Introduction to Instrumentation in Life Sciences

Introduction to Instrumentation in Life Sciences

Prakash S. Bisen
Anjana Sharma

CRC Press
Taylor & Francis Group
Boca Raton London New York

CRC Press is an imprint of the
Taylor & Francis Group, an **informa** business

CRC Press
Taylor & Francis Group
6000 Broken Sound Parkway NW, Suite 300
Boca Raton, FL 33487-2742

First issued in hardback 2017

Version Date: 20120801

ISBN-13: 978-1-4665-1240-5 (pbk)
ISBN-13: 978-1-138-44070-8 (hbk)

Library of Congress Cataloging-in-Publication Data

Bisen, Prakash S.
Introduction to instrumentation in life sciences / Prakash S. Bisen, Anjana Sharma.
p. ; cm.
Includes bibliographical references and index.
ISBN 978-1-4665-1240-5 (pbk. : alk. paper)
1. Biological apparatus and supplies. 2. Life sciences--Research--Equipment and supplies. I. Sharma, Anjana. II. Title.
[DNLM: 1. Biology--instrumentation. QH 324.42]

QH324.42.B57 2013
570.284--dc23 2012025562

Visit the Taylor & Francis Web site at
http://www.taylorandfrancis.com

and the CRC Press Web site at
http://www.crcpress.com

Contents

Preface

The current technological boom has put forth a variety of techniques being used in life science teaching and research. There is now more than ever a great deal of interaction between the physical and biological sciences. The biologist today depends on instrumentation to study the physiology and genetics of living organisms, particularly at the molecular level. A series of technologies have been developed over the years to address the challenges of different disciplines of biology. There is therefore a need for a thorough understanding of the physical principals involved in the operation of instruments and parameters of study required in using such instruments. This book is an endeavor toward that end.

This book on instrumentation includes 11 chapters that amalgamate basic techniques along with advanced instrumentation that hold the modern impetus in research and development. The book deals with the principles, concepts, techniques, and applications of optical and electron microscopy; the concepts and applications of micrometry, especially in microbial taxonomy; the principles and uses of pH meters and oxygen electrodes; the basic theoretical and practical details of chromatography to separate and purify a product from complex mixtures; the methods based on quantum principles and important spectroscopic and spectrophotometric techniques to determine the structure and underlying function of different biomolecules; various centrifugation techniques for the separation of mixtures for both preparative and analytical purposes; electrophoretic techniques and their applications for deciphering the molecular nature of cell components and also in gene technology; the principles involved in x-ray microanalysis and applications of radioactivity in biology; and the principles of fermentation and applications of fermentation technology and downstream processing to identifying the different techniques, their methodologies, and their applications regarding separation of the products of interest.

All 11 chapters have been organized from the basic to the practical with appropriate illustrations to make them comprehensive. The book will be useful for undergraduate and graduate students of life sciences, pharmacy, biotechnology, and microbiology as an introduction to the tools and techniques currently in practice. The book fine-tunes the principle behind the *modus operandi* of various instruments that illustrate the seed of prospective rationale in undergraduate students. All 11 chapters encompass a well-defined methodology that describes the instruments and their corresponding applications in different fields of life science, illuminating the young minds of graduate students knocking on the door of innovative research.

A unique facet of this book is its broad subject coverage that incorporates fundamental concepts like microscopy, micrometry, and electrochemical techniques and also advocates the important applications of modern molecular and proteomic tools that lay the basis for state-of-the-art research in the present era of metagenomics, metaproteomics, and metabolomics. This book devotes a chapter to fermentation, which emphasizes the principles, designs, and types of different bioreactors and the various separation techniques that help in purifying the products of interest for further studies. The present aspects of industrial microbiology and biotechnology will provide readers with comprehensive insight into the dynamics of the core commercial application industry, a sanguine component that has been missing in contemporary books related to biological science. The book can also help researchers by laying down the inventory of methods and their principles that could be put to practice to further their study.

About the Book

Introduction to Instrumentation in Life Sciences addresses different aspects of instrumentation that hold the keys to cutting-edge research and innovative applications. The book has been designed to serve a wide multitude of students and researchers diversified in the field of life sciences: pharmacy, biotechnology, microbiology, biochemistry, and environmental sciences. Graduate students are introduced to different aspects of basic experimental methods and instrumentation used in practical life sciences. Due emphasis has been given to techniques encountered both in practical classes and also in high-throughput technique involved in the modern era of commercialization and product inventories. As a further aid to students, well-illustrated diagrams are presented to explain the principles and theories behind these instruments.

This is essential reading for all life science, undergraduate, graduate, and medical students for whom practical molecular biology, proteomics, biochemistry, microbiology, and immunology are part of the syllabus. We hope that the book will be able to respond to the requirements of students and researchers involved in life science teaching and research.

Acknowledgments

We wish to express our gratitude to Dr. M. P. Kaushik, Dr. R. Vijayaraghavan, and Dr. S. J. S. Flora, Director, Former Director and Scientist G, respectively, of the Defense Research and Development Establishment (DRDE), Defense Research and Development Organization (DRDO), Ministry of Defense, Government of India, Gwalior; Professor G. P. Agarwal, Senior Microbiologist, Department of Postgraduate Studies and Research in Biological Sciences, R. D. University, Jabalpur; Prof. G. B. K. S. Prasad, Head, School of Studies in Biochemistry, Jiwaji University, Gwalior; R. S. Rathore, Vikrant Group of Institutions, Gwalior; and Puneet Davar, Tropilite Foods Pvt. Ltd., Gwalior, India, for providing necessary facilities and extending help in various ways. Sincere thanks to Devendra Singh and Avinash Dubey for their computational work preparing the book in a presentable form.

We express our sincere thanks to our research students, Dr. Ruchika Singh Raghuwanshi, Gulab Singh Thakur, Bhagwan Singh Sanodiya, Rakesh Kumar Baghel, Rohit Sharma, Dr. Abhishek Bhattacharya, Rani Verma, Chandan R. Bora, Varsha Shukla, Padmini Ramteke, Poonam Kumar, Ankita Shrivastava, Arunima Sharma, Shraddha Trivedi, Pankaj Parihar, and Islam Husain for extending scientific inputs and technical support.

Finally, we would like to express thanks to the Council of Scientific and Industrial Research, New Delhi, for the award of Scientist Emeritus to Professor Prakash S. Bisen.

Prakash S. Bisen
Gwalior, India
Anjana Sharma
Jabalpur, India

1 Microscopy

1.1 INTRODUCTION

A microscope is defined as an instrument that magnifies objects by means of lenses to reveal details that are invisible to the naked eye. It was only after the discovery of the first microscope around 1590 by a Dutch spectacle maker, Zacharias Jensen, that it was realized there were certain living organisms so small in size that they were invisible to the naked eye and hence never believed to be in existence. The microscope opened new doors into the living world, bringing forth a realm of microorganisms. This gave concrete evidence to Louis Pasteur's germ theory of disease. Thus began the science of microbiology.

Through the microscope, the different shapes, sizes, and even colors of microorganisms can be seen. The degree of magnification needed to see a microorganism depends on the size of the microbe. Protozoa, fungi, algae, and bacteria whose sizes range from 1–200 μm can be viewed with a light microscope, that is, a microscope that uses visible light to illuminate a specimen. It gives a magnification of about 1500×. Viewing smaller microorganisms like viruses, whose size varies from 0.015 to 2.0 μm, as well as the internal structure of bacterial cells or eukaryotic cell organelles, requires the use of a more specialized electron microscope, which has a higher magnification, that is, 200,000× (Table 1.1). It may be noted here that the extent of magnification is limited by the capacity of resolution, which we shall discuss in Section 1.2. Without resolution, magnification is called empty magnification and it is of practically no use whatsoever.

Today, microbiologists have a variety of microscopes at their disposal (Table 1.1). Together with the different techniques available for exploration, they can choose and use these microscopes for study. The choice of a particular microscope depends on the size of the object, degree of detail to be viewed, and purpose of microscopic observation.

In this chapter, we examine different type of microscopes, their advantages and disadvantages, and also different analytical techniques using microscopes. In order to understand the indispensable role played by the microscope in the study of microorganisms, it is necessary to appreciate the intrinsic limitation of the eye as a magnifying instrument. The image formed by the eye lens, L, in Figure 1.1 must appear on the retina, R, in order to be clearly seen. The ciliary muscles attached to the lens surfaces can alter the focal length of the lens. This enables the eye to focus different distances on the retina; this is a property of the eye known as its power of accommodation. However, the eye cannot accommodate objects closer than approximately 25 cm due to muscular orientation.

1.2 MAGNIFICATION, RESOLUTION, AND CONTRAST

In order to view an object closer to the eye than 25 cm, a converging lens is placed between the object (at less than 25 cm from the eye) and the eye. This produces an enlarged virtual image 25 cm from the eye (Figure 1.2). The apparent size of an object as viewed by the unaided eye is directly related to the angle subtended by the object at the eye. Magnification is achieved by increasing the angle subtended by the image at the eye so that the size of the object apparently increases from h to h' (Figure 1.2). Magnification is therefore defined as $M = \theta'/\theta$, where

M = Magnification
θ' = Angle subtended at the eye by an image at 25 cm
θ = Angle subtended at the unaided eye by the object at 25 cm

TABLE 1.1
Microscopes, Magnification, and Applications

Type of Microscope	Maximum Useful Magnification	Resolution	Useful Application
Bright field	1500×	100–200 nm	Extensively used for visualization of microorganisms and their gross morphological features; usually staining is necessary to view specimens.
Dark field	1500×	100–200 nm	Used for viewing live microorganisms particularly those with characteristic morphology, for example, spirochetes. Staining is not required, and the specimen appears bright on a dark background.
Phase contrast	1500×	100–200 nm	Used to examine cellular structures of living cells of yeast, algae, protozoa, and some bacteria; does not require staining.
Interference	1500×	100–200 nm	Used to examine structure of microorganisms; produces sharp, multicolored image with 3D appearance.
UV	2500×	100 nm	Useful for obtaining improved resolution; largely replaced by an electron microscope.
Fluorescence	1500×	100–200 nm	Used for fluorescence staining. Useful in many diagnostic procedures for identifying microorganisms.
TEM	500,000×–100,000×	1 nm	Used to view ultrastructure of microorganisms including viruses; much greater resolving power and useful magnification achieved than with light microscope.
SEM	10,000×–1,000,000×	1–10 nm	Used for viewing surface structures in detail; produces a 3D image.

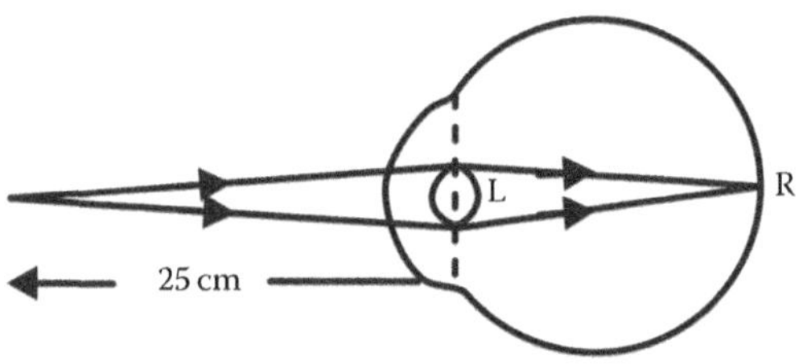

FIGURE 1.1 Image formation on the retina of the eye.

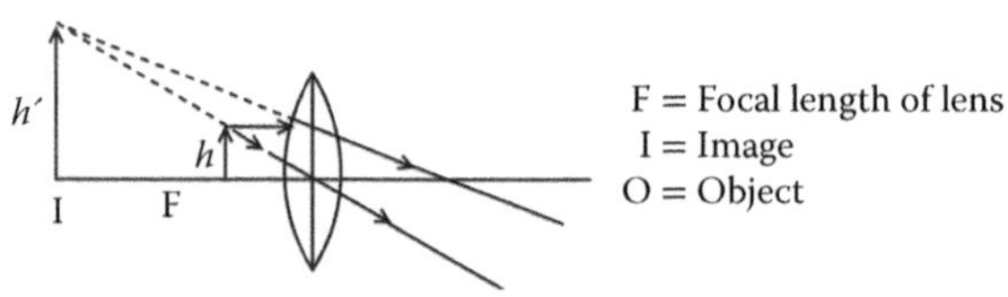

FIGURE 1.2 Use of a converging lens to view an object closer than 25 cm to the eye.

In Figure 1.2,

F = Focal length of lens
I = Image
O = Object

The principle of a compound microscope involves the use of an eyepiece to magnify an already enlarged real image produced by a first lens called an objective. In a compound microscope, magnification is brought about by a system of lenses.

Two other factors, contrast and resolution, are of great importance in microscopy. In order to be perceived through a microscope an object must possess a certain degree of contrast with its surrounding medium, and in order to produce useful magnification the microscope must have resolution, that is, the ability to discern two closely adjacent points as separate points.

1.3 LIGHT (BRIGHT FIELD) MICROSCOPY

The use of a microscope in which the final image of an object, which is illuminated by visible light (400–700 μm), is seen through glass lenses is called light microscopy. The light microscope consists of three separate but coordinated lens systems (Figure 1.3): (1) condenser, (2) objective, and (3) eyepiece. The total useful magnification produced is about 1500× and is equal to the product of the magnifications of the objective and eyepiece. The lens systems are defined as follows:

Condenser: The condenser collimates the light beam, regulates the passage of light, and eliminates peripheral rays from a source. Occasionally, there is a substage condenser, which concentrates the light beam on an object. This effectively increases the numerical aperture (NA), which, as will become clear in Section 1.3.1, increases the resolving power of the microscope.

Objective: The objective lens system magnifies an object by about 90× to 100× and produces a real image inside the microscope.

Eyepiece: The eyepiece magnifies the real image to form a virtual image on the retina of the eye, producing a total magnification of 1500×. In a compound light microscope, light rays from below the condenser are refracted through the condenser and emerge from the top surface of the slide, at the plane of the object as a cone of light with the apex pointing downward (Figures 1.4 and 1.5).

Single lenses have two inherent defects:

1. Spherical aberration
2. Chromatic aberration

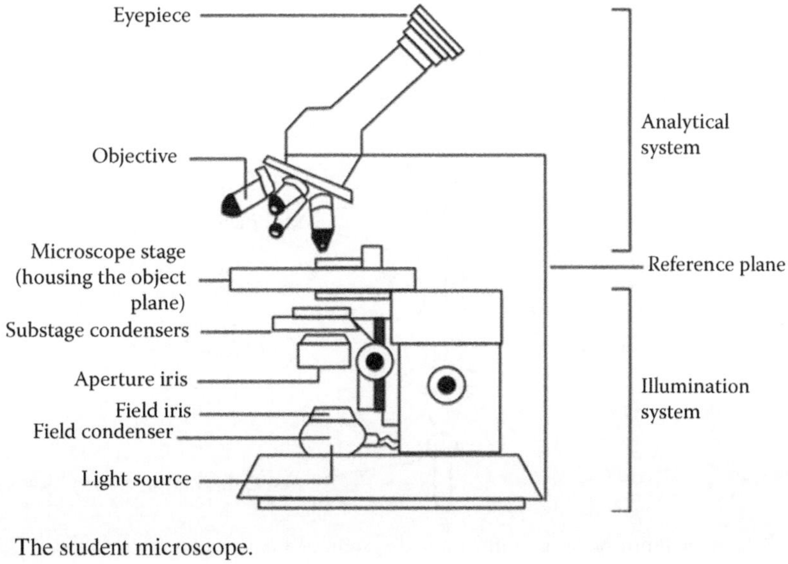

FIGURE 1.3 The student microscope.

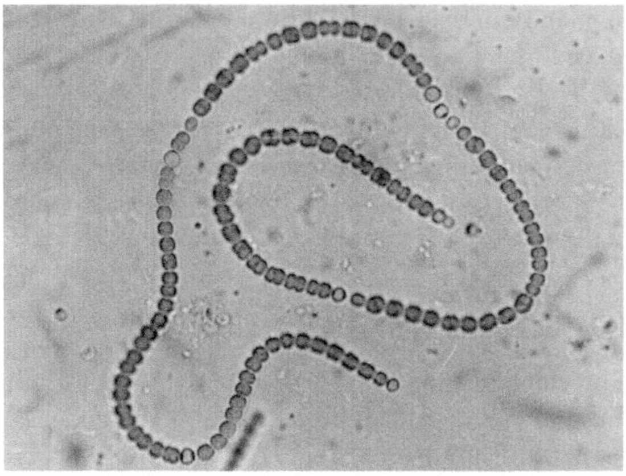

FIGURE 1.4 **(See color insert.)** Microscopic view under a compound microscope of the cyanobacterium *Anabaena variabilis*, showing cell differentiation.

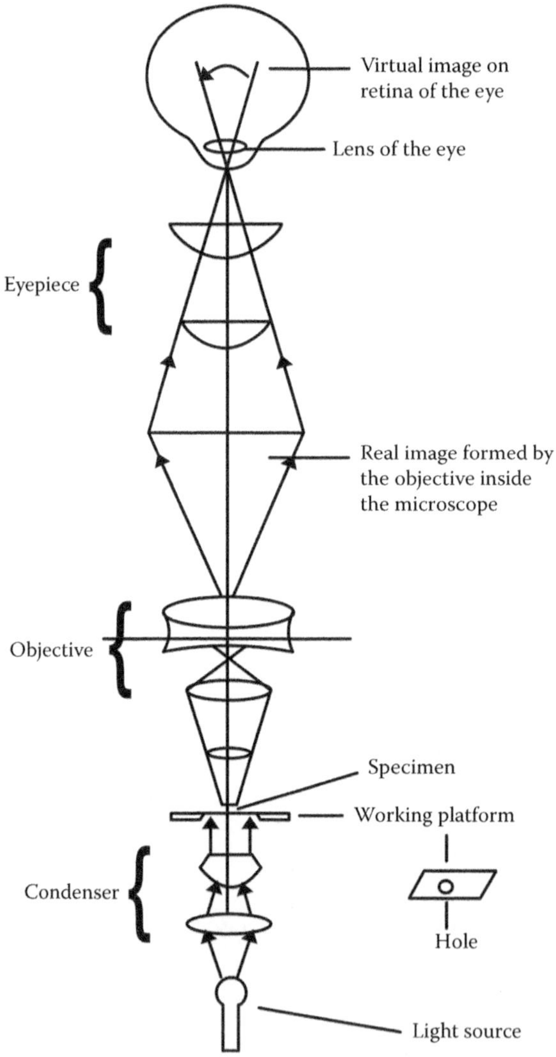

FIGURE 1.5 Schematic representation of the optical system of a compound microscope.

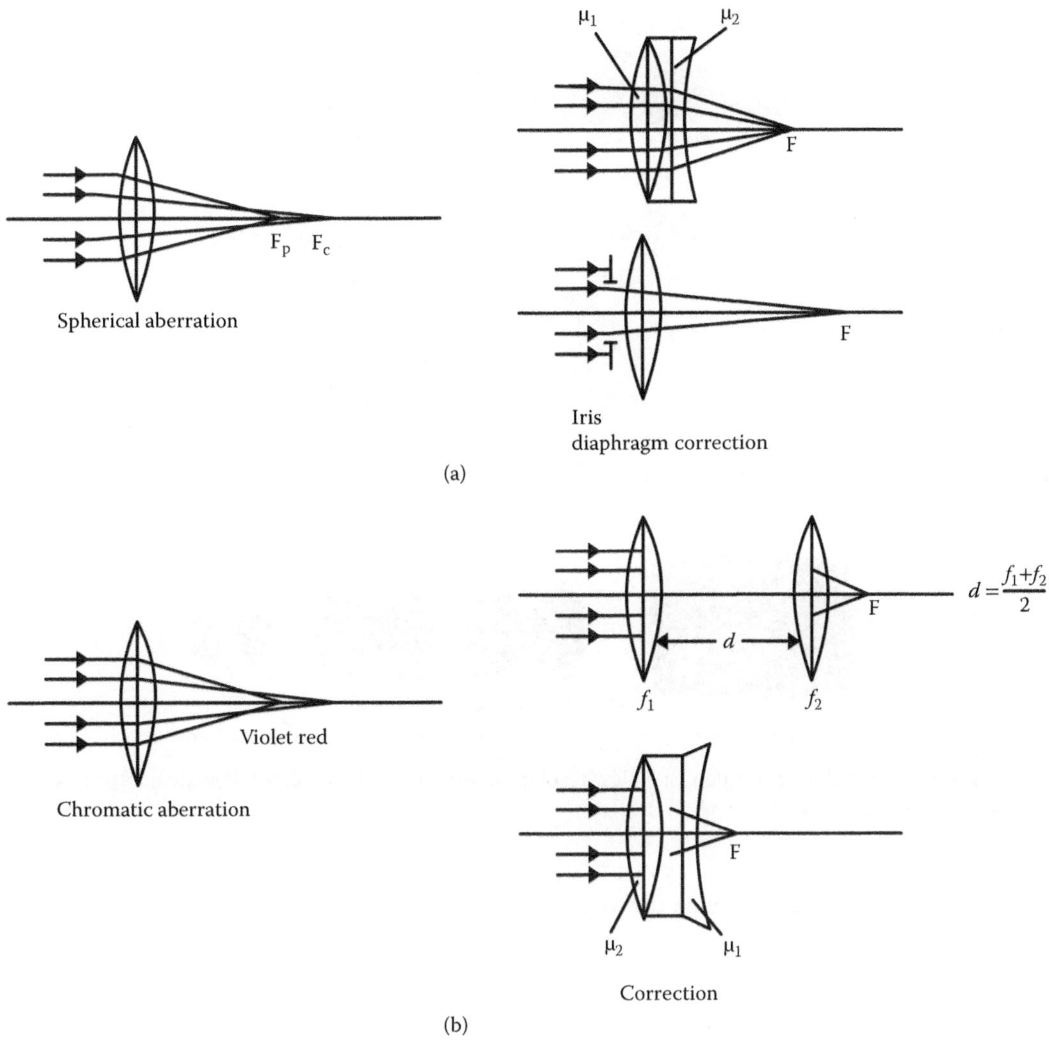

FIGURE 1.6 Defects of the lens and their correction: (a) spherical aberration and (b) chromatic aberration.

Spherical aberration is the inequalities of refraction and focus by the peripheral portions of the lens. This is due to the curved surface characteristics of the lens element. Rays at the outermost margins of the lens are refracted to a greater degree, which causes the image to be formed at a point closer to the emergent side of the lens. This can be corrected either by combining with the lens another lens of opposite diverging power or by using an iris diaphragm, which eliminates peripheral rays (Figures 1.6 and 1.7).

Chromatic aberration is the inequality of refraction and focus of all the different wavelengths of white light producing multiple colored images. This is corrected by combining two lenses that have different refractive indices.

1.3.1 Resolution

The ability to distinguish between two closely spaced objects is called resolution. It is the property of the microscope by which magnification is rendered useful and more detail can be observed (Figure 1.8).

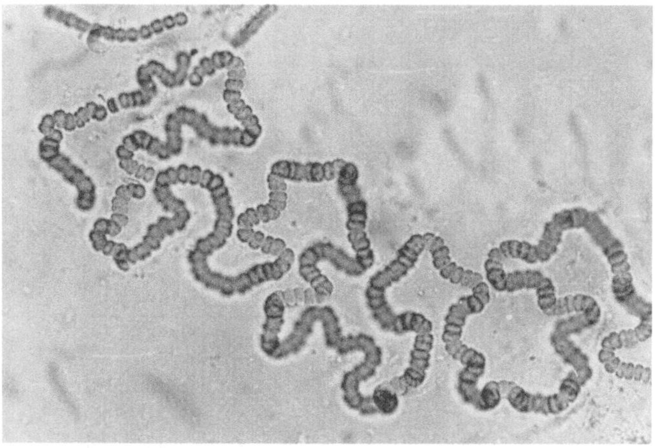

FIGURE 1.7 **(See color insert.)** Microscopic observation of a mutant *A. variabilis* under high resolution.

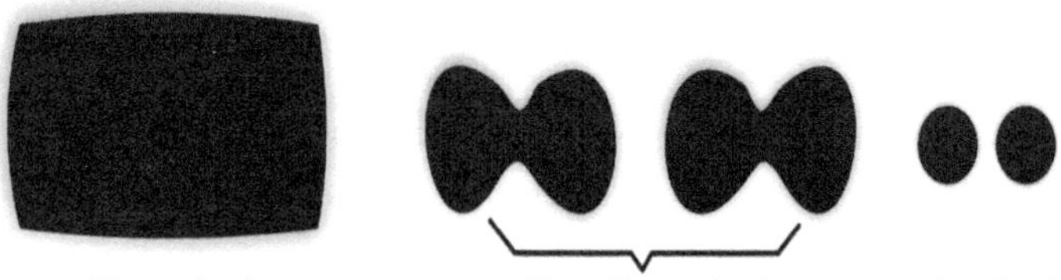

FIGURE 1.8 Resolution of two points: At low resolution, structures blur together. The greater the resolution the more detail that can be observed.

The distance between two points that can just be distinguished is called the resolving limit (*d*). The resolving limit is dependent on the wavelength of light (λ) and the NA:

$$d = 0.5\lambda / \text{NA}$$

The NA is the property of a lens that describes the amount of light that can enter it:

$$\text{NA} = \mu \sin\theta$$

where

μ = Refractive index of the medium between the specimen and lens
θ = Half the angular aperture

The angular aperture is the angle between the most divergent rays of the inverted cone of light emerging from the condenser that enters the objective (Figure 1.9). Resolving power can be increased by using a shorter wavelength and increasing the refractive index of the medium that fills the space between the specimen and the front of the objective.

Due to the physiological restriction of the human retina to perceive light between 400 and 700 nm, the resolving limit, *d*, in a light microscope is approximately 200 nm. Ultraviolet (UV) light, which still has a shorter wavelength than visible light, is preferable for increasing resolution, but because UV light does not penetrate glass lenses well and directly viewing UV light results in eye damage, it is normally not possible to depend on the improved resolving power that could be achieved using this shorter-wavelength light. Also, the advent of the electron microscope, which utilizes the wave motion of electrons (electrons have a much shorter wavelength than UV light), has made such UV microscopes obsolete.

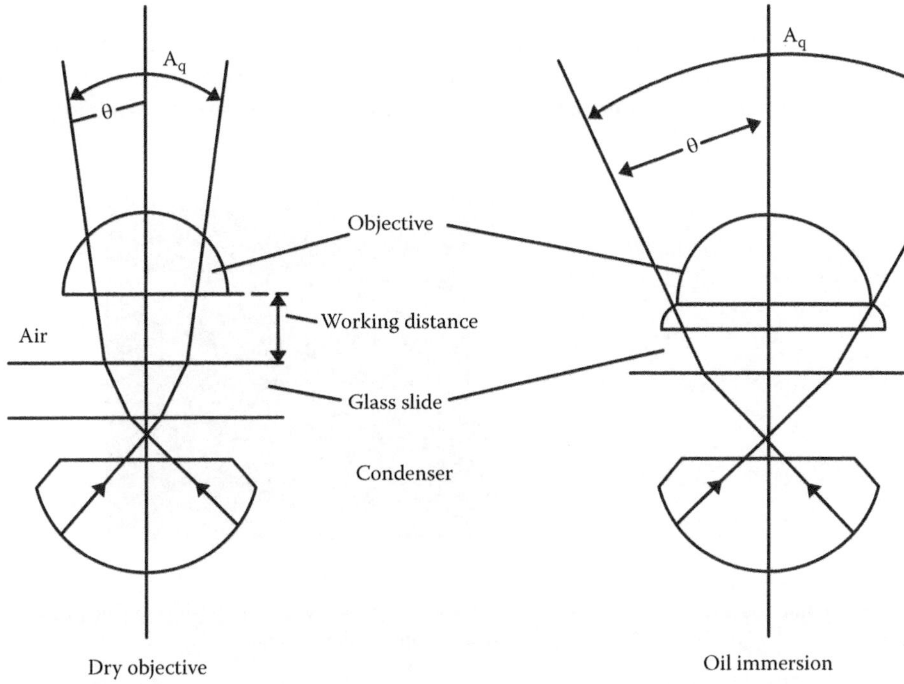

FIGURE 1.9 The NA is improved by the use of immersion oil to replace the air between the specimen on the glass slide and the objective as shown by the wider angular aperture (Aa) obtained using an oil immersion lens. It is noted that θ is the angle between the most divergent rays entering the objective and the optical axis, and it is equal to half Aa. The working distance is also reduced in an oil immersion.

The refractive index of the medium filling the space between the specimen and objective can be increased by using immersion oil, which has a refractive index of 1–5, similar to that of glass. Many of the divergent peripheral rays lost by reflection and refraction at the surface of the condenser, slide, and objective lens are refracted within the angular aperture, thereby increasing the NA and consequently the resolution. Recall that the use of a substage condenser concentrates peripheral light waves on the object, thereby effectively increasing the NA and decreasing the resolving limit, which means increasing the resolution.

The use of immersion oil also effectively decreases the focal length of the lens so that the specimen has to be very close to the objective in order to be focused. Therefore, there is a short working distance between the lens and objective. A short focal length also reduces the depth of field so that only very thin sections can be focused.

The observation of algae, fungi, and protozoa can be achieved with dry objectives, that is, air that occupies the space between the specimen and objective. The viewing of bacteria, which are smaller in size, normally requires the use of an oil immersion lens. Such lenses are specially designed for use with immersion oil and should never be used without it.

1.3.2 Contrast of a Microscope

A microbial cell largely comprises water; it is the medium in which the cell is normally suspended. In order to be seen through a microscope, the cell should have some degree of contrast with the surrounding medium. The contrast arises because less light is transmitted through the cell than through the medium. This is because some light is absorbed by the cell and some is refracted out of the optical path of the microscope by the difference in the refractive index between the cell and the surrounding medium.

Contrast can be enhanced by either staining or the use of dark field or phase contrast microscopy. In the process of staining, the specimen is treated with dyes that bind selectively either to the whole cell or to certain cell components, thus producing a much greater absorption of incidental light. Specific types of microorganisms and/or particular structures of microorganisms exhibit different staining reactions that can be readily distinguished (Figures 1.10 through 1.12).

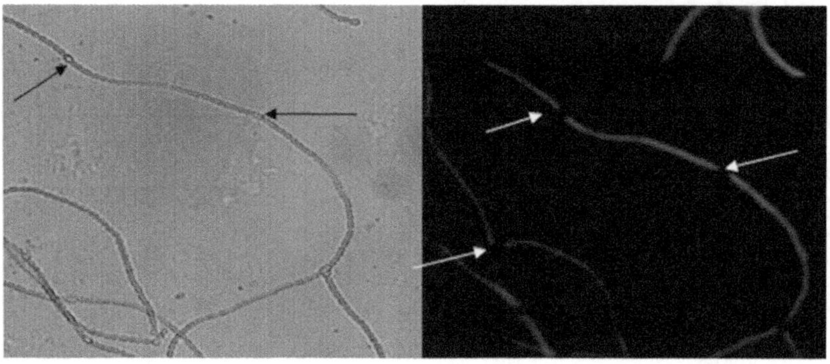

(a) (b)

FIGURE 1.10 **(See color insert.)** Photomicrograph of *A. cylindrica* showing the position of heterocysts in between the vegetative cells: (a) bright field and (b) fluorescent photomicrographs.

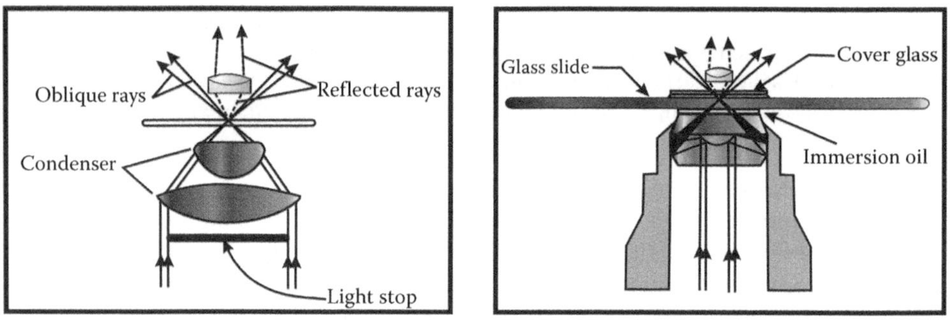

The star diaphragm allows only peripheral light rays to pass through the condenser. This method requires maximum illumination.

(a)

A cardioid condenser provides greater light concentration for oblique illumination than the star diaphragm.

(b)

FIGURE 1.11 Figure showing (a) dark field microscopy and (b) phase contrast microscopy.

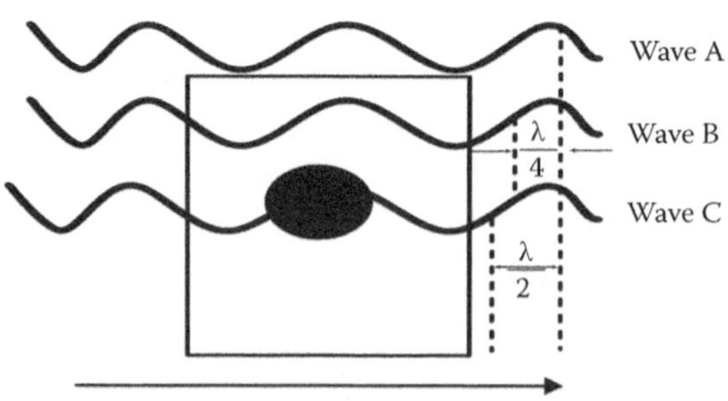

FIGURE 1.12 Figure showing retardation of phases of light waves as they pass through a transparent living cell mounted in saline. Compared to the waves that do not pass through the cell (Wave A), waves passing through the full thickness of the cytoplasm (Wave B) are retarded by $\lambda/4$ and those passing through a more refractile inclusion (Wave C) are retarded by $\lambda/2$.

1.3.3 Uses of the Light (Bright Field) Microscope

With a magnification of 1500× and a resolution of 100–200 nm, the light (bright field) microscope is a very useful tool for the gross morphological observation of microorganisms ranging 1–200 μm. This range includes bacteria, yeasts, molds, algae, and protozoa among the microorganisms and cells of tissues of several bacterial and zoological specimens.

The morphological examination of microorganisms can be put to a variety of uses. For example, the viewing of bacteria (except cyanobacteria) requires the use of special staining techniques that not only help in visualizing the bacteria but also often form the basis of their classification. This is the case of gram staining, which stains blue for gram-positive and pink for gram-negative cells. This gives information about the chemical nature of the cell wall and forms the basis of classification. It is also possible to see the characteristic sizes, shapes, and arrangements of chains, clusters, or individual unicells.

The occurrence of fission presence of endospores and other such features help to elaborate the life cycles of bacteria. The purity of bacterial cultures is established by observing a drop of bacterial suspension under a light microscope. Contamination is detected either as morphologically different bacteria or as bacteria that stain differently from what is expected of the particular culture being examined. A hemocytometer, which is explained in Section 1.11.2, can estimate the number of bacteria in a sample.

Again, by using an ocularmeter and a stage micrometer, the size of bacterium can be found. Using the hanging drop technique, the mode of motility of bacteria can be observed, for example, the vibratory movement of *Oscillatoria*, a cyanobacterium, or the twisting movement of spirochetes. Using nuclear stain, the chromatin material of a cell becomes visible. In a prokaryote, this shows the location of the nucleus or its dividing condition during fission. In a eukaryote, the different stages of mitosis in a cell of a young dividing tissue can be seen.

The morphological observation of fungi, algae, and protozoa is likewise used for classification and elaborating lifecycles, identifying modes of motility, finding the number and size of spores, and so on. As an example, the presence or absence of septa in a fungal mycelium separates Phycomycetes from higher fungi (Figure 1.13a and b).

The anatomy of plant and animal tissue specimens is also studied principally under the light microscope. By simple observation of a blood smear on a slide, the heterogeneous nature of blood tissue can be understood. Treatment with antisera and subsequent observation under the microscope shows whether or not coagulation has occurred, which aids the process of blood grouping.

The light microscope is by far the most widely used microscope due to its low cost, easy workability, and immense versatility. However, due to its low magnification and resolution as compared to the electron microscope (Table 1.2), the detailed examination of microorganisms or, for that matter, examination of small microorganisms like mycoplasmas and viruses is not possible (Figure 1.13c). Until the advent of the electron microscope, the great complexity of a eukaryotic cell was concealed because structures like endoplasmic reticulum, Golgi bodies, mitochondria, and the internal structure of chloroplasts cannot be seen through a light microscope. Also, because it has no system for enhancing the contrast, the light microscope relies on either staining or the intrinsic contrast of the specimen. A structure with a low refractive index is difficult to stain, so, for example, slime sheaths around bacteria often go unnoticed. Under such circumstances, one resorts to either dark field microscopy (Figure 1.14a) or phase contrast microscopy (Figure 1.15).

1.3.4 Care of the Microscope

Since the light microscope has so much to offer, we should take good care of it as follows:

- The lens surface of the eyepiece and objective should be wiped clean using either a lint-free soft cloth or lens-cleaning paper before and after use.
- The lens covers should be duly replaced after use.

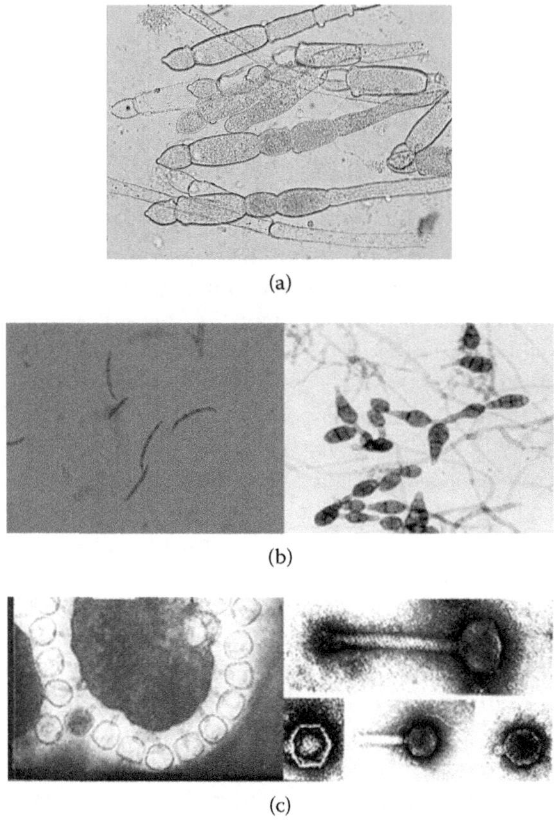

FIGURE 1.13 **(See color insert.)** Figure showing (a) sporangium and sporangiospore in a slime mold viewed under a compound microscope, (b) fungal conidia of *Fusarium* and *Alternaria* under oil immersion, and (c) electron micrograph of phycovirus attached to cyanobacteria and different types of phycoviruses.

TABLE 1.2
Depiction of Relative Sizes of Microbes, Molecules, and Atoms Together with an Indication of the Useful Range of Different Types of Microscopes

	Size	Object	Microscope	
0.1 mm	100 μm	Protozoa	Light microscope ↓	↑
	10 μm	Blood cells		
1000 mm	1 μm	Bacteria		
100 mm		Viruses	↑ UV ↓	Electron microscope
10 mm		Macromolecules		
1 mm	10 Å	Molecules		↓
	1 Å	Atoms		

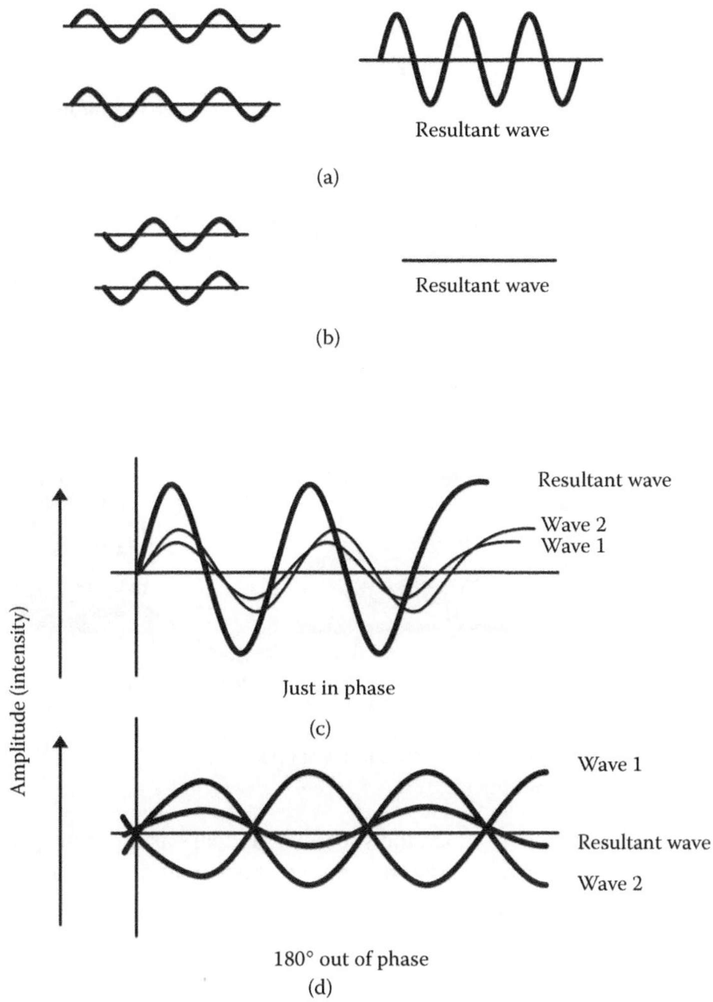

FIGURE 1.14 Additive and destructive interference of light waves: (a) amplitude objects, (b) phase objects, (c) direct and diffracted rays, and (d) coincidence and interference.

- If fungal growth occurs on the lens, it should never be scratched with nails or a sharp instrument. It should be given to specialists who will clean it by dipping it in special solutions.
- The 100× objective lens should never be used without oil.
- The microscope should be kept covered when not in use.

1.4 DARK FIELD MICROSCOPE

Contrast can be enhanced not only by staining but also by using several alternative microscope designs that do not require staining, thereby permitting the visualization of living specimens. The simplest of these microscopes is the dark field microscope. Many objects appear invisible against a bright background due to a lack of contrast, but they are visible against a dark background. In order to obtain a dark background, direct light that illuminates the objective must not enter the objective. This is accomplished by the use of a special condenser that transmits a hollow cone of light apex down and through the specimen. The diverging rays do not enter the objective. Only light scattered

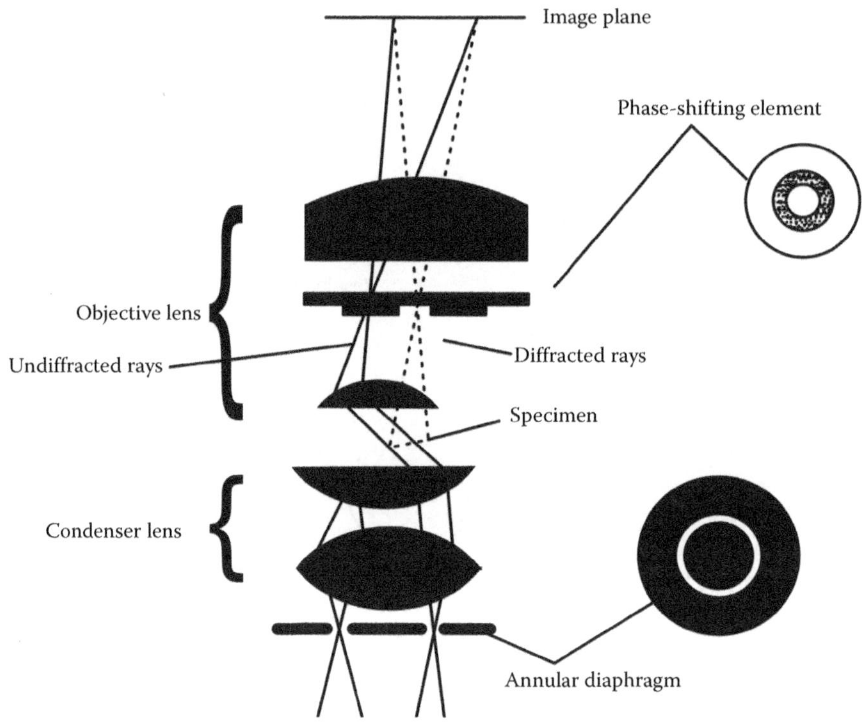

FIGURE 1.15 Schematic representation of a phase microscope.

by the specimen enters the objective. Thus, the specimen appears as a bright speck in a dark field. The special dark field condenser can be either paraboloid or cardioid. It could also be the bright field condenser used with a dark field stop (Figure 1.15).

1.5 PHASE CONTRAST MICROSCOPE

The contrast between a specimen and a surrounding medium can be increased by using a specialized microscope called a phase contrast microscope, in which the optical system is modified. The phase contrast microscope optically converts differences in the speed with which light passes through a specimen into differences in contrast that can be seen. Similar to the dark field microscope, it is useful for visualizing living microorganisms and eliminates the necessity of staining to view microbial structures.

This type of microscope relies on the fact that light passing through a cell of a higher refractive index (with a greater ability to change the direction of a ray of light) than the surrounding medium slows down relative to the light passing directly through the less dense background medium. The greater the refractive index of the cell the greater the retardation of a light wave. Thus when light passes through a microorganism, there is a slight alteration in the phase of the light wave, that is, the point of advancement within the light wave cycle. The conversion of differences in the phase of the light wave is based on interference between light waves reaching the image plane. When two waves that are out of phase with each other by a phase difference of $\lambda/2$ reach the image plane, they destructively interfere, cancel each other out, and produce darkness. On the contrary, two light waves that are in phase combine and reinforce each other on reaching the image plane to produce a wave of twice the amplitude and higher intensity, which results in brightness (Figure 1.14).

The phase contrast microscope is designed to separate direct, undiffracted background light from light passing through an object and getting diffracted, causing these two different waves to be approximately 90° out of phase with each other so that they destructively interfere at the image plane and cause changes in light intensity. To achieve this phenomenon, an annular diaphragm on the substage condenser allows a ring of light to pass through the condenser and the objective and fall on a corresponding ring-shaped area on a phase-shifting plate placed on the back focal plane of the objective.

The ring-shaped area of the phase-shifting plate is thinner than the rest of the plate so that rays passing through it are altered in phase by 90° or $\lambda/4$. Rays that pass through the specimen directly, undiffracted, fall on this thin ring-shaped area and are advanced by 90°, while those diffracted by the specimen pass through the thicker portion and are retarded by 90°. When these two waves recombine on the image plane, they interfere destructively, which greatly increases the contrast of the cells or intracellular structures that differ slightly in the refractive index from their surroundings. The difference in phase increases contrast, hence the design of a phase contrast microscope.

1.6 INTERFERENCE MICROSCOPE

Similar to a phase contrast microscope, an interference microscope is used for enhancing contrast. Both phase contrast and interference microscopes utilize the fact that light travels as waves and the addition of light waves that are out of phase with each other produces interference that alters the amplitude of light waves. Whereas phase contrast microscopes use one beam of light passing through the specimen, interference microscopes have two beams of plane-polarized light that are combined after passing through the specimen. These microscopes have higher NAs and better contrast, and they can produce colored pictures with vivid topographic relief. However, because of its expense, the phase contrast microscope is the preferred device for observing wet specimens of bacteria.

1.7 UV AND FLUORESCENCE MICROSCOPES

The resolving power of the light microscope is directly related to the wavelength of the light used:

$$d = 0.5\lambda/\text{NA}$$

A slight improvement in resolution (about twofold) can be achieved by the use of a UV light source; UV light has a shorter wavelength than visible light. This is called UV microscopy, and the microscope is modified as glass lenses are replaced by quartz lenses since glass is opaque to UV light. Also, a camera is used to record the image because the eye cannot perceive UV light; in fact, UV light causes blindness. However, its complexity and high cost have limited the use of UV microscopy. A modification called fluorescence microscopy, which has the same resolving power as UV microscopy, has recently come into the limelight.

Microscopy that involves staining with fluorescent dyes is known as fluorescence microscopy. When a fluorescent dye is illuminated by light of one wavelength, the excitation wavelength, it gives off light of another wavelength, the emission wavelength, which is always shorter than the excitation wavelength. The wavelength of light used to excite the dye may be in the UV range, but the emitted light that is to be viewed must be in the visible range (Figures 1.16 through 1.18).

Excitation light may be transmitted either from below the specimen, in which case it is called transmitted fluorescence, or to the specimen through the objective lens, in which case the system is referred to as epifluorescence (Figure 1.16a and b). Fluorescence microscopes are equipped with various excitation filters that permit the selection of the wavelength used to illuminate the specimen and barrier filters that prevent all but the emission wavelength from reaching the ocular lens.

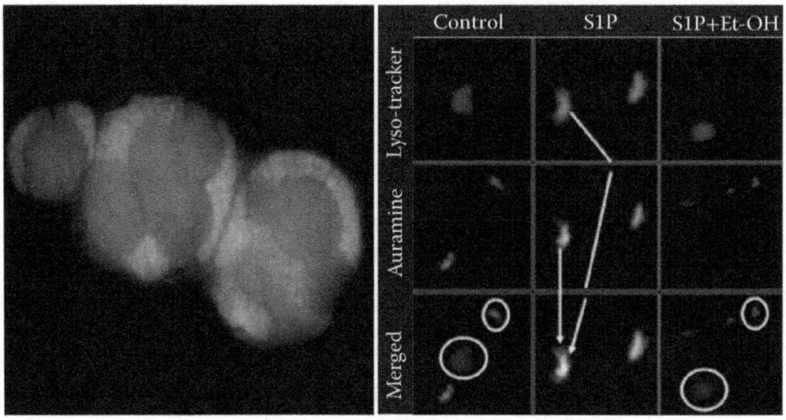

FIGURE 1.16 **(See color insert.)** Figure showing (a) fluorescence in situ hybridization of *Betaproteobacteria* and (b) immunofluorescence detection of sphingosine 1-phosphate (S1P) in tuberculosis. (Courtesy of Pernthaler et al., 2008, *Proceedings of the National Academy of Sciences* 105, 7052–7057 10.1073/pnas.0711303105.)

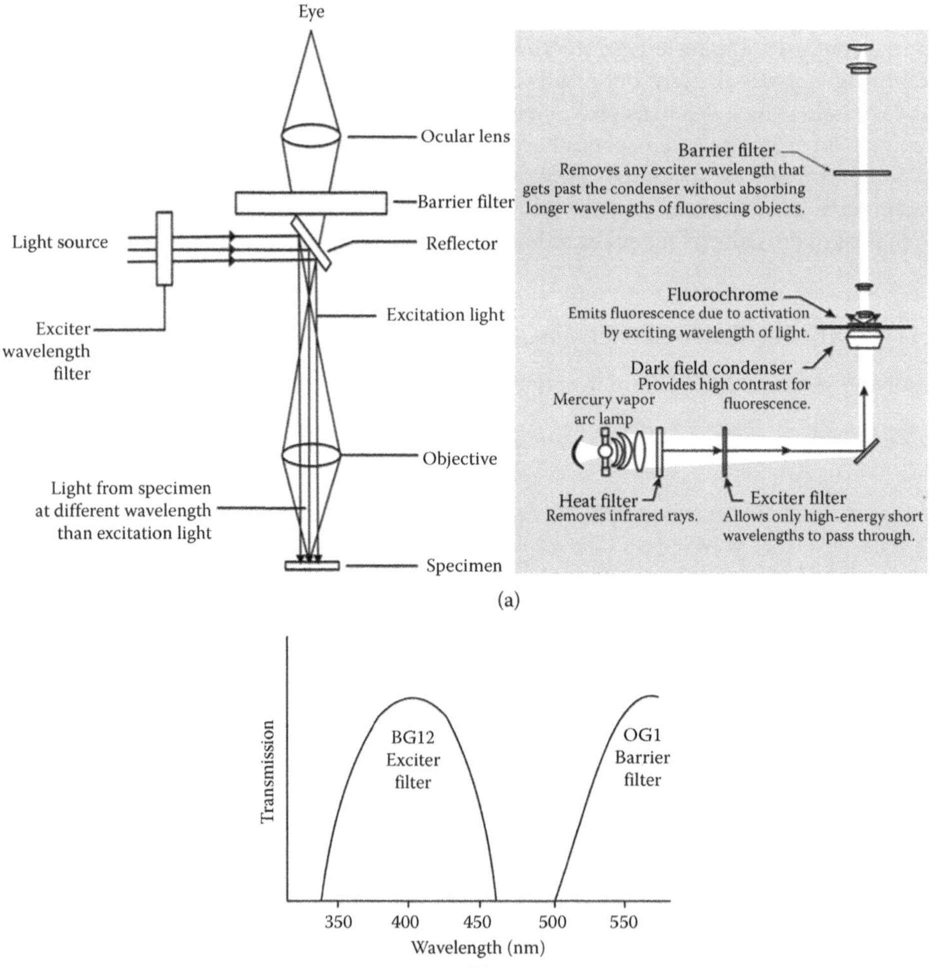

FIGURE 1.17 Figure showing (a) a diagram of an epifluorescence microscope showing the light path and (b) the spectral transmissions of Bg12 and OG1 filters.

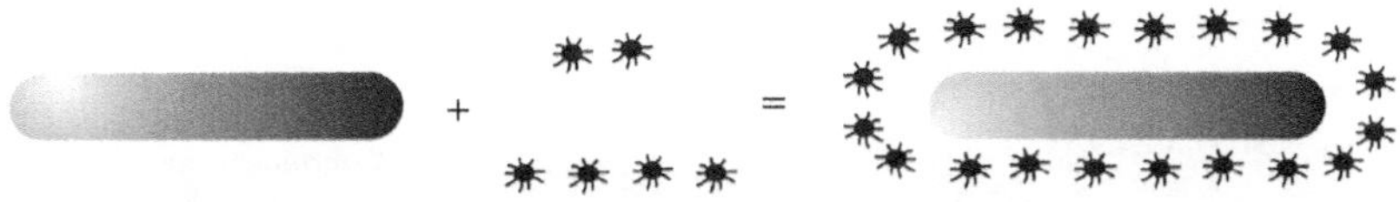

FIGURE 1.18 Fluorescence staining technique and microscopy.

1.7.1 Uses of UV and Fluorescence Microscopes

An example of direct staining of bacteria with fluorescent dye/antibodies is shown in Figures 1.16 and 1.18. One of the reasons fluorescence microscopy has become important in microbiology is that fluorescent dyes can be conjugated (linked) with antibodies (specific proteins produced against antigens as part of the immune response), providing great specificity in staining procedures. Antibodies to which a fluorescent dye is attached are referred to as labeled antibodies. These are mixed with a suspension of bacteria (antigen) and then the preparation can be examined by fluorescence microscopy. The bacterial cells that have been combined with the labeled antibody will be visible. This phenomenon is called immunofluorescence, and it is widely used in diagnostic procedures (Figures 1.16 and 1.18).

1.8 ELECTRON MICROSCOPY

Electron microscopy is based on the discovery that a circular electromagnetic field acts on a beam of electrons in a way that is analogous to the action of a glass lens on a beam of photons. The circular electromagnetic field acting as a lens is a lens coil formed by several thousand turns of wire in a soft iron casing. When a current passes through the coil, a magnetic field is developed, which directs the movement of electrons. An electron beam has the properties of an electromagnetic wave of a very short wavelength. When accelerated through an electric field, its wavelength (λ) has the following relation with the accelerating voltage:

$$\lambda \; \alpha \; 1\sqrt{\text{accelerating voltage}}$$

With an accelerating voltage of 100 KV, a wavelength of 0.04 nm (10,000× shorter than visible light) is obtained. Consequently, the resolving power increases and high useful magnification can be achieved.

There are two types of electron microscopy:

1. Transmission electron microscopy (TEM), in which electrons are transmitted through the specimen. TEM is similar in many respects to light microscopy (Figure 1.19). Both require the electron beam to pass through a vacuum. If this is not done, molecules in the air deflect the electron beam and a sharp image cannot be obtained. Extremely thin specimens must be used for TEM studies. These may be prepared in a number of different ways. The differential scattering of electrons by a specimen is viewed on a fluorescent screen or captured on a photographic plate. This was first constructed by Borries and Ruska in Berlin and Hillier and Vance in the United States in 1938.
2. Scanning electron microscopy (SEM), in which electrons are made to impinge the specimen from above. The secondary electrons ejected are collected by a positively charged plate (anode), amplified, and viewed on a cathode ray tube (CRT).

The number of secondary electrons ejected depends on the surface topography of the specimen. Hence, a three-dimensional (3D) specimen is possible.

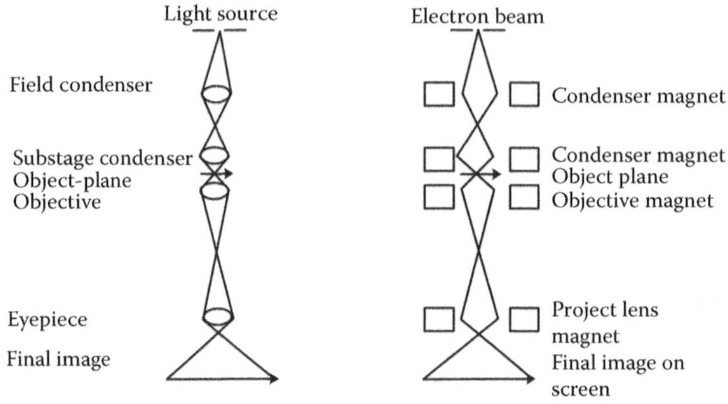

FIGURE 1.19 Comparison between a light microscope and a transmission electron microscope.

1.8.1 Operation of TEM

1.8.1.1 Parts

Source: The source of electrons is a hot tungsten filament at 30–150 KV potential. Electrons are drawn from the filament and accelerated as a fine beam past an anode by a high voltage established between the filament and the anode.

Lens system and viewing: Analogous to the glass lens system in a light microscope, there is a magnetic lens coil system in a transmission electron microscope, comprising

a. Condenser lens coil system
b. Objective lens coil system
c. Intermediate projector

The condenser lens collimates the electron beam on a specimen by varying the current to the lens. After transmission through the specimen, the objective lens coil focuses the electron beam into a first (real) image of the object enlarged about 2000×. The projector, like the eyepiece of the light microscope, magnifies a portion of the first image to about 250,000× total magnification and projects the image onto a fluorescent screen or film plate, which is necessary because the electron beam cannot be viewed directly. The operator can look into the main tube by means of portholes or magnifying binocular glasses, and can scan the images on the fluorescent screen, manipulate the objects, and adjust the alignment and field strength of the focusing magnets.

Vacuum system: Air is removed from the path of the electron beam to prevent collisions with gas molecules, which scatter the electron beam, and to reduce heat associated with the electron beam, which would otherwise destroy the biological specimen and also reduce the life of the filament. This necessitates introduction of air locks for insertion and removal of photographic plates and specimens.

1.8.1.2 Sample Preparation

The vacuum system of the microscope, lack of sufficient contrast in biological specimens, and extra thickness of microorganisms necessitate the preparation of a sample prior to microscopic examination.

Fixing and dehydration: Biological specimens containing water cannot be placed under a high vacuum because the water will boil, destroying the integrity of the specimen. Fixing and dehydration is carried out carefully in several stages (Figure 1.20). Fixation is achieved either by rapid freezing or treatment with chemicals to stabilize and cross-link the protein and lipid components of membranes. Osmium tetra oxide is a common fixative for electron microscopy. Fixed tissue is

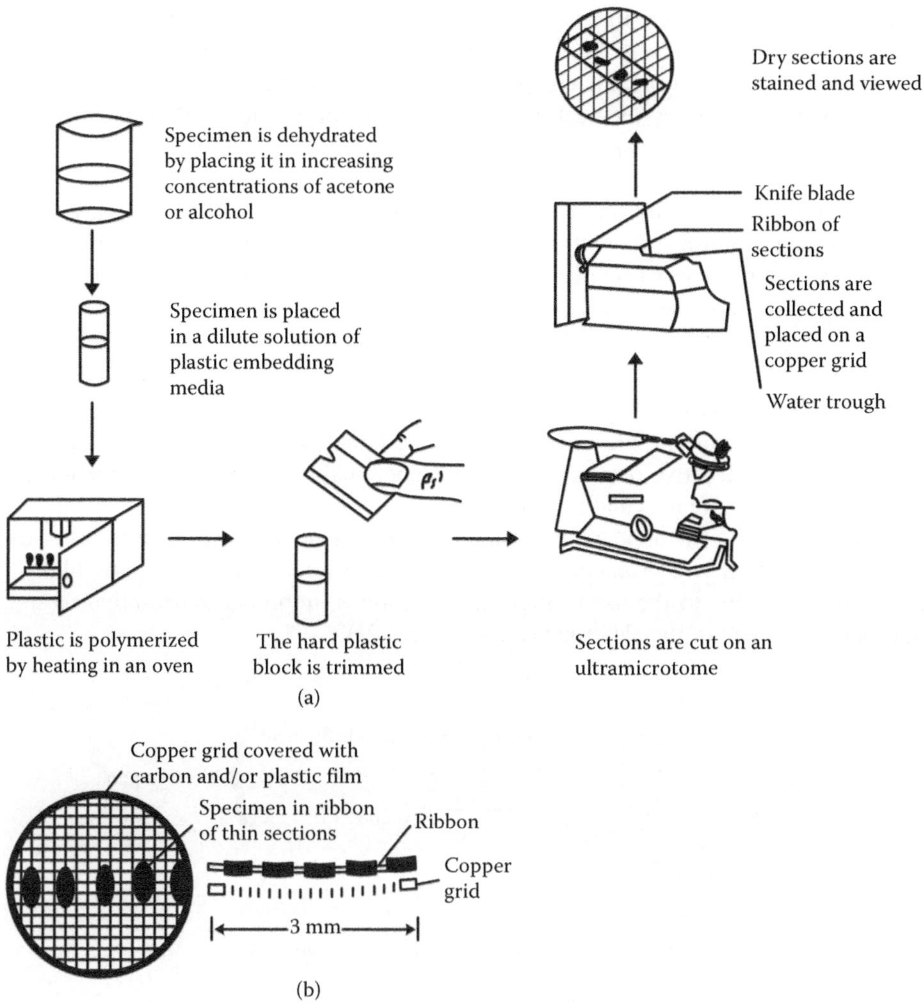

FIGURE 1.20 (a) Preparation of a specimen for viewing by TEM. (b) Inset dry section which goes for staining and viewing.

then dehydrated by passing through increasing concentrations of ethanol. A more recent technique is critical point drying, where after treatment with ethanol the specimen is immersed in pressurized liquid CO_2 and the temperature is raised to 32 °C, when liquid CO_2 vaporizes, leaving a dry, undistorted specimen.

Ultramicrotomy: Even microorganisms are too thick to be viewed under TEM; therefore, it is necessary to slice them into thin sections by a process called microtomy. In this process, the specimen after being dehydrated and fixed is embedded in a plastic material for easy handling under a microtome, which is a mechanical slicing instrument that moves a specimen across a knife that has a diamond or glass edge. The plastic material is removed with solvents. The specimen is subsequently stained.

Staining: In TEM, contrast is produced by differential scattering of electrons by a specimen. Electron images are "shadows" produced mainly by the scattering of electrons. Scattering is produced when electrons encounter atoms. Heavy atoms like Au, Pb^{207}, U^{237}, and OS^{192} produce more scattering than light atoms such as C^{12}, N^{14}, and O^{16}. Since biological specimens comprise mainly light atoms such as C, N, and O, electron scattering and therefore contrast is very slight. Contrast can be greatly enhanced by "positive" staining, that is, a combination of the organic matter in a cell with metals of high molecular weights, such as Pb, U, Os, and Au. Since they have different weights and combine differently with different organic compounds, their use permits a helpful degree of differential staining. In "negative" staining, the background is stained with an electron-opaque material, which scatters electrons, commonly phosphotungstic acid. This does not penetrate the cell but darkens the background.

Mounting: Instead of placing the specimen on a glass slide as is done in light microscopy, it is placed on a copper mesh grid within the evacuated column of the electron microscope. The specimen is sprayed onto a very thin film of electron-transparent organic material such as collodion, which is then supported on the copper mesh grid.

1.8.1.3 Operational Problems

Magnetic lenses suffer from problems similar to those found in glass lenses, although they are of electrical rather than refractive origin, for example, spherical and chromatic aberration caused by differences in electron velocities and energies.

A common problem is the appearance of artifacts that are not true representations of the specimen being viewed. This is a problem common to all microscopes, but it is particularly true for electron microscopes due to the use of high magnifications, improper dehydration of specimens, and improper adjustment of the electron beam (Figure 1.21).

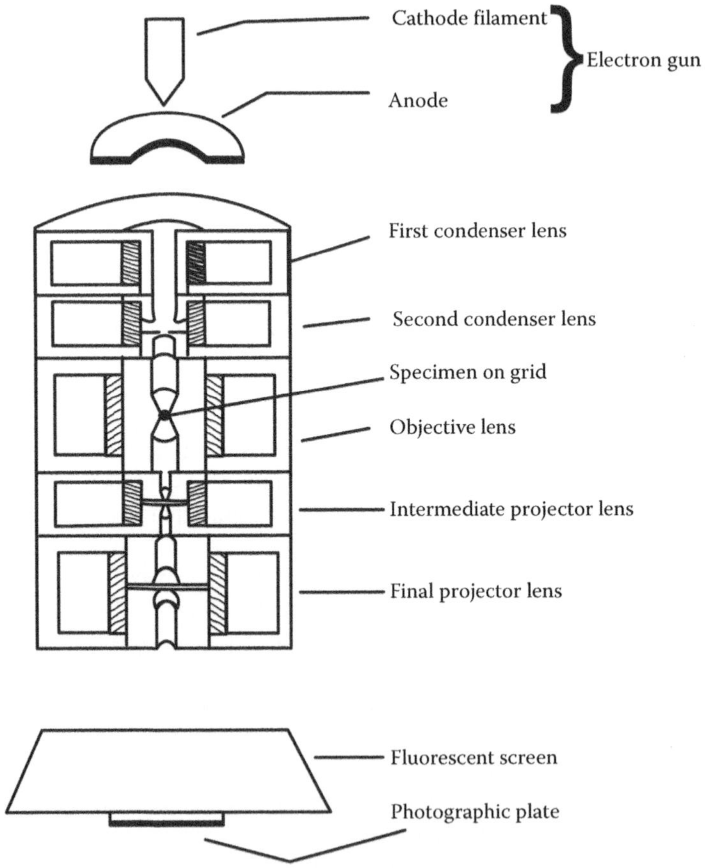

FIGURE 1.21 A transmission electron microscope allowing visualization of fine detail of a microbial cell.

1.8.2 Disadvantages of TEM

The most outstanding disadvantage for biological specimens is that they cannot be viewed live. The high-energy electron beam falling on them and the necessary process of sample preparation kill the specimen cell. Also, with the sophistication required for such a high magnification, the cost of an electron microscope becomes prohibitively expensive.

1.8.3 Operation of the SEM

1.8.3.1 Principle

The operational principle and design of an SEM are quite different from those of a TEM. The principle combines the mechanisms of electron microscopy and television. It involves the electronic amplification of signals generated by irradiating the surface of a specimen with a very narrow beam of electrons (probe). Such primary irradiation knocks off electrons from the specimen. These secondary electrons are collected on a positively charged plate called a detector and amplified and viewed on a CRT. Magnification is the ratio of the size of the image on the CRT to the diameter of the area scanned by the probe. Resolution depends on the size of the phosphorescent dots that are used to illuminate the CRT screen and on the size of the primary electron beam.

1.8.3.2 Parts

Source: An anode accelerates electrons generated on lanthanum hexabromide cathode at 30–150 KV potential (Figure 1.22).

Condenser lens coil system: The condenser lens coil system sharply focuses a fine beam of electrons on the specimen. Instead of forming an inverted cone of rays illuminating a wide field as in a light microscope/TEM, electrons are made to form a needle-sharp probe. The primary beam or probe acts as an exciter of image-forming secondary electrons that are ejected from the surface of the specimen. The number of electrons ejected depends on topography of the specimen. The probe scans the specimen in a raster pattern, similar to that on a blank TV screen.

Detector: The secondary electrons are magnetically deflected to a collector or detector, which is a positively charged plate.

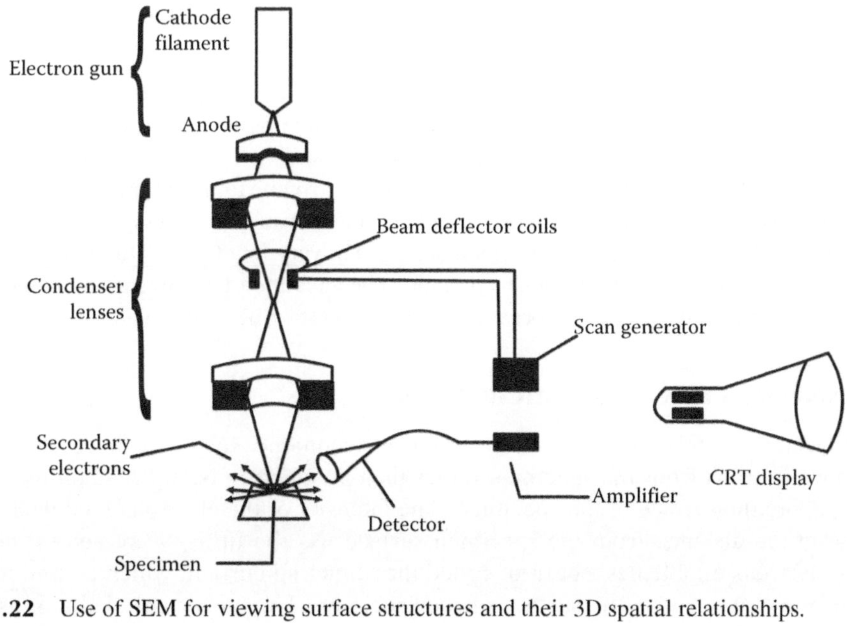

FIGURE 1.22 Use of SEM for viewing surface structures and their 3D spatial relationships.

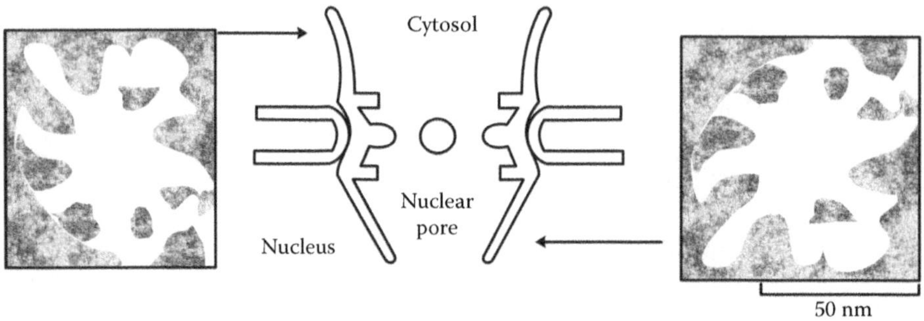

FIGURE 1.23 A nuclear pore and frozen nuclear envelopes as seen in a high-resolution SEM, which is equipped with an emission gun as the source of electrons.

Amplifier: The successive signals from the detector are amplified and transmitted to a CRT. The CRT beam and scanning beam are synchronized so that the image on the CRT is an accurate reproduction of the scanning image.

ACRT display: The final image is actually a series of pictures of different points on the specimen seen in such rapid succession as to provide the eye with a unified view of the entire surface of the specimen.

1.8.3.3 Sample Preparation

Fixing and dehydration: As in TEM, the electron fixing and dehydration beam in SEM must be transmitted through a vacuum, and therefore biological specimens must be fixed and dehydrated. The process is similar to the one followed in TEM (Figure 1.23).

Coating with metal: Once the specimen is dehydrated, it is covered by a thin film of metal, like gold or gold palladium, by vaporizing the metal under vacuum and depositing it on the specimen. Coating with metal produces a conductive surface that permits dissipation of secondary electrons, which otherwise create a surface charge on the nonconducting biological specimen. This distorts the image.

Mounting: After coating, specimens are mounted and viewed. Unlike TEM, thin sectioning of the specimen is unnecessary because only the surface structure of the specimen is viewed (Figure 1.24).

1.8.4 Advantages and Disadvantages of SEM over TEM

Since the intensity of the secondary electron beam depends on the topography of a specimen, the CRT screen gives a 3D appearance of the specimen. This is not possible in TEM. However, TEM has a better resolution (1 nm) as compared to SEM (10 nm); also, TEM allows examination of internal structures of cells, whereas SEM does not, because the electron beam is not transmitted through the specimen. However, it is possible to expose and then view subsurface layers by a technique called cryofracturing. In this technique, the specimen is frozen at very low temperatures, usually in liquid N_2 (–198°C), and then fractured with a sharp blade. The specimen fractures along planes that correspond to the internal surfaces of the organism, which can then be coated with metal and viewed.

1.9 TUNNELING ELECTRON MICROSCOPY

Tunneling electron microscopy is a relatively recent technique. In this technique, magnets are used to draw electrons from the specimen rather than an electron beam. A scanning needle tip is used to explore the surface of the specimen. The intensity of the electron cloud decreases with an increase in the distance from the specimen surface. As the tip is swept across the surface, the flow of electrons establishes a current called the tunneling current, which is used to hold the tip at a uniform height above the surface of the specimen. The movement of the tip is detected

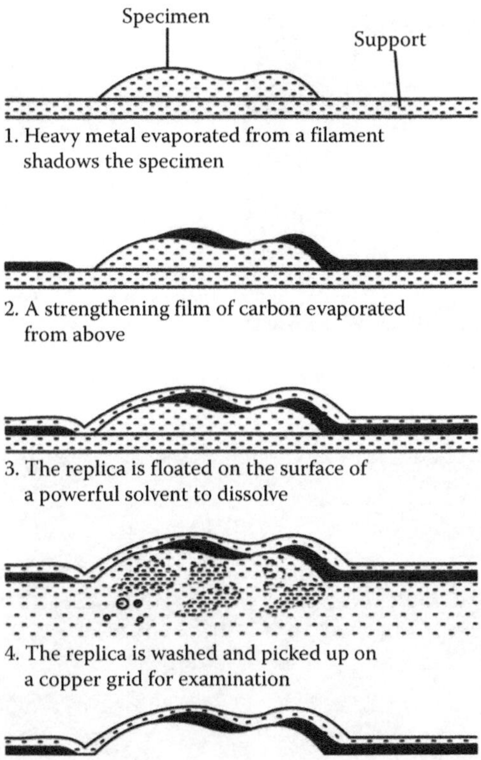

FIGURE 1.24 Preparation of a metal-shadowed replica of the surface of a specimen for SEM.

and computer processed to produce an image on the screen with high resolution and 3D view. This technique is used to view surface structures of viruses and deoxyribonucleic acid (DNA) macromolecules.

1.10 CONFOCAL MICROSCOPY

A confocal microscope creates sharp images of a specimen that would appear blurred when viewed through a conventional microscope. This is achieved by excluding most of the light from the specimen that is not from the microscope's focal plane. This image has less haze and better contrast than the image of a conventional microscope and represents a thin cross section of the specimen. Thus, apart from allowing better observation of fine details, it is possible to build 3D reconstructions of a volume of the specimen by assembling a series of thin slices taken along the vertical axis (Figures 1.25 and 1.26).

Confocal microscopy was pioneered by Marvin Minsky in 1955 while he was a junior fellow at Harvard University (Minsky 1988). Minsky's invention performed a point-by-point image construction by focusing a point of light sequentially across a specimen and then collecting some of the returning rays. By illuminating a single point at a time, Minsky avoided most of the unwanted scattered light that obscures an image when the entire specimen is illuminated at the same time. Additionally, the light returning from the specimen passed through a second pinhole aperture that rejected rays that were not directly from the focal point. The remaining "desirable" light rays were then collected by a photomultiplier and the image was gradually reconstructed using a long-persistence screen. To build the image, Minsky scanned the specimen by moving the stage rather than the light rays. This was to avoid the challenge of trying to maintain the sensitive alignment of moving optics. Minsky managed to obtain a frame rate of approximately 1 image every 10 seconds by using a 60-Hz solenoid to move the platform vertically and a lower-frequency solenoid to move the platform horizontally.

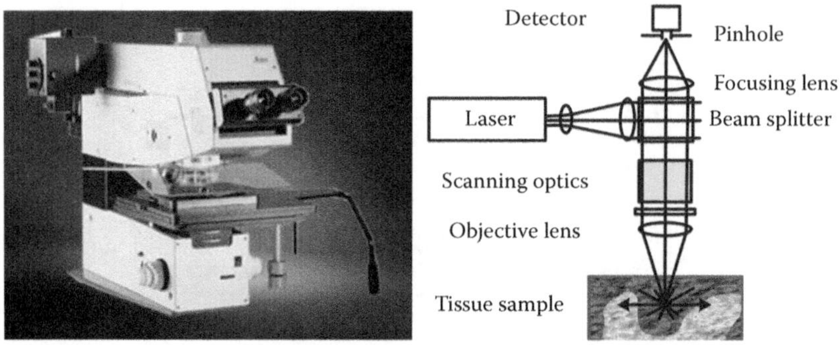

FIGURE 1.25 A modern confocal microscope.

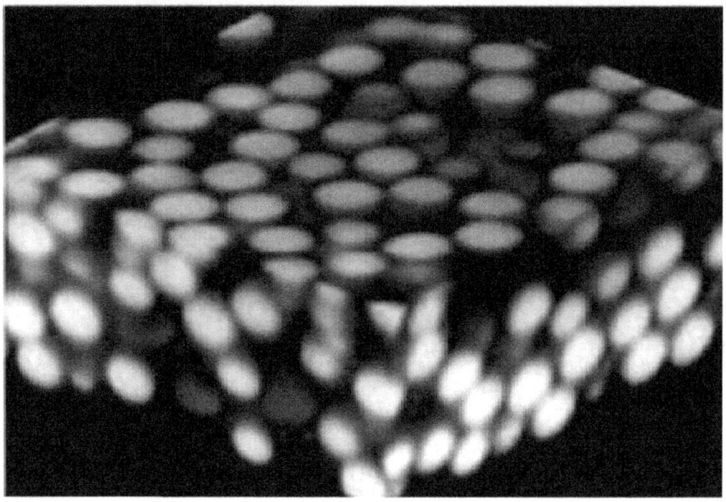

FIGURE 1.26 Point-by-point illuminations and imaging onto specimens.

1.10.1 Modern Confocal Microscopes

Modern confocal microscopes retain the key elements of Minsky's design, that is, the pinhole apertures and point-by-point illumination of a specimen. Advancements in optics and electronics have been incorporated into current designs to provide improvements in speed, image quality, and storage of generated images. Although there are a number of different confocal microscope designs, we discuss the general type here; note that the other designs are not markedly different from the general design.

The confocal microscope incorporates the ideas of point-by-point illumination of a specimen and rejection of out-of-focus light. One drawback of imaging a point onto the specimen is that there are fewer emitted photons to collect at any given instant. Thus, to avoid building a noisy image each point must be illuminated for a long time to collect enough light to make an accurate measurement. This increases the length of time needed to create a point-by-point image. The solution is to use a light source of very high intensity, which Minsky did using a zirconium arc lamp. The modern choice is a laser light source, which has the additional benefit of being available in a wide range of wavelengths.

A laser provides intense blue excitation light. The light is reflected by a dichroic mirror, which directs it to an assembly of vertically and horizontally scanning mirrors. These motor-driven mirrors scan the laser across the specimen. Recall that Minsky's invention kept the optics stationary and instead scanned the specimen by moving the stage back and forth in vertical and horizontal directions. As awkward (and slow) as the method seems to be, it does have, among others, the following two major advantages (Sheppard 1997):

1. The specimen is illuminated axially everywhere rather than at different angles as in the case of the scanning-mirror configuration, thereby avoiding optical aberrations. Thus, the entire field of view is illuminated uniformly.
2. The field of view can be made larger than that of the static objective by controlling the amplitude of stage movements.

In confocal microscopy, there is never a complete image of the specimen because at any instant only one point is observed. Thus, for visualization the detector is attached to a computer, which builds up the image one pixel at a time. For a 512-pixel image, this is typically done at a frame rate of 0.1–30 Hz. The image created by the confocal microscope is of a thin planar region of the specimen, an effect referred to as "optical sectioning." Out-of-plane unfocused light is rejected, resulting in a sharp, better-resolved image. The ability of a confocal microscope to create sharp optical sections makes it possible to build 3D renditions of a specimen. Data gathered from a series of optical sections imaged at short and regular intervals along the optical axis are used for the 3D reconstruction. Software can combine two-dimensional (2D) images to create a 3D rendition. Representing 3D information in a meaningful way out of 2D data is nontrivial, and a number of different schemes have been developed for this purpose.

Figure 1.27 shows a 3D reconstruction from slices of a suspension of 2-mm-diameter colloidal particles using "alpha blending," a technique that combines images by first making each of their individual pixels less or more transparent according to a computed weight called the alpha value (Porter 1984). The result is a 3D-like structure.

A confocal microscope provides significant imaging improvement over a conventional microscope. It creates sharper, more detailed 2D images and allows for the collection of data in three dimensions. In biological applications, it is particularly useful in measuring dynamic processes. A number of designs have been developed to achieve video-rate confocal microscopy, which enables the capture of short-timescale dynamics.

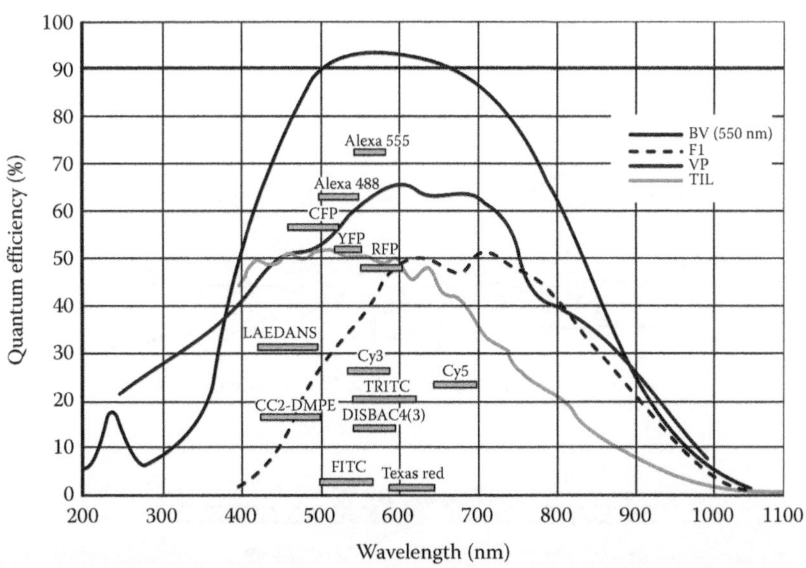

FIGURE 1.27 The 3D reconstruction of a series of 2D images of poly(methyl methacrylate) spheres suspended in a cyclohexylbromide and decalin solution. The image was created using alpha blending.

1.11 TECHNIQUES IN MICROSCOPY

1.11.1 Hanging Drop Technique

The hanging drop technique enables the viewing of the size, shape, arrangement, and motility of live microorganisms in fluid media. It requires the use of special ground slides (Figure 1.28).

In this technique, a loopful of bacterial suspension is placed in the center of a cover slip. At the four corners, tiny droplets of mineral oil are placed. The hollow ground slide is placed over the cover slip with the depression side down and the slide is inverted quickly so that water cannot run off to one side. However, the lack of contrast yields limited, although valuable, information.

1.11.2 Use of the Hemocytometer

The hemocytometer, which was originally devised for counting hemocytes, can be used for counting the number of bacteria, fungal spores, and so on, in a given volume of a sample. The hemocytometer is a glass slide with a central area partitioned off by ridges into regular cubicle chambers of exactly known volume (Figure 1.29). By counting the individual cells in each chamber under a microscope and adding them, the number of living and dead organisms may be computed.

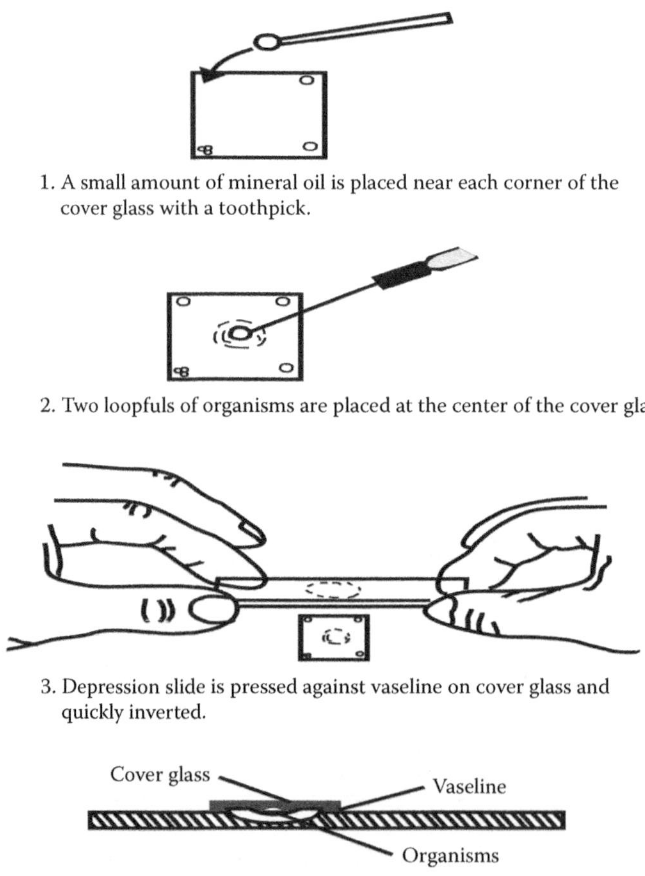

1. A small amount of mineral oil is placed near each corner of the cover glass with a toothpick.
2. Two loopfuls of organisms are placed at the center of the cover glass.
3. Depression slide is pressed against vaseline on cover glass and quickly inverted.
4. The complete preparation can be examined under oil immersion.

FIGURE 1.28 Hanging drop preparation.

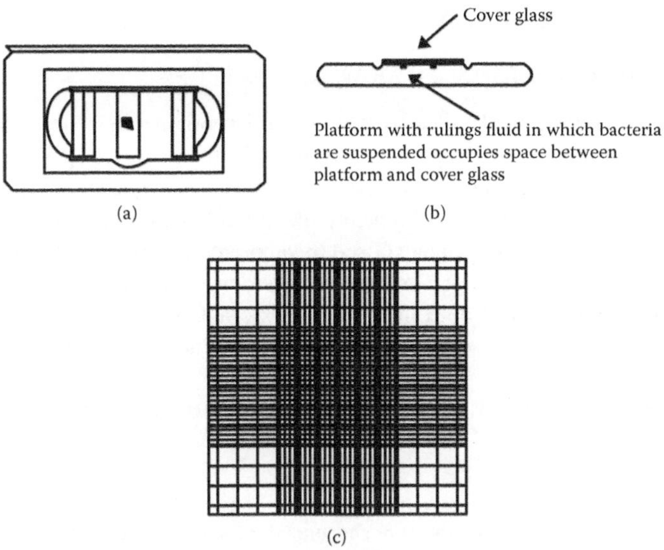

FIGURE 1.29 A hemocytometer adapted for counting bacteria and other microorganisms: (a) plan view showing a central dark square covered by ruled chambers, which is enlarged in the figure at the bottom (c), and (b) vertical section with cover glass in place.

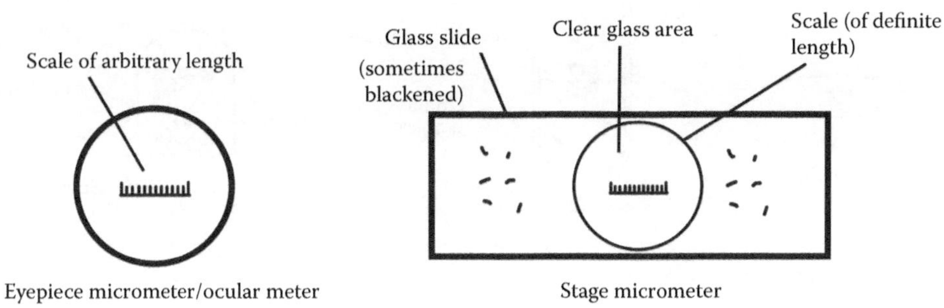

FIGURE 1.30 An ocular meter and a stage micrometer.

1.11.3 Ocular Meter and Stage Micrometer for Micrometry

This is a technique in which the microscope is calibrated so that the size of the objects being viewed can be found. It involves the use of an ocular meter or eyepiece micrometer, and a stage micrometer or an object micrometer (Figure 1.30).

The following operations are performed in sequence:

1. The eyepiece is removed from the microscope and the ocular meter is inserted between the lens and the diaphragm.
2. The stage micrometer is viewed through this eyepiece.
3. The number of divisions of the eyepiece and the stage micrometer is noted, say, x divisions of the stage micrometer = y divisions of the ocular meter and x/y division of the stage micrometer = one division of the ocular meter; x/y is called the least count.
4. The stage micrometer is then replaced by the slide with the specimen.
5. The number of divisions of the ocular meter equal to the length of object to be measured is observed.
6. This value is then multiplied with the least count to give the size of the object.

1.12 ELECTRON MICROSCOPY

1.12.1 Freeze Etching and Metal Shadowing

Freeze etching is used to reveal the detailed structures of microorganisms. In this procedure, a specimen frozen in liquid N_2 is fractured by striking it with a knife blade; the fractured specimen is then etched, that is, some of the ice is allowed to evaporate, which raises the surface layer of the specimen (Figure 1.31). The specimen is then exposed to vapors of a heavy metal while being held at a 45° angle to produce a shadow effect, after which it is rotated and exposed to vaporized carbon at a 90° angle to produce a replica of the surface. Any adhering biological specimen is removed, and the carbon replica is then viewed. This method reveals much detail of both internal and external surface structures and also eliminates some problems with artifacts that arise from chemical fixation and sectioning of biological specimens.

Experiment: Microscopic Examination of Microorganisms

Principle

Microorganisms are difficult to observe in a broth or when wet because there is very little contrast between them and the liquid in which they are suspended. In studying their properties and differentiating microorganisms into specific groups for diagnostic purposes, biological stains and staining procedures in conjunction with light microscopy have become major tools in microbiology.

Chemically, a stain may be defined as an organic compound containing a benzene ring plus a chromophore and auxochrome group. Numerous staining techniques are available for the visualization,

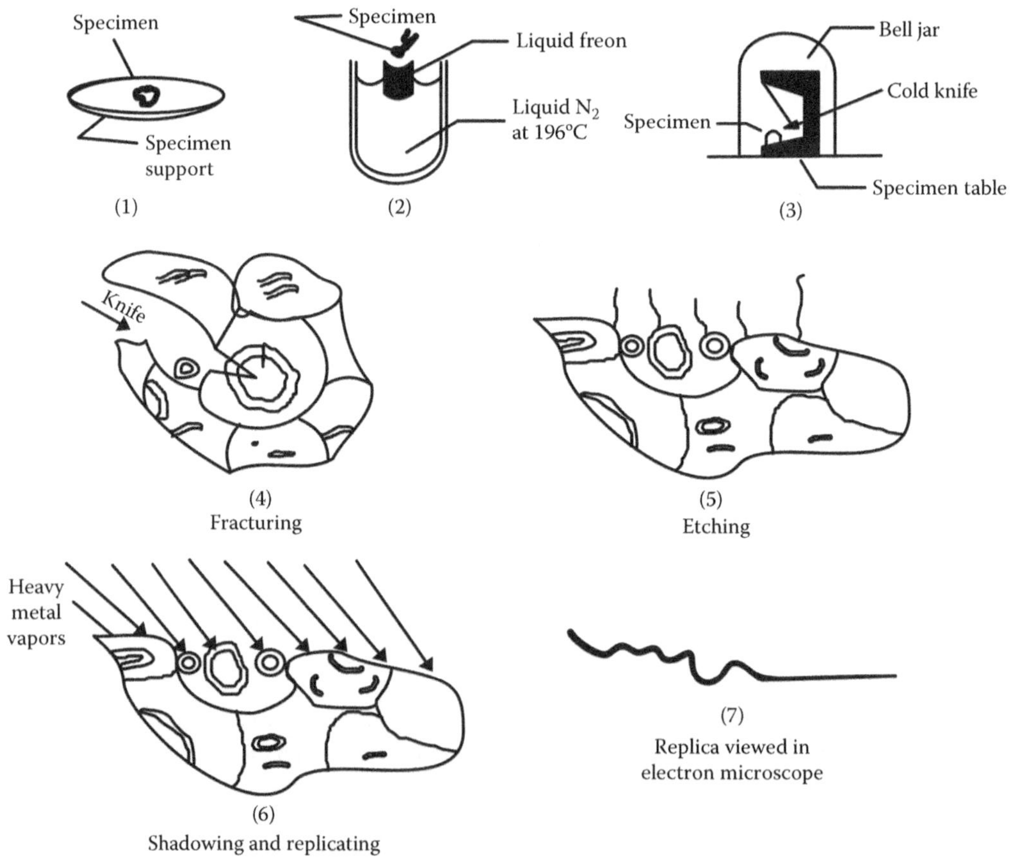

FIGURE 1.31 Diagrammatic view for the procedure for the formation of freeze-fracture replicas used for visualizing surface structure in conjunction with TEM.

differentiation, and separation of microorganisms. Different types of stains are used to differentiate between different groups of microorganisms. Two major types of staining techniques are as follows:

1. Simple staining or use of a single stain
2. Differential staining or use of two contrasting strains

A. Fungi

Place the material to be examined on a clean glass microscopic slide. Add a drop of lactophenol to the material and mix. Place a cover glass over the preparation and observe it under the appropriate microscope.

B. Algae and Cyanobacteria

Place the material to be examined on a clean glass microscopic slide. Place a cover slip over the smear preparation in simple water or 10% glycerol, and observe it under the appropriate microscope.

C. Bacteria

The success of a staining procedure depends on the preparation of a suitable smear of organisms. A good smear is one that when dried appears as a thin, whitish layer or film. A properly prepared smear withstands one or more washings during staining without any loss of organisms. The first step in preparing a bacterial smear differs according to the source of the organisms; those made from broth cultures or cultures from a solid medium require variations in technique.

Procedure

A. From Liquid Media (Figure 1.32)

1. Apply one or two loopfuls of suspended cells to a clean glass slide.
2. Spread it evenly over a small area on the slide.

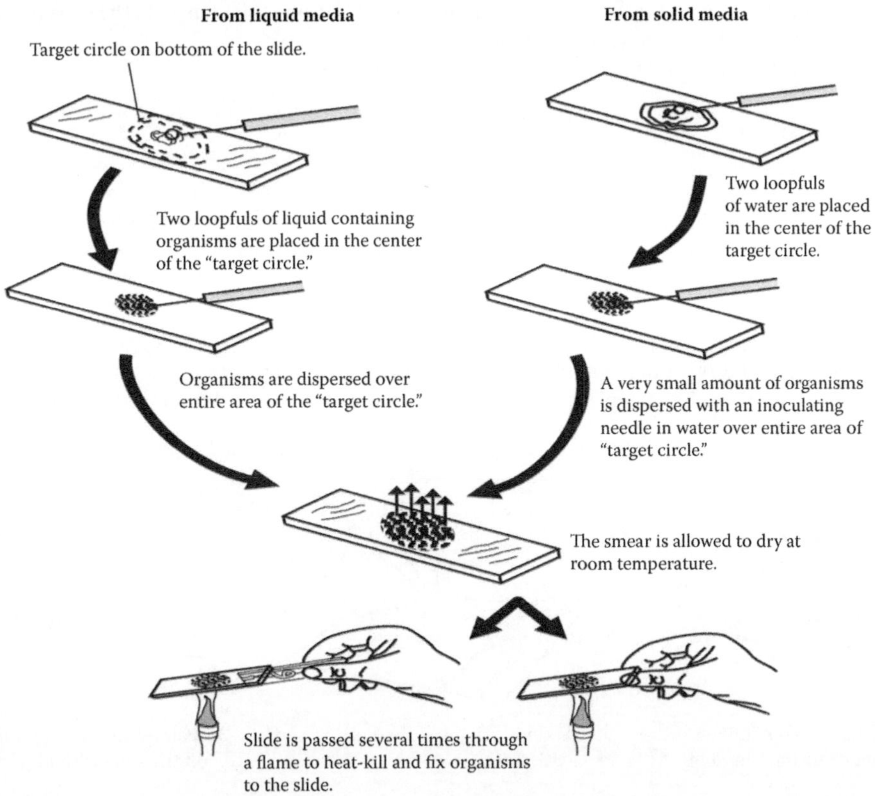

FIGURE 1.32 Procedure for preparing bacteria smear from liquid media and from solid media.

3. Allow the slide to dry by normal evaporation of water. Do not apply heat.
4. After the smear is completely dry, pass the slide over a Bunsen burner flame to heat-kill and fix the organisms to the slide.

B. From Solid Media (Figure 1.32)

1. Place a loopful of water on the slide.
2. Flame an inoculating needle and let it cool; pick up a very small amount of the organism, and mix it with the water on the slide.
3. Disperse the mixture over a small area at the center of the slide.
4. Allow the slide to dry by normal evaporation of the water.
5. Once the smear is completely dry, pass the slide over the flame of a Bunsen burner to heat fix the organism to the slide.

D. Simple Staining

Principle

In simple staining, the bacterial smear is stained with a single reagent (Figure 1.33). Basic stains with a positively charged chromogen are preferred, since bacterial nucleic acids and certain cell wall components carry a negative charge that strongly attracts and binds to the cationic chromogen. The purpose of simple staining is to elucidate the morphology and arrangement of bacterial cells. Commonly used basic stains are methylene blue, crystal blue, and carbolfuchsin.

Procedure

1. Prepare a bacterial smear as described in the smear preparation section.
2. Place the slide on the staining tray and flood with a required simple stain using the appropriate exposure time (carbolfuschin requires 15–30 seconds, crystal violet 20–60 seconds, and methylene blue 1–2 minutes).
3. Wash the smear with tap water to remove excess stain. During this step, hold the slide parallel to the stream of water to reduce the loss of cells.
4. Blot dry the slide and observe it under a light microscope.

E. Negative Staining

Principle

Negative staining requires the use of an acidic stain such as eosin or nigrosine (Figure 1.34). The acidic stain, with its negatively charged chromogen, will not penetrate the cells because of the negative charge on the surface. Therefore, unstained cells are easily visible against the colored background.

Procedure

1. Place a small drop of nigrosin close to one end of a clean slide.
2. Using a sterile technique, place a loopful of inoculums from the culture in the drop of nigrosin and mix.

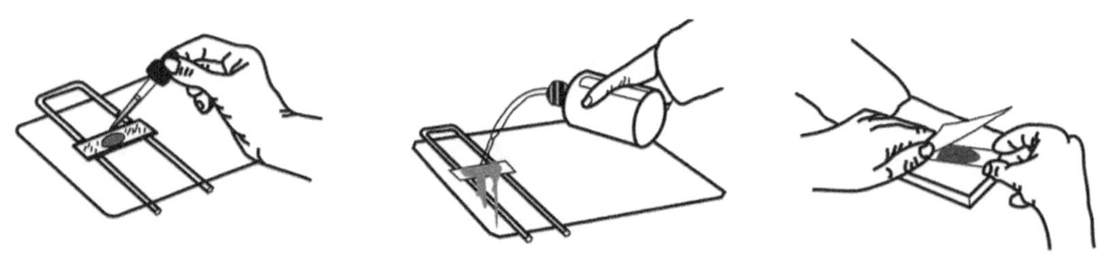

1. A bacterial smear is stained with methylene blue for 1 minute.
2. Stain is briefly washed off the slide with water.
3. Water drops are carefully blotted off the slide with bibulous paper.

FIGURE 1.33 Simple staining.

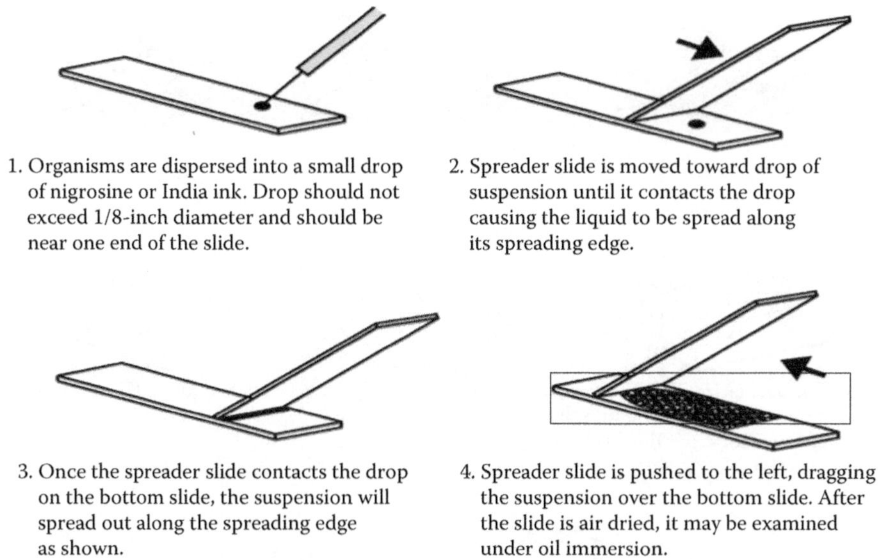

FIGURE 1.34 Negative staining of bacteria smear.

3. With the edge of a second slide held at a 30° angle and placed in front of the bacterial suspension, push the mixture to form a thin smear.
4. Air dry. Do not heat-fix the smear.
5. Examine the slide under oil immersion.

F. Gram Staining

The most important differential stain used in bacteriology is the Gram stain, which is named after Dr. Christian Gram. A Gram stain divides bacterial cells into two major groups, gram-positive and gram-negative, which make it an essential tool for the classification and differentiation of microorganisms.

Principle

In Gram staining, a bacterial smear is dried and then heat-fixed. It is then stained with crystal violet (primary stain), which is rinsed off and replaced with an iodine solution. The iodine acts as a mordant, that is, it binds the dye to the cell. The smear is then decolorized with alcohol and counterstained with safranin. In gram-positive organisms, the purple crystal violet dye, complexed with the iodine solution, is not removed by alcohol, and thus the organisms remain purple. On the other hand, the purple stain is removed from gram-negative organisms by alcohol and the colorless cells take up the red color of the safranin counterstain (Figure 1.35).

Procedure

1. Prepare a smear and heat-fix it.
2. Cover the smear with crystal violet; leave it for 20 seconds.
3. Briefly wash off the stain using a wash bottle of distilled water. Drain off excess water.
4. Flood the slide with Gram's iodine for about 30–60 seconds, and wash with water.
5. Flood the smear with 95% ethyl alcohol for 10–20 seconds or until no more purple dye runs off.
6. Immediately wash the slide with tap water.
7. Flood the smear with safranin and leave it for 20 seconds. Wash with tap water.
8. Immediately examine the slide using an optical microscope under oil immersion.

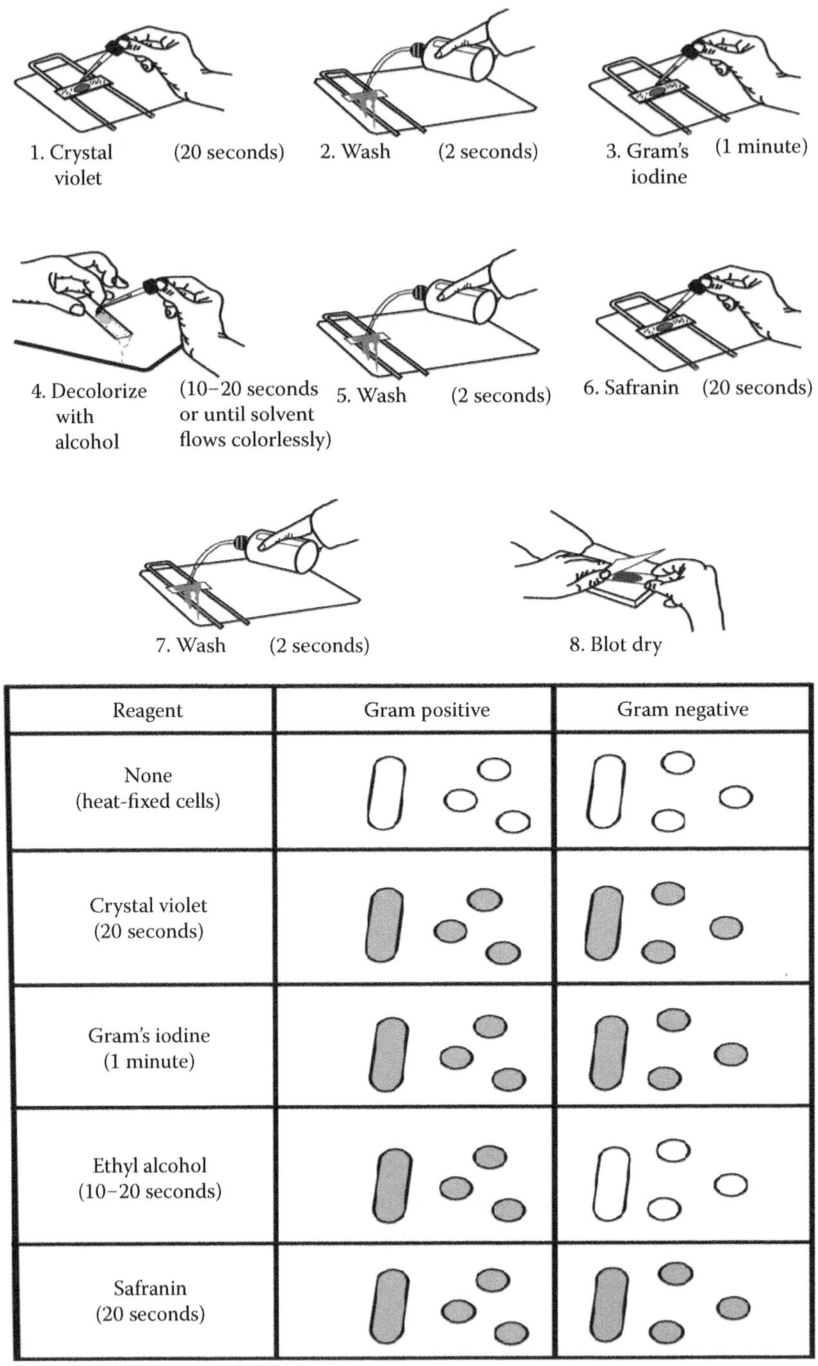

FIGURE 1.35 Steps in Gram staining.

Note

1. Preferably use a fresh culture for making smear.
2. Old cultures of gram-positive organisms lose their ability to retain the crystal violet.
3. Do not excessively decolorize with alcohol. The gram-positive organisms may also appear gram negative.
4. Use thin smears. It is difficult for the dyes to penetrate properly if thick smears are used.

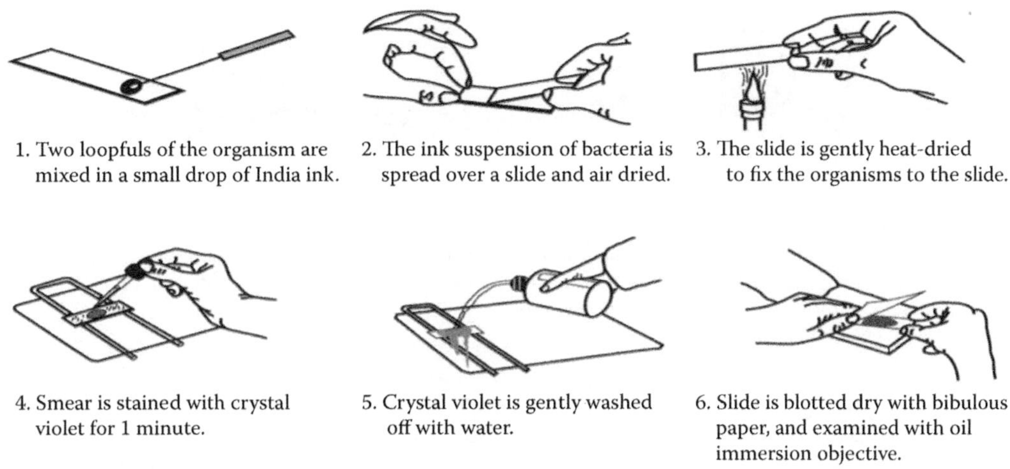

FIGURE 1.36 Capsular staining.

G. Capsular Staining

Principle

Some bacterial cells are surrounded by a pronounced gelatinous or slimy layer called a capsule. Capsules appear to be made up of glycoprotein or polypeptides. They can be observed by differential staining (Figure 1.36).

Procedure

1. Make a suspension of the organism in a drop of water on a clean slide.
2. Put a small drop of India ink next to it.
3. Spread the ink suspension of bacteria over the slide and air dry it. Caution: Do not heat-fix.
4. Stain the smear with crystal violet for 1 minute.
5. Wash off the crystal violet gently with water.
6. Gently blot dry and immediately examine the slide using an optical microscope under oil immersion.

H. Spore Staining

Species of bacteria belonging principally to the genera *Bacillus* and *Clostridium* produce extremely heat-resistant structures called endospores. Besides heat, endospores are resistant to many chemicals and starvation. The endospore is resistant to most stains, so special staining procedures are needed. There are two different methods that are mainly used for spore staining: (1) Schaeffer–Fulton and (2) Dorner.

I. Schaffer–Fulton Method

This method utilizes malachite green to stain an endospore and safranin to stain the vegetative portion of the cell. A properly formed spore will have a green endospore contained in a pink sporangium (Figure 1.37).

Procedure

1. Prepare a smear on a clean slide and heat-fix it.
2. Take a beaker with about an inch of water and bring it to boil.
3. Place the slide on the beaker.

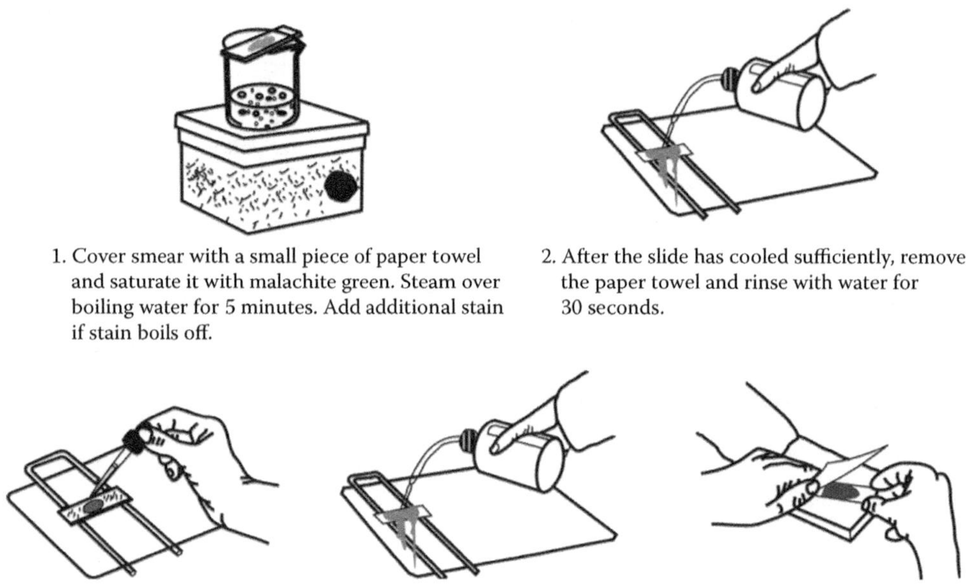

1. Cover smear with a small piece of paper towel and saturate it with malachite green. Steam over boiling water for 5 minutes. Add additional stain if stain boils off.
2. After the slide has cooled sufficiently, remove the paper towel and rinse with water for 30 seconds.
3. Counterstain with safranin for about 20 seconds
4. Rinse briefly with water to remove safranin.
5. Hot dry with bibulous paper and examine slide under oil immersion.

FIGURE 1.37 Steps in the Schaffer–Fulton method of endospore staining.

4. Flood the slide with malachite green. Steam the slide over a boiling water bath for 5 minutes. Add additional stain if stain comes off.
5. After the slide has cooled sufficiently, rinse it with water for 30 seconds.
6. Counterstain with safranin for about 30 seconds and then wash with tap water for 30 seconds.
7. Blot dry and microscopically examine the slide under oil immersion.

J. Dorner Method

The Dorner method for staining endospores produces a red spore within a colorless sporangium. Nigrosine is used to provide a dark background for contrast (Figure 1.38).

Procedure

1. Make a heavy suspension of bacteria by dispersing several loopfuls of bacteria in five drops of sterile water.
2. Add five drops of carbolfuchsin to the bacterial suspension.
3. Heat the carbolfuschin suspension of bacteria in a beaker of boiling water for 10 minutes.
4. Mix several loopfuls of bacteria in a drop of nigrosine on the slide.
5. Spread the nigrosine–bacteria mixture on the slide.
6. Allow the smear to air dry. Examine the slide under oil immersion.

K. Acid Fast Staining

Principle

Acid fast staining is useful for identifying bacteria that has a waxy lipid cell wall. Most of these organisms are members of a group of the genus *Mycobacterium*. These organisms have a gram-positive cell wall, but the lipid in the cell wall prevents staining with Gram-stain dyes (Figure 1.39). In the Ziel–Neelsen method, three different reagents are used:

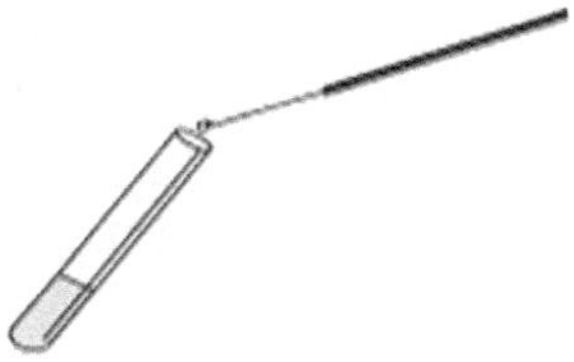

1. Make a heavy suspension of bacteria by dispersing several loopfuls of bacteria in five drops of sterile water.

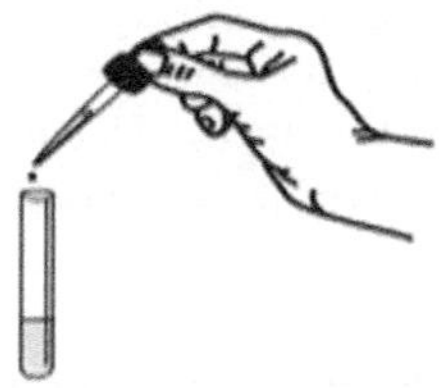

2. Add five drops of carbol fuchsin to the bacterial suspension.

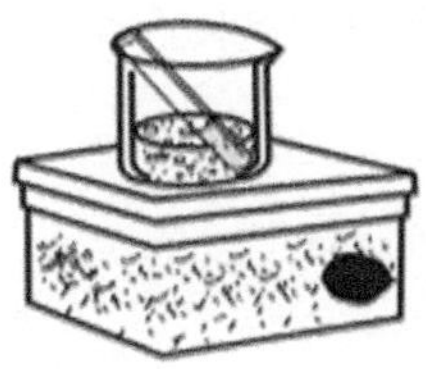

3. Heat the carbol fuchsin suspension of bacteria in beaker of boiling water for 10 minutes.

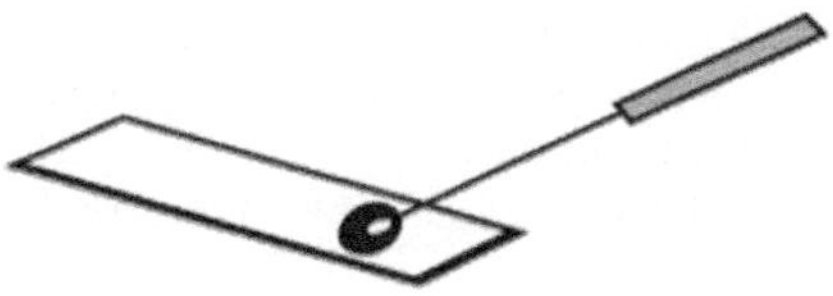

4. Mix several loopfuls of bacteria in a drop of nigrosine on the slide.

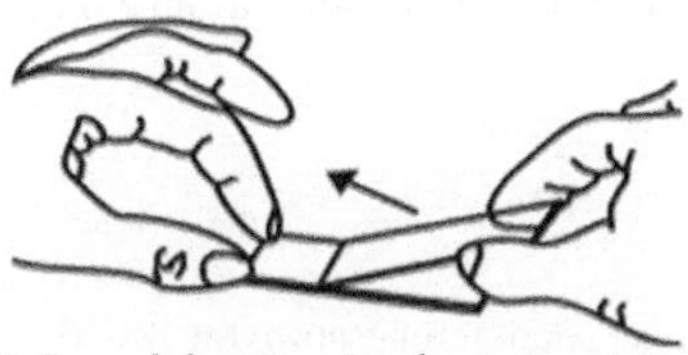

5. Spread the nigrosine-bacteria mixture on the slide.

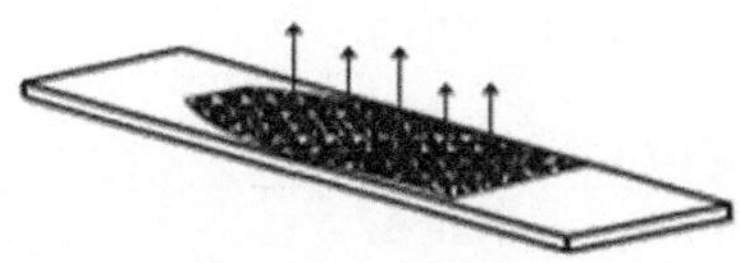

6. Allow the smear to air dry. Examine the slide under oil immersion.

FIGURE 1.38 Steps in the Dorner method of endospore staining.

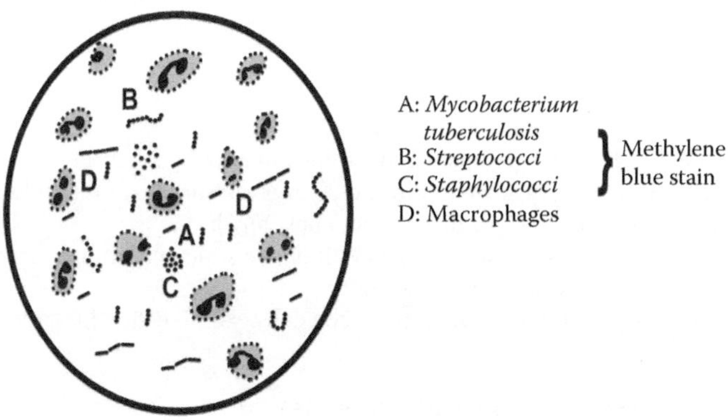

FIGURE 1.39 Various forms of acid fast stain bacteria.

1. Primary stain: Carbolfuschin, a phenolic stain, is driven into the waxy cell wall with steam.
2. Decolorizing agent: Acid alcohol (3% HCl + 9.5% ethanol) is used as the decolorizing agent. The mycobacterial cells are acid fast and not decolorized, and thus they retain the primary stain. Nonmycobacteria are, however, decolorized by acid alcohol.

3. Counterstain: Methylene blue is used as the final reagent to stain previously decolorized cells. The colorless nonmycobacteria take up the blue color; so, they contrast with the pink acid-fast bacteria that are not decolorized.

Procedure

1. Prepare a smear of the material and heat-fix it.
2. Place the slide over a beaker of boiling water and cover the slide with carbolfuschin. Cover it with a paper towel to prevent the dye from flowing out.
3. Keep the slide covered with the stain and steam for 5 minutes.
4. Remove the paper and wash off the carbolfuschin using tap water.
5. Flush all freely removable stain with acid alcohol.
6. Flood the entire slide with acid alcohol and allow it to stand for 1–2 minutes.
7. Wash with water for 5 seconds.
8. Counterstain for about 30–45 seconds with methylene blue.
9. Wash with water.
10. Blot dry carefully and microscopically examine the slide under the oil immersion lens.

L. Flagella Staining

Some bacteria have flagella for motility. As their width is below the resolving power of the microscope, they cannot be seen in a light microscope. Flagella can be viewed if they are dyed with a special stain that precipitates on them, making them appear thicker. Leifson's method accomplishes this by using a single staining reagent that utilizes pararosaniline as a staining reagent and tannic acid as a mordant. The arrangement of flagella on bacteria is usually characteristic of the organism and can aid in its identification (Figure 1.40).

Procedure

A. Culture preparation

1. Inoculate nutrient broth with the organism and incubate at room temperature for 18–20 hours.
2. Add 0.25 mL of formalin to the culture, mix by shaking, and let it stand for 15 minutes.
3. Fill the tube to within 1 cm of its top with distilled water, and mix and centrifuge for 3 minutes.
4. Pour away the supernatant fluid without disturbing the pellet.
5. Resuspend the pellet in about 2 mL of distilled water.
6. Dilute the suspension with additional distilled water until the suspension is barely turbid.

B. Staining procedure

1. Heat a clean slide in a Bunsen burner flame.
2. While the slide is still hot, mark an oval outline on the slide with pencil.
3. Place several loopfuls of the organism at the right end of the cooled slide.
4. Tilt the slide to allow the organism to flow down over the surface of the slide.
5. Allow the smear to completely air-dry. Do not apply any heat.
6. Cover the smear with Leifson's stain and leave it on the slide until all the alcohol evaporates.
7. Wash gently to remove the stain from the slide.
8. Allow the stained organisms to air-dry. Microscopically examine the slide under oil immersion.

Experiment: Scanning Electron Microscopy

To perform an experiment for preparing sample for SEM, *Anabaena cylindrica* cells growing in log phase, are harvested by centrifugation and prefix in a culture medium by the addition of an equal volume of 1% glutaraldehyde in phosphate buffer. Cells are allowed to stand for 30 minutes on ice, pelletized, suspended in phosphate buffer with 2% glutaraldehyde, and incubate for 1 hour at room temperature. Samples are then washed with phosphate buffer, post-fix in 1% osmium tetraoxide in the same buffer, and wash once in distilled water. After that the samples are kept on carbon stubs and gold coating is done with fine coat ion sputter JFC 1100. Samples are observed under an electron microscope (JEOL JSM-840; Figure 1.41).

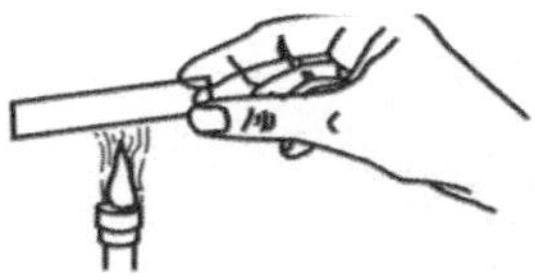

1. Heat a clean slide in blue portion of Bunsen burner flame.

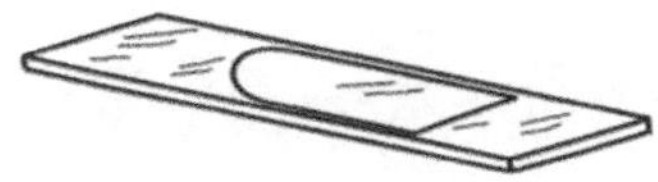

2. While the slide is still hot, mark with a China marking pencil as shown.

3. Place several loopfuls of organisms at the right end of cooled slide as shown.

4. Tilt slide to allow organisms to flow down over the surface of the slide.

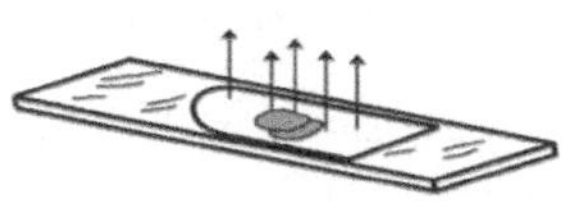

5. Allow the smear to completely air dry. Do not apply any heat.

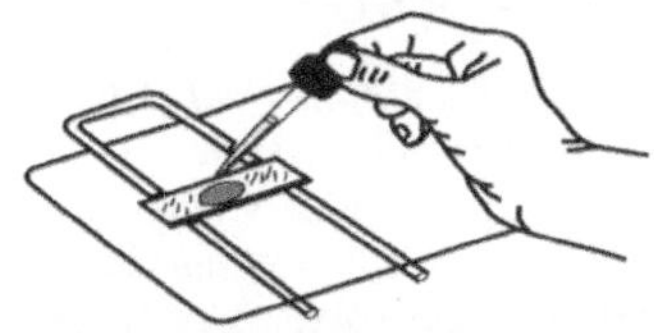

6. Cover the smear with Leifson's stain and leave it on the slide until all the alcohol has evaporated.

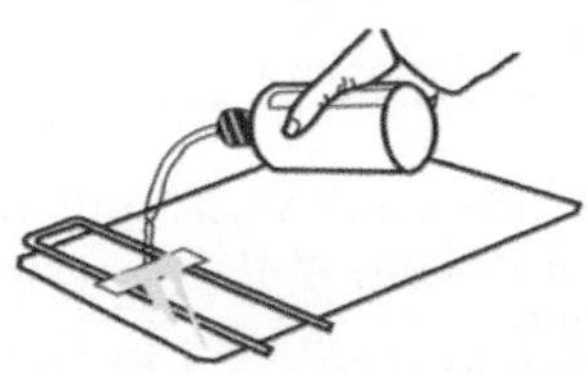

7. Wash gently to remove the stain from the slide.

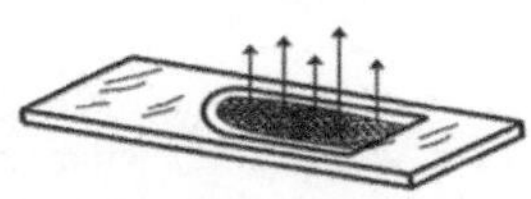

8. Allow the stained organisms to air dry. Examine under oil immersion.

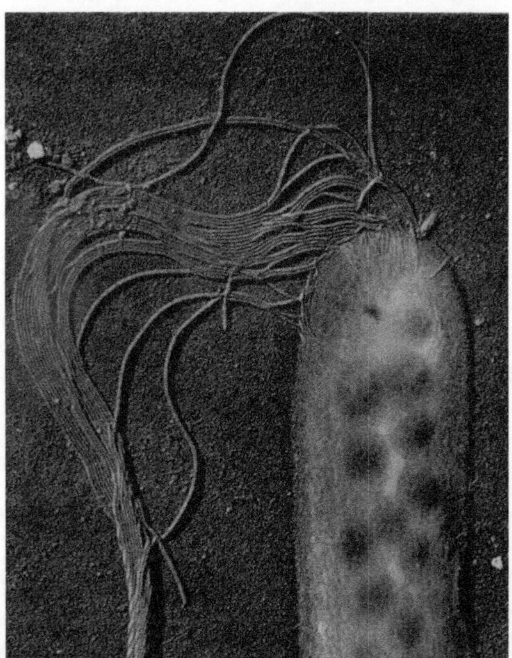

FIGURE 1.40 Steps in flagella staining and an electron microscope view of bacterial flagella.

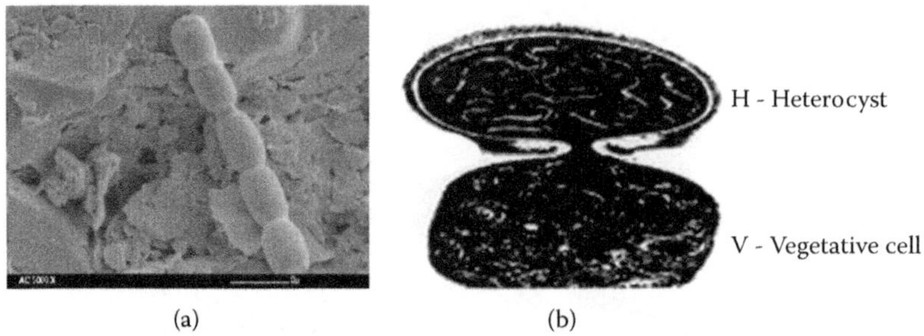

FIGURE 1.41 Scanning electron micrograph image of *Anabaena* sp.: (a) cell arrangement and (b) differentiation.

SUGGESTED READING

Benson, H. J. 1998. *Microbiological Applications Laboratory Manual in General Microbiology,* 7th ed. New York: WCB/McGraw-Hill.

Boatman, E. S., M. W. Berns, R. J. Walter, and J. S. Foster. 1987. Today's microscopy. *BioScience* 37: 384–394.

Bradbury, S. 1997. *Introduction to Light Microscopy*, 2nd ed. New York: Springer-Verlag.

Clark, G. L. 1961. *The Encyclopedia of Microscopy*. New York: Van Nostrand Reinhold.

Cosslett, V. E. 1966. *Modern Microscopy or Seeing the Very Small*. Ithaca, NY: Cornell University Press.

Gerhard, P., R. G. E. Murray, W. A. Wood, and N. R. Krieg, eds. 1994. *Methods for General and Molecular Bacteriology*. Washington, DC: American Society for Microbiology.

Hawker, L. E., and A. H. Linton. 1974. *Micro-Organisms Function Form and Environment*. London: Edward Arnold Ltd.

Minsky, M. 1988. "Memoir on Inventing the Confocal Scanning Microscope." *Scanning* 10: 128–138.

Perkins, G. A., and T. G. Frey. 2000. Microscopy, Optical. In *Encyclopedia of Microbiology*, 2nd ed., vol. 3, edited by J. Lederberg, 288–306. San Diego: Academic Press.

Pernthaler, A., A. E. Dekas, C. T. Brown, S. K. Goffredi, T. Embaye, and V. J. Orphan. 2008. "Diverse Syntrophic Partnerships from Deep-Sea Methane Vents Revealed by Direct Cell Capture and Metagenomics." *Proceedings of the National Academy of Sciences* 105: 7052–7057.

Porter, T. 1984. "Compositing Digital Images." *Computer Graphics* 18: 253–259.

Rawlins, D. J. 1992. *Light Microscopy*. Philadelphia: Coronet Books.

Rochow, T. G. 1994. *Introduction to Microscopy by Means of Light, Electrons, X-rays, or Acoustics*. New York: Plenum.

Sheppard, C. J. R. 1997. "Confocal Microscopy." XVI Meeting of the Brazilian Society for Electron Microscopy. Caxambu, Brazil, keynote.

Slayter, E. M. 1992. *Light and Electron Microscopy*. New York: Cambridge University Press.

IMPORTANT LINKS

1. Light microscope: http://www.sigmainstruments.org/laboratory-instruments.html#microscope
2. Dark field microscope: http://www.olympusmicro.com/primer/techniques/darkfieldindex.html
3. Phase contrast microscope: http://www.olympusmicro.com/primer/techniques/phasecontrast/phaseindex.html
4. Fluorescence microscope: http://microscopeinternational.com/Fluorescence-Microscopes
5. Transmission electron microscope: http://www.jeol.com/PRODUCTS/ElectronOptics/TransmissionElectronMicroscopesTEM/300kV/J EM3100F/tabid/128/Default.aspx
6. Scanning electron microscope: http://www.jeol.com/PRODUCTS/ElectronOptics/ScanningElectronMicroscopesSEM/tabid/92/Default.aspx
7. Confocal microscope: http://www.zeiss.de/c12567be0045acf1/ContentsFrame/731dfc84f15572fac12574310056687c

2 Micrometry

2.1 INTRODUCTION

Micrometry is the measurement of microorganisms. Since microorganisms can only be seen under a microscope, a suitable scale for their measurements should be located somewhere in the microscope itself. For this, an ocular micrometer serves as a scale or rule. An ocular micrometer consists of disks of glass upon which lines are etched. There are in general use two practical methods of measuring microscopic objects: (1) by means of the ocular micrometer and (2) by means of camera lucida sketches. The latter method is the best one for measuring spores and bacterial size in taxonomy (Figure 2.1a and b).

A stage micrometer and an ocular micrometer are necessary for measuring microscopic objects. A stage micrometer should be ruled in tenths and one-hundredths of a millimeter. It does not matter what the spacing is in an ocular micrometer, except that the lines must be at equal distances from one another. As a matter of fact, the ocular micrometer is generally ruled in tenths of a millimeter, but this ruling is more or less magnified by the lens of the ocular.

2.2 STRUCTURE OF AN OCULAR MICROMETER

Illustrated in Figure 2.2 is a modern microscope eyepiece (often termed an ocular) equipped with an internal reticle scale. Also presented in the figure is a stage micrometer, which contains a small, metalized millimeter ruler that is subdivided into increments of 10 and 100 μm. Juxtaposing graduations on the eyepiece reticle with those on the stage micrometer enables the microscopist to calibrate the reticle gauge and perform linear measurements on specimens.

2.3 CONJUGATE IMAGE-FORMING FOCAL PLANES

The principle of a transfer scale has been in use since the earliest days of humankind, and it can be applied to specimens studied under a microscope even though they may not be accessible for direct measurement with a standardized scale. There are various approaches for using a transfer scale in microscopy, including placing the scale on a transparent material for use with a drawing tube or conducting measurements directly on a projected image. An alternative method is to photograph or engrave the scale onto a glass element, which can be placed in the optical path at one of the image-forming conjugate planes of the microscope, so that it can be observed in sharp focus superimposed on the specimen image. Before a quantitative measurement is done, the arbitrary divisions of the transfer scale must be calibrated by comparison with the absolute graduations of a master scale, such as a stage micrometer.

The critical requirement for superimposing a graduated scale onto a specimen in such a manner that it can be imaged together with the specimen is to place the scale in a suitable conjugate plane of the microscope. Two primary sets of principle conjugate focal planes occur along the optical axis of a properly focused and aligned compound microscope. One set of planes consists of four image-forming or field planes (Figure 2.3), whereas the other set consists of four illumination or aperture planes. Each plane within a set is termed conjugate with the other planes in the set because the planes are simultaneously in focus and can be viewed superimposed on one another when observing specimens through the microscope. An object placed in one plane of a conjugate set appears in sharp focus at all the other conjugate planes of the same set. Obviously, if a scale must be visible and in focus while observing the image of a specimen, then the scale must be placed in one of the image-forming sets of planes.

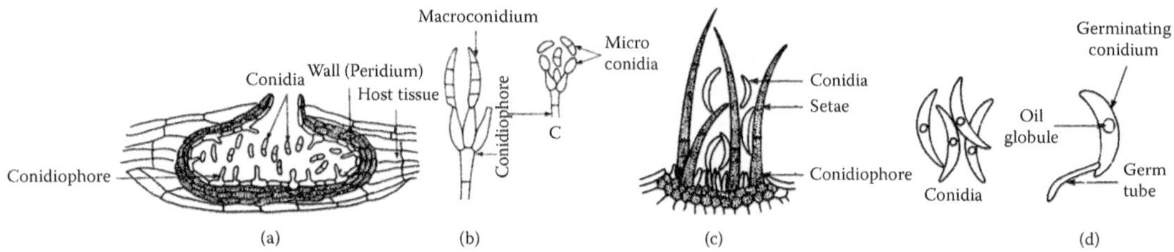

FIGURE 2.1 Camera lucida sketches of (a) the pycnidium of a fungus, (b) *Fusarium* sp. macro and micro-conidium, (c) acervulus, setae, and conidia of *Colletotrichum gloeosporoides*, and (d) the conidial structure and conidial germination of *Colletotrichum gloeosporoides*.

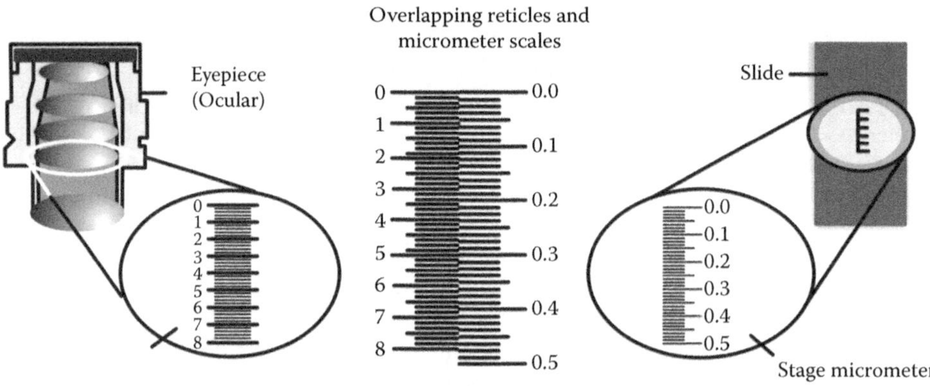

FIGURE 2.2 Eyepiece reticles and stage micrometers.

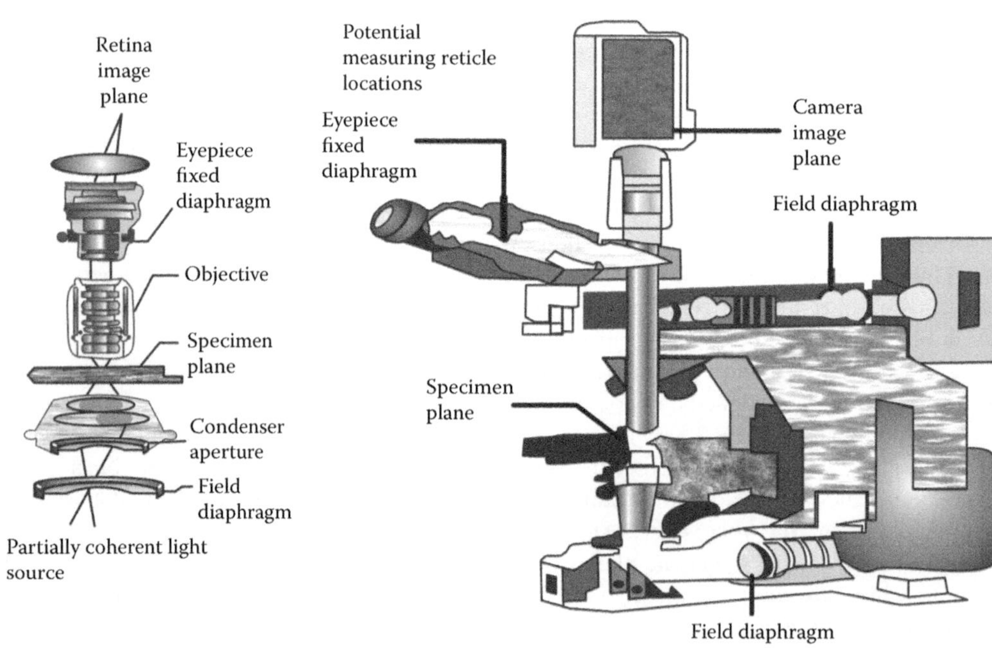

FIGURE 2.3 Image-forming conjugate planes in the optical microscope.

2.4 EYEPIECE DESIGNS

The image-forming conjugate in a set has an alternative location, that is, an intermediate image plane, at which a measuring scale can be inserted. This plane coincides with the diaphragm fixed to the eyepiece, which is generally easily accessible (Figure 2.4). Nearly any eyepiece can be fitted with a scale in the focal plane, converting the eyepiece into a measuring device for specimen features observed through the microscope. Eyepiece scales are often referred to as reticles, although the terms reticules or graticules are commonly used in the same sense and are frequently encountered in literature. The most common conventional eyepieces differ with respect to the physical location of the fixed diaphragm. Some eyepiece designs position the diaphragm at the center of the unit (between the lenses), whereas other models have a fixed diaphragm at the base of the eyepiece, beneath and external to the lens assembly. In both eyepiece styles the field diaphragm is located at the intermediate image focal plane, but the external diaphragm design is preferred for measurement because any reticle, pointer, or other scale will be outside the optical system of the eyepiece.

2.4.1 Types of Eyepieces

One of the simplest eyepiece designs, known as the Huygenian (or Huygens) eyepiece, consists of two plano-convex lenses mounted with their convex faces oriented toward the objective (Figure 2.4). The lens nearest to the eye is referred to as the eye lens, and the one closest to the objective is termed the field lens. Eyepieces of this type are uncorrected for optical aberrations and have the disadvantage of the image plane being located between the two lenses (internal diaphragm). Therefore, reticle accuracy is affected by aberrations of the eye lens alone, whereas the specimen image suffers from any optical defects arising in the field lens as well.

The Ramsden eyepiece has a construction motif similar to that of the Huygenian eyepiece except that the field lens is oriented with the plane surface facing the objective (Figure 2.4). In addition, the focal plane and diaphragm are located outside the optical system (external diaphragm design), just beneath the field lens. A reticle or similar scale placed in the diaphragm experiences less distortion than with the Huygenian design, and any optical aberrations of the eyepiece affect the specimen and reticle images equally. One of the primary applications of the Ramsden eyepiece is in micrometry.

For the introduction of infinity-corrected optical systems, compensating eyepieces are utilized to assist in the correction of chromatic aberration. These eyepieces are generally constructed with two separate lenses, one or both of which are doublets or triplets (Figure 2.4, widefield eyepiece). Compensating eyepieces can be identified by the color fringe appearing around the inside edge of the fixed diaphragm when the eyepiece is viewed in front of a bright light source (ordinary eyepieces display a blue fringe, whereas compensating eyepieces exhibit a yellow, orange, or blue fringe).

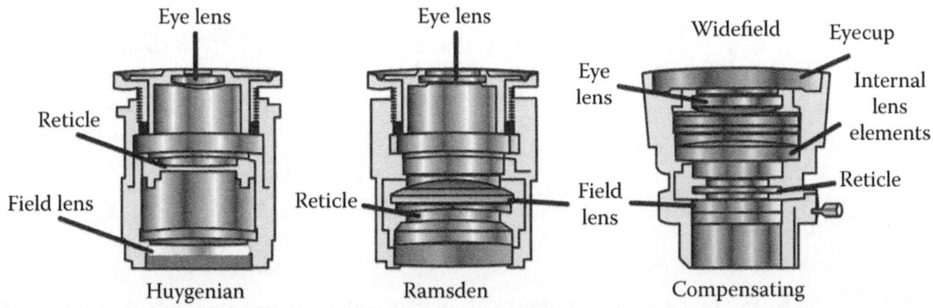

FIGURE 2.4 Microscopy eyepiece anatomy and reticle location.

Chromatic difference of magnification, an aberration common to all high-power objectives, can be corrected by coupling the optical system to a compensating eyepiece. In addition, compensating eyepieces are designed to correct image curvature to a limited extent.

2.5 STAGE MICROMETER

Linear measurement requires the comparison of the object to be measured with a standardized scale, such as a ruler. In utilizing eyepiece reticles or micrometer eyepieces for measurements in a microscope, the arbitrary units of the transfer scale (reticle), which is superimposed on the specimen image, must be converted to absolute units such as millimeters or micrometers. The calibration of reticle scale graduations is commonly performed by imaging a stage micrometer with the same objective used for specimen measurements (Figure 2.5). Proper calibration involves determining an absolute distance on the stage micrometer, imaged in place of a specimen, which corresponds to one division of the scale in the eyepiece reticle. This value is often referred to as the micrometer value or calibration factor for that particular objective. Once the value is determined, the size of any specimen feature may be calculated by multiplying the number of eyepiece reticle divisions spanned by the feature with the calibration factor for the objective in use.

Stage micrometers designed for applications using transmitted-light microscopes consist of standard-sized microscope slides (1 × 3 in.) that have scales of defined length attached directly to the surface or, preferably, sandwiched beneath a cover glass of known thickness. Micrometers commonly have a graduated scale, either 1 or 2 mm in length, subdivided into units that are one-tenth of a millimeter in length (100-μm units). Each 100-μm unit is further divided into 10 equal sections, and the smallest graduation represents 10 μm.

2.5.1 Counting Chambers Stage Micrometer

Included in the broad category of stage micrometers are calibration scales, finder reticles, and counting chambers (Figure 2.6). Finder reticles are utilized for locating a region of interest on a specimen, whereas counting chambers are designed to enable particle and cell counts in a specific volume of liquid. Counting chambers are widely used for counting blood cells and spermatozoa and consist of a thick glass slide (Figure 2.6) that has a central polished and ruled platform. The platform is positioned a short distance (typically 0.1 mm) beneath twin polished coverslip supports to create a chamber that can be filled with a precise quantity of fluid. In practice, a clean glass coverslip is placed over the chamber and centrally positioned on the polished supports. The gap between the ruled counting platform and the coverslip equals 100 μm, and the ruled (engraved) face is divided into squares of exact dimensions. As a result, the volume of the liquid placed in the chamber can be easily calculated to yield an accurate analysis of the number of particles (cells) present per unit volume in a suspension.

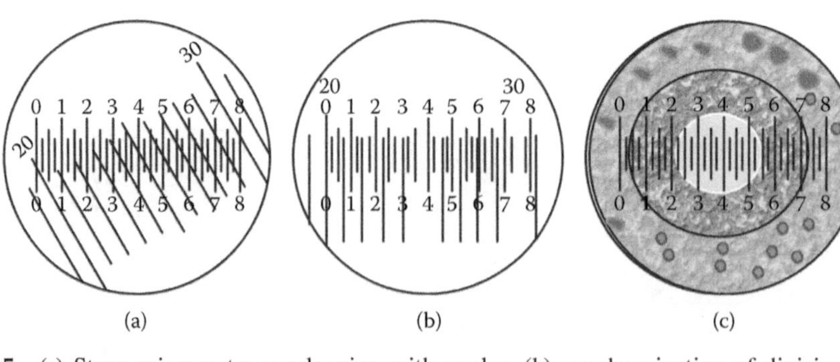

FIGURE 2.5 (a) Stage micrometer overlapping with ocular, (b) synchronization of division of stage and ocular micrometer, and (c) with object.

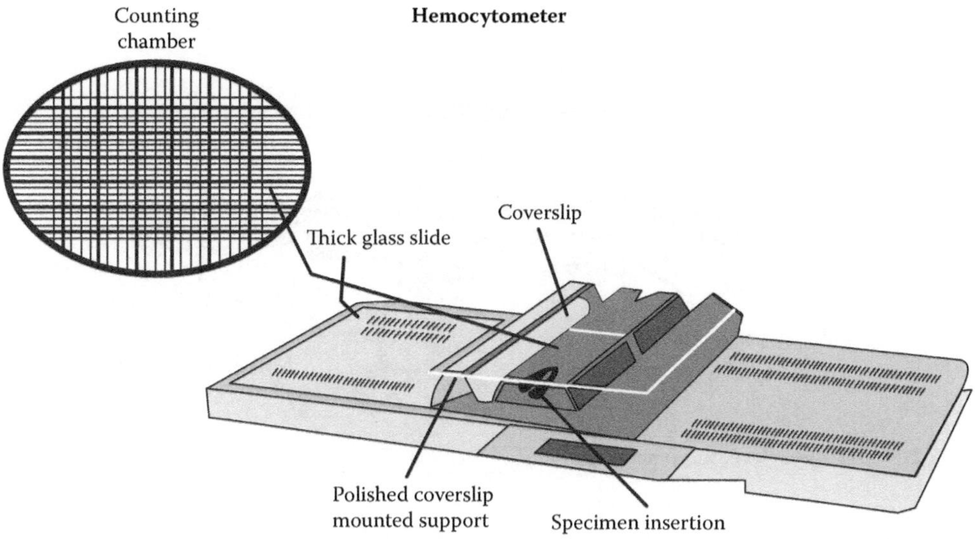

FIGURE 2.6 Counting chamber stage micrometer.

2.6 FILAR EYEPIECE MICROMETER

The standard eyepiece reticle, when combined with a precision stage micrometer, provides a rapid, convenient, and accurate means of conducting measurements in the microscope. However, for easier and more precise measurements (with greater objectivity), a specialized vernier micrometer eyepiece known as the filar eyepiece micrometer is often considered essential. This specialized eyepiece micrometer utilizes the same principle as a standard eyepiece and reticle combination but features a moveable line rule (or line rule group) in addition to a fixed or mobile graduated scale positioned in the focal plane. A filar micrometer avoids the necessity of estimating fractions of a division on a stage micrometer (a difficult and subjective maneuver), which can lead to considerable error.

The mobile line rule group in a filar micrometer is designed to translate across the field of view, traversing a fixed vernier scale, by means of a precision screw mechanism that is operated by rotating an external drum. In general applications, a single rule or other reference point (depending on the particular design) is aligned with one end of the specimen feature to be measured and a reading of the calibrated drum is noted. The drum is then rotated to move the reference line across the specimen feature, and a second reading is taken on the drum scale. The difference between the two readings yields an apparent linear dimension of the specimen feature measured and, when calibrated with a stage micrometer, enables an absolute determination of the feature size.

Some filar micrometer design variations incorporate an additional movement of the reticle scale by the external drum, which allows zeroing of the drum scale after the reference line is positioned at the first edge of the object to be measured. This feature enables each measurement to begin with the drum scale set at zero and avoids the necessity of determining the difference between the two drum readings. For most filar micrometers, the primary reticle scale has a travel distance of 10 mm. The scale is also divided into 100 graduations with each division representing 0.1 mm. The drum of the micrometer screw is also divided into 100 intervals, so that one interval of the drum division corresponds to a 0.1 interval of the eyepiece scale. One full rotation of the drum translates the measuring rule (line) across one interval of the eyepiece scale.

Several modern filar eyepiece micrometer styles contain an internal zoom lens system that eases calibration of the micrometer with different objectives. The lower portion of the eyepiece contains

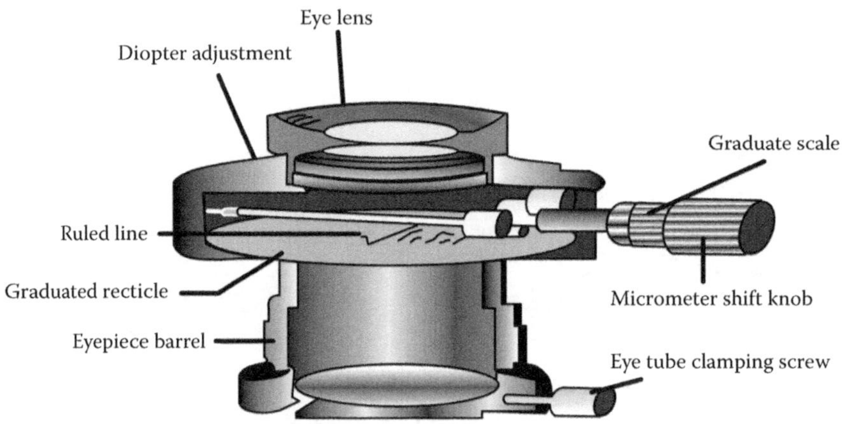

FIGURE 2.7 Filar eyepiece micrometer anatomy.

a graduated ring that can be rotated to optically alter the effective tube length in order to superimpose graduations of the stage micrometer directly on the internal scale of the filar micrometer (Figure 2.7).

2.7 PRINCIPLE OF FILAR EYEPIECE MICROMETER

The basic principle is applicable to the measurement of specimens observed under the microscope, but in practice it is often not possible with a compound microscope to place a ruler in direct contact with a specimen (although this is often done in low-magnification stereomicroscopy). Alternative mechanisms for performing measurements at high magnifications in compound optical microscopy must be used, and the most common of these is the application of eyepiece reticles in combination with stage micrometers. A majority of the measurements made with compound microscopes fall into the size range of 0.2 μm to 25 mm (the average field diameter of widefield eyepieces). Horizontal distances below 0.2 μm are beneath the resolving power of compound microscopes, and lengths larger than the field of view of a widefield eyepiece are usually (and far more conveniently) measured with a stereomicroscope (Benson 2001).

A method has been presented that uses a longitudinal mode splitting laser to sense microdisplacements and measure air refractivity with very simple construction, large measurement ranges, high sensitivity, and precision. The phase differences of two power curves of the split modes can be used to judge the movement direction, and the number of lamb dips and mode-change dips can be used to measure micrometer displacement. Because of anomalous dispersion, the relative variation of beat frequency is the function of the change of cavity length, which is the basis of measuring nanometer displacement (Zhang and Tang 1994).

2.8 WORKING

2.8.1 Components of the Ocular Micrometer

1. If eyepieces that contain ocular micrometers are available, replace the eyepiece in the microscope with one of them. If it is necessary to insert an ocular micrometer in your eyepiece, find out from your instructor whether it is to be inserted below the bottom lens or placed between the two lenses within the eyepiece. In either case, great care must be

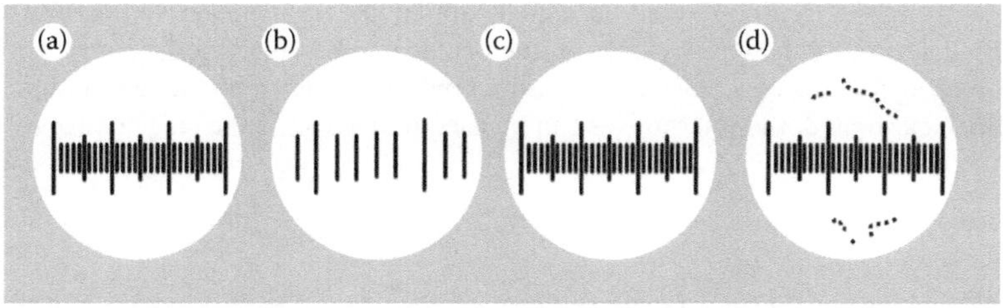

FIGURE 2.8 (a), (b), (c), and (d) show the various steps involved with the standardization of the ocular micrometer in measuring microbial cells (1μ = 1 × 10^{-6} m).

taken to avoid dropping the eyepiece or reassembling the lenses incorrectly. Eyepieces should be disassembled only with your instructor's prior approval. During the placement of the eyepiece, the graduation should be on the upper surface of the glass disk (Figure 2.8).

2. Place the stage micrometer on the stage and center it exactly over the light source.
3. With the low-power (10 μm) objective in position, bring the graduations of the stage micrometer into focus by using the coarse adjustment knob. Reduce the lighting.
4. Rotate the eyepiece until the graduations of the ocular micrometer lie parallel to the lines of the stage micrometer.
5. If a low-power objective is the objective to be calibrated, proceed to step 8.
6. If a high-dry objective is to be calibrated, swing it into position and proceed to step 8.
7. If an oil immersion lens is to be calibrated, place a drop of immersion oil on the stage micrometer, swing the oil immersion lens into position, and bring the lines into focus; then, proceed to the next step.
8. Move the stage micrometer laterally until the lines at one end coincide. Then look for another line on the ocular micrometer that coincides exactly with a line on the stage micrometer. Occasionally one stage micrometer division includes an even number of ocular divisions, as shown in the following example. In most instances, however, several stage graduations are involved. In this case, divide the number of stage micrometer divisions by the number of ocular divisions that coincide. The figure you get will be that part of a stage micrometer division that is seen in an ocular division. This value must then be multiplied by 0.01 mm to get the amount of each ocular division.

   Example: Note that 3 divisions of the stage micrometer line up with 20 divisions of the ocular micrometer:

$$\begin{aligned}\text{Each ocular division} &= 3/20 \times 0.01 \\ &= 0.0015\ \mu\text{m} \\ &= 1.5\ \mu\text{m}\end{aligned}$$

9. Replace the stage micrometer with the slides of organisms to be measured.

The micrometric technique can be used for measuring the size of protozoa, algae, fungi, and bacteria.

Experiment: Calibration of Ocular Micrometer

The measurement of minute objects seen under a microscope is called micrometry. The unit of such measurement is called a micron (μ or micrometer). 1 μm equals 1/1000 of a mm. The ocular or eyepiece micrometer consists of a diaphragm with a graduated scale unit. The ruled lines superimpose certain distance markers on the microscopic field, on which parallel lines that are approximately 0.01 mm apart are etched. By determining how many units of the ocular micrometer superimpose a known distance on the stage micrometer, the exact distance measured by each division on the microscopic field can be calculated. Once calibrated, the micrometer can be used to determine the size of various microscopic objects.

Procedure

1. Carefully, place the calibrated ocular micrometer inside the eyepiece and put it back inside the tube of microscope.
2. Put a drop of oil immersion in the centre of the prepared slide of object to be measured and focus properly.
3. Count the number of ocular divisions that occupies a single cell/object. Measure size of 5–10 cells/object and take average of it.
4. Find out the accurate size of the cell/object by multiplying with calibration value of ocular micrometer.

Results

For example, if the cell/object occupies 5 divisions of ocular micrometer and the calibration value of the ocular micrometer is 2.5 μm, the size will be 5×2.5 μm = 12.5 μm.

SUGGESTED READING

Beach, C. B., and F. M. McMurry et al. 1911. *The New Student's Reference Work*, Vol. 3. Chicago: F. E. Compton and Company.

Benson. 2001. *Microbiological Application: Laboratory Manual in General Microbiology*, 8th ed. New York: McGraw–Hill Companies.

Zhang, S., and M. Tang. 1994. Principle for measurement of micrometer and nanometer displacement and air refractivity based on laser mode split technology and lasing action. *Optical Engineering* 33: 3381–87.

IMPORTANT LINKS

1. Stage micrometer: http://www.opticsplanet.net/nikon-stage-micrometer-a-1mm-in-mbm11100.html
2. Filar eyepiece micrometer: http://www.lomoplc.com/Micrometer%20Eyepieces.htm

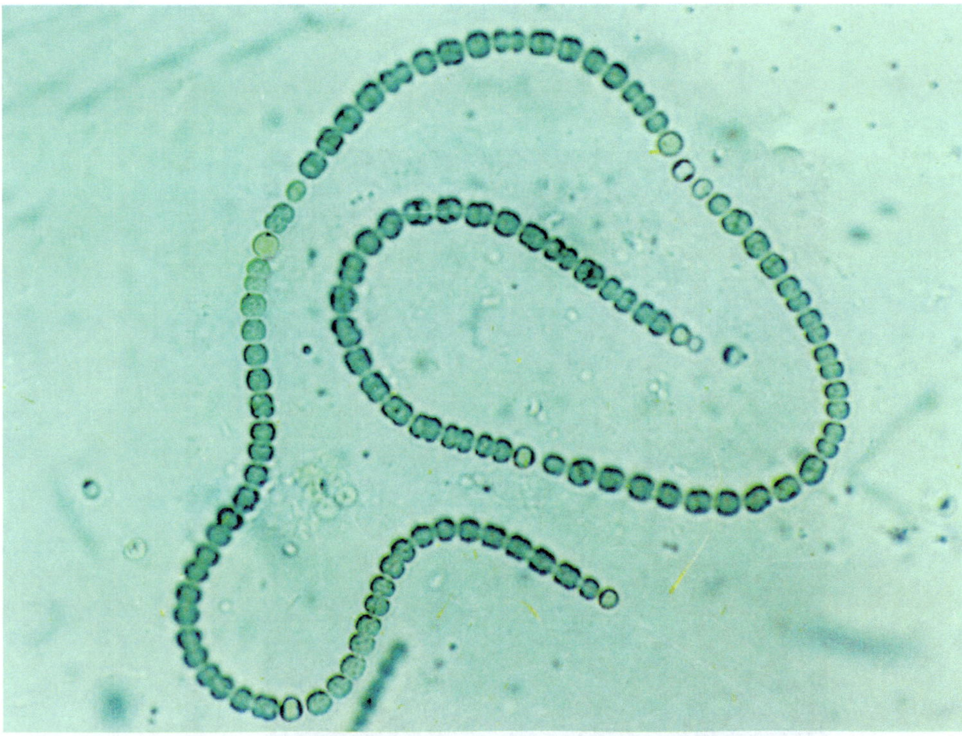

FIGURE 1.4 Microscopic view under a compound microscope of the cyanobacterium *Anabaena variabilis*, showing cell differentiation.

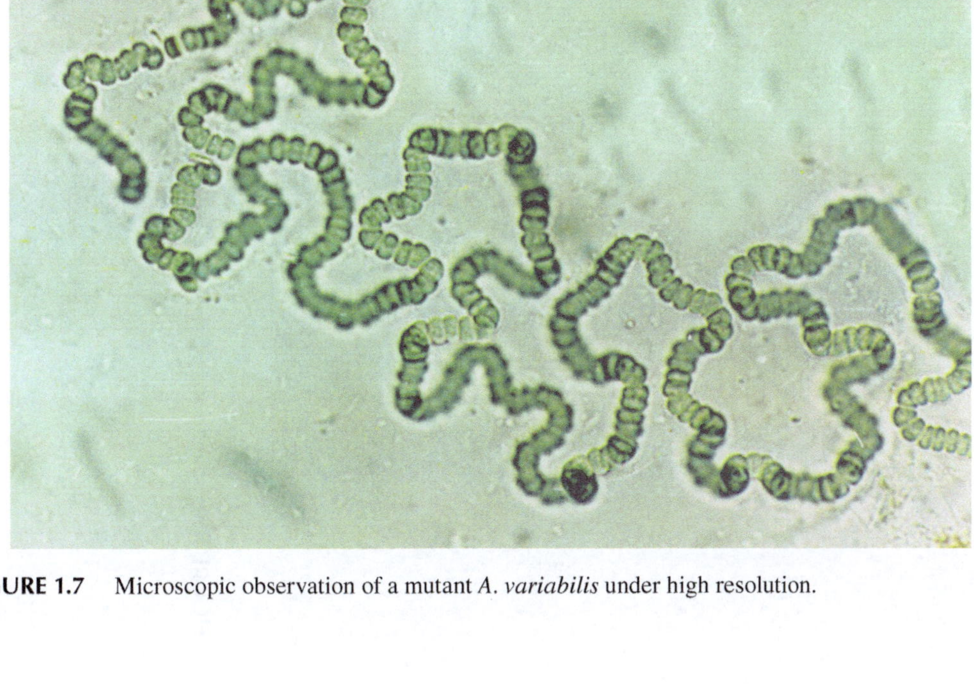

FIGURE 1.7 Microscopic observation of a mutant *A. variabilis* under high resolution.

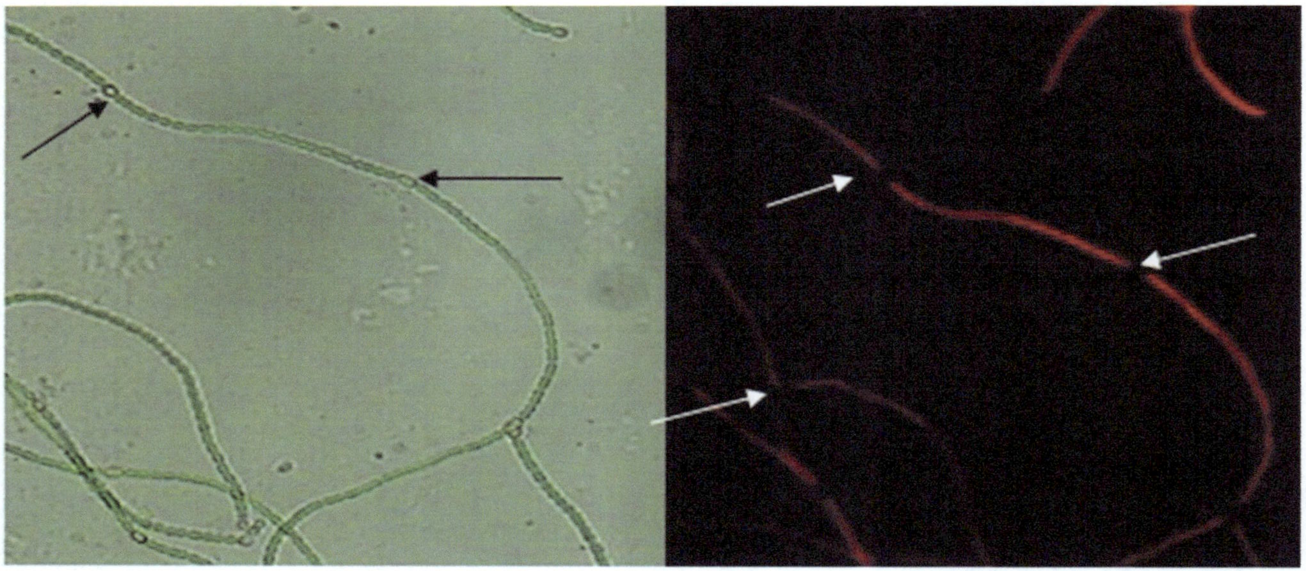

(a) (b)

FIGURE 1.10 Photomicrograph of *A. cylindrica* showing the position of heterocysts in between the vegetative cells: (a) bright field and (b) fluorescent photomicrographs.

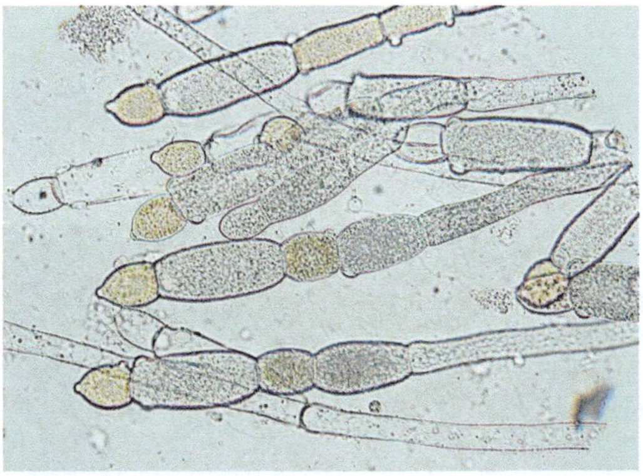

(a)

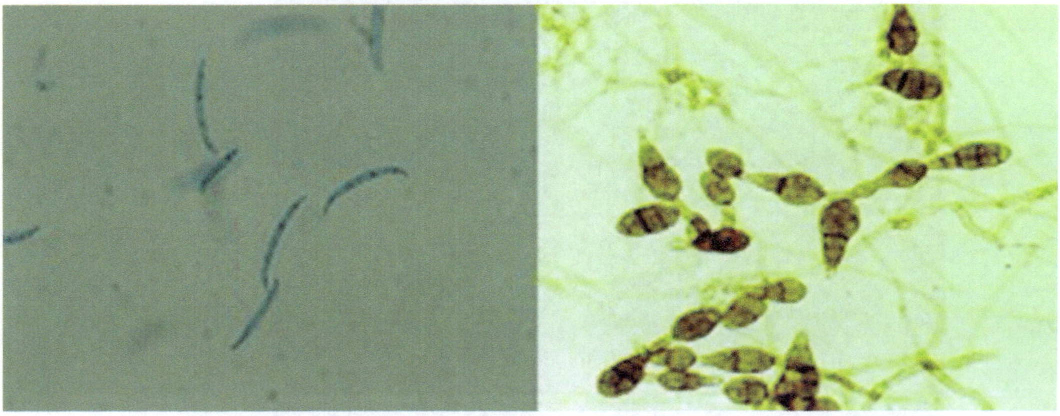

(b)

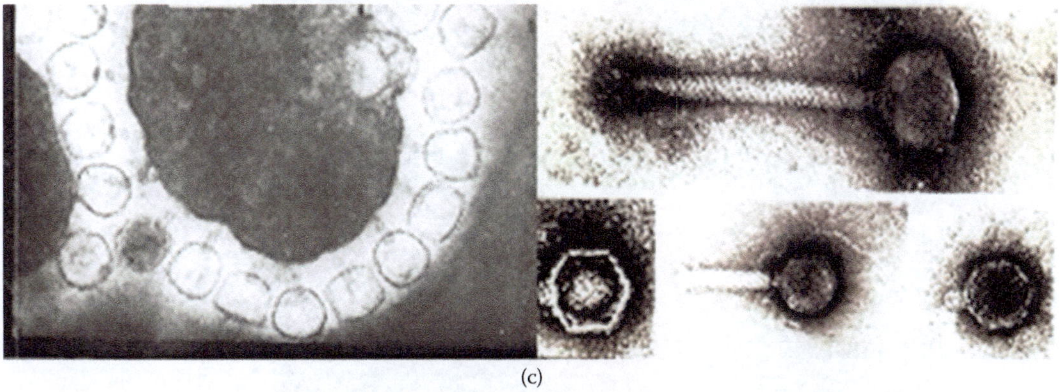

(c)

FIGURE 1.13 Figure showing (a) sporangium and sporangiospore in a slime mold viewed under a compound microscope, (b) fungal conidia of *Fusarium* and *Alternaria* under oil immersion, and (c) electron micrograph of phycovirus attached to cyanobacteria and different types of phycoviruses.

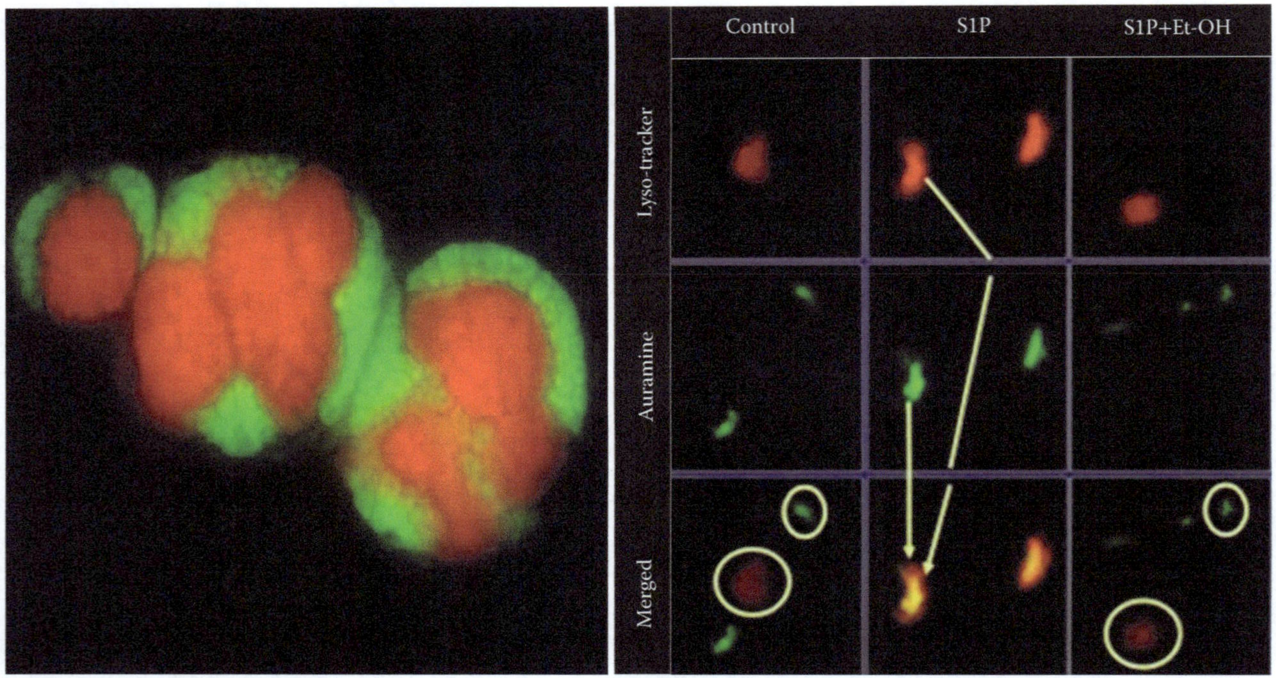

FIGURE 1.16 Figure showing (a) fluorescence in situ hybridization of *Betaproteobacteria* and (b) immunofluorescence detection of sphingosine 1-phosphate (S1P) in tuberculosis. (Courtesy of Pernthaler et al., 2008, *Proceedings of the National Academy of Sciences* 105, 7052–7057 10.1073/pnas.0711303105.)

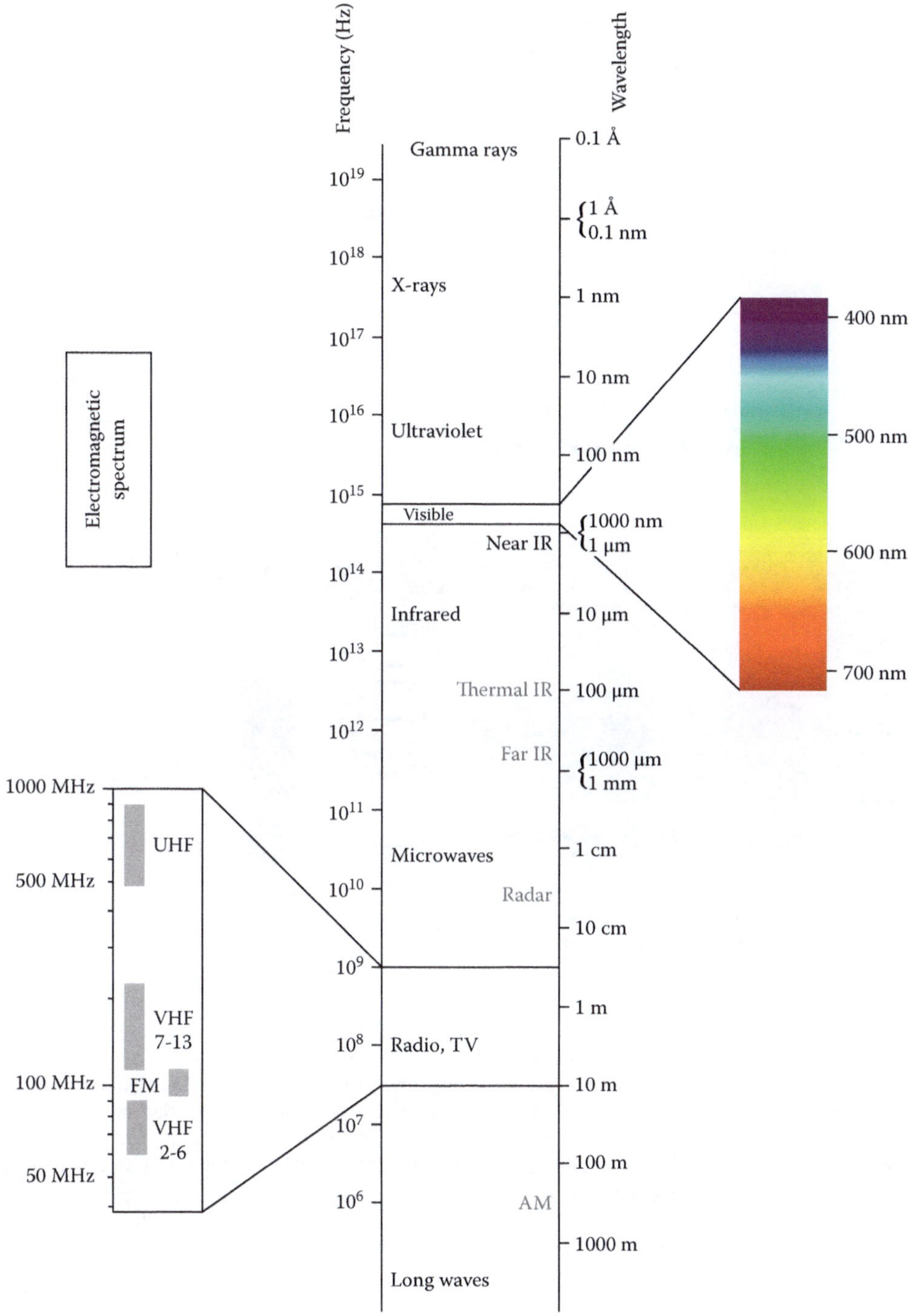

FIGURE 5.1 EM spectrum.

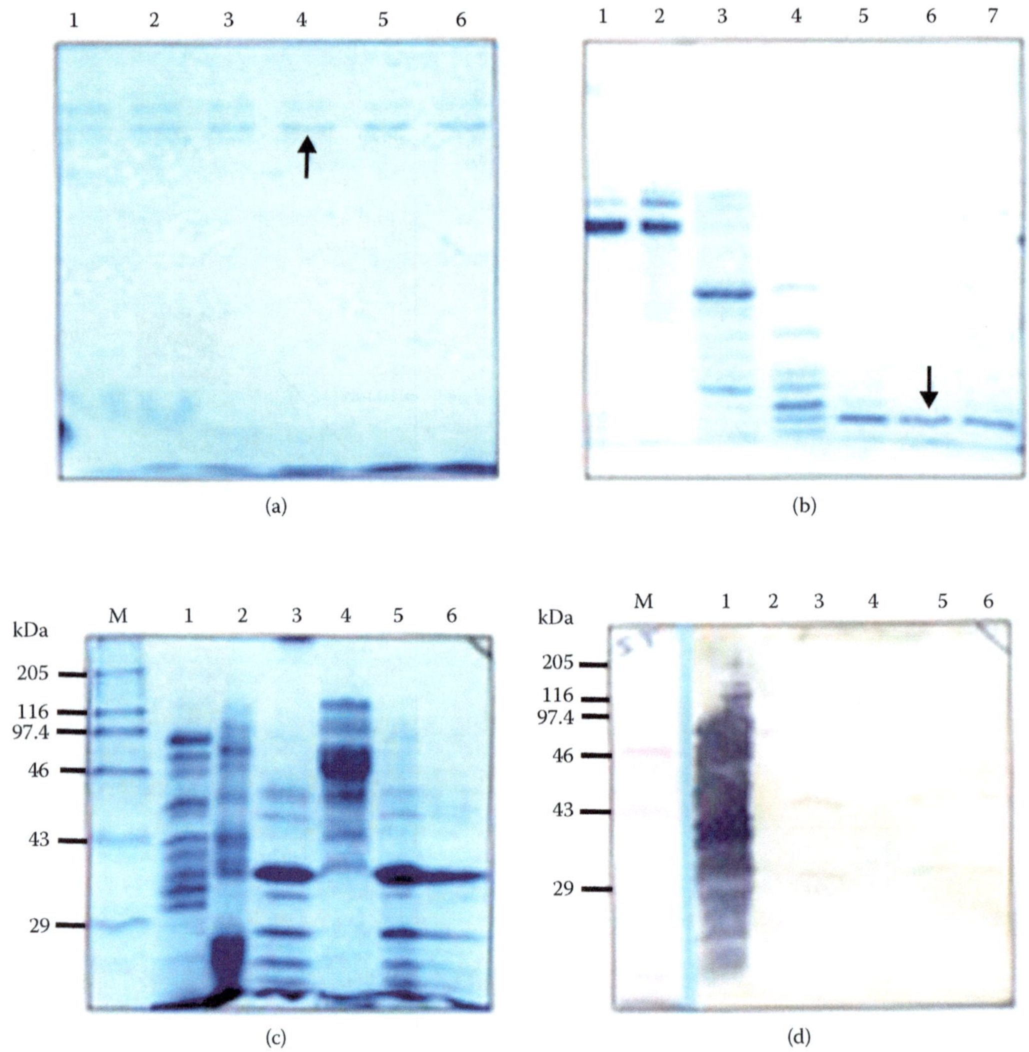

FIGURE 7.6 Top: Protein profile of different fractions of metabolic antigens of *Aspergillus fumigates* obtained by size-exclusion chromatography on 12% SDS-PAGE stained with Coomassie blue: (a) Arrow shows 70–72 kDa protein doublet bands and (b) arrow shows 18 kDa purified protein. (c) Protein profile of metabolic antigens of *Aspergillus fumigates* precipitated by graded saturation of ammonium sulfate. (d) Identification of immunogenic proteins by Western blotting. Lane M, molecular weight marker; 1-, 2-, 3-, 4-, and 5-proteins precipitated at 20%, 40%, 60%, 80%, and 100% saturation of ammonium sulfate in culture filtrate, respectively.

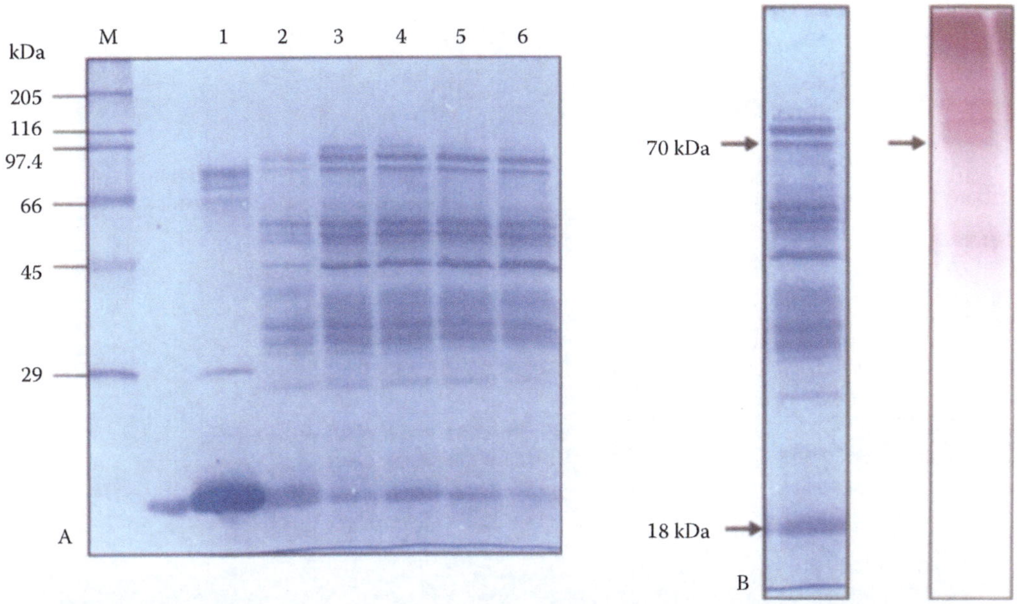

FIGURE 7.15 (b) Electrophoretic protein profile of a fungus *Aspergillus fumigates* with known-molecular-weight marker protein.

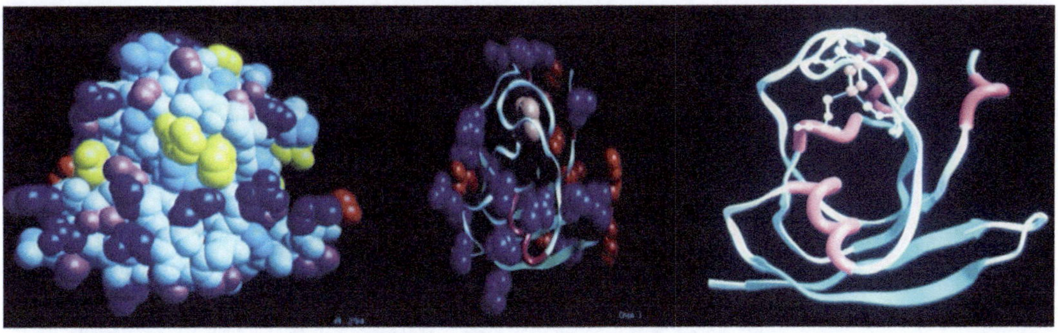

FIGURE 8.12 Computer-generated tertiary structures of crystallographic data of ferredoxin proteins FdxH1 and FDxH2 from the cyanobacterium *Anabaena variabilis*.

FIGURE 10.29 A laboratory-grade fermentor of capacity 2 L containing MRS broth.

FIGURE 10.30 A fermentor after 24 hours of incubation. The visible color difference in the media shows a high cell density of *S. thermophilus*.

3 Electrochemical Techniques

3.1 INTRODUCTION

Ordinary laboratory voltmeters cannot be used for the measurement of electron motive force (emf) of glass electrode cells because of the high electrical resistance of glass electrodes (typically, 10–200 MΩ). Special high-impedance voltmeter circuits are required, which draw 10–12 A or less from the circuit. The pH meter is a voltmeter but with several critical additional functions. It measures potential across the pH-sensing and reference electrode system, converts the potential difference measurement at a given temperature into pH terms, and provides (pH-metry) mechanisms to correct for the nonideal behavior of the electrode system. An operational amplifier not only serves as a high-impedance voltmeter but provides stability and automatic operation through the use of a feedback loop. The operational controls on a pH meter are best understood by referring to the operational manual provided by the manufacturer for an individual instrument. Modern electronic techniques permit the production of a simplified pH meter that measures pH with an accuracy of +0.1 pH unit. The microprocessor-equipped pH meters include a temperature probe to display temperature compensation, a memory to store pH values of standard buffers, a waiting period to allow draft before taking pH readings, and built-in diagnostics to alert for electronic malfunctions or defective electrodes. Along with a pH meter, we also require a reference electrode and a glass membrane electrode, which act as indicator electrodes in pH measurement.

3.2 STRUCTURE OF A pH METER

3.2.1 Glass Membrane Electrodes

The pH glass electrode, although it is somewhat mechanically fragile, resists a variety of sample media and, with the exception of hydroxide, is largely free from interferences. Moreover, pH-sensitive glass electrodes form the basis of many successful sensors for environmentally sensitive gases. Thus, glass membranes represent an important class of solid-membrane ion-selective electrodes (ISEs).

As illustrated in Figure 3.1a, these electrodes have a thin glass membrane fused to the end of a hard glass or, sometimes, epoxy body. The main body of the electrode contains an internal reference electrode, typically Ag/AgCl, and is filled with a solution that is usually aqueous HCl of concentration around 0.1 M. The selectivity coefficient of glass membrane is such that excellent dissemination is achieved against common cationic species.

A pH electrode responds to hydrogen ions as a result of the thin ion-exchange sites existing on the surface of a hydrated glass membrane. The electrode consists of a thin layer of glass, typically about 50 μm thick. Charge is transported across the membrane by sodium or lithium ions within the glass. The membranes are primarily made from lithia (lithium oxide) or sodium oxide, SiO_2, and some amount of Al_2O_3 and B_2O_3 or multicomponent glasses whose sensitivity pattern depends on the composition of the glass. The surface layer of the glass consists of silicate group associated with sodium ion ($-SiO^-Na^+$), as shown in Figure 3.1b. When this electrode is dipped in water, sodium ions are exchanged with the protons in water.

Determination of pH using glass electrodes is a very accurate and widely used method despite there being a few disadvantages. For example, the glass membrane is very fragile, so it requires great care while using. Therefore, the varying potential of a glass electrode can be compared to an external reference electrode, such as a calomel electrode, producing steady potential by joining internal and

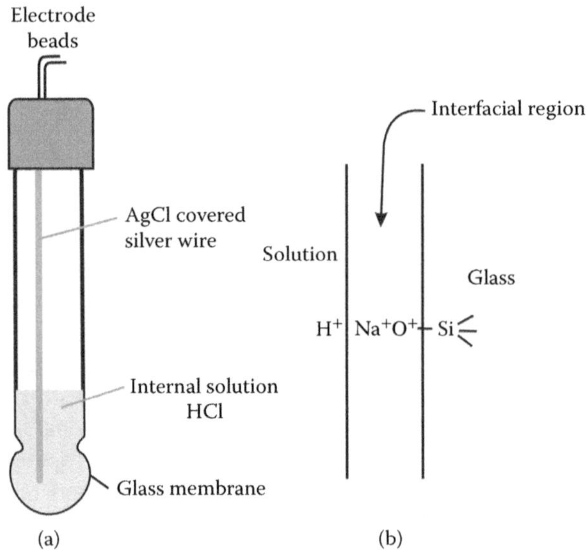

FIGURE 3.1 A glass electrode: (a) typical glass electrode consisting of both an indicator glass electrode and a silver/silver chloride reference electrode, and (b) illustration showing ion exchange.

external reference electrodes. An ordinary potentiometer cannot be used for measuring the potential of the glass electrode. Therefore, electronic potentiometers must be used; they need frequent standardization, and they cannot be used in pure ethyl alcohol, acetic acid, and gelatin. The following features of a glass electrode make it more versatile for use as an indicator electrode for pH measurement:

- A glass electrode can be used in the presence of strong oxidizing and reducing solutions in viscous media and even in the presence of proteins that interfere with the operation of other electrodes.
- It can be used for solutions with pH values of 2–10; using some special glass electrodes, measurements can be extended to pH values greater than 10.
- It is simple to operate and immune to poisoning.
- Equilibrium can be reached quickly.

3.2.2 Reference Electrode

The two most popular types of reference electrodes are saturated calomel and silver/silver chloride systems. Both types of reference electrodes exhibit many ideal characteristics, including maintaining fixed potential over time and temperature, having long-term stability, and returning to the initial potential after being subject to small currents. It may be either a separate probe (Figure 3.2a) or built around the glass electrode giving a combination electrode (Figure 3.2b):

1. A saturated calomel electrode (SCE) is composed of metallic Hg, solid mercurous chloride (Hg_2Cl_2), and a saturated solution of KCl for maintaining an equilibrium. Consequently, the potential of the SCE (+0.241/2 V versus the potential of the standard hydrogen electrode) remains constant even if some of the liquid evaporates over time. The SCE is more popular with a constant temperature bath, which gives the SCE the capacity to eliminate error caused by fluctuating temperature. The SCE can be used as a reference electrode in a sample that does not exceed 80°C.
2. The Ag/AgCl electrode includes a silver wire, which is coated on one end with the insoluble AgCl salt. When the electrode is immersed in saturated KCl solution, its potential

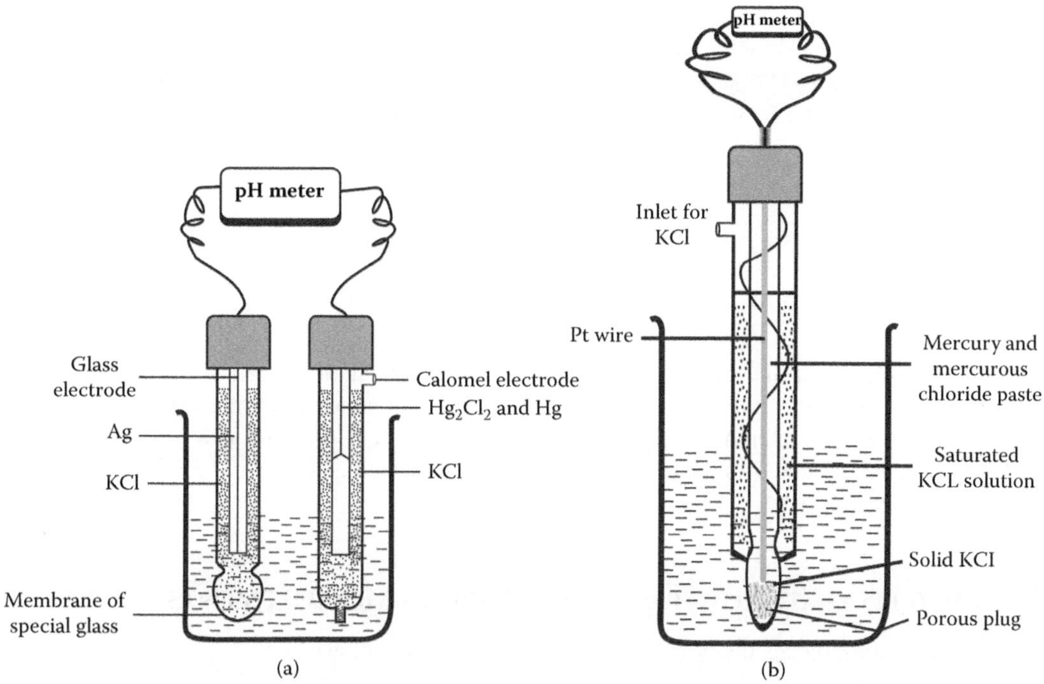

FIGURE 3.2 Reference electrodes: (a) a calomel reference electrode, and (b) a combination electrode.

at 25°C depends only on the Cl^- concentration and is +0.192 V versus the potential of the standard hydrogen electrode. The Ag/AgCl electrode should not be used in solutions in cases of experiments containing bacterial species that may precipitate or complex with silver.

If a combination electrode is in use, the level of test solution should be kept high enough to cover the liquid junction, but it should not be above the salt bridge solution in the external electrode because it is essential for KCl to diffuse slowly in the test solution. The reference electrode should be prepared and maintained so that the level of the internal liquid is kept above that of the sample solution to avoid infusion of the sample into the reference electrode. This is a commonly used precautionary measure to avoid any contamination of the sample by Cl^-, Ag^+, or Hg^{2+} ions.

3.3 PRINCIPLES

pH is a measure of the hydrogen ion activity of a solution, which defines the degree of acidity or alkalinity of the solution. Therefore, pH is the unit of measurement for acidity, as pounds are for weight. Acidity is a chemical parameter; it is a concentration. But due to the wide measurement range and unwieldy nature of chemical notations, R. L. Sorensen, a Danish chemist who developed the pH scale in 1909, defined pH as the negative logarithm of hydrogen ion concentration, which he later changed to hydrogen ion activity:

$$pH = -\log_{10}[H^+]$$

This gives a concentration of about 0.0000001 M hydrogen ions as pH 7.

The pH electrode depends on ion exchange in the hydrated layers formed on the glass electrode surface. Glass consists of a silicate network through which metal ions are coordinated to oxygen atoms, which causes the exchange of metal ions with H^+. Measuring the pH of a solution with a potentiometric cell requires the use of a pH-sensitive glass electrode, a reference electrode, and a high-input-impedance voltmeter. The output of the cell is temperature sensitive in accordance with the Nernst equation, so a temperature electrode is also included in most systems. Walther Nernst first elucidated the theory of galvanic cells in 1889, which relates the voltage of a cell to its physical and chemical properties. For pH, the equation is as follows:

E = *E*^sub 0^ 2.3*RT*/*F* log *a*^sub H

At 25°C, this becomes

$$E = E^{*} + 0.059\ \text{pH}$$

Here, *E* is the voltage of the cell; *E*^sub 0^ is the standard voltage of the cell; E^{*} includes the standard electrode potential of glass electrode, potential of the reference electrode, and liquid-junction potentials between the reference electrode and the solution; *R* is the universal gas constant; *F* is the Faraday constant; and *T* is the temperature in degrees Kelvin. The term 2.3*RT*/*F* is the Nernst number or slope.

3.4 FACTORS LIMITING THE ACCURACY OF pH MEASUREMENTS

The factors limiting the accuracy of pH measurements are as follows:

Alkaline error: It has been noted that in a high pH range (above 9 or 10 pH units), the ordinary glass electrode becomes sensitive to alkali ions and gives a low reading. The reason for this error is that, although the glass membrane is selective to hydrogen ions, it also responds to other ions. This error becomes more significant when the activity of other ions is greater than the activity of the hydrogen ions.

Note: Alkaline error is relatively more frequent in the case of sodium ions because of the higher selective coefficient of sodium ions. This occurs because sodium ions can be reduced by the use of Li_2O glass in place of Na_2O glass.

Acid error: At a low pH range (pH less than 0.5), the values determined by the glass electrode tend to be somewhat higher. This error is the result of ignoring the activity of water while writing the Nernst equation for the indicator electrode. It is assumed that the activity of water may be taken as unity, as it is in excess in the solution and behaves as a pure substance. However, in highly acidic solutions, the activity of water becomes less than unity because a good amount of water is used in hydrating the protons. Similar effects are also observed upon addition of large amounts of any dissolved salt and addition of a miscible nonaqueous solvent such as ethanol. This results in an increased pH value.

Variation in junction potential: In most cases, the compositions of the standard buffer solution and the test solution are different. In such situations, the liquid-junction potential will be different and the standard uncertainty will be ±0.015.

Error in the pH of the standard buffer: This may result from inaccurate composition of a buffer solution during preparation or storage will cause an error in subsequent pH measurements and a standard uncertainty of up to ±0.01 (Table 3.1).

Temperature: A change in temperature may affect pH measurement with an uncertainty error of ± 0.005 because the change in temperature affects the activities of ions as well as the liquid-junction potentials. Therefore, it is advised to calibrate the electrode at the temperature of the test solution.

Calibration procedures: Buffer solutions cannot be prepared with more accuracy than ±0.01 pH units. Therefore, we cannot calibrate the electrode with a better accuracy than this value.

Equipment-related errors: These errors may result from power fluctuations, parallax errors in reading analog scales, and so on.

3.5 MEASUREMENT OF pH

pH measurements are used to check and control many aspects of our day-to-day life, such as in determining potable water quality, soil usability for different plants, and water quality in aquaristics; they are performed to control industrial processes, in wine making and beer making, to check milk quality, and to check the quality of cosmetics. There is no need to mention all the laboratories throughout the world where pH measurements are performed many times a day to control reactions and analysis conditions.

There are several ways in which pH can be measured; of them, two methods are widely used. First, one of the simple, and often precise, methods is the use of colorimetric indicator methods, that is, use of pH strips (pH papers). Second, a more costly and demanding method in terms of procedure, but giving much more precise results, is the potentiometric method, which uses glass electrodes and pH meters (Figure 3.3). Colorimetric (spectroscopic) methods have never gained much popularity, although they are occasionally used in places where glass electrodes are in use.

TABLE 3.1
High-Purity Salts as Primary pH Standard

High-Purity Salts	Concentration (g/L)	pH at Stated Temperature 15°C	20°C	25°C	30°C
Potassium tetraoxolate	12.61	1.67	1.68	1.68	1.68
Potassium hydrogen phthalate	10.13	4.00	4.00	4.01	4.02
Potassium dihydrogen phosphate	3.39	6.90	6.88	6.87	6.85
Disodium hydrogen phosphate	3.53	6.90	6.88	6.87	6.85
Sodium tetraborate decahydrate	3.80	9.28	9.23	9.18	9.14
Sodium hydrogen carbonate	2.09				
Sodium carbonate	2.64	10.12	10.06	10.01	9.97

Note: The uncertainty of the tabulated pH values is estimated to be ±0.01.

FIGURE 3.3 Analog pH meter.

3.6 WORKING

3.6.1 Potentiometric Method of pH Measurement

The pH is determined by measuring the emf of a cell comprising an indicator electrode (an electrode responsive to hydrogen ions such as a glass electrode) immersed in the test solution and a reference electrode (usually an SCE); contact between the test solution and the reference electrode is achieved by means of liquid junction, which forms the part of a reference electrode. The emf of this cell is measured with pH.

3.6.2 Reagents Used in the Potentiometric Method of pH Measurement

3.6.2.1 Calibration of the Electrode System against Standard Buffer Solutions of Known pH

Buffer tablets having pH values 4.0, 7.0, and 9.2 are commercially available. Because buffer solutions may deteriorate as a result of mold growth or contamination, prepare fresh solution as needed. Alternatively, buffer solutions can be prepared by the methods discussed in Sections 3.6.2.2 through 3.6.2.4.

3.6.2.2 pH 4 Buffer Solution

Dissolve 10.2 g of anhydrous potassium biphthalate (KHC_8H_4) in boiled and cooled distilled water. Dilute to 1 dm^3.

3.6.2.3 pH 7 Buffer Solution

Dissolve 1.361 g of anhydrous potassium dihydrogen phosphate (KH_2PO_4) and 1.42 g of anhydrous disodium hydrogen phosphate (Na_2HPO_4), both of which have been dried at 110°C to 130°C, in distilled water. Use distilled water that has been boiled and cooled. Dilute to 1 dm^3.

3.6.2.4 pH 9.2 Buffer Solution

Dissolve 3.81 g of borax ($Na_2B_4O_7$ $10H_2O$) in distilled water and dilute to 1 dm^3.

3.7 PROCEDURE FOR MEASURING pH USING A pH METER

Numerous pH meters of various designs are marketed by instrument manufacturers. General-purpose pH meters are either line-operated instruments that are readable to 0.05 pH unit or battery-operated instruments suitable for field jobs. Nowadays, digital pH meters readable to 0.01 pH unit are more popular compared to scale-needle instruments. The pH of a solution can be measured with the instrument (analog meter) shown in Figure 3.3, following the procedure given here in a stepwise manner:

1. Keep the selector switch at the zero position and adjust the zero position with a screwdriver if the pointer does not indicate zero.
2. Before using the pH meter, remove electrodes from storage solutions (recommended by manufacturers) and rinse with distilled water. Electrodes should be dried by gently blotting with a soft tissue paper.
3. Mount the electrodes in the clip on the stand.
4. Connect the power cable to a 220 V alternating current (AC) supply. Switch on the instrument and wait for a few minutes until the instrument warms up.
5. Adjust the temperature/solution temperature value.
6. Put the standard buffer solution of desired range (e.g., buffer of pH 4 for acidic solutions) in a beaker. The electrode assembly is immersed in the pH reference buffer, and the solution

is agitated gently by swirling the solution in the region of the glass electrode surface so as to bring it into pH equilibrium. Ascertain that the glass electrode membrane is completely immersed in the solution. The electrodes should not touch each other, or the sides or bottom of the beaker.

7. Put the selector switch to a suitable pH range (0–7 for acidic or 7–14 for basic solutions), and adjust the buffer knob so that the pointer reads the pH of the standard buffer solution (placed in the beaker).
8. Put the selector switch back to the zero position. Remove the electrodes from the buffer solution, wash the electrodes with distilled water, and wipe them gently with tissue paper.
9. Immerse the electrodes in a second buffer with pH below 10, approximately 3 pH units different from the first one; the reading should be within 0.1 units for the pH of the second buffer. (If the meter response shows a difference greater than 0.1 pH units from the expected value, the electrodes or the pH meter may be at fault.)
10. Transfer the standard buffers back to the storage bottle and wash the beaker well with distilled water.
11. Take the sample solution in the beaker. Introduce the electrodes to the solution and swirl it gently.
12. Set the selector switch in the suitable range position and read the pH on the scale.
13. Put the selector switch back to the zero position. Remove the electrodes from the solution, wash them with distilled water, and keep the electrodes in distilled water when not in use.

3.7.1 Modified Glass and Solid-State Membrane Electrodes

Glass electrodes can be made selective for ions other than hydrogen ions with some modifications. These modifications achieved by changing the composition of the glass and the internal solution of the glass electrode. By adding aluminum oxide to sodium oxide and silicon oxide glass, and changing the internal filling solution from hydrochloric acid to sodium chloride, the glass electrode becomes selective to Na^+ ions. There is another types of glass electrode made up of Li_2O, Al_2O_3, and SiO_2, which can be also be used as sodium electrode. Sodium electrodes have many applications in the measurement of sodium in water analysis and in biological fluid analysis. For measuring potassium and ammonium ions, modified glass containing 27% Na_2O, 4% Al_2O_3, and 69% SiO_2 is used. These days, potassium/ammonium electrodes are replaced by other ISEs using more selective membranes.

3.7.2 Solid-State Membrane Electrodes

In solid-state membrane electrodes, we use a doped single-crystal membrane in place of a glass membrane. This change enables us to design electrodes which respond to a number of different anions and cations, such as F^-, Cl^+, and Ag^+ (Figure 3.4).

In this electrode system, similar to a glass electrode system, the internal solution and electrode form the internal reference. For example, fluoride ISE consists of LaF_3 membrane and an internal Ag, AgCl reference electrode immersed in an internal solution of KF and KCl. The LaF_3 membrane is highly selective and responds to fluoride ions only.

3.7.3 Storage Conditions for Glass Probes

The following conditions must be met for storing glass probes:

1. When not in use, the glass probe tip must be kept wet at all times to avoid dehydration of the pH-sensing membrane and subsequent dysfunction of the electrode.
2. A glass electrode by itself (i.e., without a combined reference electrode) is typically stored immersed in an acidic solution with a pH of around 3.0. In an emergency, acidified tap

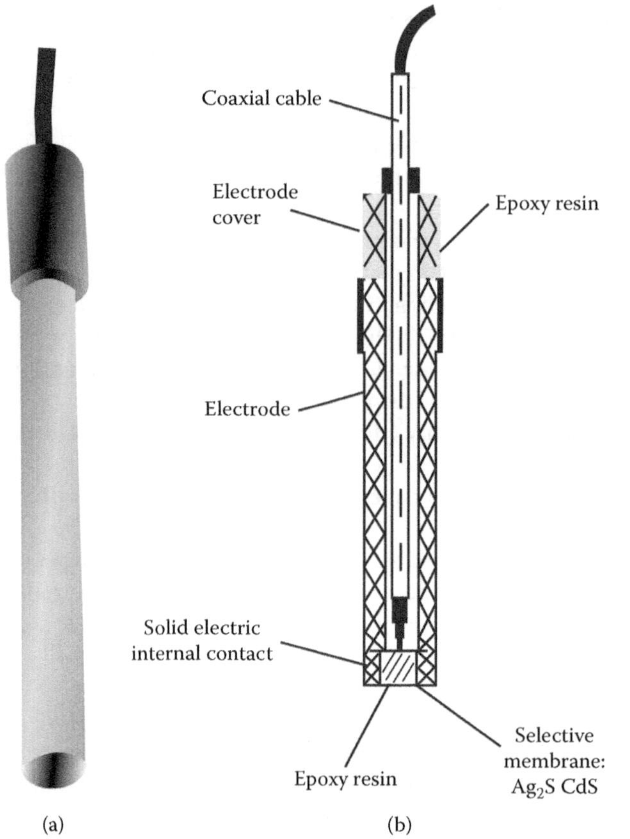

FIGURE 3.4 Typical solid state electrode with its (a) outer and (b) inner view.

water can be used, but distilled or deionized water must never be used for long-term probe storage as the relatively ionless water "sucks" ions out of the probe membrane through diffusion, which degrades it.

3. Combined electrodes (glass membrane + reference electrode) are better stored immersed in the bridge electrolyte (often 3 M KCl) to avoid diffusion of the electrolyte (KCl) out of the liquid junction.

3.8 CLEANING AND TROUBLESHOOTING OF GLASS PROBES

Cleaning and troubleshooting of glass probes must be performed at regular intervals to ensure they work properly:

1. Occasionally (about once a month), the probe may be cleaned using the pH-electrode cleaning solution; generally, 0.1 M solution of hydrochloric acid (HCl) is used, which has a pH of about 1.
2. In the case of strong degradation of glass membrane performance due to membrane poisoning, diluted hydrofluoric acid (HF < 2%) can be used to quickly etch (<1 minute) a thin damaged film of glass. Alternatively, a dilute solution of ammonium fluoride (NH_4F) can be used. To avoid unexpected problems, however, the best practice is to always refer to electrode manufacturer recommendations or a classical textbook on analytical chemistry.

3.9 APPLICATION OF pH MEASUREMENTS

pH measurements have many applications, some of which are summarized in the following list:

- pH-metric or electrometric methods help in detecting the end point of acid–base titration more accurately and precisely as compared to indicator methods.
- pH-metry is an important analytical tool in studying the acid–base equilibrium, which is controlled by the carbon dioxide–bicarbonate–carbonate equilibrium system in most natural waters.
- The estimation of alkalinity and acidity based on pH-metry serves to provide useful information regarding the buffering capacity of water.
- In water and wastewater treatment, pH measurement gives an estimate of the alkalinity required to react with the coagulant, that is, alum, ferrous sulfate, and so on, or otherwise to be supplemented with lime.
- pH measurements are used in industrial process control, especially in batch or flow-through configurations, through online pH-monitoring and chemical-dosing systems.
- In neutralization of wastewaters, pH is measured to evaluate the doses of acid or alkali to be added.
- pH measurements are useful in the development of biosensors-based pH-sensitive immobilized enzymes and in other academic studies.

Further, the flexibility in available configurations allows the ions to be monitored in a single sample solution (batch mode) or continuously in a flow-through apparatus (flow-injection analysis).

3.10 OXYGEN ELECTRODE

The Clark-type electrode is the most used oxygen sensor for measuring oxygen dissolved in a liquid. The basic principle is that there is a cathode and an anode submersed in an electrolyte. Oxygen enters the sensor through a permeable membrane by diffusion and is reduced at the cathode, creating a measurable electrical current. There is a linear relationship between the concentration of oxygen and the electrical current produced. With a two-point calibration (0% and 100% air saturation), it is possible to measure oxygen in the sample.

There are several types of oxygen electrodes, which differ in design, but the most common one is the Clark-type electrode. It utilizes an annular silver reference electrode (anode) and a glass-coated platinum electrode (cathode) connected to a small external voltage source that charges the circuit with a potential difference of 500–700 mV. This is called *polarizing voltage*. Oxygen gets reduced at the cathode giving rise to a current, which is proportional to the concentration of oxygen in the solution provided that the polarizing voltage remains constant at approximately 700 mV. Oxygen from the solution diffuses in through a thin semipermeable membrane, which separates the electrodes from the test solution (Figure 3.5). Both electrodes are immersed in the same solution at saturated KCl.

One drawback to this approach is that oxygen is consumed during pH measurement with a rate equal to diffusion in the sensor. This means that the sensor must be stirred in order to get the correct measurement and avoid stagnant water. Oxygen consumption increases with increasing sensor size, and so does stirring sensitivity. In large sensors, there also tends to be a drift in the signal over time due to consumption of the electrolyte. However, Clark-type sensors can be made very small, with a tip size of 10 μm. The oxygen consumption of such a microsensor is so small that it is practically insensitive to stirring and can be used in stagnant media such as sediments or inside plant tissue.

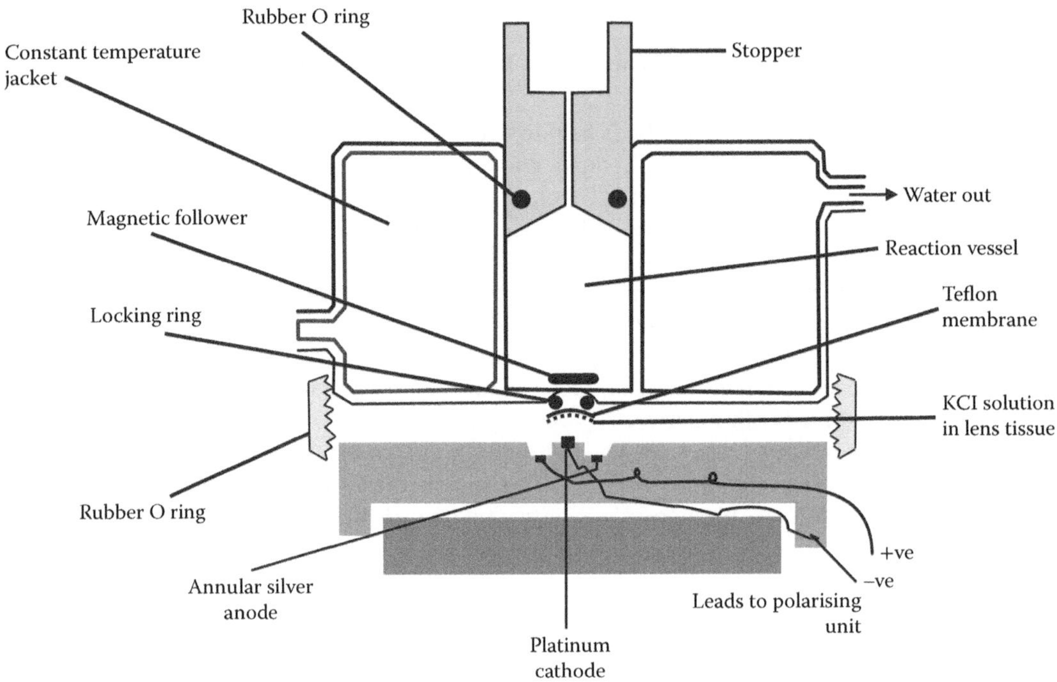

FIGURE 3.5 Diagram of the oxygen electrode.

The electroreduction of oxygen at the cathode, which gives rise to electrical current, occurs according to the following equations:

$$O_2 + 2H_2O + 2e^- \rightarrow H_2O_2 + 2e^- \rightarrow 2OH^-$$

$$OH^- + KCl \rightarrow KOH + Cl^-$$

The electrons required for this electroreduction reaction are produced at the reference electrode (anode) as follows:

$$4Ag^o + 4Cl^- \rightarrow 4AgCl + 4e^-$$

The whole unit is mounted on a magnetic stirrer and a magnetic follower (flea) placed inside the reaction vessel. This ensures that the test solution is constantly stirred so as to obtain correct values of change in dissolved oxygen levels. Since the rate of dissolution of oxygen is temperature dependent, it is important to maintain constant temperature. This is achieved by circulating water in a water jacket around the reaction vessel. Both the water jacket and the reaction vessel are made of a special type of transparent plastic so that photosynthetic studies requiring the use of light can be carried out.

The unit is connected to a control box, which takes electrical signals from the electrode and conveys them to a chart recorder. The semipermeable membrane used is generally made of Teflon. This is highly permeable, leading to quick establishment of equilibrium, but it responds differently for the same values of dissolved oxygen in gas and liquid. Less permeable membranes include polypropylene and polyethylene membranes, which show little difference in their response to liquid and gas.

3.10.1 Calibration of the Oxygen Electrode

An oxygen electrode is calibrated in the following manner:

1. Calibration is done with a liquid standard or a gas standard that has been equilibrated with water vapor. Either standard must be equilibrated to the measuring temperature.
2. The proper scale is selected; certain instruments have more than one scale to accommodate the wide ranges sometimes encountered during analysis.
3. A standard, usually distilled water, is added to the reaction vessel; the magnetic follower is introduced and the instrument is switched on for 30 minutes warm-up time. During this time, the lid of the reaction vessel is kept open. The pen recorder stabilizes at a particular point. This is taken as the dissolved oxygen concentration of the standard whose value can be obtained from Table 3.2.
4. Now, if a few crystals of sodium dithionite ($Na_2S_2O_4$) are added, all the oxygen is used up in accordance with the following reaction:

   $$Na_2S_2O_4 + O_2 + H_2O \rightarrow NaHSO_4 + NaHSO_3$$

   The pen recorder moves a certain distance and stabilizes at a particular point on the chart. This is the zero point of dissolved oxygen concentration. This point is adjusted with the zero calibration control of the recorder.
5. This process is repeated several times until the zero concentration and standard concentration need no further adjustment.
6. The number of divisions between the zero and standard concentrations correspond to the standard oxygen concentration. Hence, one division represents a certain dissolved oxygen concentration:

   $$\text{If 10 divisions} = x\ \mu\text{mol}\ 1^{-1}\ \text{oxygen}$$

   $$1\ \text{division} = 0.x\ \mu\text{mol}\ 1^{1}\ \text{oxygen}$$

7. The instrument is thus calibrated and ready to determine unknowns.

TABLE 3.2
Standard Value of Dissolved Oxygen Concentration

Temperature (°C)	O_2 (ppm)	O_2 (μmole/mL)
0	14.16	0.442
5	12.37	0.386
10	10.92	0.341
15	9.76	0.305
20	8.84	0.276
25	8.11	0.253
30	7.52	0.30
35	7.02	0.219

3.10.2 Applications of the Oxygen Electrode

The oxygen electrode is being put to increasingly wide uses because of the ease of portability of the instrument, which is also inexpensive and does not require much expertise in handling.

Clinical uses: The oxygen content of blood requires assessment during oxygen therapy. Oxygen electrodes have been designed to be small enough to be inserted into a blood vessel; however, this involves the danger of infection or blood clot formation. Hence, a transcutaneous electrode system has been devised to measure oxygen levels in the blood without having to pierce the skin. In this method, a small area on the skin, say, an earlobe or a fingertip, is heated externally to 43°C, and the electrode is held tightly to the depilated skin surface by an airtight bond. A polypropylene membrane separates the heated skin surface from the electrode assembly, which consists of a polarized platinum cathode surrounded by a silver ring-type anode held at a constant temperature, usually 43°C. Oxygen in the blood diffuses from the blood capillary through the skin and the electrode membrane, and actuates the Clark-type electrode to produce the reading.

Chloroplast and mitochondrial studies: Oxygen evolution in cyanobacteria, algae, and chloroplasts, that is, those containing photosystem II, can be studied using a sufficiently illuminated Clark oxygen electrode. The oxygen content of the suspension medium is normally reduced below 100% oxygen by bubbling nitrogen through it, so that the oxygen produced stays in the solution and is recorded. In marine biology or limnology, oxygen measurements are usually carried out in order to measure respiration of a community or an organism, but they have also been used to measure primary production of algae. There are, however, commercially available oxygen sensors that measure oxygen concentration in liquids with great accuracy. Two types of oxygen sensors are available: (1) electrodes (electrochemical sensors) and (2) optodes (optical sensors). Respiratory control studies and the effect of various inhibitors on mitochondrial respiration can also be studied using the oxygen electrode. Inhibitors of respiration slow down the rate of respiration.

Enzyme assays: The activity of enzymes that require oxygen for reaction ego, such as glucose oxidase and catalase, can be studied using the Clark-type oxygen electrode.

Microorganism studies: Bacteria that use oxygen as a terminal electron acceptor can be studied using an oxygen electrode, and the effect of electron inhibitors can be determined. The efficiency of different carbohydrates as respiratory substrates can be studied by comparing their rates of oxygen uptake. In soil respiration studies, oxygen sensors can be used in conjunction with carbon dioxide sensors to help improve the characterization of soil respiration. Typically, soil oxygen sensors use a galvanic cell to produce a current flow that is proportional to the oxygen concentration being measured. These sensors are buried at various depths to monitor oxygen depletion over time, which is then used to predict soil respiration rates. Generally, these soil sensors are equipped with a built-in heater to prevent condensation from forming on the permeable membrane, as relative humidity can reach 100% in soil.

Experiment: Determination of pH of the Given Water/Soil Sample

Principle

pH is the measure of the relative acidity or alkalinity of a solution and is represented as the negative logarithm of the concentration of free hydrogen ions in a solution. The "p" of pH denotes the power of hydrogen ion activity in moles per liter:

$$\text{pH} = -\log_{10}\left(\text{H}^+\right) = \log_{10} \times 1/\left(\text{H}^+\right)$$

When two electrodes are dipped in two solutions of different pH levels and connected with each other, a potential difference is set up between the two electrodes, which is measured using a potentiometer. This is directly related to the pH of the solution.

The pH is measured of H^+ ion concentration in an aqueous medium. Chemically, pure water at 25°C is dissociated equally into H^+ and OH^- ions with 10^{-7} H^+ per litre. If the number of free H^+ ions is greater than the number of OH^- ions, the water is acidic; otherwise, the water is alkaline. The term pH is expressed as the negative logarithm of H^+ concentration. Thus, pH 7 indicates neutral water, pH below 7 indicates acidic solution, and pH values 7–14 indicate alkaline solution.

Requirements

pH meter, standard pH buffers, and tissue paper.

Method

The pH can be measured by two different methods:

1. Colorimetric method (using indicators): Commercially available pH papers are used in this method. A color change of the paper strip on application of a drop of the test sample indicates the pH, which is matched with the provided standard color chart.
2. Electrometric method (using pH meter): A pH meter with a combined glass electrode is generally used.
   a. Wash electrode with distilled water.
   b. Standardize electrode with standard buffers in the range of pH 4–9.2.
   c. Adjust temperature knob setting according to that of the sample.
   d. Wash the electrode again with distilled water and wipe with tissue paper.
   e. Dip the electrode in the test sample and note the pH.

   Precautions to be taken when performing the electrometric method are as follows:
   a. Soak a new or dried glass electrode for several hours in 0.01 M HCl.
   b. There should be no air bubbles either at the ends of the electrode or in the salt bridge.
   c. Temperature correction should be performed before starting.
   d. After each use, the electrode should be washed with distilled water and wiped clean with tissue paper.

Experiment: Determination of the pH of Fruit Juice

Principle

The acidity and alkalinity of fruit juice is measured as a pH value.

Procedure

The pH of fruit juice can be determined colorimetrically by the use of various indicators, which are as follows:

1. pH paper: The color of the pH paper corresponds to a definite pH. A drop of fruit juice is put on the pH paper and the color of the pH paper is matched with the pH color key provided by the pH paper manufacturing company.
2. Lovibond comparator: This comparator consists of a Bakelite case with two holes at the top for tubes of standard bore and of colorless glass. Tube A contains water. The hinged door of the case holds a rotatable disk containing a series of standard colored glasses corresponding to various pH values; each glass can be brought in front of tube A in turn and viewed through aperture A. A solution of the indicator phenol red is added to tube B, which contains fruit juice, and the pH disk is rotated until a match is obtained. The pH is then read in the aperture at the bottom of the apparatus.
3. pH meter: This instrument gives an accurate pH reading. Adjust the instrument with a standard buffer solution of known pH after setting its knob at room temperature. The electrode is then washed with distilled water. Thereafter, the fruit juice is filled in a beaker and the pH reading is recorded. The electrodes should be washed with distilled water after every use. When the instrument is not in use, the electrodes should always be immersed in distilled water.

Experiment: Measurement of the pH of a Soil Sample

Principle

The pH of soil affects plant nutrient availability, microbial and fungal activity, and plant growth. It is a measurement of hydrogen ion activity in a soil/water solution. Soil pH has many primary determining factors, such as decomposition of organic material, parent material, rainfall, and anthropogenic alteration. A neutral solution has a pH of 7, whereas an acidic solution has a pH less than 7. Note that a basic solution always has a pH greater than 7, but an alkaline solution does not. This is due to the acid-neutralizing capacity of soil or buffering of a soil by strong bases.

Altering soil pH is a common practice in many fields of plant production, cultivation, and growth. Methods include addition of organic matter, hydrogen/aluminum sulfate, sulfur, fertilizers containing ammonium or urea, and liming agents. In this experiment, we will be measuring soil pH.

Reagents

Prepare all the solutions in previously boiled and cooled distilled water.

Standard buffer solutions are as follows:

1. Potassium tetraoxalate ($KH_3C_4O_{82.}H_2O$): Dissolve 3.175 g $KH_3C_4O_{82}.H_2O$ in water and make up the solution to 250 mL (pH = 1.68).
2. Potassium tartarate ($KHC_4H_4O_6$): To 250 mL water in a glass stopper bottle, add excess $KHC_4O_4O_6$; shake the bottle vigorously for about 10 minutes to obtain a saturated solution. Filter and preserve the solution by adding 0.1 g of thymol (pH = 3.55).
3. Potassium phthalate ($KHC_8H_4O_6$): Dissolve 2.552 g of $KHC_8H_4O_6$ in water and make up the solution to 250 mL (pH = 4.0).
4. Potassium dihydrogen phosphate (KH_2PO_4): Dissolve 0.30 g of KH_2PO_4, 0.8875 g of Na_2PO_4, and 0.8875 g of Na_2HPO_4 in water and make up the solution to 250 mL (pH = 6.86).
5. Borax ($Na_2B_4O_7$): Dissolve 0.9502 g of $Na_2B_4O_7$ in water and make up the solution to 250 mL (pH = 9.18).

Procedure

1. Warm up the instrument for 15 minutes.
2. Calibrate the instrument with the known buffer solutions. (Calibration is done with a buffer solution whose pH is close to that of the sample.)
3. Immerse the electrode in the unknown sample, stir for 3 minutes, and note the pH.

Note

Instead of preparing standard buffer solutions, commercially available buffer tablets can be used for calibrating the pH meter.

Experiment: Measurement of Oxygen Evolution during Photosynthesis

Principle

Oxygen enters the sensor through a permeable membrane by diffusion and is reduced at the cathode, creating a measurable electrical current. There is a linear relationship between the concentration of oxygen and the electrical current produced. With a two-point calibration (0% and 100% air saturation), it is possible to measure oxygen in the sample.

Clark-type electrodes utilize an annular silver reference electrode (anode) and a glass-coated platinum electrode (cathode) connected to a small external voltage source, which charges the circuit with a potential difference of 500–700 mV. This is called polarizing voltage. Oxygen is reduced at the cathode, giving rise to a current, which is proportional to the concentration of oxygen in the solution provided that the polarizing voltage remains constant at approximately 700 mV. Oxygen from the solution diffuses in through a thin semipermeable membrane, which separates the electrodes from the test solution. Both electrodes are immersed in the same solution of saturated KCl.

Requirements

Clark-type oxygen electrode, exponentially growing cyanobacterial culture, 0.05 M phosphate buffer with a pH of 7.8, 0.1 M KCl, magnetic stirrer.

Procedure

Estimation of photosynthetic O_2 evolution: Photosynthetic O_2 evolution was measured with a Clark-type oxygen electrode (Hansatech Instruments, Ltd., United Kingdom). Exponentially growing measured cyanobacterial cells are placed on a flat platinum cathode that is polarized at 0.6 V with reference to a large Ag/AgCl electrode. The electrodes are immersed in an electrolyte (consisting of 0.05 M phosphate buffer with pH 7.8; 0.1 M KCl). The electrodes are separated from the magnetically stirred assay medium by a Teflon membrane. The difference between the output of the electrode in water in equilibrium with air and in water in equilibrium with pure N_2 was considered to represent 0.235 mol m^{-3} in the assay medium. After injection of the same into the assay medium, the medium is illuminated from the opposite side with a projector lamp. The rate of O_2 evolution is determined from the initial slope of the electrode output as a function of time.

SUGGESTED READING

Akhtar, M. 2007. *Method of Analysis and Assay: Potentiometry. Pharmaceutical Analysis.*

Cavins, T. J., J. L. Gibson, B. E. Whipker, and W. C. Fonteno. 2000. *pH and EC meters—Tools for Substrate Analysis (Florex. 001).* Raleigh: N.C. State University.

Rupert, S. 2005. Principles and guidelines of pH measurement. *InTech.* North Carolina: Instrument Society of America NC.

Vytras, K. 1995. Potentiometry. In *Encyclopedia of Pharmaceutical Technology*, v. 12, edited by. J. Swarbric and J. C. Boylan. New York: Marcel Dekker.

Walker, D. 1990. *The Use of the Oxygen Electrode and Fluorescence Probes in Simple Measurements of Photosynthesis.* Sheffield, UK: Robert Hill Institute, University of Sheffield.

Wilson, K., and J. Walker. 2003. *Practical Biochemistry: Principles and Techniques*, 5th ed. Cambridge: Cambridge University Press.

Wilson, K., and J. Walker. 2007. *Principles and Techniques of Biochemistry and Molecular Biology.* Cambridge, UK: Cambridge University Press.

IMPORTANT LINKS

1. Membrane electrode: http://www.horiba.com/in/application/material-property-characterization/water-analysis/water-quality-electrochemistry-instrumentation/ph-guide/1/details/thick-membrane-electrode-9611-10d-9451/
2. Oxygen electrode:
   http://www.enotec.com/en/products-services/products/oxitec/oxitec-5000.Html#c9 35
   http://www.hansatech-instruments.com/
   http://www.ysilifesciences.com/index.php?page=special-methods-and-applications

4 Chromatography

4.1 INTRODUCTION

What may be called chromatographic effects today were already in use some 2000 years ago. Dyers were accustomed to judge the quality of their dyes and, in particular, to detect the presence of adulterants by letting a drop of solution spread out on a piece of cloth or paper (papyrus). The fringe of colors formed at the boundary of the spot was a useful diagnostic. This could be the root of paper chromatography (PC). Around the same time, column chromatography was born in the oil industry, where crude oils were purified by allowing them to percolate through beds of carbon. However, as with many great discoveries the exact origin of the process could not be pointed out, although a few can be appreciated for giving due importance to the process by using it for the separation, purification, and identification of compounds. The first to give convincing evidence of the power and versatility of the process was M. S. Tswett, a botanist, who in 1906 isolated the principal plant pigments by passing them in a solution of petroleum ether through a column of powdered chalk. Chromatography based on the differential partition between two immiscible solvents was first described, in 1941, by A. J. P. Martin and R. L. M. Synge. In this process, water was fixed within silica gel, the mobile phase used was chloroform, and the amino acids of a protein were analyzed. Subsequently, other classes of compounds, for example, antibiotics, were analyzed. Today, the technique has developed in all dimensions as perhaps the single most powerful analytical and preparative method available in the laboratory.

4.2 GENERAL PRINCIPLES

The basis of all forms of chromatography is the differential partition of a compound between two immiscible phases, one of which is stationary and the other, mobile. The way in which a compound is partitioned or distributed between two immiscible phases is given by the partition coefficient (K_d):

$$K_d = \frac{\text{Concentration in phase A}}{\text{Concentration in phase B}}$$

The value of K_d is constant at a given temperature. Depending on the different types of phases involved, there are different forms of chromatography (Table 4.1). Depending on the mode by which separation is achieved, there are three types of chromatography:

1. Column chromatography, in which the stationary phase is packed into glass or metal columns and the mobile phase percolates through the column.
2. PC, in which the stationary phase is supported by cellulose fibers of paper and the mobile phase moves through the interstitial spaces by capillary action. Sometimes, the cellulose fibers of the paper act as the solid stationary phase.
3. Thin-layer chromatography (TLC), in which the stationary phase is thinly coated onto glass plates and the mobile phase moves along the stationary phase.

TABLE 4.1
Types of Chromatography

	Form Phase	Mobile Phase	Stationary Phase	Principle of Separation
1.	Adsorption or solid–liquid chromatography	Liquid	Solid	Adsorption equilibrium between stationary solid and mobile liquid phases
2.	Liquid–liquid or partition chromatography	Liquid	Liquid	Partition equilibrium between stationary liquid and mobile liquid phases
3.	Gas–liquid chromatography	Gas	Liquid	Partition equilibrium between stationary liquid and mobile gaseous phases
4.	Ion-exchange chromatography	Electrolyte	Ion exchange	Ion-exchange equilibrium between ion exchangers in resin stationary phase and mobile electrolyte phase
5.	Exclusion chromatography/ gel filtration or permeation chromatography	Liquid	Liquid	Molecular sieve action of pores of gel particles
6.	Affinity chromatography	Liquid containing macromolecule	Ligand	Biological specificity and reversibility of binding between macromolecules and ligands

4.3 COLUMN CHROMATOGRAPHY

In the column chromatography technique, the components of a mixture are separated in a column as distinct zones and each zone is eventually displaced from the column as a series of fractions (Figure 4.1). The apparatus and general techniques used for column adsorption partition, ion exchange, exclusion, and affinity chromatography have much in common and are discussed in Section 4.3.1. Gas–liquid chromatography (GLC) and high-performance liquid chromatography (HPLC) have their own unique apparatus and procedures and are discussed separately.

4.3.1 Chromatography Columns

Chromatography columns are usually made up of glass and in general long columns give good resolution of components, although wide columns are better for the separation of large quantities of material. The essential features of a chromatography column are shown in Figure 4.1. The glass column should have a means of supporting the stationary phase as near to the base of the column as possible to minimize dead space below the column support in which post-column mixing of separated compounds can occur. Commercial columns possess either a porous glass plate fused onto the base of the column or a suitable device for supporting a replaceable nylon net, which in turn supports the stationary phase over it. Glass wool plugs are often used.

4.3.2 Stationary Phases

Depending on the particular form of chromatography to be carried out, there are different types of stationary phases, such as solids (alumina, silica, etc.) for adsorption chromatography, liquids for partition chromatography, gels for exclusion chromatography, and resins for ion-exchange chromatography. All materials must be equilibrated with the mobile liquid phase, called a solvent, before preparing the column. During equilibration, the solid support material is allowed to settle and fine particles remaining in the suspension are removed by decantation (Figures 4.1 and 4.2). If this is not carried out, the flow rate of the mobile phase is clogged by these fine particles. In addition, some form of pretreatment of the stationary phase is often required, for example, some gel filtration

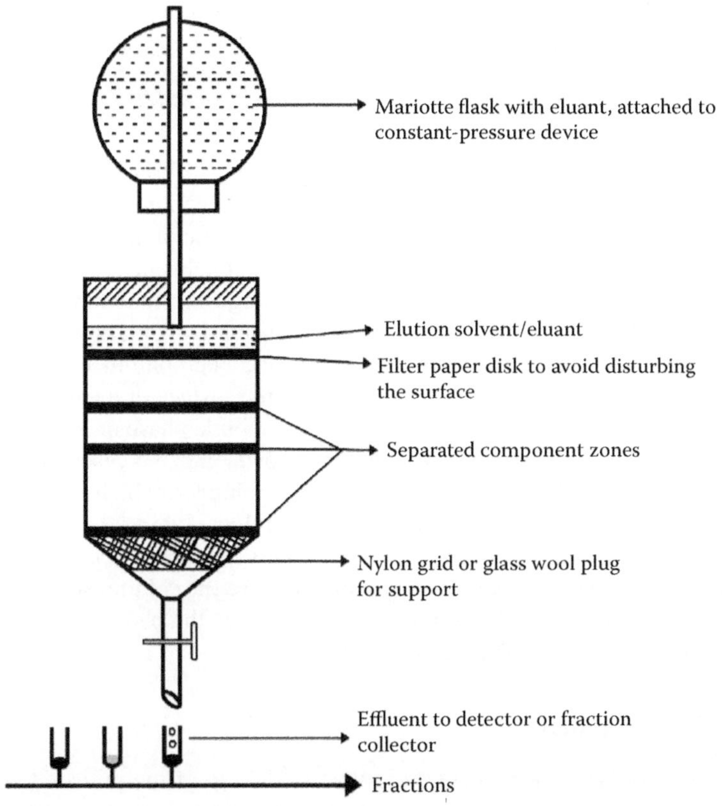

FIGURE 4.1 Column chromatography.

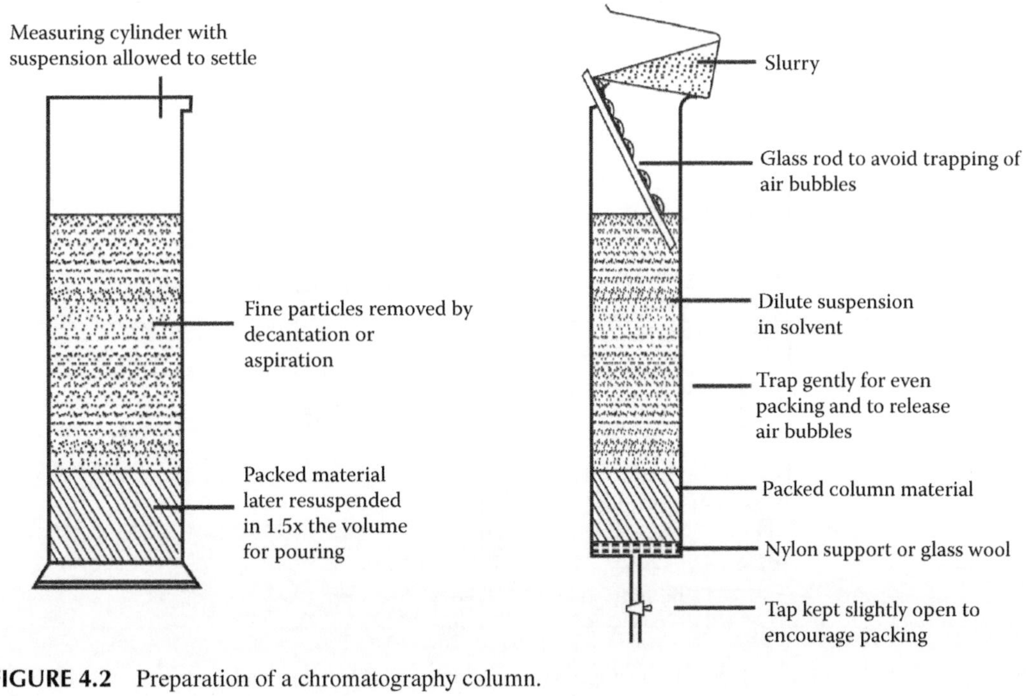

FIGURE 4.2 Preparation of a chromatography column.

materials need to swell, adsorbents need to be activated by heating or an acid treatment, or ion-exchange resins have to be obtained in the required ionized form by washing.

4.3.3 Packing of Columns

A chromatography column is packed with a stationary phase (adsorbent, resin, or gel) by pouring a slurry from the stationary phase to the mobile phase into the column with the help of a glass rod (Figure 4.2b). This prevents the trapping of air bubbles. Gentle tapping or stirring in the upper part of the slurry also ensures that no air bubbles are trapped and helps in evenly packing the column. Poor column packing gives rise to uneven flow (channeling) and reduced resolution. The slurry is added until the required height is obtained. The suspension is then allowed to settle and excess mobile phase is allowed to run off by opening the outlet. The column is then washed with the mobile phase and the level of the liquid is kept just above the surface of the material. To prevent the column from being disturbed either by the addition of the mobile phase or during the application of the sample to the column, a filter paper disk or nylon or rayon gauze is placed on the surface of the column. Some commercial columns have an adaptor or a plunger, which serves the dual purpose of protecting the surface and providing an inlet (often capillary tubing) to carry the solvent to the column surface. Once a column has been prepared, no part of it should be allowed to run dry, that is, a layer of mobile liquid should always be maintained above the column surface (Figure 4.3). The total volume of solid and liquid in the column is referred to as the bed volume, and the volume of liquid phase outside the solid is called the void volume (V_0).

4.3.4 Application of Sample

Several methods are available for applying a sample to the top of the prepared column. A simple method is to remove most of the solvent from above by suction and to just drain the remainder into the column bed. The sample is then carefully applied using a pipette and allowed to run into the column. A small volume of solvent is then applied in a similar manner to wash final traces of the sample into the bed. More solvent is then carefully added to the column to a height of 5–10 cm. The column is then connected to a suitable reservoir, which contains more solvent, so that the height of the solvent

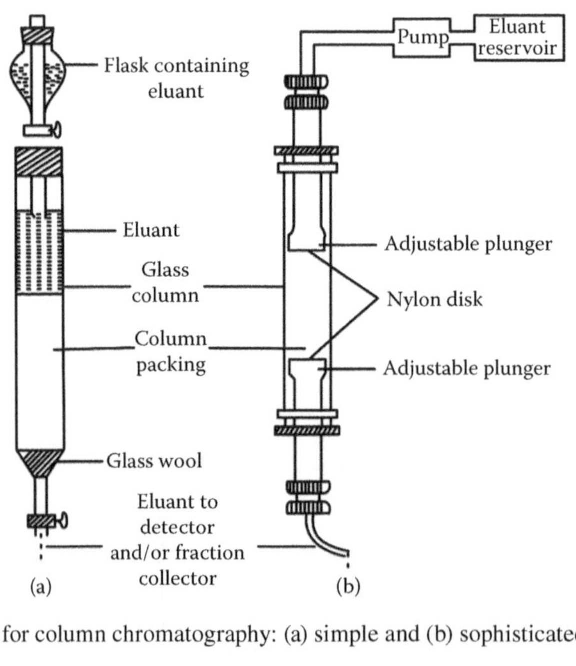

FIGURE 4.3 Columns for column chromatography: (a) simple and (b) sophisticated versions of columns.

in the column can be maintained at 5–10 cm. An alternative procedure that avoids the necessity of draining the column to the surface of the bed is to increase the density of the sample by adding sucrose to a concentration of about 1%. When this solution is layered onto the solvent above the column bed, it will automatically sink to the surface of the column and hence be quickly passed into the column. This method, however, can be used only if sucrose does not interfere with the separation and subsequent analysis of the sample. A third method involves the use of capillary tubing, a syringe, or a peristaltic pump to pass the sample directly to the column surface.

In all cases, care must be taken to avoid overloading the column with the sample; otherwise, irregular separation occurs. It is also advantageous to apply the sample in as minimal a volume as possible, as this ensures the presence of an initial tight band of material when the separation commences. In addition, to prevent anomalous adsorption effects the sample should be desalted before application to the column.

4.3.5 Column Development

Once components of the mixture have separated into zones in the column, the next step is to remove these components from the zones and collect them as separate fractions. This is carried out by continuously passing the mobile phase (eluant/solvent) through the column. This process is called column development. In displacement column development, the mobile phase interacts more strongly with the stationary phase than the component of the mixture bound to it, thereby displacing the bound molecules. However, for better resolution, elution column development is preferred. In this process, the mobile phase interacts weakly with the stationary phase than the bound molecules and overrides the bound molecules, gradually eluting them from the column. Different components can be removed from the column by changing the strength or pH of the eluting mobile phase in a stepwise fashion (gradient elution).

In order to produce a suitable gradient, two solvents should be mixed using either commercially available gradient mixers or a more simple method as follows: The two solutions are placed in separate chambers (Figure 4.4). The recipient chamber is linked to the column and the donor chamber is linked to the recipient chamber by a siphon. As the eluant enters the column from the recipient chamber, the solution in the donor replaces it and is mixed by a stirrer. The diameter of the two chambers determines whether the gradient varies with time in a linear, convex, or concave manner. Gradient elution achieves better resolution as compared to isocratic elution, which uses a single solvent as the eluant. The volume of eluant required to elute a particular solute is known as the elution volume (V_e), whereas the corresponding time for elution is known as retention time (t_r).

During elution, it is essential that the flow rate is maintained constant. The flow rate may be regulated by adjusting the operating pressure, which corresponds to the difference between the level of the solvent in a reservoir situated above the column and the level at the outlet side of the column. An ordinary open reservoir is not satisfactory because the operating pressure drops as the solvent

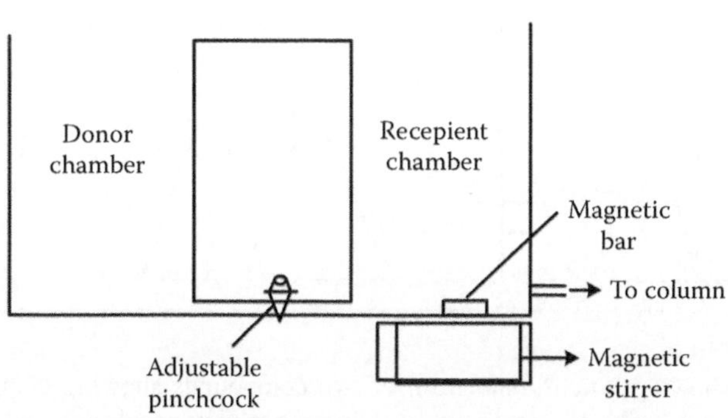

FIGURE 4.4 Gradient maker.

level drops while it runs through the column. A Mariotte flask (Figure 4.1) can keep the pressure constant. Alternatively, a peristaltic pump can pump the eluant at a predetermined rate onto the column. Care must be taken to see that the column structure does not change due to excessive pressure from the pump.

4.3.6 Fraction Collection and Analysis

As components of a mixture emerge from the column as effluent, it is necessary to detect their presence in order to enable them to be isolated for further study. This can be done in two ways. In the first method the effluent can be continuously monitored, and in the other it can be collected as separate fractions that are subsequently analyzed.

In continuous monitoring, the effluent is passed through a flow cell (8 mm^3) located in the detector. Detection may be based on UV or visible light absorption by components of characteristic wavelengths, fluorescence, changes in the refractive index of the effluent, the presence of a radioactive label, or the ease of oxidation or reduction as measured by an electrochemical detector. The signal generated by the detector is recorded on a chart recorder with each emerging compound giving a characteristic peak from which it is possible to calculate the retention time and/or the elution volume (Figure 4.5a–c).

Separate fractions can be collected either manually or by using an automatic fraction collector. The amount of effluent in each fraction can be determined in one of several ways. There may be a siphoning system to deliver a predetermined volume into each tube or there may be an electronic means of allowing a predetermined number of drops to enter each tube. This has the disadvantage that if the composition of the effluent changes during gradient elution, its surface tension and therefore the size of the drops may also change so that the volume of collected liquid changes. In another method, the effluent is allowed to enter a tube for a fixed interval of time, but if the flow rate of the column varies, the volume of the fraction collected also varies. However, once the fractions have been collected they can be detected by spectrophotometer, fluorimeter, scintillation counting, and radioimmunoassay.

Quantitative estimations of fractions collected can also be made using peak area. The area of the peak is the product of the height of the peak (h_p) and its width at half height (W_{ha}). The peak area can be shown to be proportional to the amount of the sample component present. Alternatively, the peak may be cut out, weighed, and the assumption made that area and weight are linearly related. The amount of a specific component can be found by a standard curve made by chromatographing the known sample under identical conditions and weighing the peak cutout (Figure 4.6).

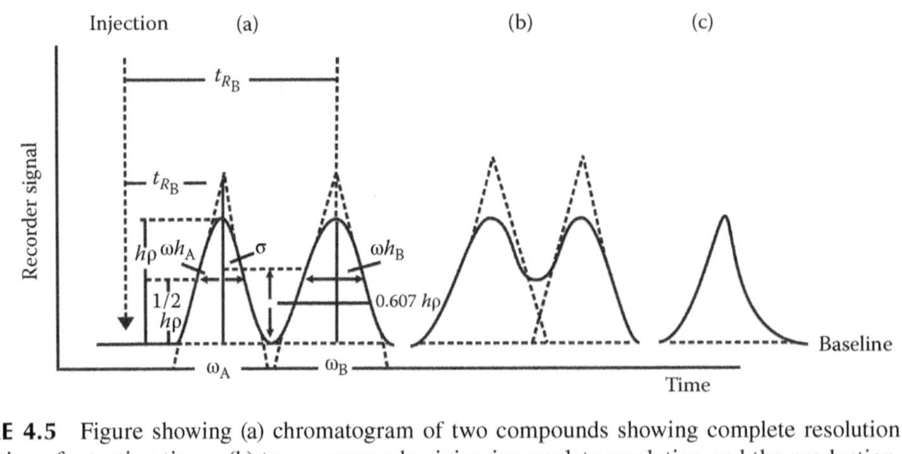

FIGURE 4.5 Figure showing (a) chromatogram of two compounds showing complete resolution and the calculation of retention times, (b) two compounds giving incomplete resolution and the production of fused peaks, and (c) a compound showing excessive tailing.

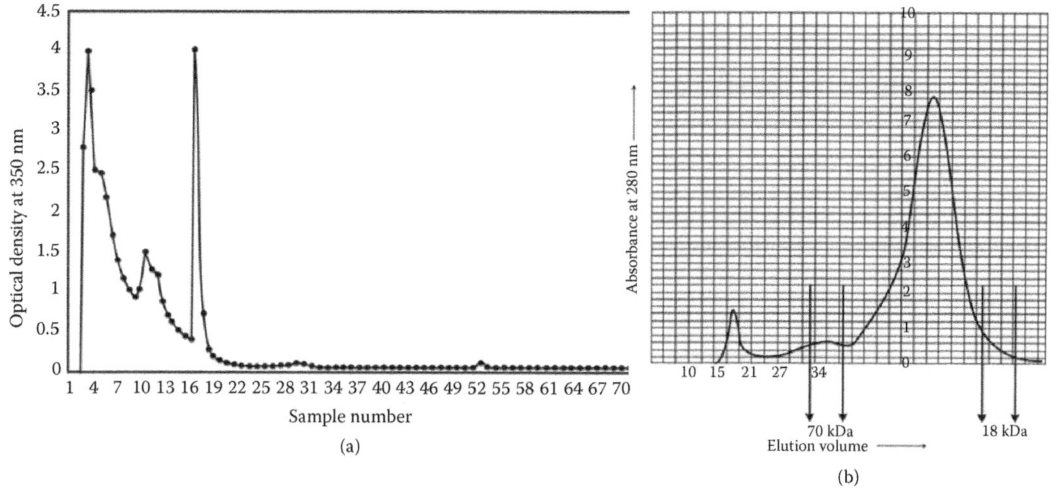

FIGURE 4.6 Figure showing (a) purification of bioactive compounds by using sepharose CL-6B column and (b) graphical presentation of fractionation of metabolic proteins of *Aspergillus fumigates* by size-exclusion chromatography.

4.4 PAPER CHROMATOGRAPHY AND THIN-LAYER CHROMATOGRAPHY

4.4.1 Principles

Unlike column chromatography, the components of a mixture after separation remain on the chromatogram in PC and TLC. On spraying the appropriate detecting agent, they appear as a series of spots. The position of the spots is recorded in terms of the R_f value (Figure 4.7). This is defined as the following ratio:

$$R_f = \frac{\text{distance moved by component}}{\text{distance moved by mobile liquid phase/solvent}}$$

The R_f values are constant for a particular component under particular conditions.

In PC, the stationary phase is supported by cellulose fibers of paper and the mobile phase moves through the interstitial spaces by capillary action. If paper merely acted as an inert support, the separation of the components of a mixture would be exclusively due to continuous partition between mobile phase flowing along the paper and water held in the paper; filter paper contains 15% of its weight in water and this may be the case in some circumstances. Usually the paper affects separation in a number of ways; it acts as an adsorbent, especially for polar molecules for which it has great affinity. They are held by hydrogen bonding and the van der Waals force; the paper also functions as an ion-exchange material due to the carboxyl groups it contains.

In TLC, a slurry of the stationary phase, generally in water, is applied to a glass, plastic, or foil plate as a uniform thin layer, by a plate spreader starting from one end and moving to the other. A variety of materials can be used as the stationary phase and, similar to PC separation, TLC is achieved in multifarious ways. On heat-activated layers of silica gel and similar substances, when nonaqueous solvent systems are used, adsorption is the predominant process, and this is useful for separating hydrophobic substances such as lipids. When aqueous solvents are used on layers of crystalline material that have a large surface area and fine particle size, a situation between these two extremes exists and the magnitude of influence of individual components is difficult to assess.

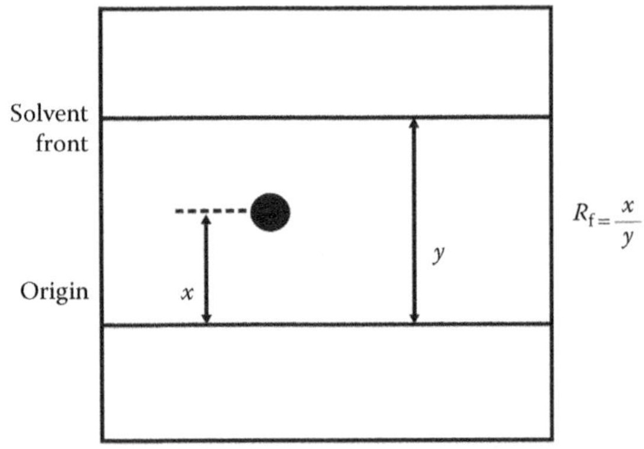

FIGURE 4.7 Finding the R_f value.

4.4.2 Types of Paper

Whatman no. 1 is the paper most frequently used for analytical purposes. Whatman no. 3 is a thick paper and is best used for separating large quantities of material. However, its resolution is inferior to that of Whatman no. 1. For rapid separation Whatman no. 4 and no. 5 are convenient, although spots are less well defined. In all cases, flow rate is faster in the machine direction, which is normally noted on the box containing the papers. The paper may be impregnated with materials like silica for adsorption, liquid for partition chromatography, and so on. Silica-impregnated papers are commercially available and used for separating lipids and hydrophobic molecules.

4.4.3 Thin Layer

Slurry of the stationary phase in water is applied evenly onto a plate. Calcium sulfate is often incorporated into the slurry to facilitate adhesion of the thin layer on the plate. For analytical purposes, a layer 0.25 mm thick is desired. For preparative separations, 5-mm-thick plates are used. With the exception of thin-layer exclusion chromatography, the plates are dried to leave a coating of the stationary phase. In the case of adsorbents, drying is carried out in an oven at 100°C–120°C. This also activates the adsorbent.

4.4.4 Sample Application

A drop of the solution containing the mixture of compounds to be separated is placed near one end of the strip by means of a syringe, capillary, or micropipette. It is allowed to dry at room temperature. The strip is then placed so that the end with the spot dips into the solvent, which is the mobile liquid phase and moves by capillary action. The mobile liquid phase is usually a mixture of solvents. It is essential that the spot is not immersed in the solvent as it would dissolve and become lost.

4.4.5 Development

It is essential to make sure that prior to development the atmosphere of the separation chamber is fully saturated with the solvent; otherwise, R_f values will vary from chamber to chamber. The process is called equilibration and is carried out as follows: About 1.5 cm of mobile phase/solvent is added to the chamber and covered with a lid for at least half an hour. Unless this is done, irregular running of the solvent will occur as it ascends the plate, resulting in poor separations. After equilibration, the lid is removed and the strip is placed inside the spot slightly over the solvent. Whenever the solvent reaches the far end of the strip or after a convenient time period, the sheet is removed and rapidly dried and spots are detected by spraying a suitable locating agent on it. When a mixture contains many components, complete separations on a strip of finite length may not be possible. In order to overcome this difficulty, a number of different solvents with different properties are used so that components running together in one solvent will probably separate in another solvent. Although many one-way chromatograms each in a different solvent can be compared, a lot more information is obtained if two solvents are used in conjunction to prepare a two-way chromatogram than if the two are used to prepare two one-way separations (Figure 4.8).

A two-way chromatogram is prepared by placing a drop of the mixture near a corner of a square or rectangular sheet of paper or layer. The solvent is then allowed to flow through the sheet with the result that a one-way separation is obtained. After drying completely, the sheet is turned at right angles and run in a second solvent, which performs a further separation and causes the components to be distributed on the sheet in two dimensions instead of the previous one dimension. This technique is called 2D chromatography.

There are two techniques that can be used for the development of paper chromatograms: (1) ascending and (2) descending (Figure 4.9). In both cases, the solvent is placed at the base of a sealed tank or glass jar to allow the chamber to become saturated with solvent vapor. In the ascending method, the end of the paper with the spot is dipped in the solvent at the base of the tank and as the solvent rises up, or ascends, by capillary action, separation is achieved. In the descending technique, the end of the paper with the spot is held in a trough at the top of the tank and the rest of the paper is allowed to hang vertically but not in contact with the solvent at the base of the tank. The solvent moves down from the trough, or descends, under the force of gravity, and separation is achieved. Although the descending technique achieves faster flow rates, it is not preferred because it is far more cumbersome to set up than the ascending technique.

FIGURE 4.8 A 2D chromatogram.

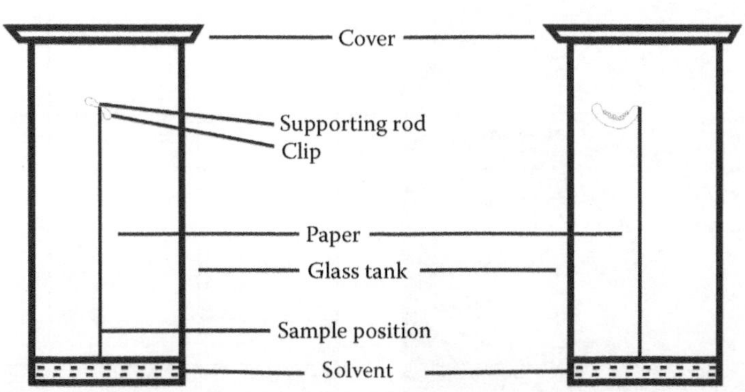

FIGURE 4.9 Methods of PC: (a) ascending and (b) descending methods.

4.4.6 Component Detection

Relatively few compounds are naturally colored or fluoresce or absorb UV light, and so after separation the majority of compounds are detected with a chemical location reagent. The term location reagent is defined as the total material applied to the dry chromatogram after chromatography in order to locate or reveal the positions of the separated substances so that a color is produced that is different from the color produced when the reagent is applied to a blank sheet that has been run in the solvent. Thus the reagent includes the chemicals (active constituents) that react to produce the color, the solvent in which they are dissolved, and any subsidiary substance added to strengthen or stabilize the colors so produced. Chemical reagents of this type suffer from the disadvantage that when compared to physical location methods (fluorescence, radioactivity, UV light, etc.), they destroy the compounds being located.

4.4.7 High-Performance Thin-Layer Chromatography

High-performance TLC (HPTLC) is an enhanced form of TLC. A number of enhancements can be made to the basic method of TLC to automate the different steps, increase the resolution achieved, and ensure more accurate quantitative measurements.

Automation is useful to overcome the uncertainty in droplet size and position when the sample is applied to the TLC plate by hand. One recent approach to automation has been the use of piezoelectric devices and ink-jet printers for applying the sample (Figure 4.10). Spot capacity (analogous to peak capacity in HPLC) can be increased by developing the plate with two different solvents. After the plate is exposed to the first solvent, the solvent is removed, the plate is rotated through 90°, and the plate is developed with the second solvent. If the two solvents show different selectivity, then the spots may be spread over the entire surface of the plate. This is obviously a form of 2D chromatography.

4.4.8 Advantages of Thin-Layer Chromatography over Paper Chromatography

A variety of supporting media can be used in TLC so that separation can be achieved in several ways, such as adsorption, partition, exclusion, or ion exchange, depending on the type of stationary phase applied to the plate. TLC is rapid, taking as little as 1 hour as compared to 7–8 hours for PC. Compounds can be detected in a much lower concentration as the spots are very compact. Separated compounds can be detected by corrosive sprays and elevated temperature, for example, spraying with 50% sulfuric acid or 25% sulfuric acid in ethanol and heating will result in most compounds becoming charred and showing up as brown spots. This is not possible with paper, which will burn and disintegrate.

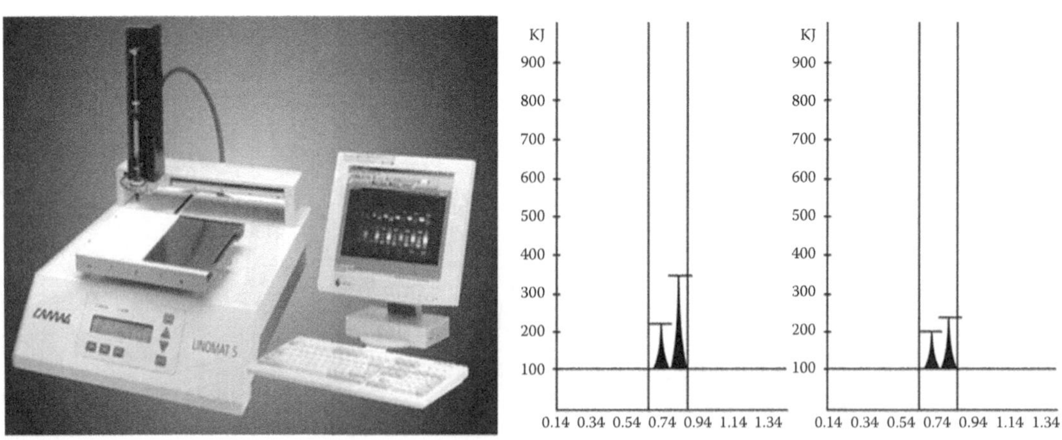

(a) (b)

FIGURE 4.10 (a) HPTLC equipment and (b) demonstration of separation and reading of intensity of separated compounds in chromatograms.

4.5 SAMPLE COLLECTION, PRESERVATION, AND PREPARATION

These are identical for all chromatographic methods as for any other biochemical investigation of biological fluids or tissues. The complex mixtures of substances present in biological fluids often result in their mutual interference when chromatography of a particular group of compounds is attempted. Either the interferences must be removed or the substances of interest selectively extracted. Care must be taken at this stage to prevent or at least minimize losses. Sometimes substances are converted to more satisfactory derivatives, which are subsequently chromatographed, for example, labile keto acids are converted to stable dinitrophenyl hydrazones or amino acids, which may be separated by TLC or PC.

4.6 ADSORPTION CHROMATOGRAPHY

4.6.1 Principle

The separation of components of a mixture depends on differences in both their degree of adsorption and their solubility in the solvent used for separation. This can be carried out in column, thin-layer, and paper modes. Compounds are adsorbed onto the stationary phase to the extent determined by the charge, van der Waal forces, dipole interactions, hydrogen bonding, and steric factors. Adsorption chromatography depends on the structure of the components due to which they have greater or lesser affinity for the stationary and mobile phases.

The mass of solute adsorbed per unit weight of adsorbent (m) depends on the concentration of the solute (c); Langmuir derived an equation on the basis that only a monolayer is adsorbed and only a proportion of the molecules in collision are adsorbed. This is known as the Langmuir adsorption isotherm:

$$m = \frac{k_1 k_2 c}{1 + k_2 c}$$

In the aforementioned equation, k_1 is a measure of the number of active adsorption sites per unit weight of adsorbent and depends on the nature of the adsorbent. The term k_2 is a measure of the affinity of the solute for the adsorbent and is affected by all the components of the system.

Langmuir assumed only one binding site, although in practice there are a number of binding sites on the surface of the adsorbent, each with a different affinity, thus giving a series of Langmuir-type isotherms. Hinshelwood therefore suggested that the following equation gives a more accurate picture:

$$m = \frac{k_1 k_2 c}{1 + k_2 c}$$

This approximates to Freundlich adsorption isotherm, which is found in practice:

$$m = Kc^x$$

In this equation, K and c^x are constants depending on the particular system used.

The difference between these two isotherms is illustrated in the graph shown in Figure 4.11a. The mixture of binding sites of different affinities is the cause of tailing observed in the elution profile in column chromatography (Figure 4.11b). This tailing can be overcome by eluting with a suitable gradient of pH, ionic strength, or polarity so that the more strongly adsorbed molecules meet a higher concentration of displacing compound than the more weakly bound molecules.

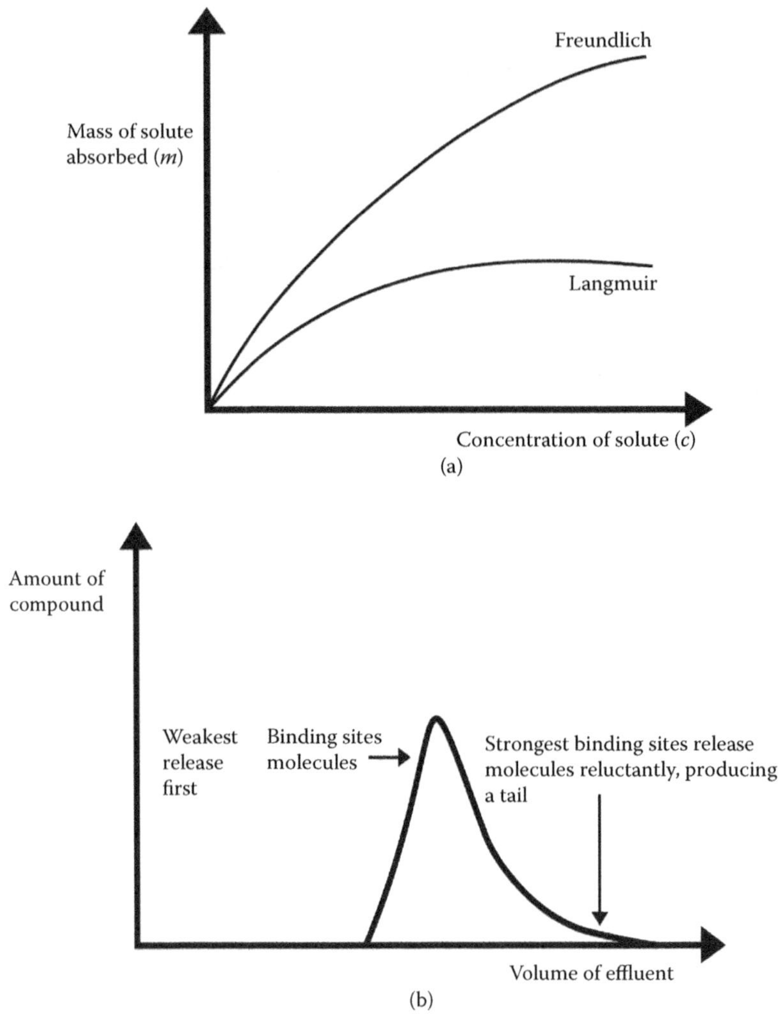

FIGURE 4.11 Figure showing (a) absorption isotherms of Freundlich and Langmuir, and (b) tailing observed in the elution of a compound from an absorption column.

4.6.2 Adsorbents

Compounds such as silicic acid (silica gel), aluminum oxide, $CaCO_3$, $MgCO_3$, $ZnCO_3$, MgO, and cellulose may be used as the stationary phase. The choice of adsorbent and solvent depends on the separation to be achieved. Hydroxyapatite ($CaPO_4$) is widely used for the separation of proteins, nucleic acids, and viruses. Unlike most other adsorbents, it has some ion-exchange properties that aid separation. In addition, in nucleic acid separation it has the advantage that it can bind to double-stranded DNA but not single-stranded DNA. Adsorbents of TLC are often impregnated with $AgNO_3$ to enhance the separation of compounds differing in the number of double bonds. This is called argentation TLC. Some adsorbents have a tendency to take up water from the atmosphere during storage and this may adversely affect their adsorption properties. In such cases, the adsorbents are activated by heating at 110°C for some time to remove any water.

Poly-E-caprolactam TLC sheets do not require activation and are semitransparent, which allows the unknown compounds and their standard to run on opposite sides to be compared. They can be reused if they are immediately cleared with ammonia-acetone and are widely used in protein sequencing. In general, the adsorption affinity of compounds increases with the polarity of molecules, the number of double bonds in alkenes, and the introduction of functional groups. In order of increasing affinity, some functional groups are as follows:

$$-O-R,\ -NO_2-COR,\ -OCOR,\ -NH_2,\ -OH,\ -CONH_2,\ -COOH$$

4.6.3 Solvents

The choice of solvents depends on the polarity of the compounds to be resolved and their distribution coefficients. Commonly used low-polarity solvents are hexane, heptane, acetonitrile, toluene, diethyl ether, and chloroform. Intermediate-polarity solvents include ethanoic acid, dichloromethane, and pyridine, whereas polar solvents are propanol, ethanol, methanol, acetone, and water.

4.7 PARTITION CHROMATOGRAPHY

4.7.1 Principles

The separation of components of a mixture depends on differences in their solubility in two immiscible liquids, one of which is kept stationary by keeping it supported by an inert matrix. The other is mobile (Figure 4.12). The component distributes itself between the two immiscible liquid phases as described by the partition coefficient K_d at equilibrium. As already mentioned,

$$K_d = \frac{\text{Concentration in solvent 1}}{\text{Concentration in solvent 2}}$$

In normal-phase partition chromatography, the stationary phase is usually water, whereas the mobile phase is an organic, nonpolar liquid that is immiscible in water and flows past the stationary phase. Thus, compounds that are soluble in both water and nonpolar organic solvents are readily separated by partition methods, as compared to adsorption chromatography, in which compounds sparingly soluble in water but insoluble in organic liquids are separated.

In the case of column chromatography, the matrix may be cellulose, starch, or silicic acid. For TLC it is most commonly cellulose, whereas for paper partition chromatography sufficient water is naturally present in paper to act as the stationary phase.

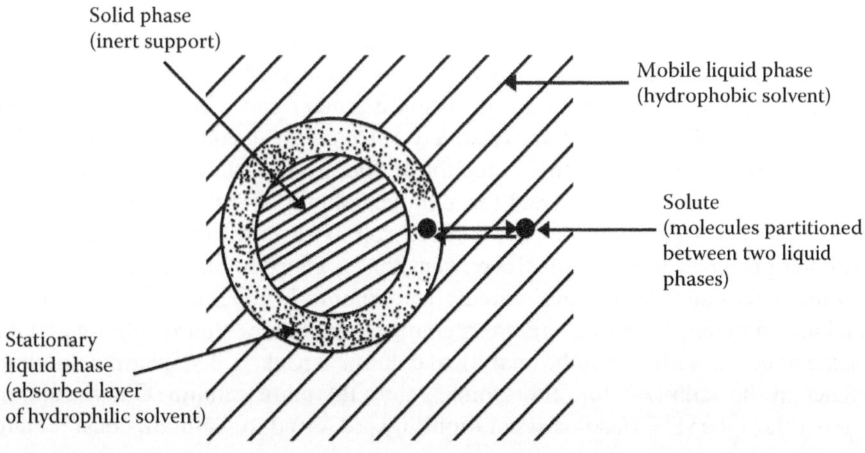

FIGURE 4.12 The principles of liquid–liquid partition chromatography.

In reverse-phase partition chromatography, the stationary phase is a nonpolar liquid, for example, liquid paraffin supported by an inert matrix as in normal phase. This phase is prepared by treating the supporting matrix with a solution of the mobile phase in a suitable nonpolar and volatile solvent; the solvent is subsequently removed by evaporation.

4.7.2 Countercurrent Chromatography

Countercurrent chromatography (CCC) is a separation process based on the distribution of a compound between two immiscible liquid phases, but neither phase is supported on an inert matrix and it differs from conventional liquid–liquid partition chromatography. The mixture components are partitioned between two immiscible liquid phases, which move relative to one another. The usual form of the required apparatus comprises a series of extraction tubes through which the less dense of the two phases moves. Each extraction is performed by shaking the tubes and, after waiting for the phases to separate, the transfer of the upper phase is effected. The components of the mixture progress through the tubes as a result of the movement of phases but at different rates. These rates depend on the K of the components in the two phases; when these coefficients are different, separation is achieved.

4.8 HIGH-PERFORMANCE LIQUID CHROMATOGRAPHY

4.8.1 Principles

HPLC is a form of chromatography in the column mode, using a small particle-size stationary phase and high-pressure eluant pumping systems to give a high performance; hence, it is named high-performance liquid chromatography.

All forms of column chromatography rely on gravity or low-pressure pumping systems for the supply of eluant to the column. The flow rates are therefore low and cannot be increased by merely increasing the pressure of the eluant supply, as this would create back pressure, which would eventually damage the column structure. The use of small particle-size stationary phases makes it possible to withstand these high pressures and, at the same time, give better resolutions. In recent years, however, special pumping systems have been designed that can give higher flow rates. These, together with the use of small particle-size stationary phases, result in quick, efficient, and better resolution in adsorption, partition ion-exchange, exclusion, and affinity column chromatography. The principal components of an HPLC system are shown in Figure 4.13a and b.

4.8.2 Column

Since glass tubing cannot withstand pressures in excess of 70 atm, stainless steel, precision-bored columns with an internal mirror finish for efficient packing are normally used. These straight columns of 15–50 cm length and 1–4 mm diameter can withstand very high pressures of up to 5.5×10^7 Pa (8000 psi) and are relatively corrosion resistant. At the end of the column, homogenously porous plugs of stainless steel or Teflon are used to retain the packing material and to ensure uniform flow of the solvent through the column. At times, the repeated application of impure samples such as urine, blood, or crude cell extracts results in clogging and the loss of the resolving power of the column. To prevent this, a short column of 1–2 cm in length and an internal diameter equal to that of the analytical column is generally introduced between the injector and the analytical column. This short column is called the guard column and it is packed with the same material with which the analytical column is packed. The guard column retains the solid particles in the sample before the sample enters the main column. Guard columns can be replaced at regular intervals. In some separations, it is required to maintain constant temperature thermostatically.

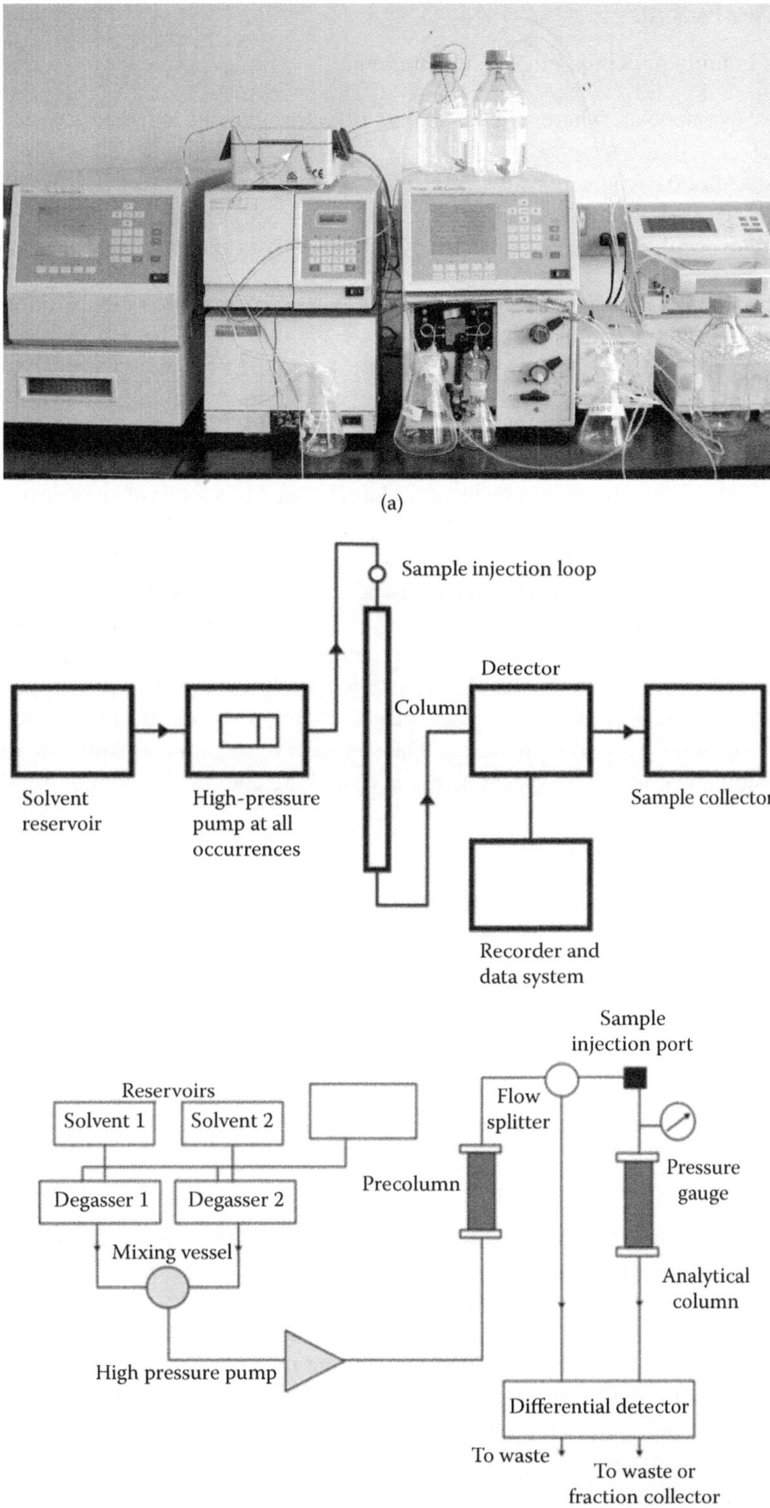

FIGURE 4.13 Figure showing (a) HPLC and (b) the diagrammatic representation of the components of an isocratic HPLC system and an HPLC column with its associated pumps, detector, and recording instruments.

4.8.3 Column Packing

Three forms of column packing material are available:

1. Microporous supports where micropores ramify through the particles, which are 5–10 μnanometers in diameter
2. Pellicular supports where porous particles are coated onto an inert solid core such as a glass bead of about 40 nm in diameter
3. Bonded phase where the stationary phase is chemically bonded onto an inert support

For adsorption chromatography, adsorbents such as silica or alumina are available in microporous or pellicular forms in a range of particle sizes. Pellicular forms have high efficiency but low sample capacity and, therefore, microporous supports are preferred where applicable. All HPLC packing materials are regular and spherical in shape, giving good packing and flow properties.

In liquid–liquid partition systems, the stationary phase may be coated onto an inert support. Both microporous and pellicular supports are used, but in these the stationary solvent gradually gets washed out with the mobile solvent. To overcome this problem, bonded phases have been developed in which the stationary liquid is covalently bonded to the supporting material (silica or a silicone polymer).

Many different types of covalently bonded phase ion exchangers are available of which the cross-linked microporous polystyrene resins (silicone network) are widely used. Pellicular resin forms are also available. These are called hard gels, and they readily withstand the pressures required during analysis.

The stationary phases for exclusion chromatography in the HPLC mode are porous silica, glass, polystyrene, or polyvinylacetate beads. These are generally used with organic eluting solvents; semirigid gels such as sephadex and nonrigid gels such as sepharose are of limited use in HPLC since they can withstand only low pressures. The supports for affinity separations are the same as those for exclusion apparatus. The spacer arm and ligand are chemically bound to the support.

4.8.4 Column-Packing Procedure

The most widely used column-packing procedure is the high-pressure slurrying technique. A suspension of the packing is made in a solvent of equal density to the packing material. The slurry is then rapidly pumped at a high pressure into a column with a porous plug at its outlet. The resulting bed of packing material can then be prepared for use by running the mobile phase through the column when hard gels are packed. It is necessary to allow them to swell first in the solvent to be used in the chromatographic separation before packing under pressure. Soft gels cannot be packed under pressure, and they are allowed to pack from a slurry in the column under gravitational sedimentation only as in conventional column chromatography.

4.8.5 Chromatographic Solvent (Mobile Phase)

The choice of mobile phase depends on the type of separation to be achieved. Isocratic separations may be made using a single solvent, two solvents, or more solvents mixed in fixed proportions. Alternatively, a gradient elution system may be used. For this system, two pumps are required to pump the two solvents to produce a gradient. In all cases, the solvent should be free from even traces of impurities as this would hamper the development and interfere in the detection of compounds. To this end, 1–5 nm microfilters are used to filter the solvent before it enters the pump, even though extra-pure HPLC solvents are commercially available.

4.8.6 Pumping Systems

The main features of a good pumping system are that it should be capable of outputs of at least 3.4×10^7 Pa (5000 psi) and there should be no pulses of flow through the system. There must be a flow delivery of at least 10 cm·min^{-1} for normal analysis and up to 30 cm·min^{-1} for preparative analysis. Several pumping systems are available that operate on the principle of constant pressure or constant displacement. Constant-pressure pumps produce a pulseless flow through the column, but any decrease in the permeability of the column will result in lower flow rates for which the pump will not compensate. These pumps operate by the introduction of high-pressure gas into the pump, and the gas in turn forces the solvent from the pump chamber into the column. The use of an intermediate solvent between the gas and the eluting solvent reduces the chances of dissolved gas directly entering the eluting solvent and causing problems during analysis.

Constant-displacement pumps maintain a constant flow rate through the column irrespective of changing conditions within the column. One form of the constant displacement pump is a motor-driven syringe type; it is a pump in which a fixed volume of solvent is forced from the pump to the column by a piston driven by a motor. Such pumps provide uniform solvent flow rates and also yield a pulseless solvent flow, which is important because certain detectors are sensitive to changes in solvent flow rate.

4.8.7 Detector Systems

Most commonly, the detector is a variable-wavelength UV spectrophotometer, fluorimeter, refractive index monitor, or electrochemical detector. A recent development is the interfacing of HPLC to a mass spectrometer. In all cases, the detector should be very sensitive since the quantity of material is very small.

4.8.8 Practical Procedure

The correct application of a sample onto an HPLC column is important in achieving successful separations. Ideally, the sample should be introduced as an infinitely narrow band into the column. Two methods are generally used. (1) The first method makes use of a microsyringe designed to withstand high pressure. The sample is injected through a system in an injection port, either directly onto the column packing or onto a small plug of inert material immediately above the column packing. (2) The second method involves the use of a loop injector. It consists of a metal loop of a small volume that can be filled with the sample by means of an appropriate valve; the eluant from the pump is channeled through the loop, the outlet of which leads to the column. The sample is thus flushed onto the column by the eluant without the interruption of solvent flow to the column. Automatic versions of the loop injector are commercially available (Figure 4.14).

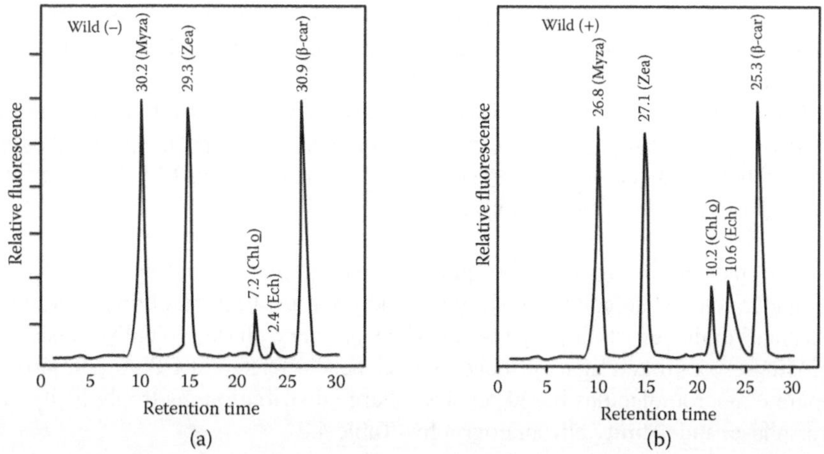

FIGURE 4.14 HPLC analysis of acetone extract of pigments of wild type (a and b) and Het⁻ Fix⁻ mutant strain (c and d) of *Anabaena variabilis*.

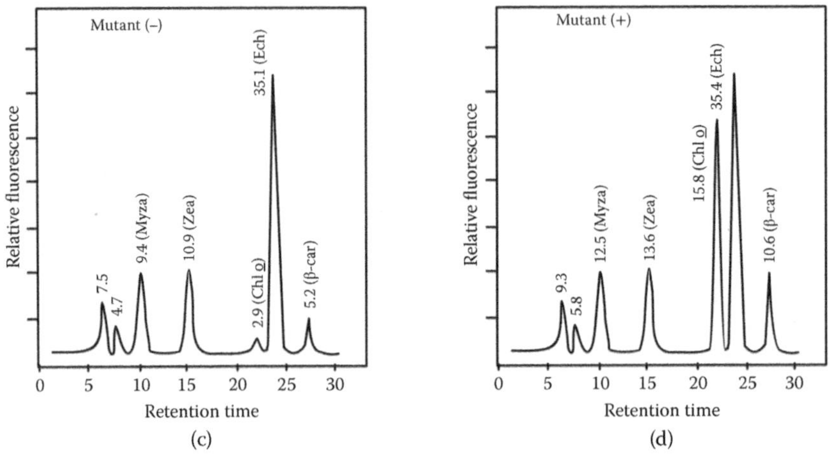

FIGURE 4.14 (*Continued*) HPLC analysis of acetone extract of pigments of wild type (a and b) and Het⁻ Fix⁻ mutant strain (c and d) of *Anabaena variabilis*.

4.9 FAST PROTEIN LIQUID CHROMATOGRAPHY

4.9.1 Principles

Fast protein liquid chromatography (FPLC) is also known as fast performance liquid chromatography. It is used to separate or purify proteins, biomolecules, and other polymers from complex mixtures.

In FPLC, the constant flow rate of solvents is controlled by a pump. The solvents (buffers) enter through tubing from an outside reservoir; the pump pushes the buffer or sample through the tubing into the injector, which is then carried into the column by the flowing solvent. As the solvent is forced into the chromatographic bed, the sample separates into various zones (bands). These fractions are collected using a fraction collector. FPLC uses tubing that has inert fluid surfaces (i.e., Teflon, titanium, and glass), which can operate at pressures up to 700 psi.

4.9.2 Apparatus and Materials

The FPLC system consists of a program controller, two high-precision pumps (one each for buffers/solvents A and B), a mixer for gradient formation, an inline filter, an automatic motor valve for injection of samples, a column, a sensitive UV monitor (fitted with a flow cell and a filter), and a fraction collector.

The FPLC system is controlled by a software program that allows automatic control of separation procedures and also systematic recording and documentation of profile generated for an applied sample with the help of a recorder. The principal components of an FPLC system are shown in Figures 4.15 and 4.16 and described as follows:

Column: Various columns are used in liquid chromatography depending on the type of separation under study. The column contains a microporous, high-resolving packing material with a small diameter (stationary phase). The stationary phase typically consists of resins comprising cross-linked agarose beads or gel beads. Different FPLC columns are used to separate macromolecules based on size, charge distribution, hydrophobicity, desalting, reverse phase, and affinity chromatography (Table 4.2).

Pumps: Pumps allow the analyte to flow through the column in a controlled manner at a constant flow rate. High-precision, laboratory-grade, peristaltic, or some other type of pump is

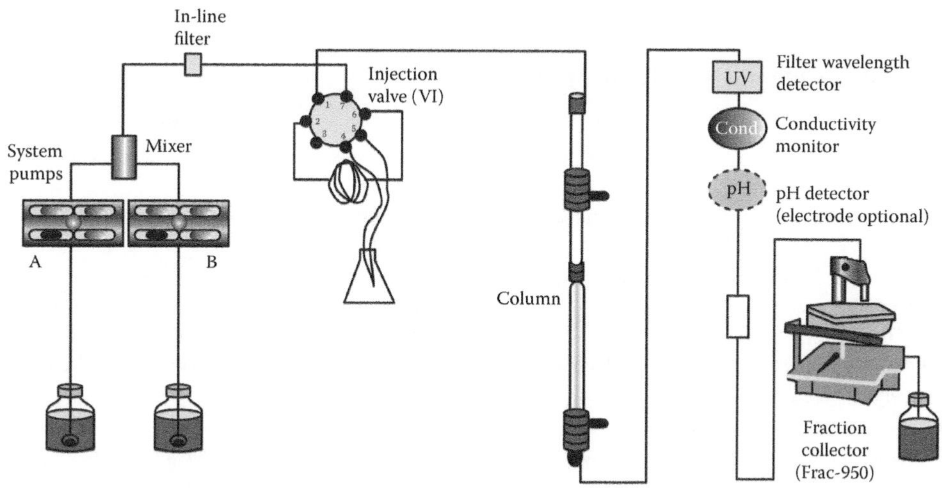

FIGURE 4.15 Schematic representation of working of an FPLC system.

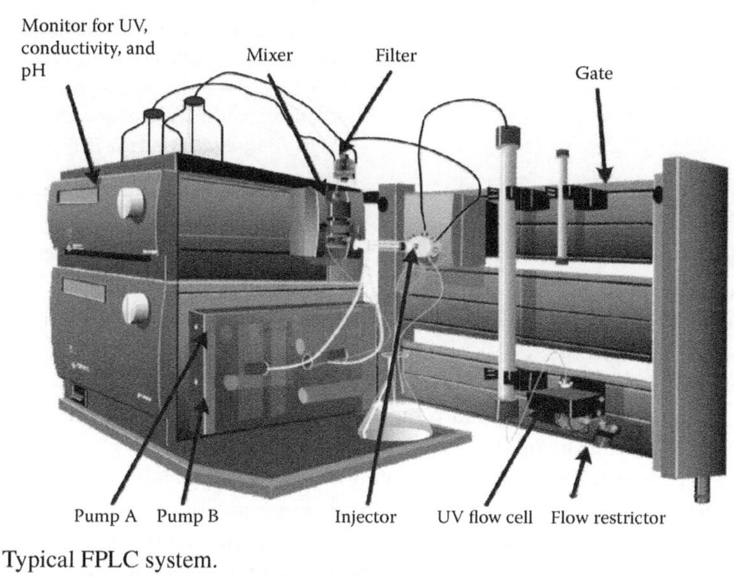

FIGURE 4.16 Typical FPLC system.

required in the FPLC system, which can generate moderate back pressures of 50–600 psi. A twin pump system provides the highest degree of accuracy, precision, and reproducibility in gradient formation. One pump delivers the binding buffer, whereas the other delivers the elution buffer. The pump controls the flow rate from a few milliliters per minute to liters per minute and substantiates the requirements for industrial scale purification.

Monitor: A monitor consists of a control unit, an optical unit with lamp assembly (Zn or Hg lamp), and conductivity and pH flow cells with a pH electrode. These monitoring modules measure traces of the analyte with the help of software interfaces.

UV and conductivity flow cells: The types of UV flow cells are selected depending on the amount of sample applied and the size of the column. The continuous recording of the absorbance of the analyte mixed with elution flow is achieved by pumping the elute through a flow-through cell placed in the UV monitor operated at 254, 280, or 405 nm. The choice of wavelength depends on the spectral properties of both the eluant and the substances to

TABLE 4.2
Different FPLC Columns and Their Applications

Columns	Type	Applications
Mono Q HR 5/5 Hi trap Q	Anion exchange	Frequently used chromatographic technique for the separation of proteins
Mono SH R 5/5	Cation exchange	Polypeptides, nucleic acids, polynucleotides, and other charged molecules
Superdex 200 HiLoad 16/60 Superdex 200 HR 10/30 Superose 6 HR 10/30	Size exclusion (gel filtration)	Include fractionation and determination of molecular weight of proteins, nucleic acid separation, plasmid purification, and polysaccharide fractionation
HiPrep 16/60 (Phenyl high sub), Phenyl Superose HR 5/5	Hydrophobic interaction	For protein, peptides, and other biomolecules with hydrophobic ligands
HiTrapChelating 1 mL HiTrapChelating 5 mL	Chelating	Allow purification of proteins with exposed histidine groups via formation of metal ion complex
HiTrap Protein A 1 mL HiTrap Protein G 5 mL	Affinity	Used for purification of immunoglobulins and tagged proteins
HiTrap Desalting Fast Desalting HR 10/10 HiPrep 26/10 Desalting	Buffer Exchange/desalting	Used for removing free low-molecular-weight labels or contaminants from proteins/peptides, substrates, and inhibitors or cofactors from enzymes; and for preparing samples for concentration, freeze drying, and storage
Polystyrene/divinyl benzene	Reverse phase	Used for high-resolution analysis and purification of peptides, proteins, and oligonucleotides, especially when high pH is needed, and use in liquid chromatography–mass spectrometry (LC-MS) applications

be detected. The FPLC system is delivered with a 5-mm cell and it also contains a 2-mm cell that is available as an accessory, which allows the desired sensitivity detection.

Mixer: The mixer is a chamber where the buffer/solvents are mixed properly. It is powered and controlled by a pump. All the eluants commonly used in ion exchange, hydrophobic interaction, and affinity and reverse-phase chromatography can be mixed in the mixer with a high degree of accuracy and reproducibility. Mixing is important when gradients between two buffers are formed.

Injection valve: A motorized valve is associated in the FPLC system for loading the sample. Different operating positions (load, inject, and purge positions) are used for loading the sample in the sample loop via an injection valve. Depending on the sample volume, sample loops of varying sizes are available (25 μL, 50 μL, 100 μL, 200 μL, 500 μL, 8 mL, etc.).

Fraction collector: Fraction collectors are made up of glass and polypropylene. They allow for efficient sample collection and prevent sample loss. There are two types of fraction collectors:

1. Fixed-volume fractionation allows for the collection of a fixed volume before elution starts.
2. Eluate fractionation or automatic peak fractionation allows for the collection of fixed volumes during elution within a set interval. It also minimizes peak dilution and cross contamination.

The choice of collection mode varies from microtiter plates to large vessels or funnels according to the fractions to be collected.

Flow restrictor: The flow restrictor maintains a steady back pressure in order to prevent the formation of air bubbles when the column is released. It is comparable to a cork on a bottle of champagne.

In-line filters: An in-line filter is fitted between the output of the mixer and position 7 of the injection valve. It creates a back pressure of maximum 0.5 MPa and filters out any dust particulates that may clog or block the fluidity of the system.

Programming of the FPLC system: The FPLC system is controlled by an automated software program. The software allows method validation and process optimization for performing a better run, along with the efficient evaluation of data. All instrument settings and functions are under the direct control of this software.

4.9.3 Calibrations

The calibrations performed on the FPLC system are as follows:

System component calibration: This involves the calibration of various components including pH electrode, conductivity flow cell, pump, temperature reading, pressure reading, flow restrictors, and fraction collectors. An automated program installed with the FPLC system calibrates the instrument components when commanded.

Sample preparation calibration: The need for sample preparation prior to performing the first chromatographic step is dependent on the sample type. Samples must be free from particulate matter. It is important to set objectives for quality and maintenance of the FPLC instrument along with quantity and biological activity of the sample.

Solvent system (buffer) calibration: Buffer calibration is important to control the pH and prevent changes in the pH in order to resolve sharp and pure peaks. They are most effective at ±0.5 pH units around their acid dissociation constant (pK_a) values.

Besides the aforementioned system calibrations, the following individual component calibrations are required before performing any run in the FPLC system:

	Component	Calibration Requirements
1.	pH electrode (optional)	Every day
2.	Pressure reading	Once a year or when required
3.	Cell constant of the conductivity flow cell	Only necessary if specific conductivity with high accuracy is measured
4.	Temperature	Must be checked when changing the flow cell
5.	Entering a new cell constant	Must be done when changing the flow cell
6.	Sample pump (optional)	Whenever the running conditions are changed

4.9.4 Maintenance

Regular maintenance is important for the safe and trouble-free operation of an instrument. The user should design a daily and monthly maintenance regime. Preventive maintenance should be performed on a yearly basis by qualified service personnel. For smooth functioning of the FPLC system, regular maintenance is necessary. The following are recommendations to maintain the system in working condition:

1. Liquid should not be allowed to spill on the instrument. The surface should be regularly wiped with a damp cloth. The system should be allowed to dry completely before use.
2. All tubing and flow paths should be properly rinsed, the pH electrode must be calibrated, and the pump should be regularly checked for any leakage. If there are signs of liquid leakage from the assembly, the components must be replaced. The pump should be rinsed with ethanol (20%), and if leakage is observed the piston must be replaced.
3. The online filters must be replaced every week. The inlet filters should be checked and replaced if necessary. To clean the titanium filter, the piece must be placed in a beaker containing 0.1N NaOH and stirred overnight or sonicated for 30 minutes.

4. The back pressure generated by the flow of the restrictor must be calculated using the user manual and the restrictor should be replaced if it is not within the given limit.
5. Every 6 months, the flow cells of the monitor and the fraction collector should be cleaned according to the user manual.
6. The valves should be checked for internal or external leakage and the distribution plate must be replaced yearly.

Some points should be kept in mind for periodic maintenance:

1. The power supply must be disconnected before attempting to replace any item on the system.
2. When using hazardous chemicals, the entire system should be flushed thoroughly with distilled water before performing service and maintenance.
3. Care should be taken while changing the location of the instrument so that the tubings are not entangled and no leakage occurs from the connections.
4. During servicing and maintenance of the system, the buffer bottles must be placed on the laboratory bench to prevent draining.
5. The pH electrode should be removed from the flow cell when the system is not in use and it should be kept in cover containing a mixture (1:1) of pH 4 buffer and 2 M KNO_3.
6. When the sample is loaded in the injector, care should be taken to ensure the needle inserted is tight; if necessary it should be tightened by hand, but the needle should not be overly tightened.

4.9.5 Troubleshooting

Common symptoms such as erratic retention times, noisy baselines or spikes in the chromatogram, and leaks at pump fittings or seals can result in poor chromatography. For the eradication of specific problems related to the apparatus, troubleshooting problems and their probable causes along with their remedial measures are listed in Table 4.3.

4.9.6 Advantages

- The use of FPLC columns allows much higher protein loadings as compared to HPLC along with the use of a wide range of aqueous and biocompatible buffer systems.
- FPLC is an intermediate between classical column chromatography and HPLC. The FPLC pump delivers a solvent with a flow rate in the range 1–499 mL/h, whereas the HPLC pump delivers a solvent flow rate in the range 0.01–10.0 mL/min.
- Since lower pressures are used in FPLC (0–40 bar) than in HPLC (1–400 bar), a wider range of column supports is possible.
- FPLC has many advantages in situations where a nonvolatile or thermally unstable sample must be separated, as is the case with many biochemicals, and it can also adapt to large-scale preparative procedures.

4.9.7 Applications

Some applications of FPLC are as follows:

- FPLC has proved to be especially useful in the purification of murine monoclonal antibodies.
- The FPLC method is useful in quantitatively measuring hemoglobin A2 (HbA2) levels.
- An efficient purification of recombinant his-tagged proteins from bacterial and mammalian and baculovirus-infected cells can be achieved by using His60 nickel resin. Tags, either N or C terminal, consisting of consecutive histidine residue, bind selectively to

TABLE 4.3
FPLC Problems and Their Probable Causes and Remedies

Problem	Probable Causes	Remedy
Noisy signal or no peaks or small peaks	1. Detector lamp off 2. Chances of impurity in buffer	1. Check that the flow restrictor generates a back pressure in its limit; if not, there may be air in the flow cell. 2. Degas buffers before use. 3. Check the connections of the optical unit and filter setting. 4. Clean the UV flow cell.
Ghost peaks observed in gradient profile	1. Air in the eluant	1. Check that the mixer is functioning properly. 2. Check loose tubing connections. Use a flow restrictor.
Baseline drift occurs	1. Column temperature fluctuation. 2. There may be air in the flow cell.	1. Control column and mobile phase temperature; use a heat exchanger before detector. 2. Check that the column is equilibrated. If necessary, clean the column. 3. Use HPLC-grade solvents, and high-purity salts and additives. Degas mobile phase before use; sparge with helium during use. 4. Flush cell with methanol or some other strong solvent. If necessary, clean the cell with 1N HNO_3 (never with HCl; never use nitric acid with PEEK tubings or fittings).
Waves on the gradient persist	—	1. Check that the pump and the valves are operating properly and are programmed correctly.
Erratic flow, noisy baseline signal, irregular pressure trace (higher than the usual)	1. Air bubbles passing through or trapped in the pump. 2. Injector, in-line filter, or tubing.	1. Check that there is sufficient eluent in the reservoirs. 2. Check all connections for leakage.
	3. Pump valves not functioning correctly	1. Clean the valves according to system manual. 2. Piston seal leaking. Replace the piston seal according to the instructions in User Manual.
	4. Blockage or partial blockage of the flow path	1. Flush the flow path to clear the blockage. 2. If necessary, replace the tubing. 3. Check the inlet tubing filter. It can get clogged if unfiltered buffers or samples are applied.
Unstable UV baseline, conductivity curve	1. Monitor and mixer	1. Check main cable (power connection). 2. Use a large mixer chamber. 3. Recalibrate the conductivity cell.
Low eluant flow and noise as piston moves; erratic pump pressure	1. Air trapped in pump cylinders 2. Partially blocked solvent filters 3. Leaking connections 4. Piston seal leakage 5. Pump valve malfunction 6. Piston damaged	1. Disassemble the pump cylinder and examine the piston seal. 2. Follow pump user manual.

(*Continued*)

TABLE 4.3 (*Continued*)
FPLC Problems and Their Probable Causes and Remedies

Problem	Probable Causes	Remedy
Variable retention times	1. Leaks, salt buildup, and unusual noises in the pump. 2. Air trapped in pump. (Retention times increase and decrease at random times.) 3. Column temperature fluctuations (especially evident in ion-exchange systems). 4. Column overloading. (Retention times usually decrease as the mass of the solute injected to column exceeds column capacity.) 5. Sample solvent incompatible with mobile phase. 6. Column problem. (Not a common cause of erratic retention. As a column ages, retention times gradually decrease.)	1. Change mobile phase composition. (Small changes can lead to large changes in retention times.) Change pump seals if necessary. 2. Check makeup of mobile phase. If mobile phase is machine mixed using proportioning values, hand mix and supply from one reservoir. 3. Purge air from pump head or check valves. Change pump seals if necessary. Be sure mobile phase is degassed. 4. Use reliable column oven. (Note: Higher column temperatures increase column efficiency. For optimum results, heat the eluant before introducing it to the column.) 5. Inject smaller volumes (e.g., 10 μL vs. 100 μL) or inject the same volume after 1:10 or 1:100 dilutions of a sample. 6. Adjust solvent. Whenever possible, inject samples in mobile phase. 7. Substitute new column of the same type to confirm the column is at fault. Discard old column if restoration procedures fail.
Loss of resolution	1. Mobile phase contaminated/deteriorated	1. Prepare fresh mobile phase. 2. If problem persists, the column may be clogged with strongly retained contaminants; hence, change or replace the column.
Split peaks	1. Sample solvent incompatible with mobile phase	1. Adjust solvent whenever possible and prepare sample in mobile phase.
Tailing peak	1. Mobile phase is contaminated or deteriorated and there is an interfering component in the sample	1. Check mobile phase. 2. Check column performance with standards.

immobilize Ni^{2+} ions. The purification of recombinant his-tagged proteins consists of cell lysis/extraction, binding, washing, and elution. The His60 nickel resin allows one-step protein purification under either native or denaturing conditions. Purified recombinant his-tagged proteins can be eluted with imidazole or by a reduction in pH.

- The FPLC method is used for glutathione-S-transferase (GST)-tagged protein purification. The GSTs are a family of enzymes that modify toxic substances, such as nitro and halogenated compounds, by adding sulfur from glutathione, leading to their excretion in urine. Many mammalian GSTs can be purified by affinity chromatography using the immobilized cofactor glutathione, which is followed by competitive elution with an excess of reduced glutathione.

The GST is the second most commonly used protein tag. The GST gene fusion system uses pGEX plasmids for inducible, high-level intracellular expression in *Escherichia coli* of genes or gene fragments fused with *Schistosoma japonicum* GST. The GST-tagged protein can be purified using a GSTrap high pressure (HP) column, and pure, untagged protein can be obtained. The activity of the tag can be easily measured by an enzyme assay or immunoassay. If necessary, the tag can be cleaved during the purification process to generate native proteins with high purity, which is suitable for further studies.

- FPLC is used for antibody purification. Affinity purification of monoclonal antibodies has been largely confined to the use of protein A and protein G chromatography. The antibodies obtained are sufficiently pure for radiolabeling, conjugations (e.g., fluorescein), or preparation of immunoaffinity columns.
- The FPLC system is used in biotin binding. Nowadays, biotinylated proteins/enzymes binding to avidin/streptavidin columns are frequently used. The high affinity ($K_d = 10^{-15}$ M) between avidin and biotin is one of the strongest known noncovalent interactions, which can be used in pretargeting by coupling avidin to an antibody and using an effector molecule comprising biotin and a radionuclide.
- FPLC is used in phosphoprotein enrichment. Protein phosphorylation is probably the most common and important form of protein modification. Protein phosphorylation provides detailed methods designed to detect and identify phosphorylated proteins and to analyze their specific sites of modification using both isotopic and nonisotopic approaches.
- Another application of FPLC is in protein desalting. Desalting columns apply size-exclusion chromatography to facilitate the removal of low-molecular-weight components (MW < 1000 Da) from DNA, proteins, or peptides (MW > 5000 Da). Protein desalting columns desalt or exchange buffers of protein samples and have exceptional desalting characteristics, with a ≥95% retention of salts and small molecules while providing excellent recovery of proteins greater than 7000 Da. Multiple samples can be processed in less than 5 minutes without cumbersome column preparation steps.
- An improved ribonucleic acid (RNA) purification method using FPLC size-exclusion chromatography allows the preparation of milligram quantities of pure RNA in a single day.
- FPLC is used for the separation of urinary protein and several plasma proteins and for the diagnosis of β-thalassemia. It is also used to identify protein profiles or variability within a single protein of clinical significance.
- FPLC in conjunction with nano-HPLC-ESI-tandem mass spectrometry as a new integrated methodology is suitable for the proteomic analysis of human lipoprotein fractions.
- FPLC is used for measuring levels of tubular proteinuria. Tubular proteinuria is a medical condition in which a person's kidney secretes more than 150 mg of urine proteins daily. FPLC is used for cases of tubular proteinuria to isolate specific proteins from urine to measure the levels of excreted proteins.
- FPLC is used in the profiling of plasma proteins in cerebrospinal fluid. Using the anion-exchange column in the FPLC process, plasma proteins, transport proteins, hemoglobin, and isoenzymes in cerebrospinal fluid are profiled. Protein profiling and monitoring contributes much to research on various diseases and medical conditions, including Alzheimer's disease, leukemia, cerebral hemorrhages, and renal failure.
- FPLC is used in analyzing pancreatitis. FPLC determines the specific protein-related cause of pancreatitis (presence of pancreatic juice) that potentially leads to disease.

4.10 GAS–LIQUID CHROMATOGRAPHY

4.10.1 Principle

Gas chromatography (GC) as the name suggests is particularly suited for the separation of gases and volatile liquids or solids in their gaseous state. Compounds of low polarity are best separated by GC. The technique is highly sensitive and reproducible and has a high speed of resolution. When the stationary phase is an active solid such as silica, the method is referred to as gas–solid chromatography (GSC). However, if the stationary phase is a liquid such as polymers of silicone coated on the surface of an inert granular solid, then the technique is known as GLC. The stationary phase, whether a solid or a liquid coated as a thin film on the surface of a solid support, is packed in a glass or stainless steel column, which is narrow, coiled, and 1–3 m long with a 2–4 mm internal diameter. An inert carrier gas (mobile phase) such as nitrogen, helium, or argon is made to flow through the column. The temperature of the column is maintained high in an oven to keep the compounds separated in their volatile states. These volatilized compounds get partitioned between the liquid or solid stationary phase and the gaseous mobile phase, and thus they get separated because of differences in their partition coefficients. After leaving the column, the separated compounds pass through the detector, are sensed, and are recorded by the recorder. The essential components of GC equipment are presented in Figure 4.17.

Capillary columns with internal diameters of 0.03–1.0 mm and lengths of up to 100 mm made of glass or steel are used for performing GC. Two types of capillary columns, that is, wall-coated open tubular columns (WCOTs) and support-coated open tubular columns (SCOTs), are available for this purpose. In WCOTs, as the name suggests, the walls of the capillary column are coated with the stationary phase. Since the stationary phase is a liquid and is directly coated on the walls of the capillary column, only a small amount of the stationary phase is present in this system. Accordingly, only a small amount of the sample can be applied onto the WCOT column. In SCOTs, the stationary phase is in the form of a thin layer on the surface of a solid support, which is in turn packed into the capillary column. Hence, the capacity of the SCOT system is much higher than that of the WCOT system and a larger amount of sample can be applied to the former (Figure 4.18a and 4.18b). GC can be carried out only in the column mode. It has high sensitivity, reproducibility, and speed of resolution and it has proved to be the most versatile of all chromatographic methods. The only limiting factor is that the components of the mixture must be vaporized to give heat-stable vapors up to a temperature of 300°C.

4.10.2 Solid Support and Stationary Phase

The purpose of the solid support is to provide a large, uniform, and inert surface area for holding a thin layer of the liquid stationary phase. The support should be inert and have high mechanical strength, a large surface area, regular shape, and uniform size. The most commonly used support is celite, the OH groups of which are modified by silanization with hexamethyl disilazane to minimize the support's interaction with the sample.

The correct choice of the stationary phase is perhaps the most important parameter in GC. Ideally, the stationary phase must be nonvolatile and thermostable at the temperature used for analysis. It should be chemically inert toward the solutes of interest at the column temperature. The high boiling points of organic compounds such as polyethylene glycols; methyl phenyl and methyl vinyl silicon gums; esters of adipic, succinic, and phthalic acids; polyesters; and polyethylene glycols are

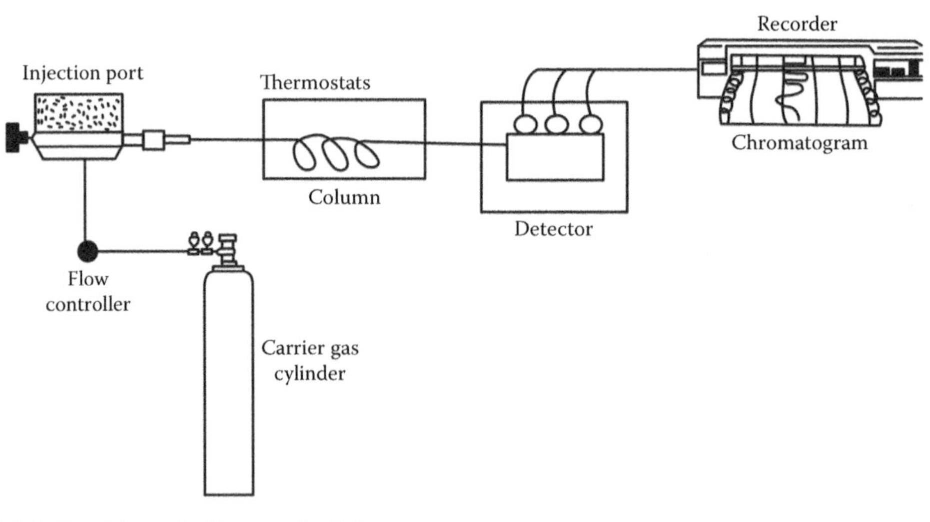

FIGURE 4.17 Schematic diagram of a GC system.

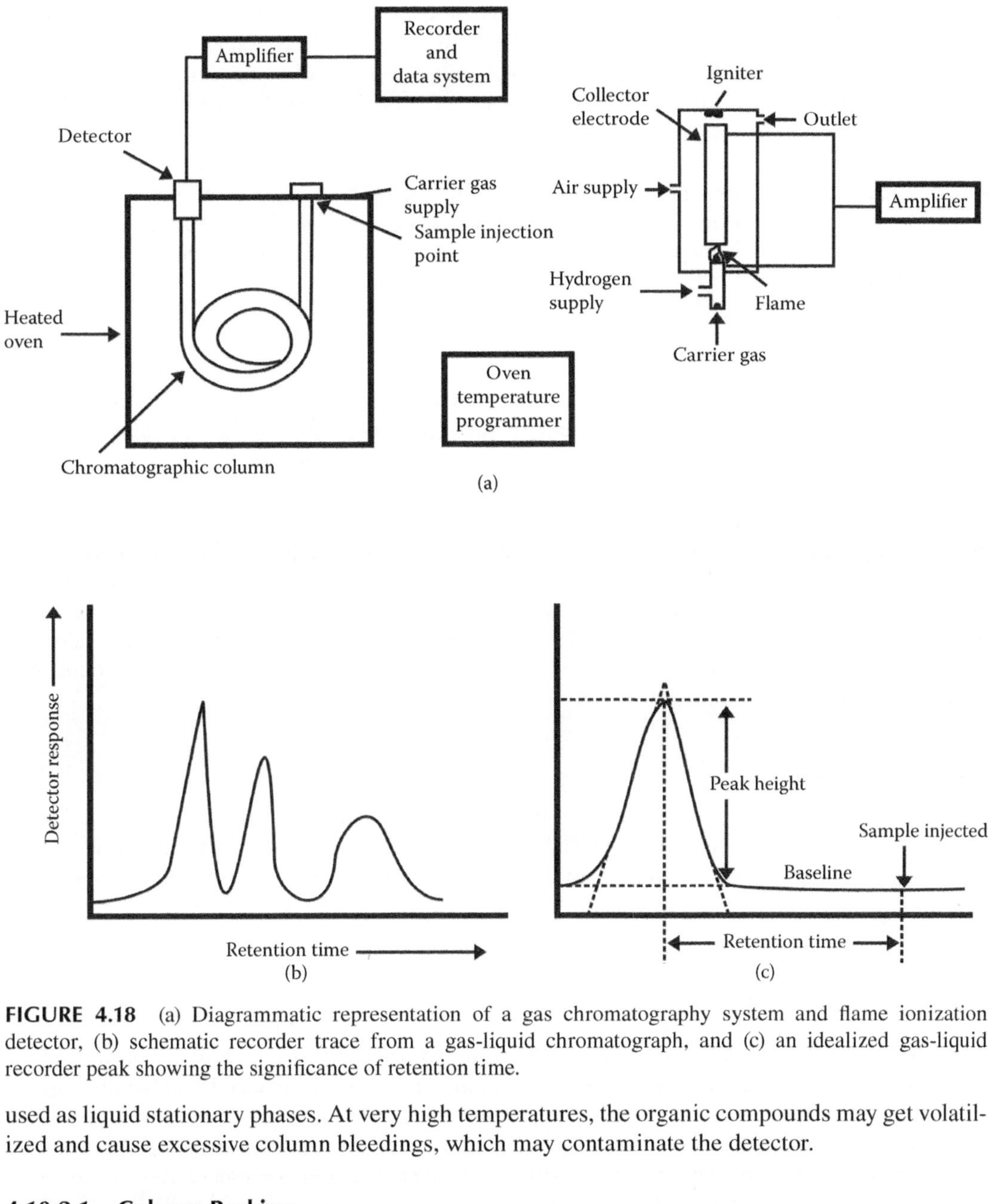

FIGURE 4.18 (a) Diagrammatic representation of a gas chromatography system and flame ionization detector, (b) schematic recorder trace from a gas-liquid chromatograph, and (c) an idealized gas-liquid recorder peak showing the significance of retention time.

used as liquid stationary phases. At very high temperatures, the organic compounds may get volatilized and cause excessive column bleedings, which may contaminate the detector.

4.10.2.1 Column Packing

The columns are generally dry packed under a slight positive gaseous pressure. Prepacked columns are also available. After packing, the column is kept in an oven for 24–48 hours at a temperature near the upper working limit. This is done to condition the column. During conditioning, the carrier gas is passed through the column at normal flow rates. The column is not connected to the detector; otherwise, the detector may get contaminated.

4.10.3 Sample Preparation and Application

The sample should be prepared in such a way that it does not get retained on the column for an excessive period of time. This will lead to poor resolution and peak tailing. Polar groups such as

$-NH_2$ $-COOH$, and OH are derivatized by methylation, silanization, and trifluoromethyl silanization to increase the volatile character and distribution coefficient of the compounds.

Solvents such as ether, heptane, and methanol are used to dissolve the sample, which is then injected with the help of a microsyringe onto the column through a rubber septum in the injection port. The temperature of the injection port is generally maintained higher than the temperature of the column to ensure the rapid and complete volatilization of the sample. As a rule of thumb, the temperature of the injection port should be 50°C higher than the boiling point of the sample. Too high a temperature of the injection port may cause the decomposition of the sample. Therefore, the temperature of the injection port should be such that it causes rapid vaporization of the sample without decomposing it.

4.10.4 Carrier Gas

The primary function of a carrier gas is to carry the volatile components through the column. The gas used should be inert and should not react with either the sample or the stationary phase. Its secondary purpose is to carry the separated components to the detector, so the carrier gas should be suitable for detector use. It should be readily available in extra-pure form and inexpensive.

Normally, nitrogen, helium, and argon are the three most commonly used carrier gases. The column temperature must be high enough that analysis can be accomplished in a reasonable length of time. The retention time doubles with every 30°C decrease in column temperature. The lower the temperature the better the resolution and the longer the analysis time. Therefore, a balance has to be struck between peak retention time and resolution. Chromatographic separation can be achieved isothermally where a constant temperature is used or by temperature programming where the temperature is increased gradually.

4.10.5 Detectors

Selectivity, sensitivity, response, noise, and minimum detectable quantity and linear range should be considered when choosing a detector. The detector should be simple to operate, inexpensive and, as far as possible, insensitive to changes in flow rate and temperature. Some of the commonly used detectors are as follows:

Flame ionization detector (FID): It responds to almost all organic compounds and is the most widely used detector. It has a wide linear response range (10^6) and can detect even at low concentrations. The detector consists of two electrodes: one of the electrodes is the jet of the flame produced by introducing a mixture of hydrogen and air into the detector, whereas the second electrode is made of brass or platinum wire, which is mounted near the tip of the flame. When the carrier gas carrying the sample components emerges from the column, the sample components are ionized in the flame and sensed by the detector giving the signal, which is recorded by the recorder. An FID has an upper temperature limit of 400°C, and the minimum quantity that can be detected by this detector is on the order of 5×10^{-12} g. A typical FID is shown diagrammatically in Figure 4.19.

Nitrogen phosphorus detector (NPD): As the name suggests, NPD responds efficiently to detect compounds containing N or P or both. It shows poor response to compounds that contain neither of these elements. This detector has an upper temperature limit of 300°C, a narrow linear response range of 10^4, and detection limits of 10^{-11} $g{\cdot}s^{-1}$. Its principle of operation is the same as that of FID, but NPD has the sodium salt fused onto the electrode system. The NPD is widely used in the analysis of organophosphorous pesticides.

Electron capture detector (ECD): Compounds that have the capacity to capture electrons are best detected by this detector. Here, a radioactive source (^{63}Ni) ionizes the column gas and produces electrons, which give a current across the electrodes to which suitable voltage

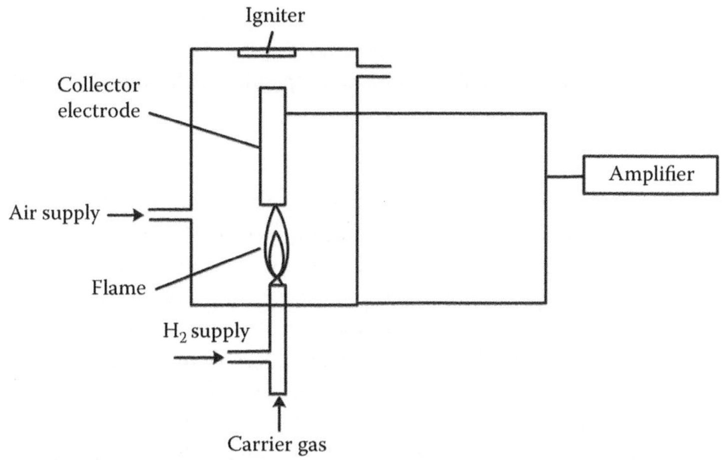

FIGURE 4.19 An FID.

is applied. When the carrier gas carrying the electron-capturing substance emerges from the column, it captures the ionized electrons. This results in a drop in the current, which is traced on chart paper by the recorder. The detector has an upper temperature limit of 300°C and high detection sensitivity (10^{-12} g·s^{-1}) but a low linear range (10^2–10^4). This detector is best suited for halogen-containing compounds and is widely used in the analysis of polychlorinated compounds such as the pesticides DDT, dieldrin, and aldrin.

Thermal conductivity detector (TCD) or Katharometer detector: This detector measures the change in the thermal conductivity of the carrier gas as a component emerges. This is measured by means of change in the resistance of a platinum wire. All components of a mixture, whether organic or inorganic, are detected up to a limit of 10^{-8} g·s^{-1}.

4.10.6 Amplifiers and Recorders

When components leave the column and pass through the detectors, small and weak electrical signals are produced, which are amplified by an amplifier before they are fed to the recorder. Recorders generally consist of two basic parts, that is, a chart paper and a pen; the pen moves on the chart paper and traces the signals being activated from the amplifier in the form of peaks.

In cases where the identity of a compound is unknown, the GLC system is connected to a mass spectrometer. Special separators separate the carrier gas from the sample, which is introduced into the mass spectrometer. The GLC system can also be linked to an infrared (IR) spectrometer and nuclear magnetic resonance (NMR) spectrometer in order to identify the compounds.

4.11 ION-EXCHANGE CHROMATOGRAPHY

Ion-exchange chromatography is a type of adsorption chromatography in which the retention of a solute occurs due to its reversible electrostatic interaction with the oppositely charged group on an ion-exchange substrate. This technique is useful in the separation of compounds that bear a net electric charge, such as proteins, amino acids, and nucleic acids. Ion exchangers are prepared from either certain synthetic resins, which are insoluble porous organic molecules, or naturally occurring biopolymers such as cellulose to which various groups known as fixed ions are covalently attached. These fixed ions are balanced by equal and oppositely charged ions from the solution, referred to as counterions. Depending on the nature of the counterions, ion exchangers are of two types: (1) In cation exchangers counterions are cationic or positively charged ions, and (2) in anion exchangers counterions are negatively charged ions. Counterions are mobile and can be easily exchanged by other similarly charged molecules in the sample.

The nature of the resin matrix remains unchanged during this exchange process. Generally, resin-based ion exchangers are used for the separation of low-molecular-weight biomolecules and cellulosic ion exchangers are more suitable for the isolation of macromolecules such as proteins and nucleic acids.

4.11.1 Principle

Ion-exchange chromatography depends on the degree of attraction between oppositely charged particles. Hence, it is used exclusively for the separation of ionic species. These occur mainly in inorganic systems. However, ionic species occur in organic systems too, for example, amino acids, proteins, and nucleic acids components have ionizable groups that can carry positive or negative charge depending on their pK_a and the pH of the solution. Separation may be carried out in column, paper, or thin-layer modes with the help of ion exchangers, which are either positively charged (anion exchangers) or negatively charged (cation exchangers). The positive and negative charges in the exchangers are loosely bound to oppositely charged groups. When a group of a stronger charge is introduced into the matrix structure of the exchanges, they displace the loosely bound groups and bind strongly as a result of the higher degree of attraction, for example:

$$\textbf{Cation exchanger}: RSO_3 \ldots \ Na^+ \ + N^+H_3R' = RSO_3 \ldots \ NH_3R' + Na$$

exchanger counter ion to be exchanged: Charged molecule: bound molecular ion: exchanged ion

$$\textbf{Anion exchanger}: (R)_4\, N^+ \ldots \ Cl^- \ + OOCR \ = (R)_4\, N^+ \ldots \ OOCR' \ + Cl$$

The exchange molecules are recovered by selective desorption by the eluant and diffusion of the molecule to the external solution. Selective desorption is brought about by changing the pH or ionic concentration, or by introducing an ion that has a greater affinity for the exchanger than the bound molecules (Figure 4.20).

4.11.2 Ion-Exchange Materials

The common ion-exchange materials are resins insoluble in water. These are produced by the copolymerization of styrene and divinyl benzene. Copolymerization results in cross-linkages, which renders the polymers insoluble. By varying the amount of the two polymers, the amount of cross-linking is controlled so that the system swells in water and is susceptible to water molecules and ionic species. Sulfonic acid groups are introduced after polymerization. This results in the formation of the anionic group, which binds mobile cations, that is, cation exchangers (Figure 4.21). Anion exchangers are made by copolymerizing styrene with chloromethyl ether and allowing the chloro groups to react with tertiary amines, which provides the cationic group that binds mobile anions. Modified cellulose is now available as an alternative to polystyrene-based anion exchangers with

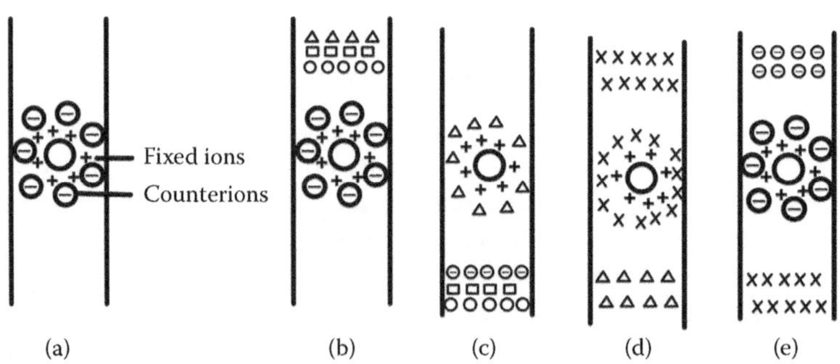

FIGURE 4.20 Separation process of solute during ion-exchange chromatography. Demonstration of various steps (a to e) in separation of solute during ion exchange chromatography where Δ is a negatively charged solute and ○ and □ are positively charged and neutral solutes.

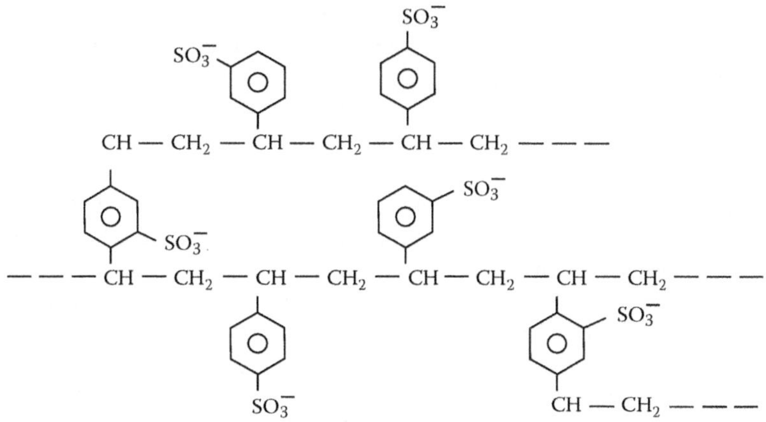

Sulphonated polystyrene resin

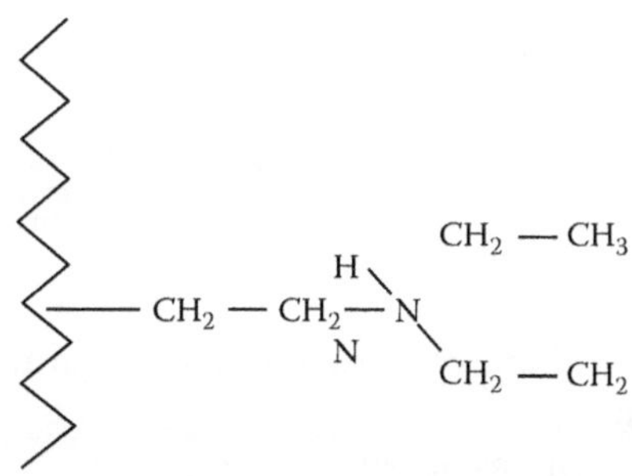

Cellulose
DEAE cellulose
(diethylaminoethyl cellulose)

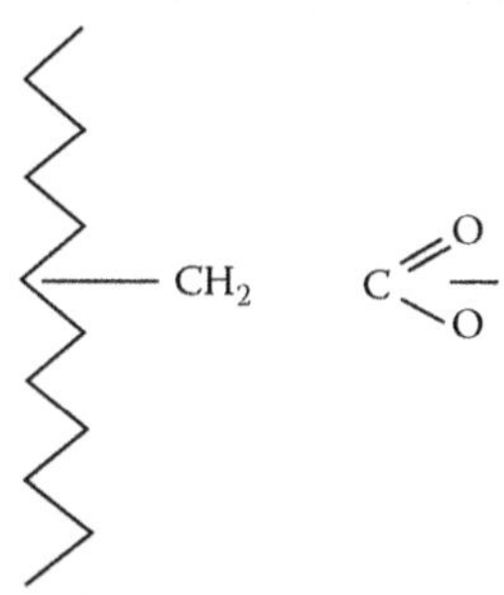

Cellulose
CM cellulose

FIGURE 4.21 Carboxymethyl (CM) cellulose.

better flow and exchange properties are now available, which possess good flow and exchange properties. In addition, separation is aided by the sieving action of gels and beads of these materials.

4.12 EXCLUSION CHROMATOGRAPHY

4.12.1 Principle

The separation of molecules on the basis of their molecular size and shape is achieved by utilizing the molecular sieve action of gels. The term gel filtration is used to describe the separation of molecules of varying molecular sizes utilizing gel materials. Porous glass granules have been used for this purpose, and the term controlled pore glass chromatography has been coined for this process. The term exclusion or permeation chromatography is used to describe all separation processes that utilize the molecular sieve action of gels.

Gels are insoluble hydrophilic semisolid colloids, which swell in buffer to form a 3D network of pores. In order to explain the molecular sieve actions of gels, consider a mixture of large, medium, and small particles being eluted through a gel (in equilibrium with a suitable solvent for the molecules to be separated; see Figure 4.22a). Large molecules, which are completely excluded from the pores of the gel particles, pass through the interstitial spaces first, whereas small molecules are distributed between the solvent inside and outside the gel particles and then pass through the column at a slower rate. Due to variation in the pore size of gel particles, molecules of medium size penetrate partially, fully, or not at all (Figure 4.22b). Hence the distribution coefficient, Kd, of a particular solute between the solvent inside and the solvent outside a particular gel system varies between Kd = 0 for large molecules completely excluded and Kd = 1 for small particles completely reaching the inner solvent. The intermediate values will be for molecules of medium size. These variable K values are responsible for the complete separation of the components of a mixture based on differences in molecular weight or size (Figure 4.22c).

The elution volume for a given solute depends on several parameters given by the following expression:

$$V_e = V_o + K_d V_i \tag{4.1}$$

where

V_o = void volume (volume of solvent outside gel particles)
V_e = elution volume
V_i = volume of solvent inside gel particles
K_d = partition coefficient

It is noted that V_i can be calculated from W_r, or water regain value, which is the amount of solvent taken up by 1 g of dry gel:

$$V_i = aW_r$$

where a = dry weight of gel.

Thus from Equation 4.1,

$$K_d = \frac{V_e - V_o}{V_i} = \frac{V_e - V_o}{aW_r}$$

For a molecular species, completely exclude $V_e = V_o$. Then,

$$K_d = 0$$

The molecular weight of a given unknown species can be determined by comparing its ΔV (arbitrary volume increments) value with that of compounds of known molecular weight eluted under the same conditions (Figure 4.23).

4.12.2 Materials and Methods

Gel filtration is usually carried out in the column mode. However, the thin-layer mode can also be used. The swollen gel is spread on a glass plate without the addition of a binding material. This is called thin-layer gel filtration (TLG). Similar to TLC, the solvent in the interstitial spaces is the

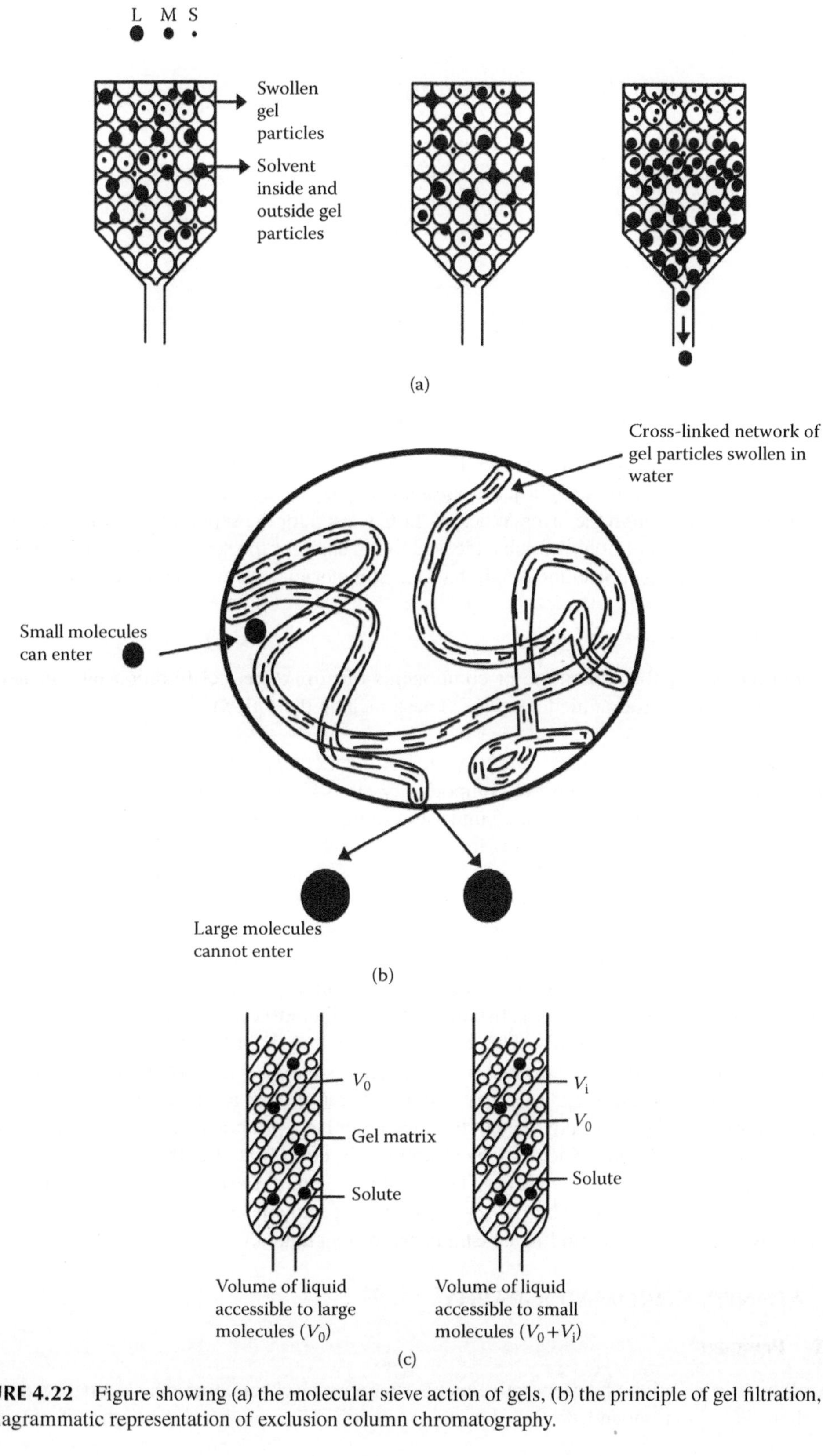

FIGURE 4.22 Figure showing (a) the molecular sieve action of gels, (b) the principle of gel filtration, and (c) the diagrammatic representation of exclusion column chromatography.

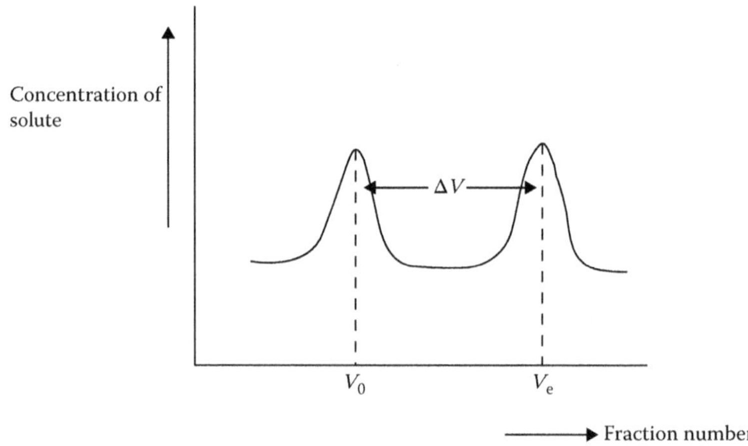

FIGURE 4.23 Elution diagram for exclusion chromatography.

mobile phase, but unlike TLC the layer in TLG is never dried; hence, there is no solvent front in TLG. The plate is kept in an airtight jar at an angle of 20° to the horizontal position to facilitate movement of solvent through the layer. Whereas TLC is used for the separation of amino acids, sugars, oligosaccharides, and lipophilic substances, TLG is used for the separation of hydrophilic substances, particularly high-molecular-weight biological material such as proteins and nucleic acids.

4.12.3 Applications

Apart from its use for the separation of components of a mixture, gel filtration has further uses unlike any other chromatographic technique. These include the following:

Determination of the molecular weight of an unknown compound: For determination of the molecular weight of an unknown compound, V ($V_e - V_0$) values are compared with the ΔV values of known compounds eluted under the same conditions.

Concentration of solute: Water and low-molecular-weight substances are absorbed by the swelling gel, whereas high-molecular-weight substances remain in the solution. After 10 minutes, gel is removed by centrifugation, leaving behind high-molecular-weight material in the solution whose concentration has increased without altering pH and ionic strength.

Desalting solutions of high-molecular-weight compounds: High-molecular-weight substances move with the void volume which is the volume of mobile phase (V_0) in a column, whereas low-molecular-weight substances are distributed between stationary and mobile phases and hence move slowly. This method of desalting is faster and more efficient than dialysis, which is used for the same purpose. This is used for removing monosaccharides from polysaccharides, NH_4SO_4 from protein preparation, and amino acids from proteins.

Protein binding studies: To study the binding of a ligand to a protein, a protein/ligand mixture is applied to a column, which has been previously equilibrated with ligand of the same concentration as that in the mixture. The sample is eluted. Early fractions contain ligand and protein-bound ligand. Thus by repeating the experiment for a series of ligand concentrations, the appropriate binding constants can be calculated.

4.13 AFFINITY CHROMATOGRAPHY

4.13.1 Principle

Unlike all other separation and preparative techniques such as chromatography, centrifugation, and electrophoresis, which depend on the physical properties (size, molecular weight) of molecules,

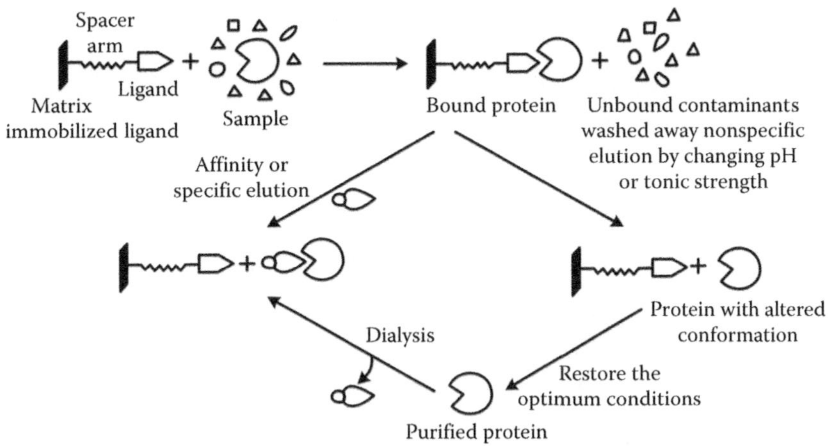

FIGURE 4.24 Principle of affinity chromatography applied in the purification of a protein.

affinity chromatography depends on the biological interactions of molecules to achieve purification. It is therefore very sensitive and extremely efficient in achieving absolute purity in a single process without destroying the molecules. This factor makes it superior to other forms of chromatography for which repeated fractionation is necessary to achieve a high degree of purity. It is well known that an enzyme can bind specifically to a substrate inhibitor or activator. These binding media are called ligands.

A column is prepared in which a ligand of the required enzyme is covalently bonded to the inert insoluble matrix when a solution containing the enzyme is passed through this column. It alone binds to the ligand while all other molecules pass through. Even slightly modified or denatured molecules do not bind and are thus eliminated. The enzyme is then eluted by changing either the pH or the ionic strength of the solvent. Originally the technique was developed for the purification of enzymes, but it has now been extended to the purification of nucleotides, nucleic acids, immunoglobulins, membrane receptors, and even whole cells and cell fragments. Any molecule is capable of reversibly binding to a specific ligand, which is attached to an insoluble matrix.

The main limitation found so far is the difficulty in simulating the natural affinity between macromolecules in an artificial system. It requires detailed knowledge of the interactions and conditions (pH, ionic strength, temperature, etc.) in which binding can take place. Also, the matrix in which the ligand is bound should have spherical gel particles for a good flow of the unbound molecules to pass through. Further, the ligand should be attached such that the attachment site for the macromolecule is well exposed. For this reason, the ligand is not bound directly to the matrix; instead, a spacer arm is used to separate the two (Figure 4.24).

4.13.2 Materials and Methods

The three important components of an affinity chromatograph are the matrix, ligand, and spacer arm. The matrix should have a good network of pores through which the unbound molecules can easily pass. It should be chemically inert and have suitable functional groups to which the ligand can bind. It should not interact (or interact only weakly) with other molecules so that nonspecific adsorption does not occur. The most commonly used materials for the matrix are agarose, polyacrylamide gels, cellulose, porous glass, and silica. Ligands may be bonded to the matrix by a cyanogen bromide treatment of the matrix at pH 11 (Figure 4.25), which activates the matrix for ligand attachment. The length of the spacer arm is usually 6–10 carbon atoms or their equivalent. It should possess two functional groups, one at either end; one group binds to the matrix and the other binds to the macromolecule, respectively.

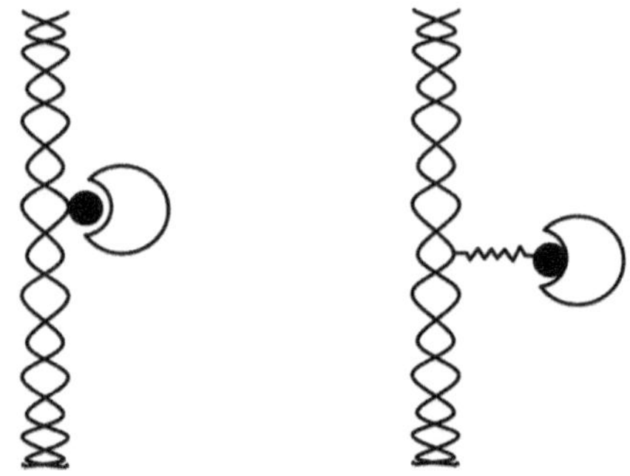

FIGURE 4.25 Representation of possible arrangements (a and b) of the role of spacer arms in properly binding a macromolecule.

Experiment: Separation and Identification of Amino Acids by Descending Paper Chromatography

Introduction

Amino acids in a given mixture or sample aliquot are separated on the basis of the differences in their solubility and hence differential partitioning coefficients in a binary solvent system. The amino acids with higher solubility in a stationary phase move slowly compared with the amino acids with higher solubility in a mobile phase. The separated amino acids are detected by spraying the air-dried chromatogram with ninhydrin reagent. All amino acids give off a purple or bluish-purple color on reaction with ninhydrin, except proline and hydroxyproline, which give off a yellow-colored product.

Safety Guidelines

1. Do not touch the paper with naked hands because sweat contains a significant amount of amino acids, which may interfere with your result.
2. The spots of the applied sample should be as compact as possible. The larger the spot the poorer the resolution.
3. At the time of fixing paper in the chromatography chamber, ensure that the baseline on which the sample has been applied does not dip into the solvent; otherwise, the sample might get washed away in the solvent.
4. Allow sufficient time for the filter paper to absorb sufficient water (which will act as a stationary phase) before pouring the solvent into the trough/tank. Inadequate conditioning or equilibration will result in improper or poor-quality resolution.
5. The solvent front should advance in a straight line and should not zigzag or slip; it should be parallel to the baseline.
6. Dry the paper thoroughly before spraying with the detection reagent. Wet paper may interfere with the appearance of evenly shaped compact spots.
7. Chromatography should be carried out in a temperature-controlled room because any fluctuation in temperature will cause the uneven flow of the solvent and may alter the R_f value.

Experimental Outline

Saturate the chromatographic chamber with BAOH (n-butanol: acetic acid: water). Develop the chromatogram in the solvent. Take the chromatogram out of the solvent after the solvent front reaches the bottom. Dry the chromatogram completely and spray it with the ninhydrin reagent.

Materials

1. Whatman no. 1 filter paper, which should be cut as per the chromatographic chamber size; standard amino acid solutions; ninhydrin; n-butanol; acetic acid; and distilled water
2. Micropipette/microsyringe

3. Hair dryer
4. Sprayer
5. Oven set at 105°C
6. Chromatographic chamber saturated with solvent vapors

Prelaboratory Preparations

1. Solvent: Take n-butanol, acetic acid, and water in the ratio 4:1:5 in a separating funnel and mix them thoroughly; saturate the mixture by incubating for 3–6 hours. Allow the phases to separate out completely. Use the lower aqueous phase for saturating the chamber. The upper organic phase is used as the mobile phase.
2. Ninhydrin spray reagent: Prepare a fresh solution by dissolving 0.2 g of ninhydrin in 100 mL of acetone.
3. Standard amino acids: Prepare solutions such as methionine, tryptophan, alanine, glycine, and threonine (1 mg/mL in 10% isopropanol).
4. Prepare a sample containing a mixture of unknown amino acids.

Method

1. Take Whatman no. 1 filter paper and lay it on rough filter paper. Throughout the experiment, care should be taken not to handle the filter paper with naked hands; for this purpose gloves should be used, or filter paper should be handled with the help of a folder piece of rough filter paper.
2. Fold the Whatman no. 1 filter paper about 2–2.5 cm from one edge. Reverse the paper and again fold it 2 cm down from the first fold.
3. Draw a line across the filter paper with a lead pencil at a distance of about 2 cm from the second fold. Put circular marks along this line at a distance of 2.5 cm from each other.
4. With the help of a micropipette or microsyringe, apply 20 μL of solution of each standard amino acid on a separate mark. Also apply a spot of the sample or mixture to be analyzed, preferably on the mark at the center of this baseline. The size of the spot should be as small as possible so that the developed spots are compact and do not overlap. If necessary, the wet sample spot should be dried with the hair dryer before applying additional aliquot.
5. Hang filter papers in a line in a chromatographic chamber, which has previously been saturated with the aqueous phase of the solvent system. This is done by keeping petri plates containing the aqueous phase at the bottom of the chamber. The paper is hung from the trough/tray and a glass rod is kept to hold it in place. Care should be taken to ensure that the baseline is not submerged when the mobile phase is added to the trough; otherwise, the spotted material will dissolve in the solvent.
6. Close the chamber firmly so that it is airtight. Allow sufficient time for the cellulose fibers of the paper to fully hydrate.
7. Pour the mobile phase through the holes provided on the lid of the chamber into the trough. Replace the rubber bungs in the hole and allow the mobile phase to run down the paper until the solvent front reaches about 5 cm from the opposite edge.
8. Remove the paper and mark the solvent front using a lead pencil, and let it dry at room temperature.
9. Spray the filter paper (chromatography) with a ninhydrin reagent; after drying it at room temperature, transfer it to an oven at 105°C. Keep it in the oven for 5–10 minutes.
10. Blue- or purple-colored spots should appear on the paper. Mark the boundary of each spot with a lead pencil.
11. Measure the distance between the center of the spots and the distance of the solvent front from the baseline.
12. Calculate the R_f value of the standard amino acids as well as of those in the given mixture or sample as follows:

$$R_f = \frac{\text{Distance traveled by unknown amino acid}}{\text{Distance traveled by the solvent system}}$$

13. Identify the amino acids in the mixture or sample by comparing their R_f values with those of the reference standards (Figure 4.26).

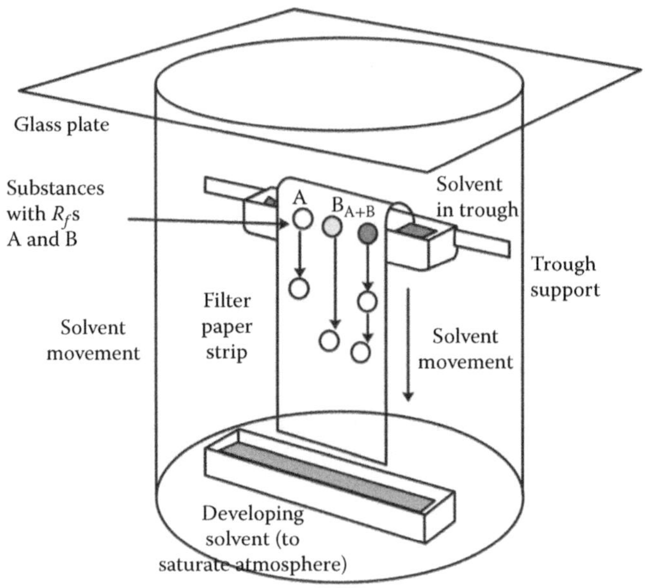

FIGURE 4.26 Chromatographic chamber for descending PC. A denotes sample 1, B denotes sample 2, and A + B denotes the mixture of 1 + 2.

Note: It is advisable to carry out chromatography in three different solvent systems before the identity of the amino acid in the mixture or sample can be established with any degree of certainty.

Observations

Distance traveled by the solvent = x cm from baseline
Distance traveled by glycine = a cm from baseline
Distance traveled by alanine = b cm from baseline
Distance traveled by threonine = c cm from baseline
Distance traveled by methionine = d cm from baseline
Distance traveled by spot no. 1 in = a cm sample from baseline

Calculations

R_f value of glycine = a/x
R_f value of alanine = b/x
R_f value of threonine = c/x
R_f value of methionine = d/x
R_f value of spot no. 1 = a/x

Conclusion

The sample contains glycine since the R_f value of spot no. 1 is identical to the R_f value of authentic glycine standard.

Experiment: Separation and Identification of Amino Acids by Ascending Paper Chromatography

Introduction

Same as given in the first experiment of this chapter.

Safety Guidelines

Same as those given in the first experiment of this chapter.

Materials

Same as those given in the first experiment of this chapter, except that cylindrical chromatography chambers are needed for this experiment.

Experimental Outline

Experimental outline is similar to the one given in the first experiment of this chapter.

Prelaboratory Preparations

Same to that given in the first experiment of this chapter.

Method

1. Take a Whatman no. 1 filter paper sheet of appropriate size so that it can be rolled into a cylinder and accommodated in a cylindrical chromatographic jar.
2. Draw a baseline 2 cm from one of the breadthwise edges of the paper. Put small circular marks along the baseline in such a way that the distance from the edge of the paper to the first spot and the distance between adjacent spots are not less than 2.5 cm.
3. Apply 20 μL aliquots of the standard amino acids and the sample at different spots. The diameter of the spotted material should be as small as possible and, if required, the applied solution may be dried prior to loading additional volume.
4. Roll the paper into a cylinder and fasten its edges with a paper clip. Pour a sufficient volume of the mobile phase into the chromatographic jar, which has been previously saturated with water vapors by lining the tank with filter paper saturated with the aqueous phase of the solvent system.
5. Gently place the rolled filter paper upright in the jar after ensuring it does not touch the sides of the chamber; at the same time, take care that the baseline where the spots have been applied does not dip into the solvent.
6. Close the tank with an airtight lid or a glass plate to which a sufficient amount of silicon grease has been applied.
7. Leave the setup undisturbed and allow the solvent to move up until it reaches about 5 cm from the upper edge.
8. Remove the chromatogram from the chamber and air dry it.
9. Spray the paper with a ninhydrin reagent and let it dry again at room temperature prior to transferring it to an oven at 105°C for 5–10 minutes. Locate the position of the amino acids from the bluish or purple-colored spots on the chromatogram.
10. Calculate the R_f values of the standard amino acids and the amino acids in the sample or mixture as described in the first experiment of this chapter.
11. Identify the amino acids in the mixture or sample by comparing the R_f values with those of applied standard amino acids.

Experiment: Separation and Identification of Amino Acids in a Given Mixture by 2D Paper Chromatography

Introduction

Amino acids that have very close R_f values in a particular solvent system may appear as single or overlapping spots in a one-dimensional (1D) chromatogram and may be mistaken as a single component. They can be separated into individual components by developing the chromatogram again in a direction perpendicular to the first run in a second solvent system in which they have different R_f values. The main limitation of this method is that only one spot of either the sample or a standard amino acid can be applied on each filter paper sheet, necessitating the running of a large number of chromatograms for standard amino acids.

Safety Guidelines

Same as those given in the first experiment of this chapter.

Experimental Outline

Develop the chromatographic paper in one solvent after it has been dried; turn round and develop it in another direction with a second solvent. Calculate the R_f values of the amino acids (Figure 4.27).

Materials

Same as the ones listed in the first experiment of this chapter, except that an additional chromatographic chamber for the second solvent system is required.

Prelaboratory Preparations

Prelaboratory preparations required are the same as those given in the first experiment of this chapter. Additional preparations required are as follows:

1. Solvent system no. 2: Phenol (distilled) in water (80:20 w/v) is used as the second solvent system. Add 125 mL of water to 500 g of phenol, and add a few drops of ammonia (0.88%) to this mixture just before use. (Caution: Phenol is corrosive and can cause burns on the skin.)
2. Standard amino acids: Prepare a 1% solution of standard amino acids such as asparagine, glycine, serine, and arginine in 10% isopropanol (w/v).

Method

1. Lay the chromatographic paper sheet flat on the rough filter paper using gloves.
2. Draw a baseline 5 cm from one of the edges of the paper.
3. Draw another line perpendicular to the first line, again 5 cm away from the adjacent edge.
4. Apply 60 μL of the sample solution or a given mixture containing unknown amino acids at the point of intersection of the two lines. The sample should be applied in small volumes at a time with the help of a micropipette with intermittent drying to ensure that the zone of the applied solution is as small as possible.
5. Repeat the same procedure for a mixture of three standard amino acids using a separate chromatographic sheet for each amino acid. The composition of the mixture of standard amino acids should be such that each amino acid is present in at least two different mixtures so that its identity can be established.
6. Hang the paper in the chromatographic tank whose interior has been previously saturated with the aqueous phase of solvent system no. 1 (n-butanol: acetic acid: water mixture in the ratio 4:1:5).
7. After allowing an equilibrium period of half an hour, pour solvent no. 1 into the trough of the chamber and let it run until it reaches about 10 cm from the opposite edge of the paper.
8. Take the paper out, air dry it, and turn it at an angle of 90°C; now develop the paper in the second chromatographic chamber using solvent system no. 2 (phenol: ammonia: water).
9. Remove the paper when the solvent has traveled up to about 10 cm from the opposite end.
10. Dry it at room temperature and spray it with the ninhydrin reagent. After air drying it, keep the chromatogram in an oven at 105°C for 10 minutes. Mark the blue- and/or purple-colored zones that appear on the paper.

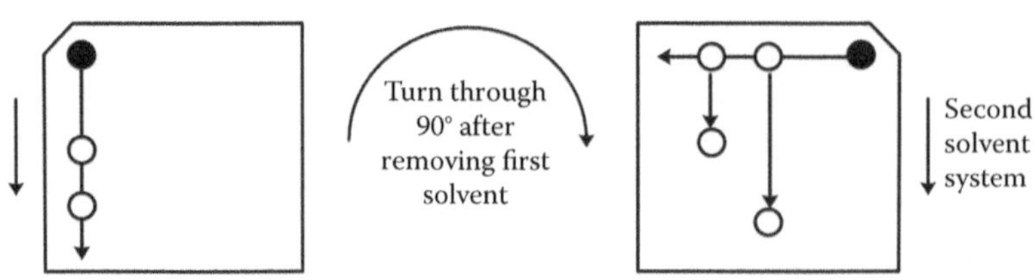

FIGURE 4.27 Scheme of 2D chromatography.

Observations

Calculate the R_f value of the standard amino acids and those in the mixture as given in both the solvents. From these values, identify the amino acids in the given mixture.

Experiment: Separation and Identification of Sugars by Adsorption TLC

Introduction

Sugars are separated on the basis of differential adsorption onto silica gel. The sugars that have higher affinities for stationary phases are adsorbed more strongly, and they migrate slowly when the mobile phase moves over them. On the other hand, those having lower affinities for stationary phases are weakly adsorbed and more easily carried by the mobile phase. The separated sugars are then located as colored zones by spraying TLC plates with an aniline diphenylamine phosphate reagent.

Safety Guidelines

1. Only thoroughly cleaned glass plates free of any greasy spots or finger marks should be used.
2. The thickness of the layer should be uniform throughout the length of the plate.
3. The slurry of the chromatographic media should be free of any clumps. This can be ensured by vigorously shaking it in an Erlenmeyer flask or by gently preparing the slurry using a pestle and mortar to ensure uniform mixing.
4. The layer on the chromatographic media should not get scraped off when putting marks on samples.
5. The size of the applied spot should be as small as possible. If a large volume of the sample has to be spotted, then it should be done in small aliquots with intermittent drying. Overloading of the sample should be avoided.
6. The chromatographic tank should be airtight and chromatography should be performed under temperature-controlled conditions.

Experimental Outline

Preparation of TLC Plates with Silica Gel-G

Develop silica gel-G plates by running glass plates in a chromatographic chamber with ethyl acetate: isopropanol: water: pyridine in the ratio 26:14:7:2. Dry the plates after they are developed, spray with aniline diphenylamine phosphate reagent, and calculate the R_f values of each spot (Figure 4.28).

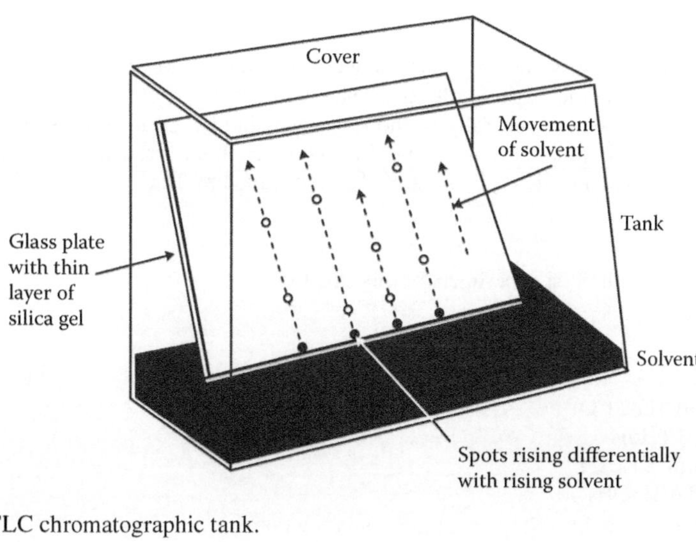

FIGURE 4.28 TLC chromatographic tank.

Materials

A TLC chromatographic tank, glass plates (20 × 20 cm), a spreader, micropipettes/microsyringe, an oven maintained at 105°C, a hair dryer, and a sprayer or an automizer.

Prelaboratory Preparations

1. Solvent system: Take ethyl acetate: isopropanol: water: pyridine in the ratio 26:14:7:2.
2. Standard sugar solution: Prepare a 1% solution of standard sugars such as glucose, ribose, fructose, and sucrose in 10% isopropanol.
3. Aniline diphenylamine phosphate reagent: Mix 5 volumes of 1% aniline, and 5 volumes of diphenylamine in acetone with 1 volume of 85% o-phosphoric acid.

Method

1. Place thoroughly cleaned and dried glass plates (20 × 20 cm) side by side on a flat plastic tray, leaving no gap between the two adjacent plates.
2. Prepare a slurry of the stationary phase (silica gel-G) free of clumps in water or an appropriate buffer.
3. Spread a uniform layer of slurry of 250 μm thickness on the glass plates with the help of a spreader or an applicator by moving it from one end to the other end of the tray. Note that ready-made thick silica gel–coated aluminum or plastic sheets are also available.
4. Activate the plates by keeping them at 105°C for 30 minutes; allow the plates to cool in a desiccator before use.
5. Gently put marks in a straight line with the help of a pin at a distance about 2 cm from one edge of the plate. The adjacent marks should be plotted carefully in such a way that the silica does not scratch off while putting on the spots.
6. Carefully apply the solution of individual standard sugars and the mixture or alcoholic extract of the sample on separate marked spots.
7. Gently put marks in a straight line.
8. Develop the TLC plates in the solvent (ethyl acetate: isopropanol: water: pyridine in the ratio 26:14:7:2).
9. Take out the plates after the solvent reaches 2 cm from the bottom.
10. Air dry the plates for at least 12 hours.
11. Identify the sugar species by spraying the plates with an aniline diphenylamine phosphate reagent.

Experiment: Isolation and Identification of Lipids in a Given Sample by Thin-Layer Chromatography

Introduction

Lipids, which are a heterogeneous group of biological compounds, are insoluble in water but soluble in ether, chloroform, and other organic solvents. The hydrocarbon content of lipids contributes to their hydrophobic nature. Lipids are generally bound to proteins in biological samples and cannot be efficiently extracted with nonpolar solvents alone. In such cases, lipids are extracted with a mixture of chloroform and methanol and are easily separated and identified by TLC.

Safety Guidelines

Same as those given in the first experiment of this chapter.

Reagents

1. Silica gel
2. Acetone ((CH3)2 CO)
3. Benzene (C6H6)
4. Chloroform (CHCl3)
5. Methanol (CH3OH)
6. Ammonium hydroxide (NH_4OH) or ammonia solution

Experimental Outline

A. Extraction of Lipids

1. Centrifuge a known volume of homogeneous algal suspension at 5000 rpm for 10 minutes.
2. Wash the pellet twice in distilled water.
3. Sonicate or grind with glass powder a known weight of the algal pellet with a known volume of chloroform: methanol (2:1) (v/v).
4. Filter the extract using the filter paper and ensure that the filter paper is free from lipids.
5. Add 1/3 volume of distilled water to the filtrate and vortex thoroughly to remove water-soluble impurities (upper layer).
6. To remove moisture content in the filtrate, add a small amount of sodium sulfate crystals and centrifuge (nonclumping of crystals indicate that the filtrate is free of moisture).
7. Transfer the filtrate to a preweighed bottle (weight is A).
8. Dry the filtrate either under nitrogen or in a rotary evaporator.
9. Reweigh the bottle (new weight is B).
10. Weight of lipid (in milligrams per gram) = $B - A$.

Comments

For storage up to a few hours, the tube is covered with aluminum foil and chilled on ice to minimize the evaporation of the solvent.

B. The TLC Method for *Separation* of Lipid Classes

1. Draw a pencil line on a precoated silica gel plate (20 × 20 cm) parallel to and approximately 2 cm from one edge.
2. Redissolve the total lipid extracted earlier in 1 mL of chloroform: methanol (2:1) (v/v).
3. Apply a lipid solution (10–50 μL) as a spot on the pencil line with a microsyringe (evaporate the solvent quickly with an air blower).
4. Develop the plate with either of the solvent systems:
   a. $(CH_3)_2$ CO/C6H6/H2O (91:30:8) (v/v)
   b. CHCl3/CH3OH/(28%) NH4OH (13:7:1) (v/v)
5. Dry the plate in the fume hood for about 20 minutes.
6. Keep the plate in the iodine chamber until the colored spots develop.
7. Mark the spots, and scrape into a flask for extraction.
8. Calculate the resolution front (R_f) and identify the sample by comparing the standards.

Experiment: Separation of Pigments from Leaves or Flowers by Adsorption Column Chromatography

Introduction

Different pigments are adsorbed to alumina to different extents. They can be selectively desorbed by using a mobile phase of increasing polarity in a stepwise manner.

Safety Guidelines

The safety guidelines outlined in the first five experiments of this chapter may be followed.

Experimental Outline

Prepare a glass column; homogenize leaves in benzene: methanol (26% ratio); and run the extract in an alumina-packed glass column with appropriate solvents. Different pigment fractions are collected and identified by their colors.

Materials

Leaves or flowers, pestle and mortar, a glass column or a burette, Whatman no. 1 filter paper, alumina or icing sugar, sodium sulfate (anhydrous), acetone.

Prelaboratory Preparations

The solvent used is benzene: methanol (2:1).

Method

A. Preparation of Extract

1. Homogenize the leaves or flowers (5 g wet weight) in a pestle and mortar, using sand as an abrasive, in 20 mL of benzene: methanol (2:1); add a small amount of this extracting medium at a time.
2. Filter the extract through Whatman no.1 filter paper and transfer the filtrate to a separating funnel.
3. Add 10 mL of water to the filtrate and after shaking the contents and allowing the phases to separate out, drain out the lower aqueous methanol layer. Repeat this step; avoid vigorous shaking.
4. Collect the benzene layer in a beaker and add a small amount of solid anhydrous Na2SO4 to remove the traces of moisture.
5. Decant the clear benzene layer to another beaker and concentrate the extract by evaporating the solvent over a boiling water bath.

B. Preparation of Column

1. Mount a burette or glass column vertically on a burette stand with the help of clamps.
2. Lightly place a plug of glass wool at the base of the burette and close the stopcock or outlet at the bottom of the column.
3. Take 5 g of alumina or icing sugar (adsorbent), which has been previously dried at 120°C for 8 hours, and prepare its slurry in benzene. Pour the slurry carefully into the column or burette by gently tapping the column or burette so that no air bubbles get trapped in the adsorbent.
4. Allow the adsorbent to settle by opening the outlet. After the adsorbent has completely settled, add 20 mL of benzene and let it pass out of the column. Care should be taken to not let the adsorbent dry.

C. Application of Sample

1. Allow the solvent at the surface of the column to drain out slowly, and transfer the leaf or flower extract with the help of a pipette without disturbing the surface of the column adsorbent. Let it enter the column, and then add a few drops of benzene to wash away the traces of the extract sticking to the wall of the column.
2. Add 20 mL more benzene to wash any unabsorbed material out of the column.

D. Column Development

For the desorption of the adsorbed substances, change the polarity of the solvent in a stepwise manner. After 20 mL of benzene passes through the column, add 10 mL of 5% acetone (v/v) in benzene and let it percolate through the column; collect 1–2 mL fractions of the effluent from the outlet. Continue increasing the concentration of acetone in benzene at every succeeding step. Finally, pass pure acetone through the column.

Result

Note the change in the color of the collected fractions. In the case of the leaf extract, the initial fractions are colorless; they are followed by yellow-colored and then green-colored extracts. The colorless fractions do not contain any pigments, but it is quite possible that these fractions contain some UV-absorbing materials.

Experiment: Desalting of Protein Sample by Gel Filtration

Introduction

Separation is based on the fact that proteins are macromolecules, whereas salts are low-molecular-weight substances. When the sample is passed through a column packed with sephadex G-10 or G-25, proteins remain totally excluded from the gel and move with the void volume, while salts enter the gel particles and take a longer time to get eluted.

Safety Guidelines

Standard laboratory guidelines may be followed in this experiment for protein purification.

Experimental Outline

Prepare the column with either sephadex G-10 or sephadex G-25; elute the column with buffer. Collect 2 mL fractions for the eluant and measure the protein content of each fraction.

Materials

Glass column; sephadex G-10, G-25, or G-75; protease; RNAase; sodium phosphate.

Prelaboratory Preparations

1. Prepare 0.1 M Tris-HCl buffer (pH 7.0).

Method

1. Suspend 5 g of sephadex G-25 (coarse) in 0.1 M Tris-HCl buffer (pH 7.0), and allow it to swell by keeping it for 3–4 hours at room temperature with intermittent stirring.
2. Decant the excess buffer along with any suspended fine particles to obtain slurry of reasonable thickness.
3. Fix the column upright on a burette stand with the help of clamps.
4. Keep the outlet of the column closed, place a plug of glass wool at the base of the column, and pour a small volume of the buffer or water into the column.
5. Pour the slurry gradually into the column along the inner surface of its wall and, if necessary, gently tap the column to expel any air bubbles.
6. Allow the chromatographic media to settle down evenly and then open the outlet to drain excess liquid from the column.
7. Place a filter paper disk or nylon gauze on the surface of the packed bed to prevent a disturbance of the upper layer while loading the sample or feeding the eluant into the column.
8. Prepare a mixture of 10 mg of bovine serum albumin and 40 mg of sodium phosphate in 2 mL of 0.1 M Tris-HCl buffer (pH 7.0).
9. Apply the mixture to the chromatography column by either of the following two methods:
   a. Drain out the mobile phase at the top of the peak until the bed surface is exposed. Close the outlet and gently apply the sample uniformly over the bed surface using a pipette; then allow the loaded sample to just enter the column by opening the outlet. Add a small amount of mobile phase (or buffer) to wash the traces of the sample into the column.
   b. In the second method, add sucrose or glycerol up to a concentration of 1% to the sample to increase its density. Apply this sample just above the surface of the bed directly through the layer of the buffer in the column bed. Since the sample has a higher density, it automatically settles on the surface of the gel. Then open the outlet to facilitate the entry of the sample into the column. When using this procedure, it is advisable to ensure that the addition of glycerol or sucrose does not interfere with the separation and subsequent analysis of the separated compounds.
10. Add a sufficient amount of buffer to the top of the column and connect it to the buffer reservoir.
11. Collect fractions (2 mL each) either manually or using an automatic fraction collector. Determine the protein content either by monitoring the absorbance at 280 nm or by applying Lowry's method and the phosphate ions method in each of the fractions.
12. Plot a graph of the concentration of protein and phosphate versus the fraction number or elution volume (Figure 4.29).

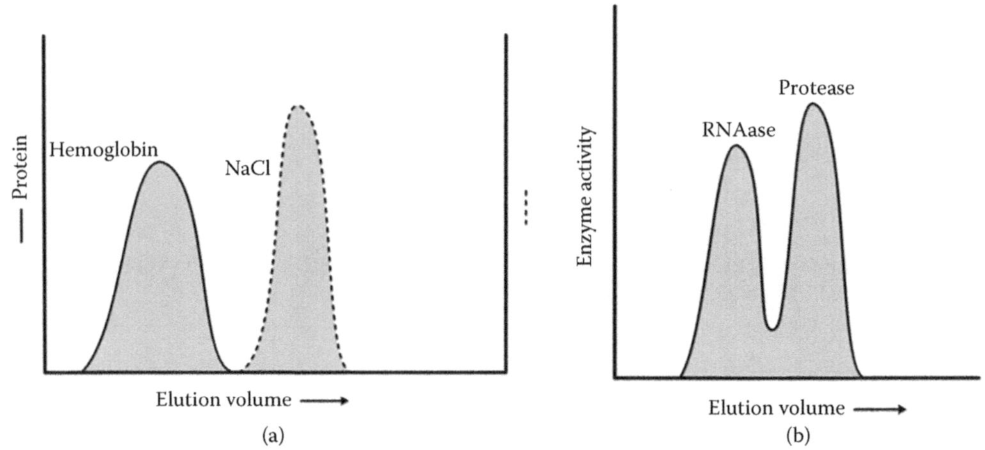

FIGURE 4.29 Figure showing (a) typical elution pattern on sephadex G-25 to separate hemoglobin from salt and (b) separation of RNAase from a protease in pancreatic extract on G-75.

Experiment: Separation of Amino Acids by Ion-Exchange Column Chromatography Using Cation Exchanger

Introduction

Ion-exchange chromatography can be used for the separation of substances that possess a net electrical charge. Anion exchangers reversibly bind negatively charged compounds through electrostatic forces, whereas positively charged molecules interact with cation exchangers. Different compounds are held by ion exchangers with varying strengths depending on charge. At very low pH values of 1.0, almost all amino acids (including acidic amino acids) exist as cations; hence, they can be separated on a cation exchanger.

Safety Guidelines

The safety guidelines outlined in the first six experiments of this chapter may be followed.

Experimental Outline

Prepare the glass column by packing it with an ion-exchange resin. Elute the amino acids with a buffer, collect the eluant, and identify the amino acids.

Materials

Chromatographic column (2.5 × 25 cm), colorimeter

Prelaboratory Preparations

1. Dowex-50 resin in 0.05 M citrate buffer, pH 3.0
2. Citrate buffer 0.05 M, pH 3.0
3. Citrate buffer 0.05 M, pH 6.0
4. Citrate buffer 0.05 M, pH 9.0
5. Amino acids (aspartic acid, alanine, lysine, and histidine); 2 mg of each amino acid per milliliter of 0.1 M HC1, pH 1.0
6. 4N HCI
7. Ninhydrin reagent
8. 0.1N HCl, pH 1.0

Method

A. Preparation of Ion Exchanger

Dowex-50, a cation exchanger, must be fully saturated with H+ first; this H+ form can be converted to Na+ form so that the resin can function as a cation exchanger.

1. Suspend 10 g of Dowex-50 into sufficient volume of 4 N HCI for 15 minutes to ensure that the resin is saturated with H+ ions.
2. Filter the suspension and wash it repeatedly with distilled or deionized water until the pH of the filtrate is neutral.
3. Transfer the resin into 2 N NaOH and keep for 15 minutes to get the Na+ form. Wash it until the pH of the filtrate is neutral.

B. Equilibration of the Resin

1. Suspend the resin in citrate buffer (of pH 3.0) and allow it to stand for 1 hour.
2. Mount the column upright and pour the suspension with the help of a glass rod while tapping the column gently.
3. Allow the suspension to settle down, open the outlet, and pass two to three bed volumes of citrate buffer (0.05 M, pH 3.0) through the column. This will fully equilibrate the resin to pH 3.0. When only a thin layer of buffer remains at the top of the resin, stop the flow by closing the column outlet.

C. Sample Loading

1. Open the outlet and let the buffer at the top drain into the column surface. Close the stopcock.
2. Load the column with 1 mL of a mixture of amino acids in 0.1 M HCl that have a pH value of 1.0.
3. Add a small amount of buffer to wash away traces of the sample from inside the walls of the column, and when the level just reaches the surface, close the stopcock.

D. Development of the Column

All the amino acids are in cationic form at pH 1.0 and, thus, they are bound to the cation exchanger. Gradient elution using increasing pH and ionic strength facilitates sequential elution of the bound amino acids. Test each eluant with ninhydrin.

EXPERIMENT: DETERMINATION OF THE MOLECULAR WEIGHT OF A GIVEN PROTEIN BY GEL FILTRATION

Introduction

During gel filtration, solutes are separated primarily on the basis of their molecular size. Due to the molecular sieving effect, large molecules are eluted from the column first, followed by compounds of smaller molecular mass. A plot between Kd or the elution volume versus logarithm to the base 10 of molecular weight gives a straight line. The molecular weight of a given protein can be established from its elution volume through the gel filtration column, which has previously been calibrated with standard marker proteins of known molecular weight.

Safety Guidelines

The safety guidelines are outlined in the first six experiments of this chapter; pack the glass column with sephadex G-100, avoiding the entrapment of any air bubbles.

Experimental Outline

1. Preparation of the glass column with sephadex G-100
2. Elution with buffer solution
3. Collecting the protein fractions
4. Calculating the molecular weights

Materials

Glass column (2.5 × 70 cm), sephadex G-100, MgC12, dithiothreitol (DTT), glycerol, standard protein markers, blue dextran

Prelaboratory Preparations

1. The HEPES [4-(2-Hydroxyethyl)piperazine-1-ethanesulfonic acid]–NaOH buffer (20 mM; pH 8.0)

Method

1. Suspend 15 g of sephadex G-100 in 20 mM HEPES–NaOH (pH 8.0) buffer containing 5 mM $MgCl_2$ and 5 mM DTT for 5 hours in a boiling water bath.
2. Allow it to cool and pack it into the glass column.
3. Equilibrate the column by passing buffer equivalent to 2–3 volumes of the bed volume.
4. Find out the void volume (V_o) of the column by determining the elution volume of blue dextran solution (2 mg/mL) through the column. Again pass 2 bed volumes of the starting buffer.
5. Apply the mixture of the standard marker proteins of known molecular weight and elute the column with the buffer.
6. Collect fractions of 2 mL each and determine the protein content in each fraction either by Lowry's method or by recording absorbance at 280 nm.
7. Determine the elution volume of standard proteins and prepare a graph of logarithm to the base 10 of molecular weight versus V_e or K_d, again pass 2 volumes of the starting buffer.
8. Layer the sample containing the protein whose molecular weight must be determined. Elute it and collect fractions of 2 mL each. Test each fraction for the presence of protein and determine its V_e.
9. Determine the molecular weight of the given protein from the calibration curve prepared in step 7 as shown in Figure 4.30.

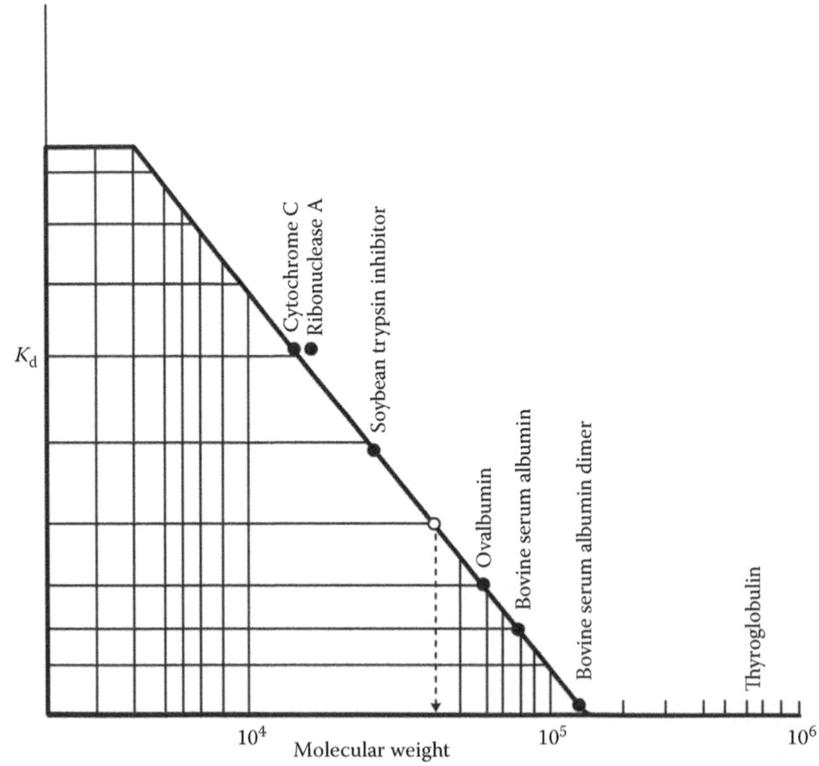

FIGURE 4.30 Relationship between K_d and logarithm of the molecular weight of proteins.

Experiment: Identification and Quantification of Amino Acids by High Performance Liquid Chromatography (HPLC)

Introduction

The derivatized amino acids have greater resolutions as well as detectability at picomole levels. There are two types of reagents used for derivatization: (1) precolumn and (2) postcolumn. Precolumn derivatization is commonly used, and the reagents used are phenylisothiocyanate (PITC) and ortho-phtalaldehyde (OPA). Derivatives could be measured above 250 nm. A fluorescence detector could also detect OPA derivatives.

A. Phenylthiocarbamyl Method

Principle

PITC reacts with primary and secondary amino acids, and the reaction yields phenylthiocarbamyl (PTC) amino acids derivatives. These have strong absorbances between 240 and 260 nm.

Reagents

1. PITC.
2. 6N HCl.
3. Acetate buffers (0.025 M) of pH 5 and pH 7: Prepare 0.25 M of acetic acid (0.23 mL in 100 mL of distilled water) and 0.25 M sodium acetate (3.4 g in 100 mL of distilled water). Add sodium acetate to 0.25 M acetic acid until the pH of the solution reaches 5 and 7, respectively, for acetate buffers of pH 5 and 7.
4. Acetonitrile.
5. Coupling buffer: Prepare the coupling buffer by mixing 10 mL of acetonitrile, 5 mL of pyridine, 2 mL of trimethylamine, and 3 mL of water.

Procedure

1. Take commercially available protein hydrolysate and prepare ethanolic solution.
2. Place the samples in the vacuum desiccator and completely dry them under a vacuum.
3. To the dried sample, add 200 µL of coupling buffer and evaporate to dryness.
4. To the sample, again add 200 mL of coupling buffer; this is followed by the addition of 20 µL of PITC.
5. Incubate the sample at room temperature for 5 minutes and evaporate to dryness.
6. Store the dried sample in the freezer until analysis.
7. Prior to analysis, dissolve the dried sample in 100 µL of acetate buffer adjusted to a pH value of 7.
8. Filter the sample before injection.

Column Conditions

Column: Teledyne Isco, Inc. P.O. Box 82531, Lincoln, Nebraska, 68501 USA (ISCO) C18, 4.6 × 250 mm, 5 µm pack ink
Mobile phase A: Sodium acetate buffer, pH 5.0
Mobile phase B: 50% sodium acetate buffer/50% acetonitrile (before use, filter and degas)
Flow rate: 1.5 mL·min–1
Detector: ISCO V4 variable wavelength, 5 mm flow cell, and wavelength 254 nm

Result

Run the standard PTC amino acids. From the retention time and peak area, identify and calculate the amount of individual amino acids in the sample and express the amount as nanomoles per milligram dry weight of sample or protein.

B. OPA Method

Principle

Peptides with N-terminal primary amines are derivatized by OPA. One advantage of the OPA derivative is its enhanced sensitivity; further, it can be detected in a fluorescence detector without interference.

Reagents

1. Borate buffer (0.25 M, pH 9.5): Prepare 0.25 M of boric acid (15.457 g in 100 mL of distilled water) and 1 M of NaOH (4 g in 100 mL of distilled water). Add NaOH solution to boric acid solution until the pH reaches 9.5, and make up to 100 mL.
2. Phosphate buffer (0.1 M, pH 6.5): Prepare 0.1 M of K2HPO4 (2.720 g in 100 mL of distilled water) and 0.02 M sodium acetate (164 mg in 100 mL of distilled water). Take 50 mL of 0.1 M K2HPO4 and titrate with 0.02 M sodium acetate until the pH reaches 6.5; make up to 100 mL.
3. OPA stock solution: Dissolve 0.26 g of OPA in 20 mL of methanol and make up to 200 mL with borate stock solution and store the solution in a refrigerator.
4. Derivatizing solution: Add 30 μL of mercaptoethanol to 20 mL of OPA solution (prepare this on the day of use).
5. Tetrahydrofuran.
6. Acetic acid.
7. Methanol.

Procedure

1. Take commercially available protein hydrolysate and prepare ethanolic solution.
2. Mix equal amounts of ethanolic hydrolysates and derivatizing solution thoroughly before injection (prior to sampling, filter the solution through a 0.45 μ nylon filter).
3. Load the mixture into the HPLC injection loop and allow the reaction to continue (do the mixing and injection within 1 minute; for reproducible results, the timing of the injection is critical).

Column Conditions

Column: ISCO C18, 4.6 × 250 mm, 5 μm packing
Mobile phase A: 10% tetrahydrofuran/90% phosphate buffer (pH 6.5), K2HPO4 (0.1 M), sodium acetate (0.2 M)
Mobile phase B: 80% methanol/20% acetic acid (0.1 M)
Flow rate: 1.5 mL·min–1
Detector: ISCO FI-2, 9 μL flow cell
Excitation filter: 305–395 nm
Emission filter: 430–470 nm

Calculations

Run the standard OPA amino acids. From the retention time and peak area, identify and calculate the amount of individual amino acids in the sample and express the amount as nanomoles per milligram dry weight of sample or protein.

Experiment: Isolation and Identification of Carbohydrates by GC

A. Hydrolysis of Polysaccharides

Principle

Carbohydrates that occur as trioses (CH2O)3, tetroses (CH2O)4, pentoses (CH2O)5, and hexoses (CH2O)6 are hydrolyzed to simple sugars by strong acids because of the dehydration of hydroxyl groups.

Reagents

1. Hydrolyzing agent: H2SO4/trifluoroacetic acid (TFA)
2. Neutralizing agent: Barium carbonate [$Ba(CO_3)_2$]

Procedure

1. Centrifuge a known volume of homogeneous bacterial/fungal/cyanobacterial suspension at 5000 rpm for 10 minutes.
2. Wash the pellet twice in distilled water.

3. To a known weight (10 mg) of the pellet in a tube, add 1 mL of 0.70% H2SO4 and seal. Hydrolyze at 105°C for 6 hours.
4. Neutralize the hydrolyzed contents with barium carbonate and centrifuge at 5000 rpm for 10 minutes.
5. Evaporate to dryness over a boiling water bath.

B. Identification by Gas Chromatography

Principle

For GC analysis of carbohydrates, the trimethylsilyl (TMS) derivative is recommended to increase the volatility of the carbohydrates. The chloride present in trimethylchlorosilane acts on hydroxyl groups in carbohydrates and forms Me3Si-O-carbohydrates in the presence of a pyridine base.

Reagents

1. Anhydrous pyridine (pyridine dried over KOH pellet is suitable)
2. Hexamethyldisilazane (HMDS)
3. Trimethylchlorosilane (TMCS)

Procedure

1. Prepare a trimethylsilation reagent by mixing anhydrous pyridine (5 mL), HMDS (2 mL), and TMCS (0.5 mL).
2. To the hydrolyzed sugar sample, add 1 mL of the pyridine–silane mixture and shake well until the sugar dissolves completely.
3. Allow the sample to stand at room temperature for 5 minutes and then perform the injection using the following column conditions:

    Column: 3% OV-225
    Column temperature: 190°C
    Injection port temperature: 240°C
    Detector temperature: 240°C
    Gas flow rates:
        H_2: 15 mL·min^{-1}
        N2: 15 mL^{-1}
        Air: 300 mL·min^{-1}

   or

    Column: 3% OV-275
    Column temperature: 230°C
    Injection port temperature: 275°C
    Detector FID
    Gas flow rates:
        H2: 15 mL·min^{-1}
        N2: 15 mL^{-1}
        Air: 300 mL·min^{-1}
4. Run standard sugars with TMS derivatives. From the known retention time (RT), find the sugars of the unknown sample and quantify them using the peak area; express their amounts as nanogram per milligram dry weight of sample.

SUGGESTED READING

AKTA Design Purification Method Handbook. Amersham Biosciences. Catalog number 18-1124-23. Amersham Biosciences 2006.

AKTA FPLC, System Manual, Amersham Pharmacia Biotech. http://www.hhmi.umbc.edu/toolkit/aktadesign.pdf, 18-1140-45 Edition AB.

ÄKTAFPLC, System Manual 18-1140-45 Edition AB. Amersham Pharmacia Biotech AB 2000.

Buffington, R., and M. K. Wilson. 1987. *Detectors for Gas Chromatography—A Practical Primer*. Hewlett-Packard Corporation, Part No. 5958-9433.

Chromatography, Theories, FPLC and Beyond. http://www.mnstate.edu/biotech/chrom_fplc.pdf

Cooper, E. H., R. Turner, E. A. Johns, H. Lindblom, and V. J. Britton. 1983. "Applications of Fast Protein Liquid Chromatography TM in the Separation of Plasma Proteins in Urine and Cerebrospinal fluid." *Clinical Chemistry* 29: 1635–40.

Desai, S. N., R. B. Colah, and D. Mohanty. 1998. "Comparison of FPLC with Cellulose Acetate Electrophoresis for the Diagnosis of Beta-Thalassaemia Trait." *Indian Journal of Medical Research* 108: 145–48.

Ettre, L. 1993. "Nomenclature for Chromatography." *Pure and Applied Chemistry* 65: 819–72.

Feinberg, J. G., and I. Smith. 1972. *Paper and Thin Layer Chromatography & Electrophoresis*, 2d ed. (with amendments and corrections). London: Longman.

Ferreira, P. O., and M. A. Ferreira. 2002. "Solid Phase Micro-Extraction in Combination with GC/MS for Quantification of the Major Volatile Free Fatty Acids in Ewe Cheese." *Analytical Chemistry* 74: 5199–204.

Gogou, A. I., M. Apostolaki, and E. G. J. Stephanou. 1998. "Adsorption Chromatography." *Journal of Chromatography A* 799: 215.

GST Gene Fusion System Handbook, *Amersham Biosciences Corp*. 800 Centennial Ave, P.O. Box 1327. Piscataway, NJ 08855-1327. Company bulletin 18-1157-58 (2002).

Hage, D. S., ed. 2006. *Handbook of Affinity Chromatography*, 2nd ed. New York: Taylor & Francis.

Hage, D. S., and P. F. Ruhn. 2006. "An Introduction to Affinity Chromatography." In *Handbook of Affinity Chromatography*, 2nd ed., edited by D. S. Hage. New York: Taylor & Francis.

Hill, H. H., and D. G. McMinn, eds. 1992. *Detectors for Capillary Chromatography*. New York: John Wiley & Sons.

Jeppsson, J. O., P. Jerntorp, G. Sundkvist, H. Englund, and V. Nylund. 1986. "Measurement of Haemoglobin A1c by a New Liquid-Chromatographic Assay: Methodology, Clinical Utility, and Relation to Glucose Tolerance Evaluated." *Clinical Chemistry* 32: 1867–72.

Lee, W.-C., and K. H. Lee. 2004. "Applications of Affinity Chromatography in Proteomics." *Analytical Biochemistry* 324: 1–10.

Marz, W., R. Siekmeier, H. Scharnagl, U. B. Seiffert, and W. Gross. 1993. "Fast Lipoprotein Chromatography: New Method of Analysis for Plasma Lipoproteins." *Clinical Chemistry* 39: 2276–81.

Mehler, A. H. 1993. "Glutathione S-Transferases Function in Detoxification Reactions." *In Textbook of Biochemistry*, 3rd ed., edited by T. M. Devlin, 523–24. New York: Wiley-Liss.

Parikh, I., and P. Cuatrecasas. 1985. "Affinity Chromatography." *Chemical Engineering News* 63: 17–29.

Gustavsson, P.-E., and Larsson, P.-O. 2003. "Fast Chromatography of Proteins." *In Isolation and Purification of Proteins*, edited by Rajni Hattokoul and Bo Mattiasson. Marcel Dekker, Inc.

Sambrook, J., and W. Russell David. 2001. Protein Interaction Technologies. Molecular Cloning A Laboratory Manual. New York: Cold Spring Harbor Laboratory Press.

Sandra, J. F. 2002. *Gas Chromatography. Ullmann's Encyclopedia of Industrial Chemistry*. Weinheim: Wiley-VCH VerlagGmbH.

Scouten, W. H. 1981. *Affinity Chromatography: Bio-Selective Adsorption on Inert Matrices*. New York: Wiley.

Sheehan, D., and S. O'Sullivan. 2003. "Fast Protein Liquid Chromatography." *Protein Purification Protocols* 244: 253.

Shibasaki, T., H. Gomi, F. Ishimoto, and T. Miyahara. 1990. "Urinary N-Acetyl-Beta-D-Glucosaminidase Isoenzyme Activity as Measured by Fast Protein Liquid Chromatography in Patients with Nephrotic Syndrome." *Clinical Chemistry* 36: 102–03.

Smith, I. 1960. *Chromatographic and Electrophoretic Techniques. Vol 1. Chromatography. Vol. 2. Zone Electrophoresis*. New York: Interscience Publishers, Inc.

Smith, I., J. Seakins, and T. William. 1976. *Chromatographic and Electrophoretic Techniques*, 4th ed. London: William Heinemann Medical Books.

Stahl, E., and M. R. F. Aisworth. 1969. *TLC: A Laboratory Handbook*, 2nd ed. New York: Springer Verlag.

TC information ab, Uppsala. Printed in Sweden by T. K. i Uppsala AB. Amersham Pharmacia Biotech AB 1999.

Turkova, J. 1978. *Affinity Chromatography*. Amsterdam: Elsevier.

Walters, R. R. 1985. "Affinity Chromatography." *Analytical Chemistry* 57: 1099A–114A.

Wilson, K., and J. Walker. 2003. *Practical Biochemistry: Principle and Techniques*, 5th ed. Cambridge: Cambridge University Press.

IMPORTANT LINKS

1. Column chromatography: http://www.repligen.com/bioprocessing/products/prepackedcolumn?gclid=ckogjcjzoksFQl76wodARLC-fw
2. HPTLC: http://www.camag.com/v/products/tlc-ms/
3. HPLC: http://www.hplc.com/Eicom/index.html
4. GLC: http://www.makarandelectronics.com/gas_chromatograph.html
5. Ion-exchange chromatography: http://www.gelifesciences.com/aptrix/upp01077.nsf/Content/Products?OpenDocument&parentid=5179&moduleid=165889&zone=Labsep
6. Exclusion chromatography: http://www.pall.com/main/Laboratory/Literature-Library-Details.page?id 47506
7. Affinity chromatography: http://www.jenabioscience.com/cms/en/1/browse/1450affinitychromatography.html

5 Spectroscopy

5.1 INTRODUCTION

5.1.1 Definition and General Principles

As a useful working definition, spectroscopy can be defined as the interaction of electromagnetic radiation (EM) with matter, although this does not include mass spectroscopy. Several factors have led to the branching of spectroscopy in different directions. Most significant is the order of magnitude of the energies involved, but additional factors such as the presence of a magnetic field and instrumentation considerations have led to the techniques of ultraviolet (UV), infrared (IR), nuclear magnetic resonance (NMR), and electron spin resonance (ESR) spectroscopy. Breakup and analysis of the above definition will be useful before delving into the details of the different spectroscopic techniques mentioned. The definition includes

1. Electromagnetic radiation
2. Interaction of EM with matter
3. Matter

EM radiation consists of an electric field perpendicular to a magnetic field and both at right angles to the direction of propagation of light (Figure 5.1). A fundamental property of EM radiation is that it can behave as though it exists as discrete quanta or packets of energy:

$$E = h\upsilon$$

where

E = energy
h = Planck's constant = 6.63×10^{-34} J·s
υ = frequency of radiation in Hertz

There are two ways in which EM radiation interacts with matter: absorption and emission. Absorption occurs when incident radiation increases the energy of a system. An increase in energy is manifested as a decrease in intensity of emergent radiation. Emission occurs when there is a decrease in the energy of a system. Decrease in energy may be due to

1. Thermal energy loss: Energy loss by molecular or submolecular motion like a collision, vibration, or rotation.
2. EM emission: This results in phosphorescence and fluorescence.
3. Photochemical reactions: There is a competition between the three processes, also called relaxation processes.

Matter is composed of atoms and molecules. The electrons are constrained to certain energy levels called K, L, M, and so on orbitals. When atoms combine to form molecules, electrons occupy new energy levels called molecular bonding orbitals. Atoms in the molecule can vibrate and rotate about the bond axis, giving rise to vibrational and rotational energy sublevels. Electrons usually remain in the ground state but after receiving energy, for example, from incident EM radiation,

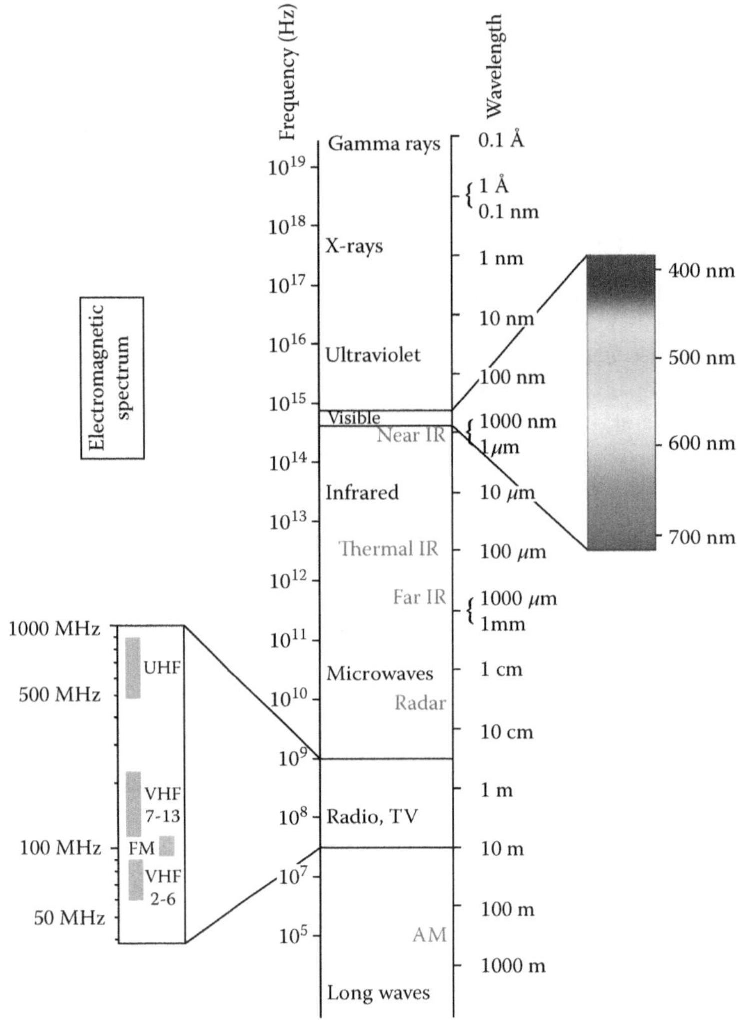

FIGURE 5.1 **(See color insert.)** EM spectrum.

they get excited to higher states. This is the absorption of energy. When electrons come back to their original ground state, the emission of energy takes place. Sometimes all the energy absorbed is emitted, while at other times some energy is lost in one or all of the relaxation processes. The amount of energy absorbed or emitted is given by

$$E = E_1 - E_2 = h$$

where

E_1 = energy of electron at original level
E_2 = energy of electron at final level
h = Planck's constant = 6.63×10^{-34} J·s
c = velocity of light = 3×10^{8} m·s^{-1}
υ = frequency of radiation in Hertz = c/λ
λ = wavelength of radiation, usually in nm

5.1.2 Beer–Lambert's Law

When a beam of light passes through a solution, a part of it is absorbed by the components of that solution. In addition, certain portions (wavelengths) of the light beam are selectively absorbed. The mathematical relationship between the concentration of a substance and the absorption of light is provided by the Beer–Lambert, or Lambert–Beer, law. The basis of this law comes from the work of Lambert on the transmission of monochromatic light by homogeneous solid substances. Beer applied the law to solutions and found that both the concentration and thickness of the solution affect light transmission through it.

Lambert's law may be expressed as follows:

$$\log\left(\frac{I_o}{I}\right) = K_1 b \tag{5.1}$$

where

I_o = the original intensity of the light beam
I = the intensity of the beam after passing through the homogeneous substance
h = the thickness of the layer of solution through which the light has passed (usually expressed in centimeters)
K_1 = proportionality constant, the value of which depends on the absorption characteristics of each compound, units of thickness, temperature, and wavelength

Beer's law may be expressed as:

$$\log\left(\frac{I_o}{I}\right) = K_2 c \tag{5.2}$$

where

I_o = the original intensity of the light beam
I = intensity of the beam passing through solution
c = the concentration of the solution in moles per liter
K_2 is similar to K_1

Combining the above two formulas, we get

$$\log\left(\frac{I_o}{I}\right) = Kcb \tag{5.3}$$

The log quantity above is called the optical density (OD), whereas K is called the extinction coefficient and may be written as E for convenience. It is specific to a certain wavelength and is large when a particular wavelength is absorbed efficiently by a compound. It is possible to use l instead of b to obtain

$$\text{OD} = Ecl \tag{5.4}$$

In short, the quantity and quality of light absorbed or transmitted by a solution depends on the absorption characteristics of the solute and solvent and is directly proportional to the concentration of the solution and the thickness of the layer used (length of the light path). The Beer–Lambert law may, therefore, be used to determine the extinction coefficient of substances in a solution and the concentration of one or more such substances in a specific solution. Examples 5.1 and 5.2 on next page will illustrate this point.

Example 5.1

Given the following absorption data for two compounds, A and B, at the indicated wavelengths, calculate the molar extinction coefficients for the two compounds at the two wavelengths. The concentration of A in both cases is 1.5×10^{-4} mol/L, and the concentration of B in both cases is 8×10^{-4} mol/L. In all cases, the light path is 1 cm.

CALCULATIONS

Compound A		Compound B	
Wavelength (μm)	**OD**	**Wavelength (μm)**	**OD**
400	0.55	400	0.07
600	0.04	600	0.75

Compound A

At 400 μm:

$$0.55 = E(1.5 \times 10^{-4})(1.0) \tag{5.5}$$

$$0.55 = E(1.5 \times 10^{-4}) \tag{5.6}$$

$$E = \frac{0.55}{1.5 \times 10^{-4}} = 3667 \tag{5.7}$$

where 0.55 is the OD, E is unknown, 1.5×10^{-4} is the concentration c in molar, and 1.0 is b or l in centimeters.

At 600 μm:

$$0.04 = E(1.5 \times 10^{-4})(1.0) \tag{5.8}$$

$$0.04 = E(1.5 \times 10^{-4}) \tag{5.9}$$

$$E = \frac{10.04}{1.5 \times 10^{-4}} = 267 \tag{5.10}$$

where 0.04 is the OD, E is unknown, 1.5×10^{-4} is as above, and 1.0 is also as above.

Compound B

At 400 μm:

$$0.07 = E(8.0 \times 10^{-4})(1.0) \tag{5.11}$$

$$0.07 \times E(8.0 \times 10^{-4}) \tag{5.12}$$

$$E = \frac{0.07}{8.0 \times 10^{-4}} = 87.5 \tag{5.13}$$

where 0.07 is the OD, E is unknown, 8.0×10^{-4} is the concentration c in molar, and 1.0 is as above.

At 600 μm:

$$0.75 = E(8.0 \times 10^{-4})(1.0) \tag{5.14}$$

$$0.75 = E(8.0 \times 10^{-4}) \tag{5.15}$$

$$E = \frac{0.07}{8.0 \times 10^{-4}} = 87.5 \tag{5.16}$$

where 0.75 is the OD, E is unknown, 8.0×10^{-4} is as above, and 1.0 is also as above.

Example 5.2

Using the molar extinction coefficients obtained above, calculate the concentration of A and B in a mixture of the two, which gives the following data: The light path is 1 cm.

- OD at 400 µm is 0.349
- OD at 600 µm is 0.7324

Suppose that:

- Concentration of A = a
- Concentration of B = b

Then at 400 µm:

$$\text{OD (A)} = (3667)(a)(1.0)$$

$$\text{OD (B)} = (87.5)(b)(1.0)$$

$$\text{OD (mixture A} + \text{B)} = \text{OD (A)} + \text{OD (B)} \tag{5.17}$$

Therefore,

$$0.349 = 3667a + 87.5b \tag{5.18}$$

and at 600 µm:

$$\text{OD (A)} = (267)(a)(1.0)$$

$$\text{OD (B)} = (938)(b)(1.0)$$

$$\text{OD (mixture A} + \text{B)} = \text{OD (A)} + \text{OD (B)} \tag{5.19}$$

Therefore,

$$0.7324 = 267a + 938b \tag{5.20}$$

Simultaneously solving Equations 5.18 through 5.20:

$$a = 0.000024 \text{ mol/L} = (2.4 \times 10^{-5}\text{M}) \tag{5.21}$$

$$b = 0.00077 \text{ mol/L} = (7.7 \times 10^{-4}\text{M}) \tag{5.22}$$

A spectrum is a plot of absorption or emission versus wavelength. The spectra of atoms are line spectra since the electrons are present in discrete energy levels. Electrons are present in discrete energy levels in molecules as well (Figure 5.2). However, a group of molecules exists in a number of different vibrational and rotational states, each state differing from another by

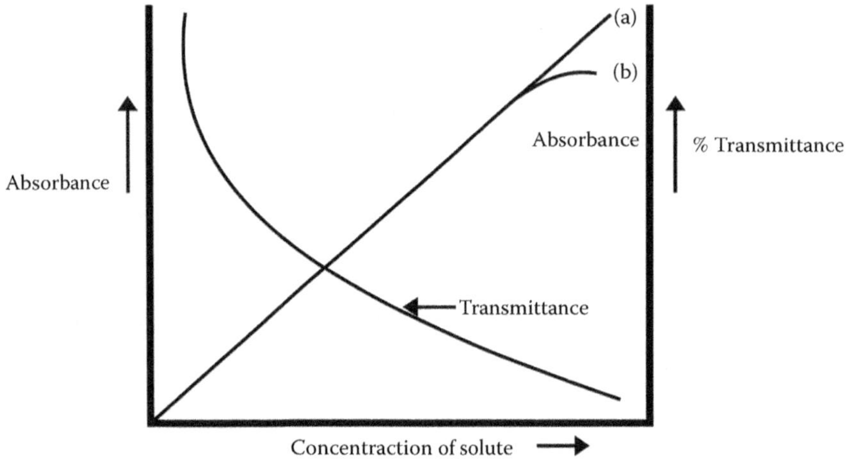

FIGURE 5.2 (a) Beer's law plot and (b) deviation from Beer's law.

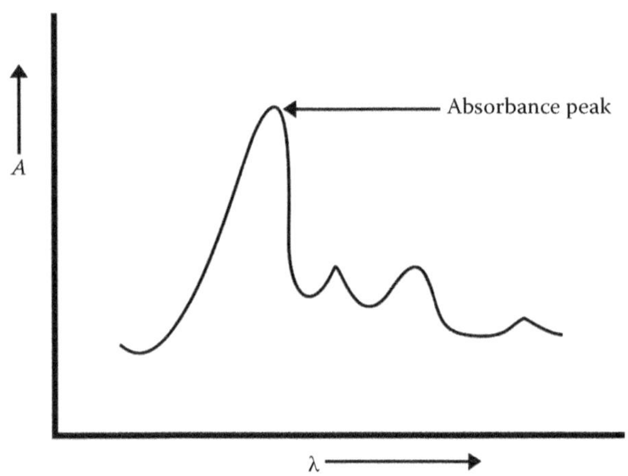

FIGURE 5.3 An arbitrary absorption spectrum showing an absorption peak.

a relatively small amount of energy. Thus, a group of molecules absorbs energy over a small range to give rise to an absorption band over certain wavelength ranges, and the absorbance rises to a maximum and the transmittance falls to a minimum (Figure 5.3). This is referred to as an absorption peak.

In quantitative analysis, the values of the absorption peak are used so that the measurements have the greatest sensitivity with respect to solute concentration. Also, the value obtained is the least sensitive to errors in the wavelength setting of the spectrophotometer.

5.1.3 Mechanics of Measurement

The basic requirements of a spectrophotometer are as follows:

Source: The source provides radiation in the range in which the absorbing species will absorb or is expected to absorb. A tungsten lamp is used to obtain light in the visible range. In the near-UV region, a hydrogen lamp, deuterium lamp, or xenon lamp may be used. In the IR region either a Nernst glower, which is a hollow rod of yttrium and zirconium oxides heated to about 1450°C, or the globar, which is rod of silicon carbide heated to about 1200°C, may be used.

Monochromator: This allows only one particular wavelength of light to pass through and be incident on the solution. These are optical filters, diffraction gratings, or both.

Absorption cell: The solution containing the absorbing species is placed in this absorption cell.

Detector: The detector measures absorbance, and it is made to compare the intensity of light transmitted by a cell containing a solvent (I) with that transmitted by a solution containing the absorbing species (I_0). In the visible and UV regions, photoelectric detectors are used. In the IR region, thermal detectors like bolometers are used.

Meter: This is calibrated in terms of absorbance and transmittance.

Absorbance (A) = log I_0/I

Transmittance (T) = I/I_0

[Transmittance is given as a percentage (0–100%). The range of absorbance commonly recorded is 0–2.0. The most accurate range for measurement is 0.1–0.8 (20–85% T).]

5.2 UV–VISIBLE SPECTROSCOPY

5.2.1 Definition

Interaction of EM radiation in the range of 200–400 nm (UV) and 400–700 nm (visible) with matter gives rise to UV and visible spectroscopy, respectively.

5.2.2 Principle

The absorption of light energy by compounds in the visible and UV regions involves the promotion of electrons in the σ, π and ν molecular orbitals to an antibonding orbital which is at a higher energy level. Organic molecules containing several types of molecular orbitals σ and π are called bonding orbitals and are occupied by a pair of electrons in the ground state. There are corresponding σ* (sigma star) and π* (pi star) antibonding orbitals at higher energy levels, which are unoccupied in the ground state. A third type of "ν" orbital or nonbonding orbital also occurs in molecules that contain lone pairs of electrons like oxygen and nitrogen. The electrons here are not directly involved in bonding and are therefore known as nonbonding orbitals.

The electronic transitions (→) that are involved in the UV and visible regions are of the following types, in order of energy associated with them:

$$(\sigma \rightarrow \sigma^*) > (\nu \rightarrow \sigma^*) > (\pi \rightarrow \pi^*) > (\nu \rightarrow \pi^*)$$

The energy associated with σ and σ* transitions is so high that the corresponding λ falls short of the visible range. Hence, fully saturated compounds do not show any significant absorption above 200 nm and are therefore colorless compounds that contain nonbonding electrons on oxygen, nitrogen, sulfur, or halogen atoms, which are capable of showing absorptions owing to ν → σ* transitions involving lower energy.

In unsaturated and delocalized systems such as benzene and porphyrins, the π → π* transition is of sufficiently small energy (long λ) to produce an absorption band in the near-UV (benzene) or visible (porphyrin) range. This increase in absorbing wavelength due to delocalization is called a bathochromic shift, whereas a decrease in delocalization caused, for example, by protonating a ring nitrogen atom causes a hypsochromic shift, which leads to a decrease in absorbing wavelength. Hyperchromic and hypochromic shifts refer to an increase and decrease in absorbance, respectively. Electronic transitions within a molecule may be associated with a given group in the molecule called a chromophore.

5.2.3 Instrumentation

5.2.3.1 Colorimeter

A colorimeter is used to measure the absorbance of particular wavelengths of light by a specific solution. This instrument is most commonly used to determine the concentration of a known solute in a given solution by the application of the Beer–Lambert law, which states that the concentration of a solute is proportional to its absorbance.

The output from a colorimeter may be displayed by an analog or digital meter and may be shown as transmittance (a linear scale from 0 to 100%) or absorbance (a logarithmic scale from zero to infinity). The useful range of the absorbance scale is from 0 to 2, but it is desirable to keep within the range 0–1 because, above 1, the results become unreliable due to the scattering of light. In addition, the output may be sent to a chart recorder, data logger, or computer. Changeable optical filters are used in the colorimeter to select the wavelength of light which the solute absorbs the most, in order to maximize accuracy. The usual wavelength range is from 400 to 700 nm. If it is necessary to operate in the UV range (below 400 nm), then some modifications to the colorimeter are needed. In modern colorimeters, the filament lamp and filters may be replaced by several light-emitting diodes of different colors.

The most important parts of a colorimeter are (Figure 5.4)

- A light source, which is usually an ordinary tungsten lamp
- An aperture, which can be adjusted
- A set of filters in different colors
- A detector that measures the light that has passed through the solution

5.2.3.1.1 Filters

Different filters are used to select the wavelength of light that the solution absorbs the most. This makes the colorimeter more accurate. Solutions are usually placed in glass or plastic cuvettes. The usual wavelengths used are between 400 and 700 nm. If it is necessary to use UV light below 400 nm, then the lamp and filters must be changed.

5.2.3.1.2 Output

The output of the colorimeter may be shown in graphs or tables by an analog or digital meter. The data may be printed on paper or stored in a computer. It either shows the amount of light that is absorbed by the solution or the amount of light that has passed through the solution.

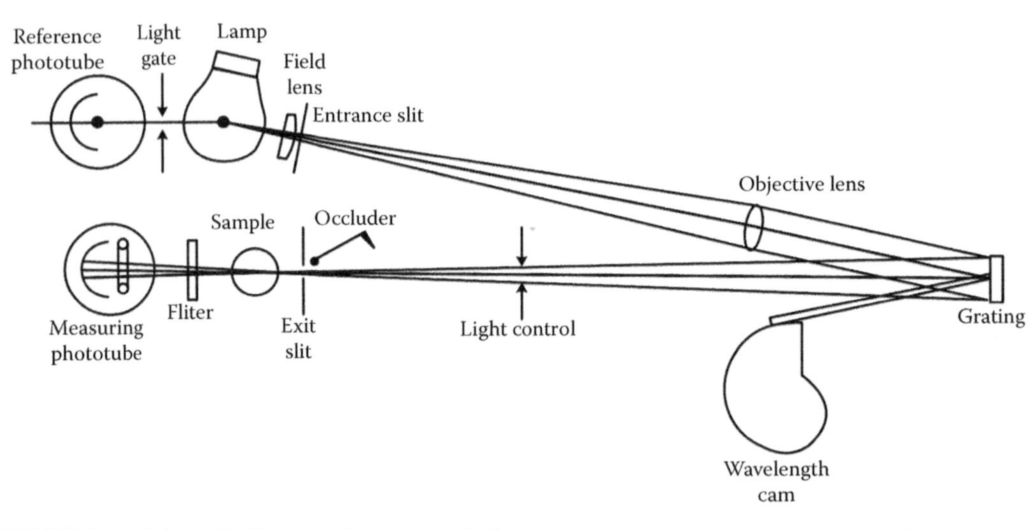

FIGURE 5.4 Schematic diagram of a spectrocolorimeter.

5.2.3.2 Spectronic 20 Spectrocolorimeter

The Spectronic 20 spectrocolorimeter is a routinely used instrument in most laboratories and educational institutions. It is therefore described in some detail.

5.2.3.2.1 Description of the Optical System

White light emanating from the tungsten lamp passes through the entrance slit and is focused by the field lens onto the objective lens. The objective lens focuses an image of the entrance slit at the exit slit after it has been reflected and dispersed by the diffraction grating. To obtain various wavelengths, the grating is rotated by means of an arm which rides on the wavelength cam. In setting the wavelengths, the cam rotates the grating so that the desired wavelength passes through the exit slit. The monochromatic light passes through the exit slit contained in a test tube or cuvette placed in the light path (the light also goes through the red filter in the IR wavelength range) and finally terminates at the measuring phototube, where the light energy is converted to an electric signal. Whenever the sample is removed from the instrument, an occluder automatically falls into the light beam so that the zero may be set without further manipulation. A light control is provided to set 100% transmittance or zero absorbance with a reference or standard solution in the sample compartment. The optics of the lens tube provides an extended range, which goes to 340 nm.

A wall separates the optical system from the electronics, thus preventing dust and dirt from spoiling the efficiency of the optical system. At the same time, the wall shields the meter to prevent erroneous readings and fluctuations from stray light entering through the meter face or permitting error-free operation in bright light or even fluctuating sunlight.

5.2.3.2.2 Description of the Electrical System

The measuring circuit consists of a direct current (DC) wheatstone bridge-type differential amplifier in which variations in characteristics, nonlinearities, and drift of one triode are cancelled by another. A current gain of 5000 V is obtained from this amplifier.

A wheatstone bridge circuit similar to that in the detection–amplification system is used to balance and compare the monitoring output with the constant voltage from the power supply voltage regulator tubes. This differential output is further amplified by a transistor cascade (powered by a silicon hill-wave rectifier circuit), transformed into conductance variation of the two transistors operating on the center-tapped secondary transformer. Thus, a change in transformed load is reflected back to the transformer primary and to the 75 Ω and 50 W resistors, causing readjustment of the primary voltage. The net effect is that an increase or decrease in supply voltage is cancelled out through the feedback system, providing a constant voltage at the primary of the supply transformer. Since this transformer supplies the lamp and amplifier tube filament, the lamp output and filament current are independent of the line voltage. In this way, not only is amplifier stability achieved by a compensating circuit for variation in tube characteristics, but any second-order errors, due to individual tube filament temperature dependence, are removed by keeping the filament at constant operating conditions.

5.2.3.3 Choice of Instruments for Colorimetry

When the absorption characteristics of a compound are unknown, a spectrophotometer or a recording spectrophotometer should be employed. By scanning the compound through the entire light spectrum, an absorption curve is obtained. This curve or some of its portions may be utilized in identifying the compound or in the development of analytical techniques. When a narrow absorption peak is utilized for such purposes, a spectrophotometer should be employed due to its narrow wavelength "window." In instances where peaks are broad or may shift slightly, a colorimeter with its broader "windows" is the more-useful apparatus.

5.2.3.4 UV–Visible Spectrophotometer

Colorimeters are relatively simple instruments, which are designed to function only in the visible range. On the other hand, spectrophotometers are equipped to operate both in the visible and UV ranges. They are fitted with deuterium or hydrogen (for UV light) and tungsten (visible range) as the light sources. Several models of spectrophotometers of varying degrees of sophistication are available. These include single-beam, double-beam, recording, and multibeam instruments. In dual-beam spectrophotometers, the incident light is split into two beams of equal intensity, one of which passes through the reference cuvette and the other through the sample cuvette. The multibeam instrument is designed to simultaneously record absorbance changes at two or more predetermined wavelengths. Recording spectrophotometers can both scan the absorption spectrum of the sample in the desired range and also record the change in absorbance with time at a fixed wavelength. Reflectance spectrophotometers are used for determining the spectra of pastes and suspensions like those of microorganisms. The latter type of spectrophotometers enable measurement of the radiation absorbed when the light beam is reflected by a sample, which is too opaque to allow transmission of light (Figure 5.5).

The spectrophotometer has the following parts (Figure 5.6). The light source provides EM radiation in the UV and visible regions. For UV, a hydrogen or deuterium lamp is used. An ordinary tungsten lamp provides visible radiation as used for the colorimeter. The slit S_1 allows a thin beam of light to pass through and reach the monochromator (Figure 5.6).

A monochromator, as the name suggests, produces light of only one particular wavelength from a multiwavelength source of radiation. The instrument has a provision for selecting the required wavelength by turning a knob located on the exterior of the instrument. Monochromators are usually optical filters, diffraction gratings, or both. Often prisms are used, by which refraction produces light of different wavelengths. Glass prisms are used for visible wavelengths, whereas quartz is used for UV because glass absorbs radiation below 400 nm. The light emerging from any monochromator does not consist of a single wavelength in practice. It consists of a group of wavelengths called bandwidth. Bandwidth is usually defined as twice the half-intensity bandwidth, which is the range of wavelengths for which transmitted intensity is greater than half the intensity of the chosen wavelength. Bandwidth varies from 5 to 35 nm depending on the quality of the instrument.

The slit S_2 allows only a thin beam of bandwidth light to pass through. A sample is placed in a cuvette, which is an optically transparent cell made of glass and quartz for visible and UV light,

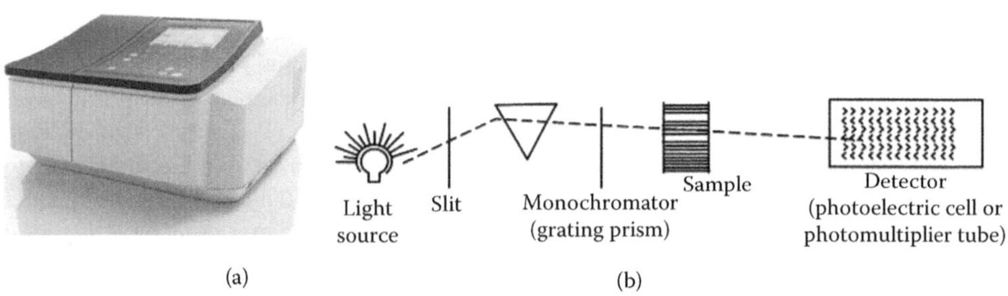

FIGURE 5.5 (a) A spectrophotometer and (b) schematic representation of a spectrophotometer.

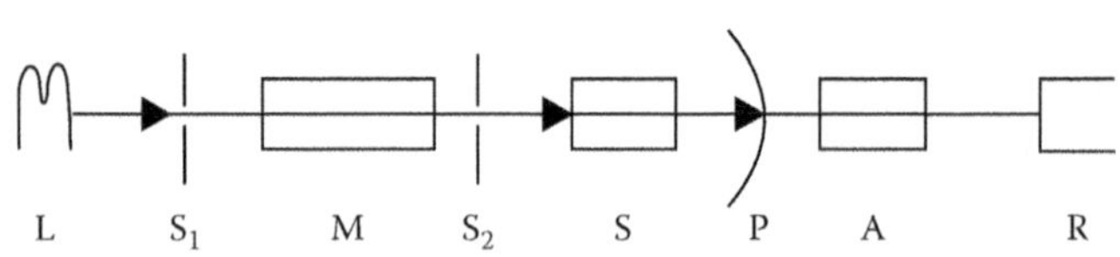

L = Light source
S_1 = Slit
M = Monochromator
S_2 = Slit
S = Sample
P = Photocell
A = Amplifier
R = Recorder

FIGURE 5.6 Parts of a spectrophotometer.

respectively. Commonly used cuvettes have an optical path length of 1 cm. A sample of 2.5–3 mL is required for accurate measurement; there are also microcuvettes requiring a 0.3–0.5 mL sample that are useful when valuable sample and reagents are being used.

The emergent beam from the cuvette reaches the photocell, which converts radiation to electrical energy which is amplified, detected, and recorded. In a photocell, photons impinging on the metal surface in a vacuum cause the emission of electrons. These are attracted by a positive electrode and hence a current flows, which causes a potential difference across a resistor present in the system. This potential difference is recorded on the potentiometer, which is calibrated to read absorbance and transmittance. Absorbance is calibrated as log I_0/I, whereas transmittance is calibrated as I/I_0, where I_0 and I are the intensities of emergent radiation from solution and pure solvent, respectively. The corresponding current produced is recorded. Photomultiplier tubes are more sensitive than simple photocells. In photomultiplier tubes, the emitted electrons are accelerated by high potential and produce secondary electrons by collision with gas molecules present in the tube. This results in a large current and thus very small changes can be measured.

5.2.4 Applications

UV-Visible Spectrophotometer is mainly used to provide valuable information on

- Concentration measurement
- Growth kinetics
- Structural studies
- Enzyme kinetics
- Effect of pH, ionic strength, and so on using difference spectroscopy
- Testing purity and homogeneity of sample
- Identification of compounds such as nucleic acid, carbohydrate, lipid, protein, organic acids, pigment, and so on.

5.2.4.1 Concentration Measurement

This is by far the most important application of this technique. Concentration can be calculated by using the equation $A = Ecl$ at a specific wavelength (λ maxima) for particular compound since l is also known (1 cm). However, it is normal to construct a calibration or standard curve at the time the samples are being analyzed and also at the same conditions. To construct a calibration curve, the absorbance of known concentrations of the substance is read on the spectrophotometer and a graph of concentration versus absorbance is plotted. Concentration of test samples may then be simply read off the graph after the measurement of absorbance values. The calibration curve should, however, embrace all values of concentration to be measured and should be measured under exactly the same conditions as the samples.

For a mixture of chromophore, quantitative measurement of the constituent chromophore is possible, provided they exhibit different absorbance at certain wavelengths. For example, if A_1 and A_2 are absorbance values at two wavelengths of two chromophore whose molar concentrations are given by [B] and [C] and molar absorption coefficients are e_1^{B}, e_2^{B} and e_1^{C}, e_2^{C} at wavelengths I_1 and I_2 then the concentration values [B] and [C] can be solved using the simultaneous equations.

$$A_1 = E_1^{\mathrm{B}}[\mathrm{B}] + E_1^{\mathrm{C}}[\mathrm{C}]$$

$$A_1 = E_2^{\mathrm{B}}[\mathrm{B}] + E_2^{\mathrm{C}}[\mathrm{C}]$$

There are certain naturally occurring chromophores, such as egg proteins (280 nm), nucleic acids (260 nm), carotenoids (455 nm), and tetrapyrroles (400 nm). However, for species that do not absorb in the visible region, a derivative is used. They are made to react quantitatively with some other reagent, which after reaction gives a colored derivative whose concentration is equal to that of the original species, for example, amino acids (280 nm) react with ninhydrin to produce a colored complex whose l_{max} = 570 nm. Proteins are treated with Folin Ciocalteau reagent to produce a colored complex whose l_{max} = 650 nm. Absorbance is measured against a reagent blank, which contains all reagents but not the substance to be measured. A calibration curve is constructed and concentrations read off it. This technique for producing a colored derivative is called colorimetry. The only drawback with the technique is that it is destructive. The compound being assayed is destroyed because it is made to form a complex with another compound.

5.2.4.2 Growth Kinetics

In situations where light scattering is the predominant factor causing a loss of intensity, measurements of absorbance are actually measurements of turbidity. This reflects the number of particles per unit volume. This is useful for constructing growth curves for bacteria.

5.2.4.3 Structural Studies

Protein structural studies: The spectrum of a chromophore depends on the polarity of its environment. Change in the polarity of a solvent changes the spectrum of a constituent amino acid chromophore without a change in the conformation of a protein. This is called solvent perturbation. As another example, if denaturation exposes a tyrosine present in an internal (hydrophilic I to an external hydrophobic) environment, the effect of pH, temperature, and ionic strength on protein denaturation may be studied (Figures 5.7 and 5.8).

Nucleic acid structural studies: The absorbance at 260 nm of double-stranded DNA in a solution increases due to denaturation on heating (hyperchromicity). Hypochromicity occurs on renaturation. Thus, the effects of pH, temperature, and ionic strength on the secondary structure of DNA can be studied. Solvent perturbation studies can be made, for example, by replacing normal water with 50% deuterium oxide (DO) in a solution of nucleic acids. Since DO only changes spectral components due to unpaired nucleotides, the fraction of unpaired bases like t-RNA can be estimated (Figure 5.9).

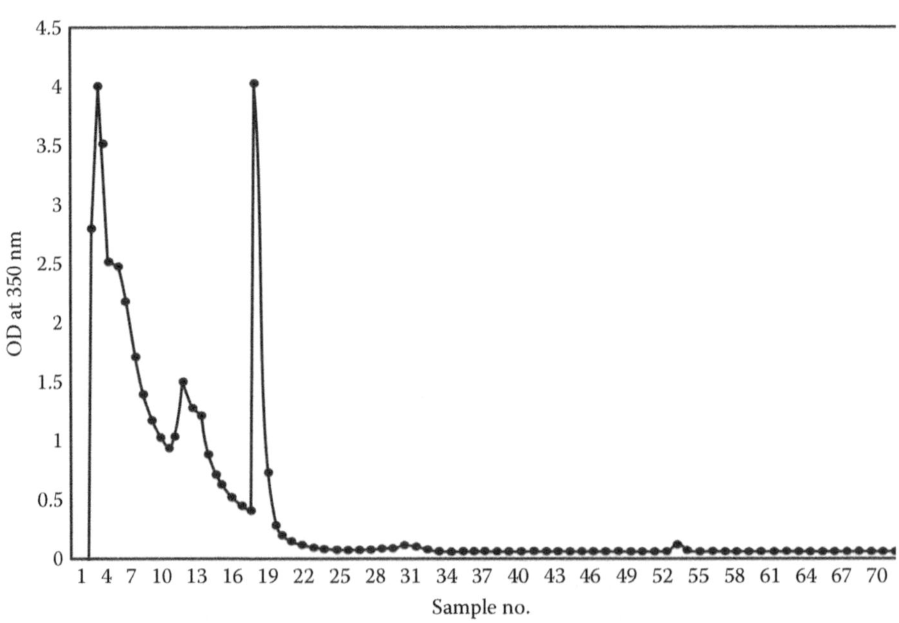

FIGURE 5.7 OD measurement of protein samples during different fractions.

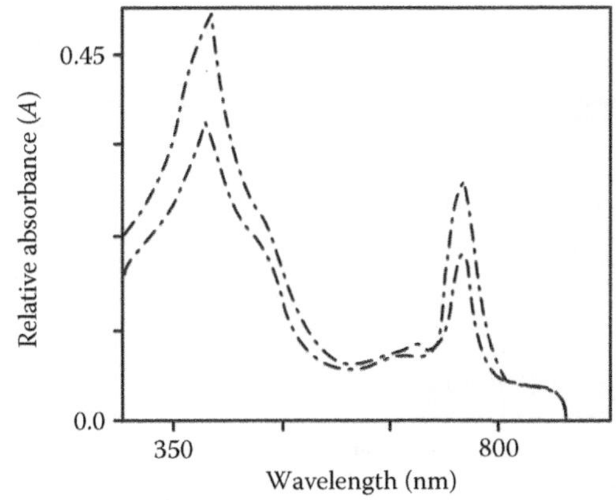

FIGURE 5.8 Absorbance spectra of pigments of cyanobacterium *Anabaena sp.*

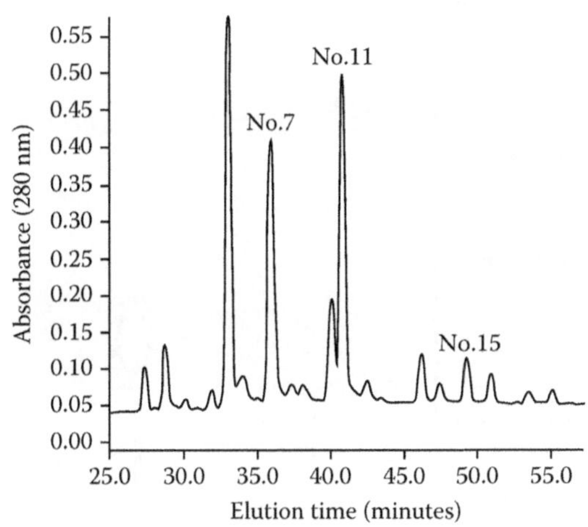

FIGURE 5.9 The elution profile of sample at 280-nm absorbance.

5.2.4.4 Enzyme Kinetics and Assays

These are carried out via the estimation of change in absorbance per unit of time with changes in the concentration of either a substrate or product, for example, the binding of a drug (substrate) to a liver microsomal monoxygenase causes a blue shift of the cytochrome P450 component of the enzyme from 420 to 390 nm.

5.2.4.5 Difference Spectra

These are produced by a double-beam spectrophotometer. Here, there are two sample cells, one for the reference solution and the other for the sample. Two beams of the same wavelength pass through the sample cells and the difference of absorption by the solutions in the two cells is measured. This enables the detection of small changes in absorption.

5.2.4.6 Purity and Homogeneity

The characteristic absorption maxima of different chromophores help in the identification of unknown compounds in both pure state and biological preparations of proteins, nucleic acids, chlorophylls, and so on. The technique may also be used to detect chemical structures and intermediates occurring in a system by comparing them with the spectrum of pure a compound under similar conditions. However, for really precise analysis, IR spectroscopy is required (Figure 5.10).

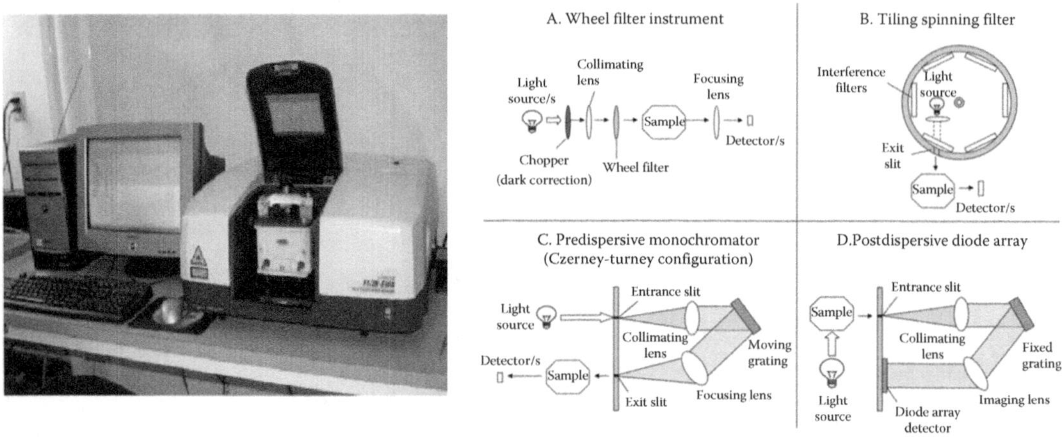

(a)

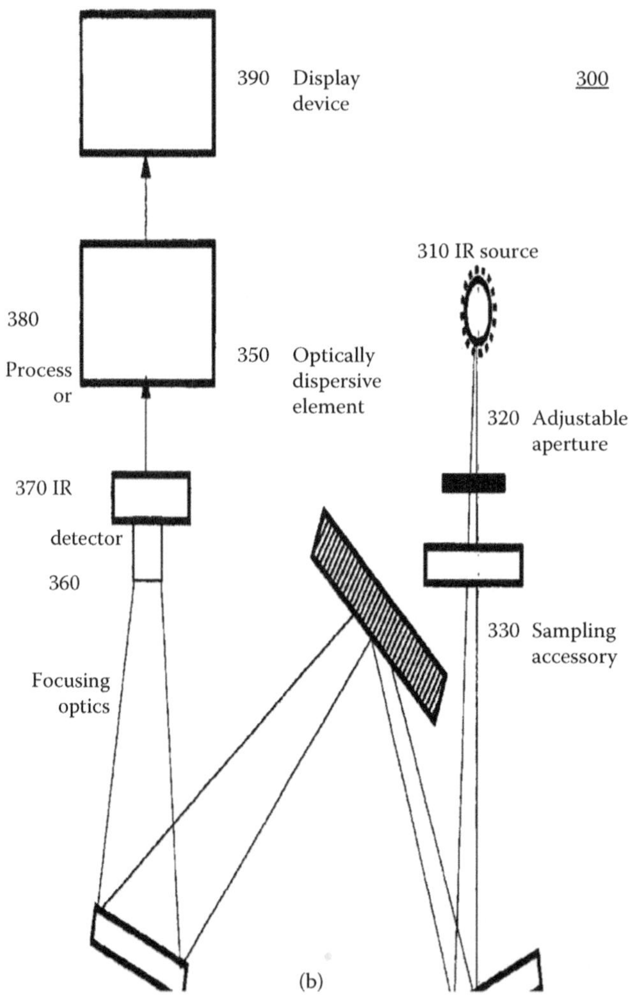

(b)

FIGURE 5.10 (a) External view of an IR spectrophotometer and different internal parts of the instrument depicted from A to C. (b) Schematic operational diagram of IR spectrophotometer.

5.3 IR (VIBRATIONAL) SPECTROSCOPY

5.3.1 Principle

IR spectra originate from different modes of vibration of a molecule. The absorption of IR energy (102–105 nm) by a compound causes the excitation of molecules between vibrational energy levels. The excitation of molecules from the lowest energy vibrational level to the first excited level gives rise to absorption bands that are referred to as fundamental bands. Additional (nonfundamental) absorption bands may occur because of the presence of overtones (or harmonics) of reduced intensity at 1/2, 1/3,…the wavelength (twice, thrice the wavenumber).

The spectral position is given in terms of wavenumber or Hertz because it deals with vibrations (wavenumber = 1/wavelength). When a light wave passes through an atom carrying an electrical charge, it is pushed first to one side and then to another. In a molecule that has a dipole moment (e.g., H–Cl), the electrical field of a light wave tends to set the charges into oscillation by stretching and compressing the HCl bond alternately. The bond has a natural frequency of vibration depending on the masses of the two atoms and the restoring the force of the bond. An incident light wave of the same frequency greatly increases the natural frequency of vibration—resonance is said to occur—and molecules absorb maximum energy at this resonant frequency. This is indicated by an absorption peak.

Bond vibration modes are one of two types:

1. Stretching—periodic stretching of bond along the bond axis
2. Bending or deformation type—displacement at right angles to *b* and the axis (Figure 5.11)

Different vibration modes lead to corresponding energy patterns. For a polyatomic molecule, there are 3–6 modes of vibration of which n–1 are stretching and 2n–5 are of the bending type, where n is the total number of atoms in the molecule. Thus, C–H has 30 fundamental vibrations. However, only those that have transient dipole moments are excited by the radiation. That is to say that in order for a particular vibration to result in the absorption of IR energy, that vibration must cause a change in the dipole moment of the molecule. Thus, the IR spectrum of benzene contains

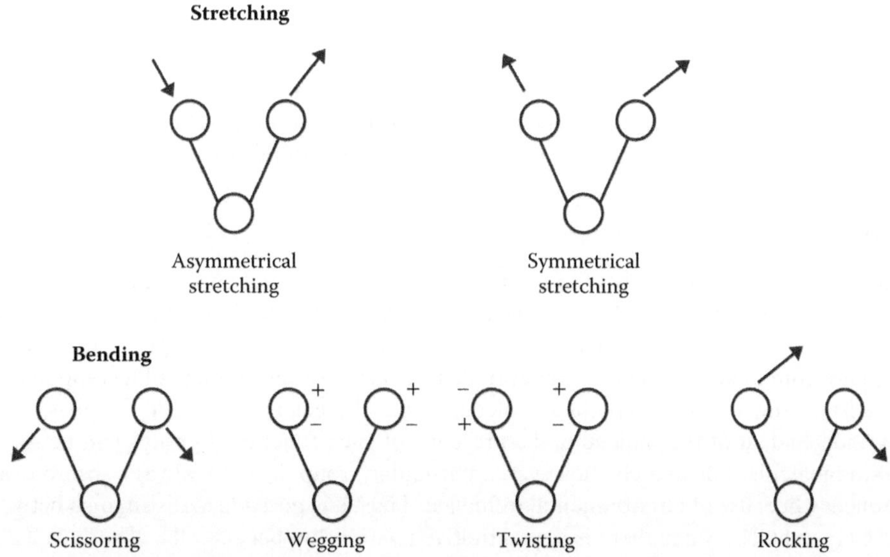

FIGURE 5.11 Bond vibration modes.

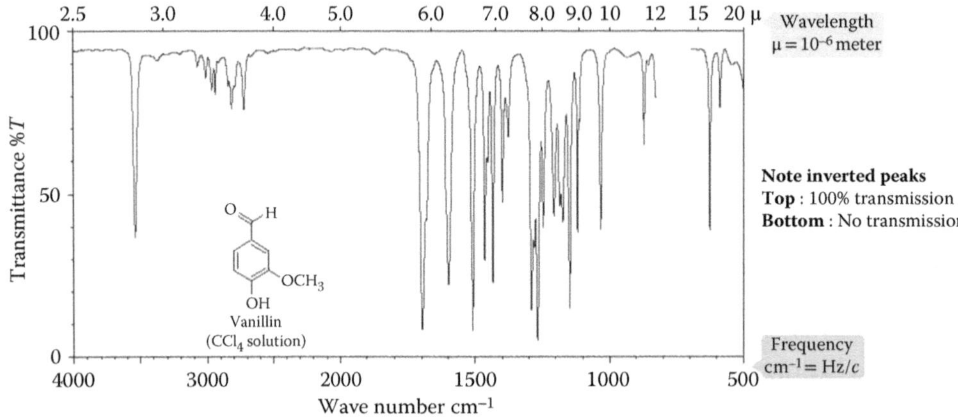

FIGURE 5.12 IR spectrum.

less than 30 absorption peaks (Figure 5.12). Although IR spectrum is characteristic of a compound, the spectra of all molecules contain three common characteristics:

1. In the region 3600–1500 cm, absorption is due to the bonds of the type X–H (e.g., N–H, O–H, C–H).
2. Below 1600 cm^{-1}, absorption is due to the bonds of the type C–C, C–N, C–O, C–halogen.
3. 1300–650 cm^{-1} is characteristic of a molecule (this is called the fingerprint region).

5.3.2 Instrumentation

A common light source for IR radiation is the Nernst glower, as a molded rod containing a mixture of zirconium oxide, yttrium oxide, and erbium oxide that is heated to around 1500°C by electrical means. Either optical prisms or gratings are used to obtain approximately monochromatic light. The beam is split into two beams: one is made to pass through the sample and one through the reference cell. Since glass and quartz absorb IR radiation, metal halides like NaCl are used as containers of absorbing species. About 1 mg of substance and 100–200 mg of alkali halide are ground together finely and pressed under high pressure to a small disk of 1–2 nm in thickness. When equal light intensity is transmitted by both the sample and reference beams, no signal is produced. The absorbance of the sample beam results in the inequality of the two transmitting beams falling on the detector, which produces a pulsating electrical signal. The detector is a bolometer or a thermocouple calibrated to give absorbance or transmittance (Figure 5.10a through c).

5.3.3 Applications

The IR spectrum of a compound is a "fingerprint" of that compound; hence, it can be used to identify a pure compound (Figure 5.13). Compound charts relate molecular structure to absorption bands. Impurities in a compound may be detected by the appearance of extra absorption peaks in the spectrum of a pure compound. IR spectra can even distinguish between isomers of a compound.

Since certain functional groups in a molecule have characteristic natural frequencies that are relatively independent of the molecule, the presence of such functional groups can be ascertained in a certain molecule. Fortunately, however, a particular group does not always absorb exactly the same frequency because of environmental influence. Thus, it is possible to distinguish between C–H bonds of CH_2 and CH_3. Since there is a quantitative relationship between the absorbance and number of absorbing molecules, quantitative chemical analysis is also possible (Figure 5.10d).

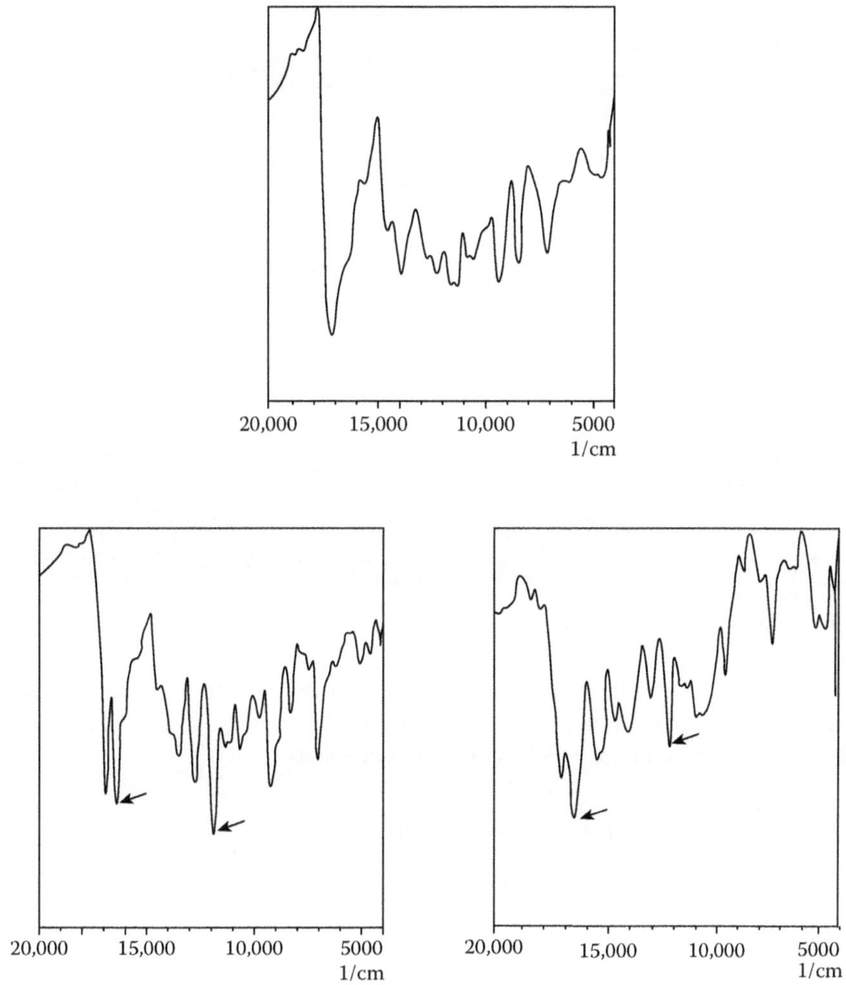

FIGURE 5.13 IR spectroscopic analysis of different preparations of heme with plasma of *Plasmodium yoelii* infected mice. The characteristic peaks of hemozoin are indicated by arrows.

Interfacing IR spectroscopy with gas chromatography is a powerful technique for analyzing drug metabolites. Its most important application is to study carbon dioxide metabolism during photosynthesis and respiration in plants and microorganisms.

5.4 FLAME/ATOMIC ABSORPTION SPECTROSCOPY

5.4.1 Principle

On absorbing energy of the order of 4 × 10 to 4 × 10 kJ·mol, the valence shell electrons of atoms get excited to higher energy levels, giving rise to absorption line spectra. On returning to the ground state, energy given off appears as emission line spectra. Atoms give rise to line spectra and not band spectra–like molecules because electronic transitions take place within discrete energy levels characteristic of that atom. Atomic line spectra are, therefore, said to be fingerprints of atoms and can be used to identify an element. Quantitative estimations are also possible since the amount of radiation absorbed or emitted is proportional to the concentration of the element.

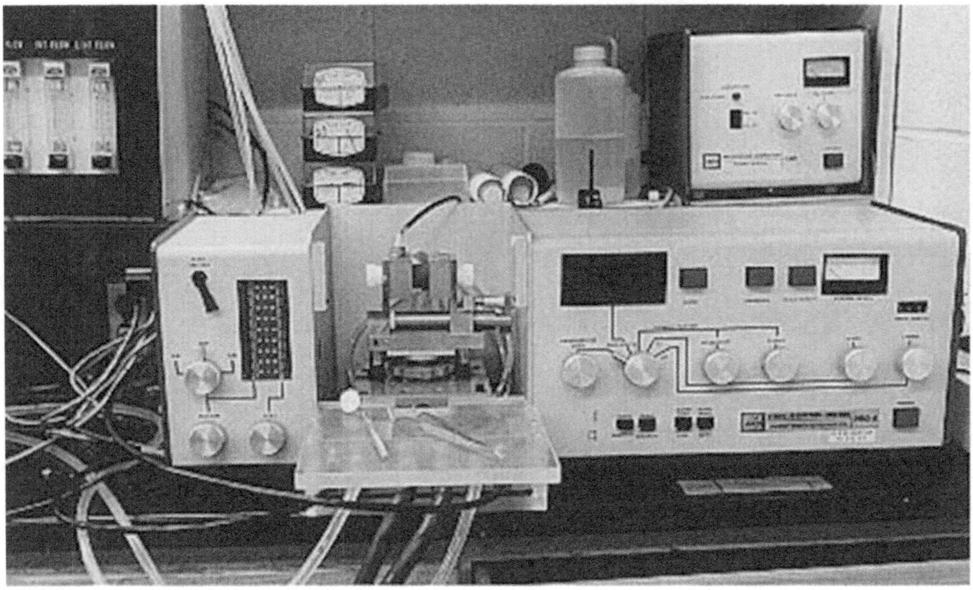

FIGURE 5.14 Atomic absorption spectroscope.

Flame spectroscopy uses a flame to provide energy for the excitation of atoms. The amount of radiation emitted or absorbed is measured as in UV–visible spectroscopy. This is used to quantify the element present in the sample, which is dissolved in water.

5.4.2 Instrumentation for Emission Flame Spectroscopy

The instrumentation needed for emission flame spectroscopy is as follows:

- The aspirator, which by vacuum action sucks up the sample solution and takes it to the nebulizer.
- The nebulizer or atomizer produces small droplets of solution that are sprayed into the flame.
- The flame serves to volatilize and excite every atom present; temperature can be controlled by gas and air pressure controls. The gas used is usually oxy-acetylene flame (2000°C). For higher temperatures of 3500°C, a mixture of oxygen and nitrous oxide is used.
- Emitted light is made to pass through a monochromator that is adjusted to the characteristic wavelength, which will be emitted by the element.
- A photoelectric detector is the same as in UV–visible type.
- The photoelectric detector is connected to a galvanometer calibrated to read concentrations. Alternatively, the instrument is calibrated with known concentrations of solution.

The amount of radiation emitted is proportional to the number of atoms excited by the flame. However, not all atoms present are excited; hence, quantitative measurements are not very precise. The technique of atomic absorption spectroscopy provides more sensitivity and precision (Figure 5.14).

5.4.3 Instrumentation for Atomic Absorption Spectroscopy

The element is excited by means of its characteristic spectra produced from a cathode tube whose cathode is made of the element being assayed. The tube contains an inert gas, usually neon, at low

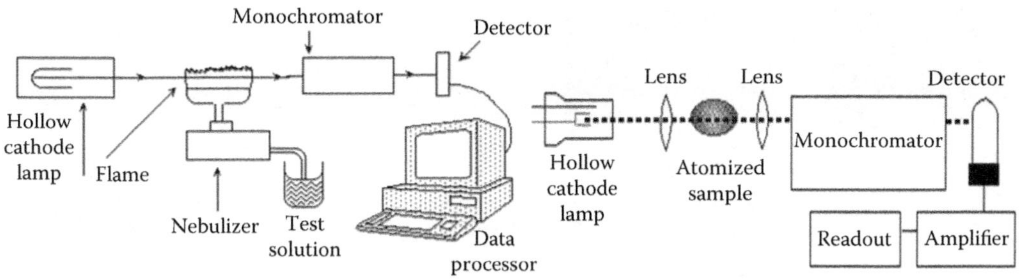

FIGURE 5.15 Functional view of atomic absorption spectroscope.

temperature. A high voltage is used to produce an arc spectrum of the element. The nebulizer, detector, and recorder are the same as in emission spectroscopy. To increase the optical path length of the sample, burners producing 10 cm flame are used (Figure 5.14).

In recent years, the flame has been replaced by electrothermal heating in a graphite furnace. The sample to be analyzed is deposited on a graphite tube in the presence of an inert gas and the temperature raise to 3000°C by electric current. The element in the sample gets volatilized and excited (Figure 5.15).

5.4.4 Applications

Since this technique can detect elements as small as less than 1 ppm, it is widely used in biochemical research for assay of various samples. More than 20 elements can be detected. These include sodium, calcium, potassium, iron, manganese, copper, nickel, chromium, zinc, cadmium, lead, lithium, and silver.

A diagnosis of certain clinical conditions can be made by observing the departure from the usual composition of elements from urine, milk, blood, saliva, and cerebrospinal fluid. In food chemistry, foodstuffs and beverages can be analyzed for the presence of trace elements or contamination by pesticides. When assaying biological samples like cells and tissues, ashing is carried out to remove organic molecules.

5.5 FLUORESCENCE SPECTROSCOPY

5.5.1 Principle

Fluorescence is a phenomenon whereby a molecule, after absorbing the radiation of a particular wavelength, emits radiation of a larger wavelength. This is called the Stokes shift. When the emitted wavelength falls in the visible region, a glow can be seen. Measurement of the intensity of this glow with respect to the intensity of incident radiation is called fluorescence spectroscopy, or simply fluorometry (Figure 5.16).

Absorption and emission are almost instantaneous, with a time lag of only seconds, approximately during which a molecule exists in an excited state. Most organic molecules in their ground state are singlets (paired); on absorbing radiation, they are excited to a higher energy state without a change of spin. These are called excited state singlets. An excited state with the lowest energy is the first excited singlet (Figure 5.17).

Fluorescence occurs when the first excited singlets relax to the ground state. The intensity of fluorescence (I_f) is related to incident radiation (I_0) by

$$I_f = I_0 2.3\mathrm{e}\lambda cdQ$$

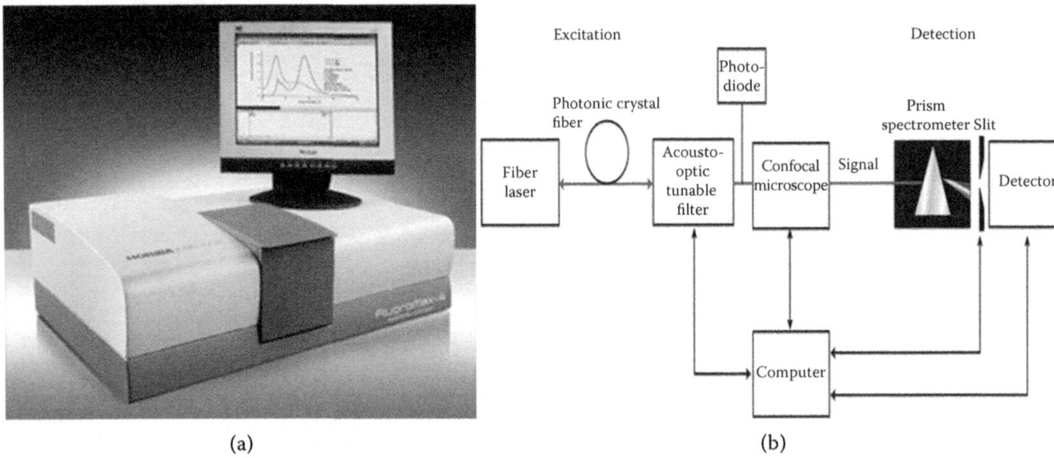

FIGURE 5.16 (a) Fluorescence spectrophotometer and (b) functional view.

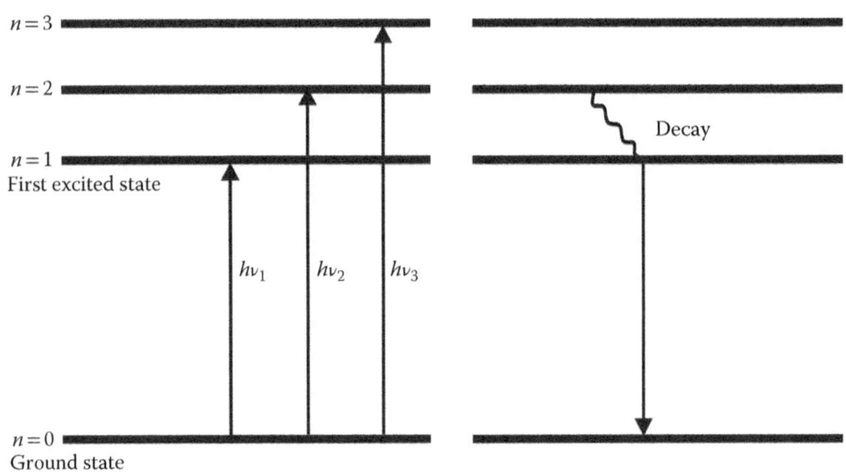

FIGURE 5.17 The emission process.

where

- c = concentration of fluorescing solution (M)
- d = light path in the fluorescing solution (cm)
- ε = molar absorptivity coefficient for the absorbing material at wavelength λ ($dm^3 \cdot mol^{-1} \cdot cm^{-1}$)
- Q = quantum efficiency that is equal to the number of quanta fluoresced divided by the number of quanta absorbed

If the initial absorption generates a higher excited state, this will relax quickly and nonradiatively to the first excited singlet (decay), which may then fluorescence. The nonradiative relaxation mechanisms include:

- Thermal relaxation transfer of energy to molecular and submolecular motions like collision, rotation, and vibration.
- Photochemical reactions. When such processes win the competition with radiative energy loss, quenching of fluorescence is said to occur.

The efficiency of these nonradiative processes depends on the environment of the molecule; therefore, so does fluorescence intensity. Thus, the fluorescence of an emitter is a probe (indicator) of its environment.

5.5.2 Instrumentation

Figures 5.18 and 5.19 show the main components of a spectrofluorimeter. The source is a mercury lamp or xenon arc. The M_1 monochromator is for selecting a chosen wavelength of irradiation. The M_2 monochromator enables determination of the fluorescence spectrum of a specimen. The photocell detector and recorder are the same as in a UV–visible spectrophotometer. The fluorescence from a sample is emitted in all directions but is examined at right angles so that the transmitted light does not interfere.

5.5.3 Pre- and Postfilter Effects

Prefilter absorption reduces the amount of incident radiation reaching fluorescent molecules furthest from the light source, and postfilter effects reduce the amount of fluorescence escaping from the cuvette. Use of (1) microcuvettes (Figure 5.20a) and/or (2) front-face illumination (Figure 5.20b) reduces both pre- and postfilter effects.

5.5.4 Applications

5.5.4.1 Concentration Measurement

The intensity of fluorescence is directly proportional to the concentration of fluorophore the (substance emitting fluorescence) according to the relation

$$I_f = I_0 2.3cIQ$$

that is,

$$I_f \propto c$$

Hence, concentration can be found as in absorption visible spectrophotometry. Since fluorescence has high absolute sensitivity, that is, even very small amounts can be detected, it is far better than absorption visible spectrophotometry at concentrations too low for absorption spectral analysis.

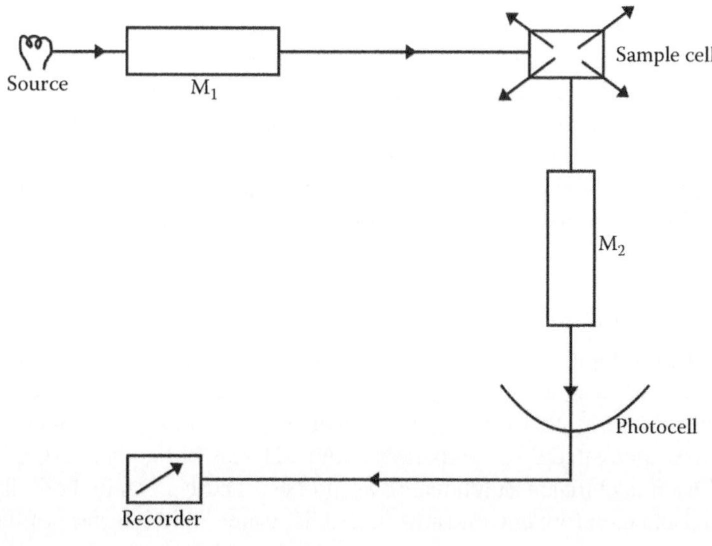

FIGURE 5.18 Main components of a spectrofluorimeter.

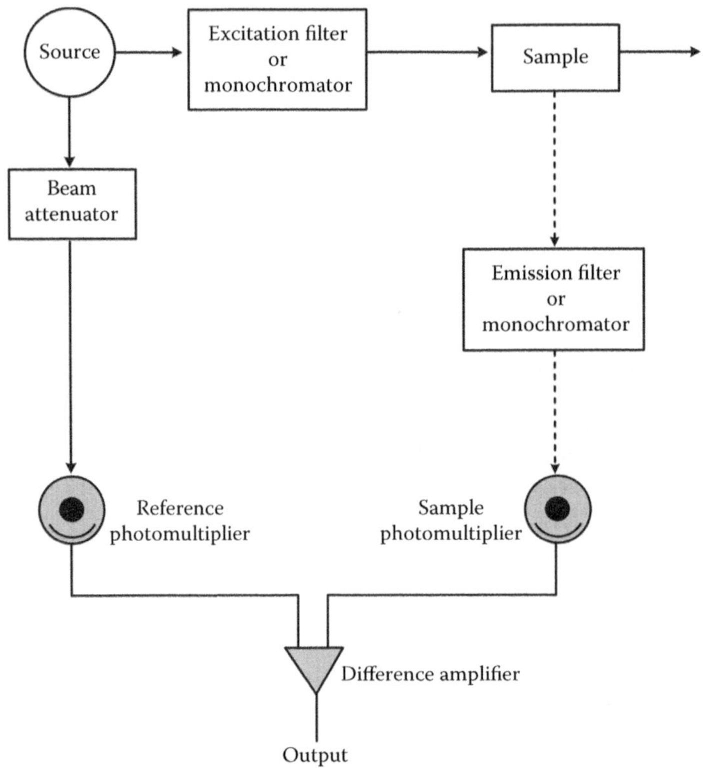

FIGURE 5.19 Functions of spectrophotofluorimeter.

If the sample lacks intrinsic fluorescence, it can be made to bind to a fluorophore or probe and the so-called extrinsic fluorescence can then be measured. Some extrinsic fluorophores include dansyl chloride, 1-anilinonaphthalene-8-sulphonate (ANS), fluorescein, ethidium bromide, and so on.

5.5.4.2 Compound Identification/Excitation Spectrum

The comparison of both the fluorescence and excitation spectra of a compound may help to identify it. An excitation spectrum is obtained by keeping the emission monochromator of the fluorimeter fixed at a particular wavelength and then successively changing the wavelength of the excitation monochromator and recording the photocell output. The spectrum produced is similar to an absorption spectrum, but it has the added advantage that it enables a fluorescent material to be detected and quantified in the presence of a nonfluorescent material that absorbs at the same wavelength. Hence, an absorption spectrum of such a mixture would give overlapping peaks (Figure 5.21a).

5.5.4.3 Kinetic and Structural Studies

Use is made of extrinsic fluorophores to label any biological structure under study (Figure 5.21b). For example, the fluorimetric assay of β-galactosidase enzyme is made using fluorescein di-(β-D-galactopyranoside) as a substrate. Even a single molecule of enzyme can be detected by this process. Membrane structure and effects of temperature and pH can be studied using ANS and MNS (*N*-methyl-2-anilino-6-naphthalenesulphonate) as probes. These contain both hydrophobic and hydrophilic groups and therefore become attached to the water–lipid interface of the membrane.

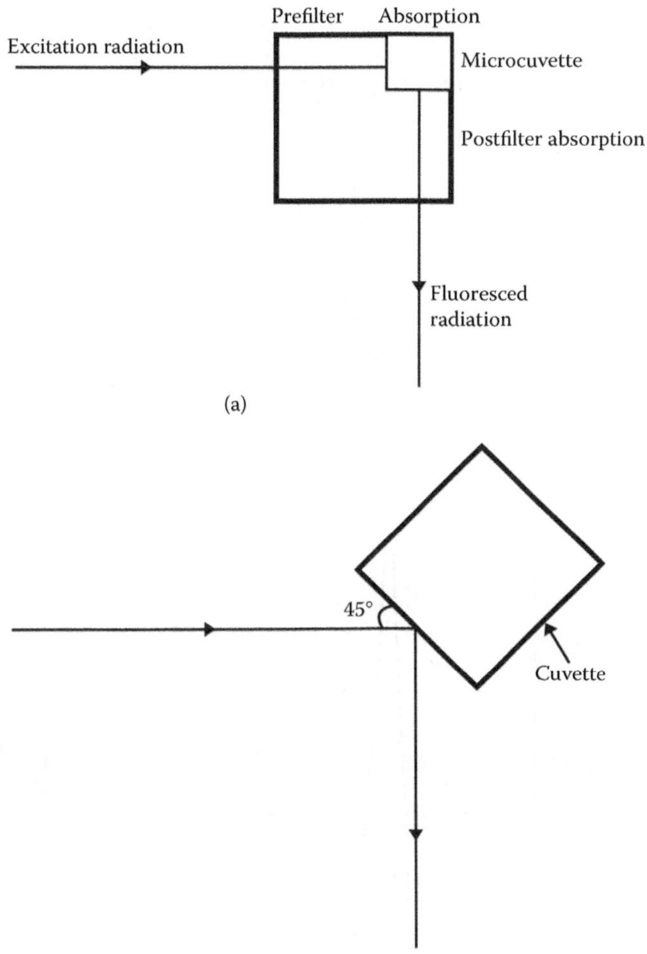

FIGURE 5.20 Reduction of filter effects (a) using microcuvette and (b) using front-face illumination.

5.6 ESR SPECTROSCOPY

5.6.1 PRINCIPLES

ESR is the interaction of an unpaired electron with a microwave field (−1010 Hz). Different from other forms of spectroscopy, energy levels arise from the application of a static magnetic field. Also, only molecules in which there are unpaired electrons can be detected. These include:

- Free radicals like CH_3O, C_2H_5O, and C_6H_5O
- Odd electron molecules such as paramagnetic molecules NO, N_2O, and O
- Paramagnetic ions and complexes such as those of transition metals

An electron not only moves around the nucleus of an atom but also rotates about its own axis either clockwise or counterclockwise, giving rise to a spin quantum number, $S + 1/2$ or $-1/2$, depending on the direction of spin. This motion may be likened to the flow of an electric current through a loop. Such a flow creates a magnetic field. A similar magnetic field is created due to the motion of the electron, which therefore has a magnetic moment. This field can interact with an external magnetic

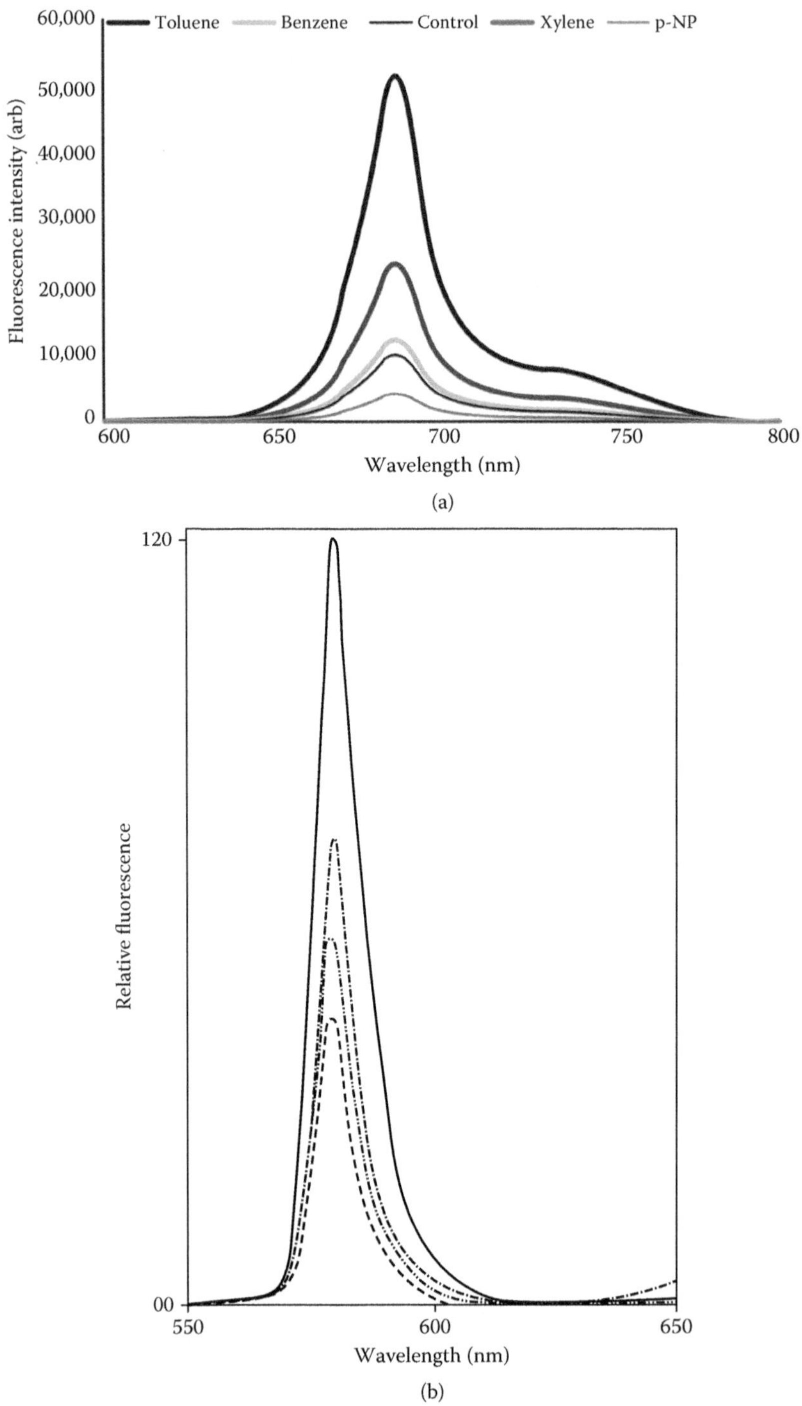

FIGURE 5.21 (a) Laser-induced chlorophyll florescence of *Synechococcus elongates* PCC 7942 under organic stress. (b) Fluorescence spectra of wild type and Het$^-$ Fix$^-$ mutant strain of *Anabaena variabilis* in the presence and absence of nitrate at an excitation wavelength of 435 nm.

field. When an external magnetic field H is applied, the electron is aligned either parallel to it in a low-energy state or antiparallel to it in a high-energy state, depending on the spin of the electron (Figure 5.22). Thus, there are two energy levels after the application of the magnetic field.

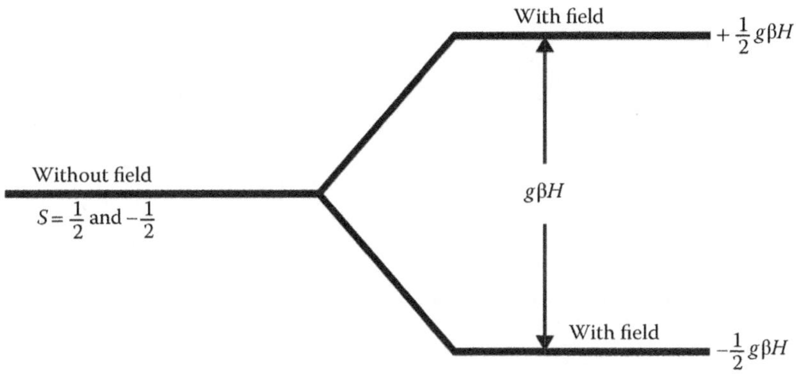

FIGURE 5.22 Energy levels after the application of magnetic field.

The unpaired electron can absorb energy from the microwave region of the EM spectrum and change from a low-energy state (spin parallel to *H*) to a high-energy state (spin antiparallel to *H*). This spin reversal (resonance) occurs if the energy, *E*, absorbed is equal to

$$E = h\upsilon = g\beta H$$

where

h = Planck's constant
υ = frequency of wavelength absorbed
g = constant called spectroscopic splitting factor
H = applied magnetic field
β = magnetic moment of electron called Bohr magneton

From the above relationship, it follows that the frequency of absorbed radiation depends on β and *H*. In practice, however, it is usual to keep β constant and vary *H*. This gives rise to an absorption peak when the magnetic field is enough to cause resonance. Such a peak corresponds to a paramagnetic species in the sample. The area under the peak is a measure of the concentration of that species, which may be quantified if a standard containing a known amount of unpaired electrons is available. In practice, ESR spectra contain many peaks and a fine structure due to hyperfine splitting. This is due to the interaction of the electron with the magnetic nuclei within the molecule.

5.6.2 Instrumentation

The requirements of an ESR spectrophotometer are as follows:

- Klystron source of monochromatic microwave (3×10^{-2} m/9000 MHz) radiation.
- Sample cell. Samples must be in a solid state so biological samples are usually frozen in liquid nitrogen.
- Magnetic field of 50–500 mT surrounding the sample. The magnetic field is generated by electromagnets. An auxiliary sweep of 10–100 mT is also present.
- Detector used to determine the resonance condition when the sample absorbs microwave radiation. The detector is a bolometer or a crystal detector.
- Pen recorder that records d*A*/d*H*, that is, the change in absorption *A* with the change in magnetic field *H*. Therefore, unlike the absorption spectrum, there is a nonsymmetrical peak adjacent to a nonsymmetrical trough and together they are called an ESR line (Figure 5.23), which is characteristic of the molecule in which the unpaired electron is present.

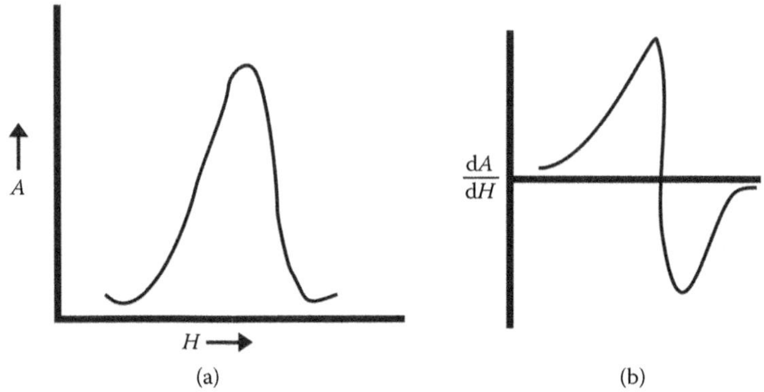

FIGURE 5.23 (a) ESR absorption curve and (b) ESR line.

5.6.3 Applications

Several transition metals constitute prosthetic groups of a number of enzymes and other proteins. Since the transition metals show intense paramagnetism, this technique is particularly useful in studies on metalloproteins such as those containing copper (e.g., cytochrome oxidase), iron (cytochromes, ferredoxin, iron-sulphur proteins, nitrite reductase, etc.), and molebdenum (aldehyde oxidase, xanthine oxidase, nitrate reductase, nitrogenase, sulfite oxidase, etc.). These metal ions possess ESR peaks in one of their oxidation states. Thus by monitoring their ESR signals their role in the activity of such metals containing isolated enzymes can be studied. ESR spectroscopy has also provided useful insight into the functioning of complex multienzymic systems such as mitochondria and chloroplast electron transport systems. The ESR spectroscopic data provides information about the environment around the metal component which contributes towards understanding structure of the molecule.

Since free radicals can be detected by ESR spectroscopy, this technique can also be extended to study macromolecules that do not contain unpaired electrons by using spin-labels. In spin-labeling, a stable and unreactive free radical is attached to the biomolecule. For example, the later movement and the "flip" rates of glycerophosphatides in the lipd bilayers of biomembranes has been examined by spin-labeling glycerophosphatides with nitroxide free radicals. ESR spectroscopy has also been employed quite extensively for investigations on generation of free radicals following irradiation of biological materials.

5.7 NMR SPECTROSCOPY

5.7.1 Principle

NMR is the absorption of energy from a radiofrequency (approximately $\sim 10^8$ HZ) of EM radiation by a system containing unpaired nuclear spins in a strong static magnetic field. The separation of energy levels and hence the frequency of absorption depends on the strength of the magnetic field.

Like electrons spinning about their own axes, protons in a nucleus also spin either clockwise or counterclockwise about their axes. The positive spinning charge gives rise to a magnetic field with a magnetic moment. Pairs of protons have net magnetic moments of zero. However, an odd proton in the nucleus imparts a magnetic moment to the molecule, which can interact with an applied magnetic field. In an applied magnetic field *H*, it can exist either in a low-energy state aligned parallel to the direction of the applied magnetic field or antiparallel to it in a high-energy state. Upon absorbing energy from the radiowave region of the EM region, a proton can change from a low-energy state to a high-energy state, causing resonance to occur and giving rise to NMR, which is therefore also known as proton magnetic resonance (PMR). For reasons already explained, NMR can occur only in atoms containing an odd number of protons, for example, protons like ^{1}H, ^{13}C, ^{15}N, ^{19}F, and ^{31}P.

The resonance condition is given by

$$\upsilon = \mu H / h\mathrm{I}$$

where

υ = frequency of EM radiation absorbed
μ = nuclear magnetic moment
H = magnetic field strength
h = Planck's constant
I = nuclear spin quantum number characteristic of an atom

The above relationship shows that the frequency of radiowaves absorbed during NMR depends on both the atom being studied (as described by I) and the strength of the magnetic field. It is common practice, however, to vary the magnetic field, H, and keep frequency constant in the radiowave region, rather than vice versa.

5.7.2 Instrumentation

The instrumentation needed for NMR spectroscopy is as follows:

- Source is a radiofrequency transmitter to irradiate the sample.
- Sample cell containing a sample dissolved in a solvent that lacks the atom containing the unpaired electron that is going to be analyzed. For example, in PMR, D_2O or $CDCl_3$ is used.
- Electromagnets providing fields of 1–10 T, in conjunction with auxiliary sweep coils to vary the magnetic field over 1–100 mT.
- Radio receiver that serves as a detector of the absorption signal.
- Recorder that plots energy absorbed against the magnetic field strength applied (Figure 5.24).

5.7.3 Chemical Shifts

The nuclear resonance of a particular atom is not always the same in a given applied static magnetic field. This is because the adjacent electron clouds interact with the applied field to give rise to small

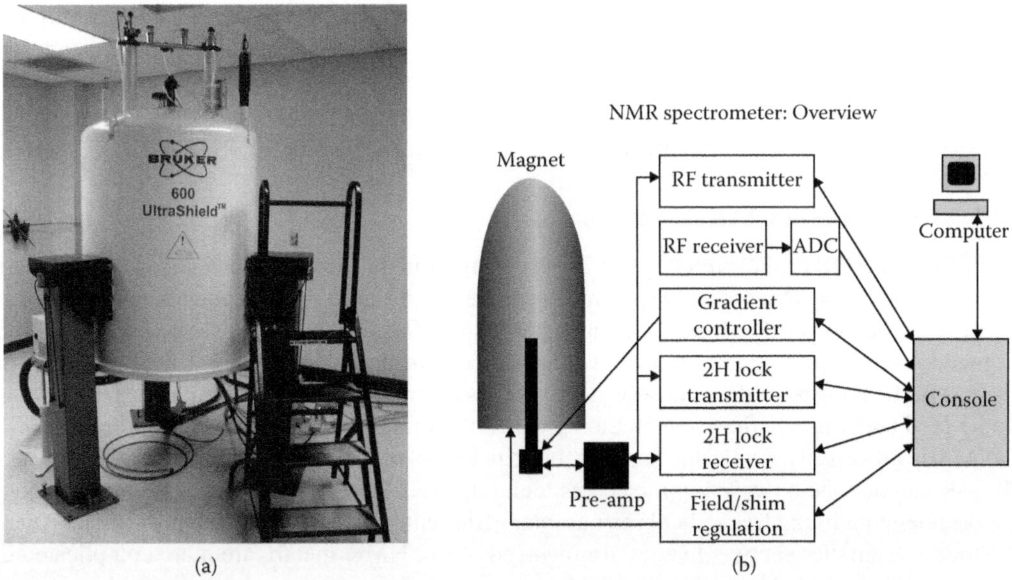

FIGURE 5.24 (a) NMR spectroscope and (b) functional view.

induced magnetic fields that alter the effective magnetic field felt by each nucleus. Thus, the actual field experienced by a proton depends on its molecular environment; for example, the NMR spectra of ethyl alcohol show that it has three different types of protons: (1) those in the CH_3 group, (2) those in the CH_2 group, and (3) those in the OH group. If the induced field opposes the applied field, a higher applied field is required to make the nucleus resonate. Such nuclei are said to be shielded. On the contrary, nuclei are deshielded if the induced field augments the applied field. Such spectral shifts in different structural environments are called chemical shifts. The extent of the chemical shift is measured relative to the PMR of tetramethylsilane (TMS) and is called the tau (τ) value:

$$\tau = 10 - \frac{(\text{frequency difference from TMS})}{\text{instrument frequency in Hertz}} \times 10^6$$

The chemical shift is influenced by

- The electronic configuration around the nucleus and therefore the molecular structure.
- The solvent in which the sample is dissolved. This is because the process of dissolution involves the bonding electrons of the solute and solvent.
- The temperature, in the case of molecules with hydrogen bonding, because it affects the strength of the hydrogen bond.

5.7.4 Applications

NMR spectroscopy is of great importance due to chemical shifts, which make the spectra very precise and the ultimate in structural analysis. Chemical shifts are utilized for studying the following situations: The presence of neighboring aromatic rings can be detected due to abnormal shifts of a particular nucleus. Specific probes can be introduced to cause changes in the shifts making it possible to study the effect of the probes on molecular structure.

In a covalent bond, the electronic magnetic moment is zero because electrons are paired. However, the nuclear magnetic moment causes the electrons to be polarized slightly. This effectively transmits the direction of the spin of one nucleus to another. Such an interaction between like or different spins through the bonding electrons, called spin–spin interactions, cause the splitting of the NMR absorption peak already separated by chemical shifts. This splitting is called hyperfine splitting, and it is used to detect and identify the number and kind of chemical groups, bond angles, and isomers present. These can be extended to biomolecules such as nucleotides, hormones, peptides, and so on. Just like other spectroscopic techniques, NMR spectra is mainly used for studying the structure of molecules, conformational changes in macromolecules, qualitative and quantitative analysis, and at times for kinetic investigations (Figure 5.25).

5.7.5 Comparison of ESR and NMR

Both ESR and NMR provide information about the structure of a molecule at the atomic level, which is not possible by any other form of spectroscopy. The two differ in the instrumentation involved and practical applications. ESR requires a magnetic field of 0.1–1 T, whereas NMR requires a higher magnetic field (10 fold higher) of 1–10 T. ESR absorbs in the microwave region of the spectrum, whereas NMR absorbs in the radiowave region of lesser energy. Both forms are nondestructive and can be used under nearly the same physiological conditions.

NMR is universally applicable since all biological molecules contain protons and most contain ^{31}P. ESR can only be observed in molecules with unpaired electrons, which restricts the technique to molecules containing free radicals on paramagnetic centers. NMR is, however, less sensitive than ESR because smaller energy changes are involved. Also, NMR spectra are more complicated due to chemical shifts and spin–spin interactions. In contrast, ESR spectra are simpler, with only one or two peaks corresponding to paramagnetic centers.

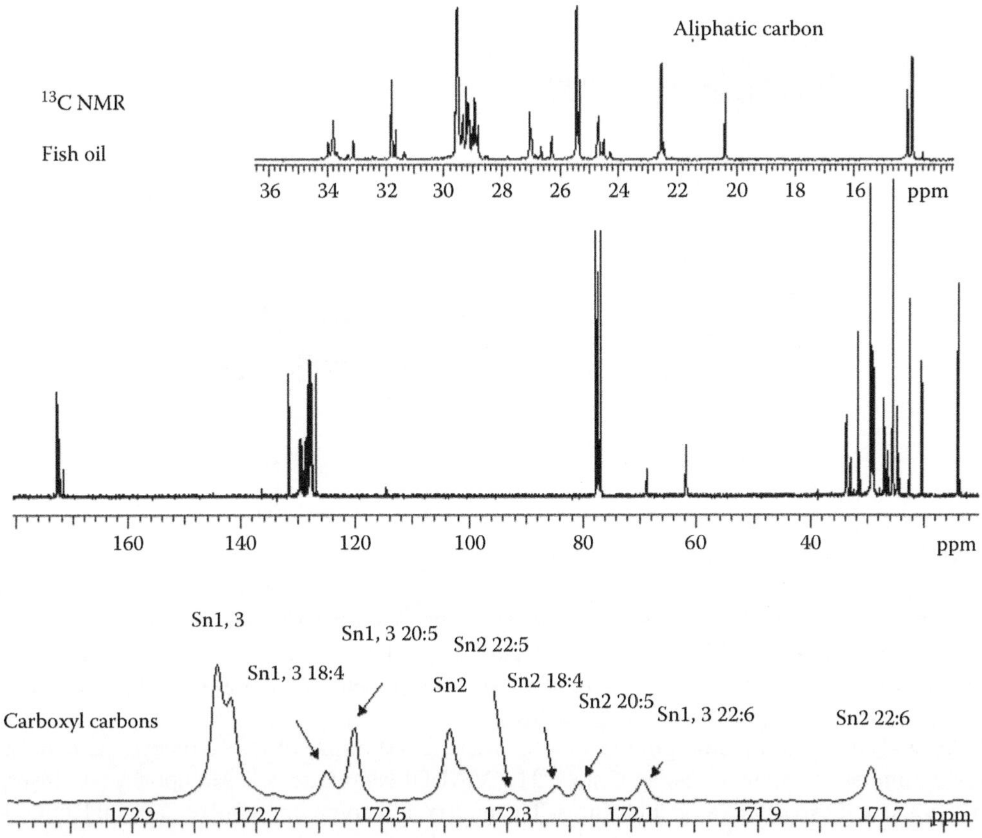

FIGURE 5.25 NMR spectra of fish oil. (From Aursand et al., 2007, *J Agric Food Chem*, 55:38-47.)

5.8 MALDI-TOF MASS SPECTROMETRY

5.8.1 Introduction

The characterization of polymeric materials is vital for predicting and elucidating polymer properties and morphology. Characterization typically involves (1) molecular mass analysis utilizing gel permeation chromatography, light scattering, osmometry, or viscometry, (2) a sequence of repeat units utilizing NMR spectroscopy, (3) end-group analysis utilizing titration, NMR spectroscopy, or Fourier transform infrared (FT-IR) spectroscopy, and (4) purity examination utilizing NMR spectroscopy, elemental analysis, and FT-IR spectroscopy. Until recently, no single technique could completely describe these characteristics of a polymer sample. The powerful capabilities of matrix-assisted laser desorption/ionization time-of-flight (MALDI-TOF) mass spectrometry are realized with the fast and accurate determination of molar masses, sequencing of repeat units, and recognition of polymer additives and impurities.

5.8.2 MALDI-Mass Spectrometry in Chemical Identification

Mass spectrometry (MS) has been appropriately used for the analysis of the molar masses of molecules for the past 50 years. However, the application of MS to large biomolecules and synthetic polymers has been limited due to the low volatility and thermal instability of these materials. These problems have been overcome to a great extent through the development of soft ionization techniques such as chemical ionization, secondary ion mass spectrometry, field desorption, fast

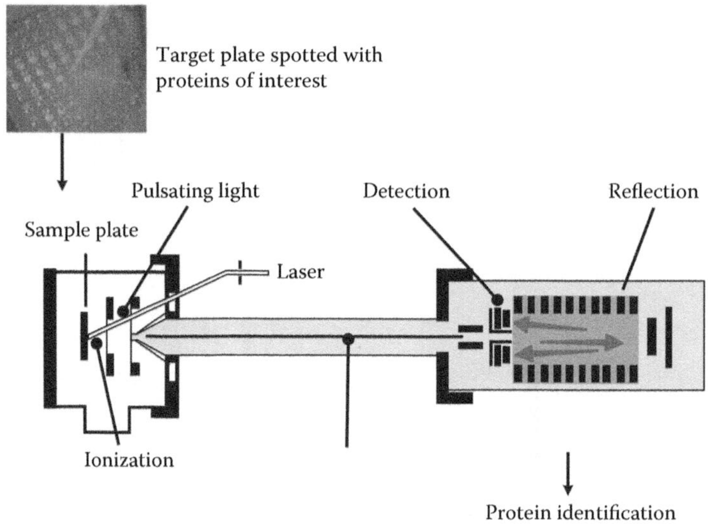

FIGURE 5.26 Functional view of MALDI-TOF.

atom bombardment (FAB), and matrix-assisted laser desorption/ionization mass spectrometry (MALDI-MS). The MALDI-MS technique, in particular, allows for the mass determination of large biomolecules and synthetic polymers with molar masses greater than 200,000 Da by ionization and vaporization without degradation (Figure 5.26).

MALDI-TOF MS is an emerging technique offering a promise of fast and accurate determination of a number of polymer characteristics. The MALDI technique is based upon a UV-absorbing matrix pioneered by Hillenkamp and Karas. The matrix and polymer are mixed at a molecular level in an appropriate solvent with a ~104 molar excess of the matrix. The solvent prevents the aggregation of the polymer. The sample/matrix mixture is placed onto a sample probe tip. Under vacuum conditions, the solvent is removed, leaving cocrystallized polymer molecules homogeneously dispersed within matrix molecules. When the pulsed laser beam is tuned to the appropriate frequency, the energy is transferred to the matrix that is partially vaporized, carrying intact polymer into the vapor phase, and charging the polymer chains. Multiple laser shots are used to improve the signal-to-noise ratio and the peak shapes, which increases the accuracy of the molar mass determination. In the linear time-of-flight (TOF) analyzer (drift region), the distribution of molecules emanating from a sample are imparted with identical translational kinetic energies after being subjected to the same electrical potential energy difference. These ions will then traverse the same distance down an evacuated field-free drift tube; the smaller ions arrive at the detector in a shorter amount of time than the more massive ions.

Separated ion fractions arriving at the end of the drift tube are detected by an appropriate recorder that produces a signal upon the impact of each ion group. The digitized data generated from successive laser shots are summed, yielding a TOF mass spectrum (Figures 5.26 and 5.27). The TOF mass spectrum is a recording of the detector signal as a function of time. The TOF for a molecule of mass m and charge z to travel this distance is proportional to $(m/z)^{1/2}$. This relationship, $t \sim (m/z)^{1/2}$, can be used to calculate the ion's mass. Through the calculation of the ion mass, conversion of the TOF mass spectrum to a conventional mass spectrum of the mass-to-charge axis can be achieved (Figures 5.28 and 5.29).

MALDI is a "soft" ionization technique in which the energy from the laser is spent volatilizing the matrix rather than degrading the polymer. The preparation of an appropriate polymer/matrix mixture is one of the critical limiting factors for the universal application of MALDI to the synthetic polymers. With the advent of MALDI in 1992, the challenge has been to discover appropriate matrix materials for use with synthetic polymers since previous efforts were

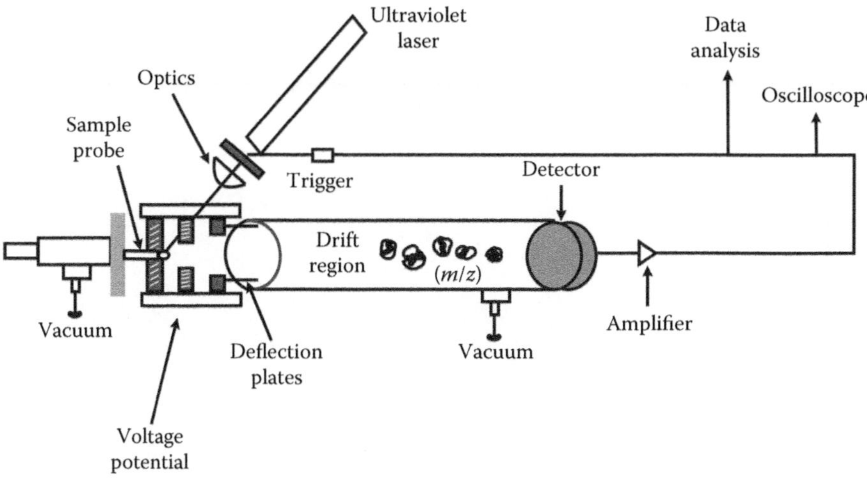

FIGURE 5.27 Schematic of a MALDI-TOF MS.

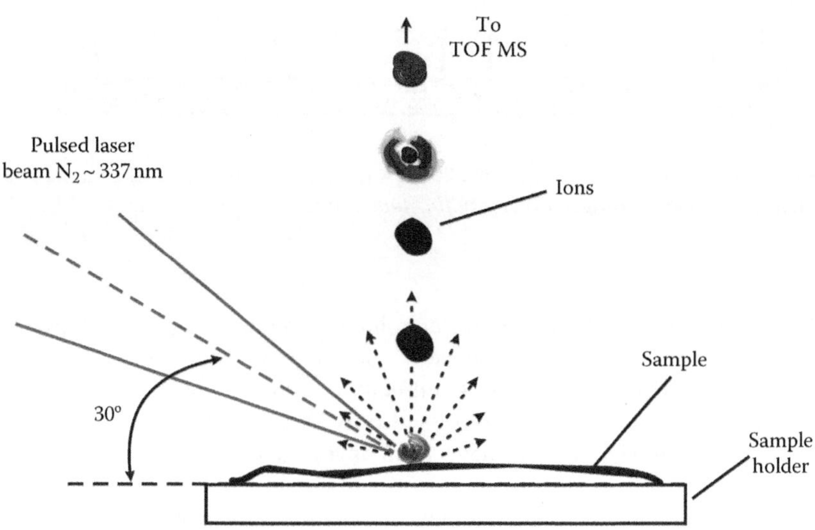

FIGURE 5.28 MALDI-TOF MS sample ionization.

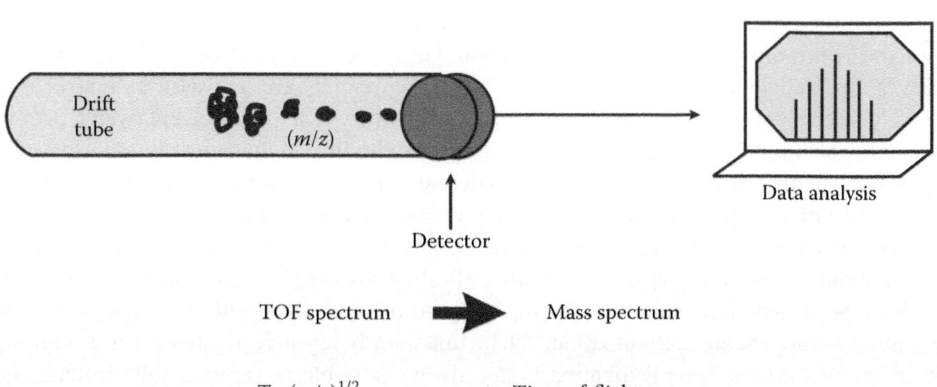

FIGURE 5.29 Conversion from TOF spectra to conventional spectra.

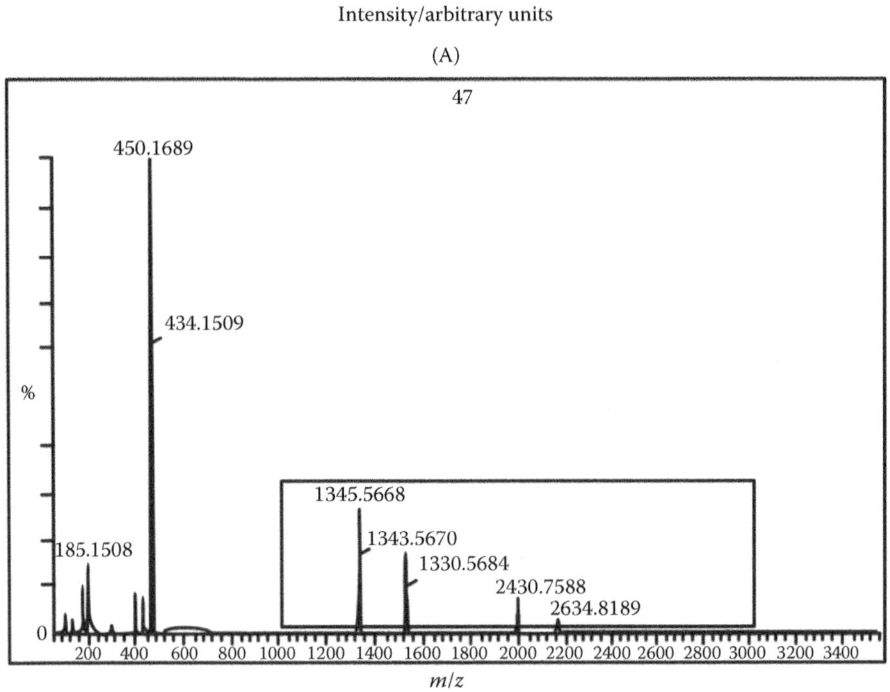

FIGURE 5.30 Peptide mass fingerprinting of a 70 kDA protein spot showing different peptide peaks of different molecular mass from an antigen of *Aspergillus fumigatus*.

centered around biopolymers. Synthetic water-soluble polymers have been shown to be capable of analysis using similar conditions to those of biopolymers. Synthetic, organic-soluble polymers, however, have exhibited analysis complications due to their seeming incompatibility with the matrix materials. Because of this fact, only structurally simplistic, synthetic, water-soluble and organic-soluble polymers have been investigated to date using MALDI analysis (Figures 5.30 and 5.31).

The purpose of the matrix material, as alluded to previously, is twofold: (1) absorption of energy from the laser light, thus preventing polymer decomposition, and (2) isolation of the polymer molecules from one another. Matrices for biopolymers have traditionally utilized just the biopolymer and the matrix material. Synthetic polymers, particularly organic-soluble polymers, have differing solubilities in the common solvents and often do not have large concentrations of ionized species. Most of the commonly used matrices are 2,5-dihydroxybenzoic acid derivatives, sinapinic acid derivatives, and indoleacrylic acid derivatives. Few compounds are as useful as matrix materials due to the numerous stipulations involved: common solubility in a given solvent (water, acetonitrile, ethanol, etc.), absorption, reactivity, and volatility are conditions that must be considered before an appropriate matrix might be found for a particular synthetic polymer. In addition to the matrix material, a cationizing species is often added to increase the concentration of ionized species. Some linear homopolymers and condensation polymers have been shown to yield adequate spectra for analysis without a cationizing species, but often alkaline salts (LiCl, NaCl, KCl) or silver trifluoroacetate have been included as the cationizing agent to increase the yield of cationized species and allow a more homogeneous cationization. Surfactants are being investigated for use with organic-soluble polymers, where homogenization is not always possible or reproducible. Enhancement of spectra is expected where the surfactant can potentially play a dual role as both a matrix emulsifier and a cationization agent.

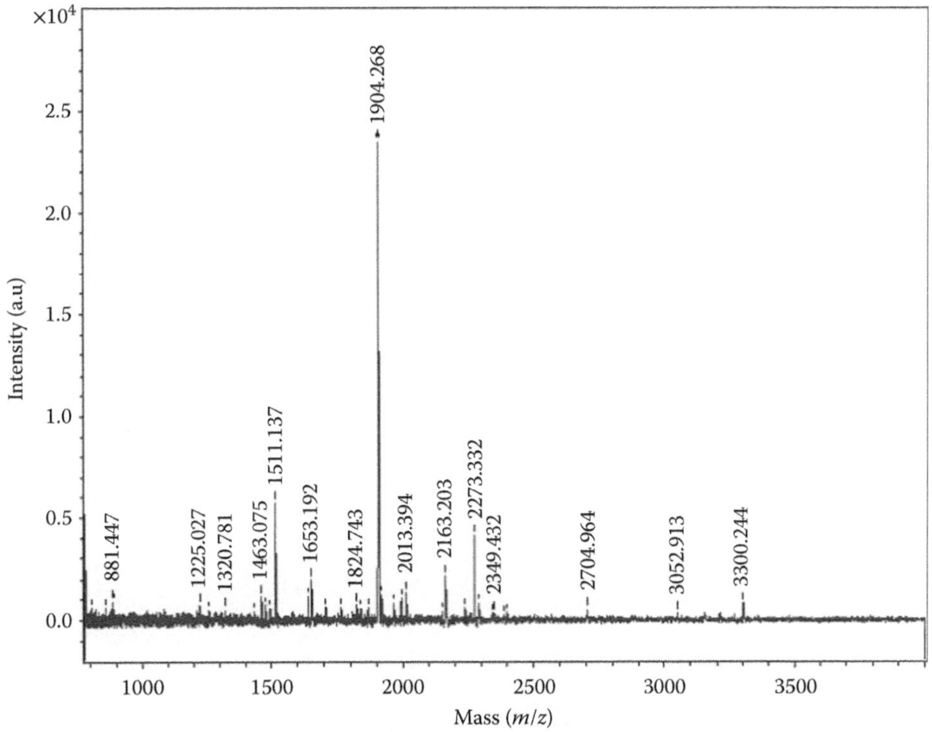

FIGURE 5.31 Representative MALDI-TOF spectrum of Spot 8114 antigen fragment (SWISSPROT accession number Q8SNCO; MHC class II antigen fragment of *Mycobacterium tuberculosis*).

5.8.3 Synthetic Polymer Analysis

The structural investigation of homopolymers using MALDI has been limited to a discrete number of water-soluble and organic-soluble systems. MALDI studies have been shown to be applicable to polymers of a broad range of chemistry, from water-soluble polymers such as poly(ethylene glycol) (PEG), poly(propylene glycol) (PPG), poly(styrene sulfonic acid), and poly(acrylic acid) to organic-soluble polymers such as poly(styrene) and poly(butyl methacrylate). Other MALDI studies in the homopolymer, condensation polymer realm have included fluorinated polymers, polymer blends, and polymer additives. MALDI was originally designed for the analysis of architecturally specific synthetic polymer systems such as copolymers, grafted polymers, living block copolymers, and dendrimers for which no standards exist. Statistical models for the mass spectra of the composition and microstructure of copolymers were also developed. These studies utilized principles of laser desorption, FAB, field ionization, and electron impact, methods in which fragmentation is observable. Actual studies utilizing MALDI analysis with copolymers have had limited investigations. Copolymers under investigation have been poly(butyleneadipate-co-butylenesuccinate) and poly[(N-vinylpyrrolidone)-co-(vinylacetate)]. The results of copolymer studies cannot be compared with other MALDI analyses since MALDI was used strictly as a detector after polydisperse copolymers were segmented with gel permeation chromatography (GPC)-MALDI.

The investigation of synthetic polymers utilizing MALDI techniques included the studies of PPG and PEG. Mixing occurs between the polymer and matrix materials on a molecular level in the solvent followed by homogeneous vacuum cocrystallization. Figure 5.32 shows the resultant spectra for one of the first successful MALDI analyses. The spectra are for a low-molar-mass (5300 g/mol) PPG sample. Due to the low molar mass of the polymer, the molar mass distribution is easily derived as 58 g/mol from the peak-to-peak mass increments. The molar masses obtained through MALDI, listed as Mn, Mw, and Mp, agree quite well with the value specified by the manufacturer of PPG-5300.

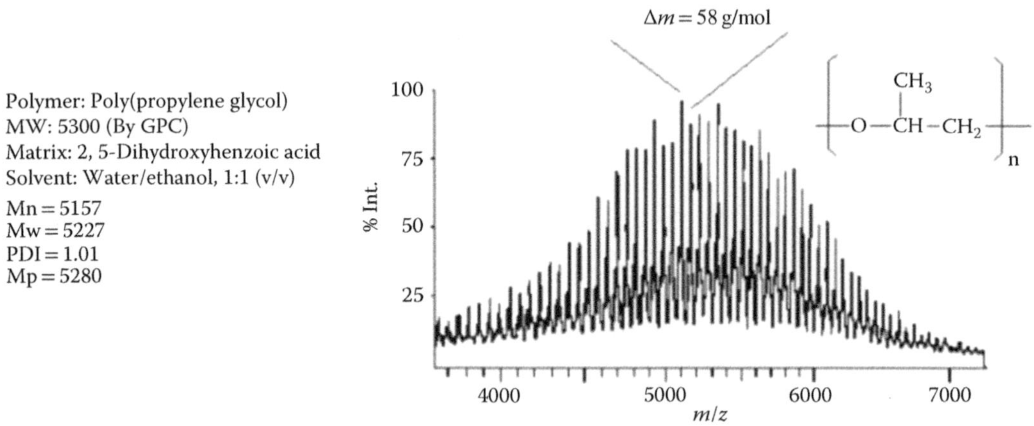

FIGURE 5.32 MALDI-TOF spectra for PPG.

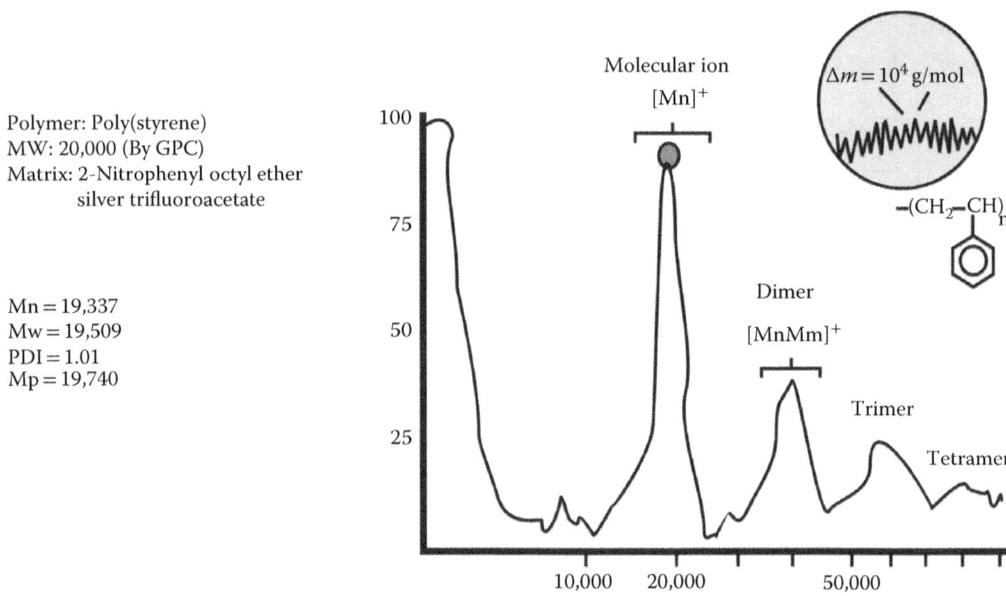

FIGURE 5.33 MALDI-TOF spectra for poly(styrene).

The MALDI technique was investigated for a PEG sample having a higher molar mass of 23,000 (Figure 5.32). At *m/z* 23,000, the mass resolution is not sufficient to resolve adjacent oligomer molecular ions with a difference of 44 mass units. Therefore, the resultant spectra show a convolution of the molecular ions, which results in a continuous distribution. Typically, there is a limit to the peak-to-peak resolving capabilities of MALDI, which has been observed by researchers to be around 20,000 g/mol. However, a MALDI study showed samples of PEG-23,600 g/mol that had the same molecular convolution of the spectrum; they were able to discern the peak-to-peak mass resolution of 44 g/mol due to the improved resolving power of more recently manufactured MALDI instruments. Even with the continuous distribution of the spectra, the data for the PEG-23,000 g/mol sample shows that the polymer distribution is 20,000–25,000 g/mol, possessing a maximum of the distribution at 22,930 and the centroid mass at 22,950. This data still exhibits a good agreement with the manufacturer's value of 23,000. Mixing between the polymer and matrix materials occurs on a molecular level in the solvent followed by homogeneous vacuum cocrystallization. Figure 5.33 shows the resultant spectra for MALDI analysis of poly(styrene).

In the low mass range, only peaks for the matrix ions appear. These spectra have a continuous distribution due to convolution of the molecular ions; however, the molar mass for individual oligomers is apparent, as shown in the inset, with a peak-to-peak mass resolution of 104 g/mol. This sample utilized 2-nitrophenyl octyl ether, a viscous liquid, as the matrix material because using 2,5-dihydroxybenzene matrix resulted in separation of the matrix and polymer. No results can be obtained from sample preparations with inhomogeneity between the matrix and polymer because the polymer is either directly ionized, resulting in the degradation of the sample (fragmentation), or the polymer is not fully ionized, leaving ion levels below the limits of detection. Spectra obtained from studies displayed good homogenization between the matrix and the polymer and do not show the dimer, trimer, or tetramer oligomer ion distributions. Here, the matrix material is possibly better in absorbing energy from the laser light, thus shielding the polymer from over ionization.

5.8.4 Impurity in Oligocarbons

Certain condensation polymers have been investigated using MALDI. Contrary to most polymerization products, condensation polymers tend to possess low molar masses. However, it is the complex polymer structure of most condensation polymers that renders them interesting for MALDI studies. MALDI studies of condensation polymers have included phenolic resins, epoxy resins, and polycarbonates, all of which are important technical products. MALDI samples of the two oligocarbonates, containing the matrix material of dithranol and LiCl and the oligocarbonate, were mixed on a molecular level in a THF solvent solution and homogeneously vacuum cocrystallized. LiCl was added as part of the matrix to increase the formation of cationized species. Spectra A and B in Figure 5.34 are the resultant distributions derived from MALDI-TOF MS analysis.

Spectra A have straightforward molar mass distributions with peak-to-peak mass increments of 254 g/mol, equaling exactly the mass of the repeating unit of bisphenol A-based oligocarbonates. The end-group can be derived through multiple subtraction of the repeat unit from any one of the mass peaks. The end-group of spectra A has, therefore, been derived as 228 g/mol, which corresponds to oligomers with hydroxyl end-groups. Spectra B include the same molar mass distribution with peak-to-peak mass increments of 254 g/mol, equaling exactly the mass of the repeating unit of bisphenol A-based oligocarbonates. However, observance of further oligomeric series in the spectrum is significant because it is representative of inhomogeneity in the sample. The peaks in the denoted oligomer series are offset by 134 g/mol from the major oligomer sample (Figure 5.35). It has been suggested that these oligomers contain cresol, which would be present as a purposefully added chain terminator or an impurity in the reaction mixture, as shown in Figure 5.36.

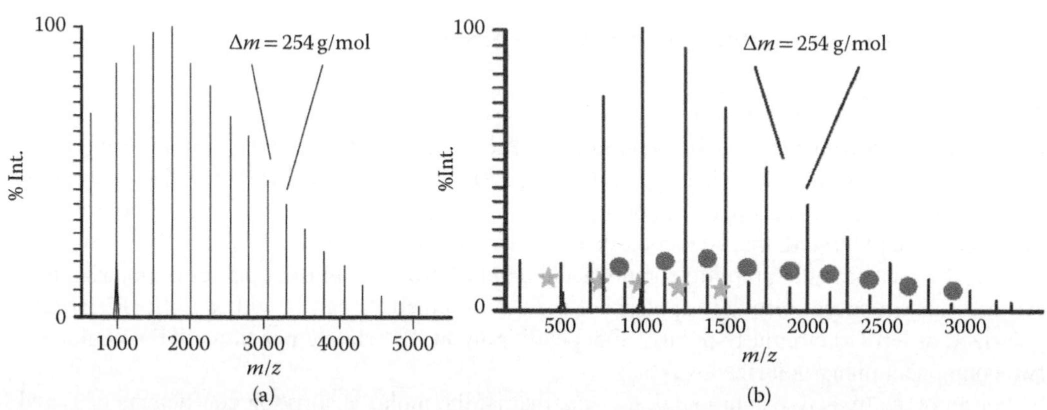

FIGURE 5.34 Oligocarbonate studies utilizing MALDI-TOF MS: (a) molar mass distribution with peak to peak mass increments and (b) mass of repeating unit of bisphenol A-based oligocarbonates.

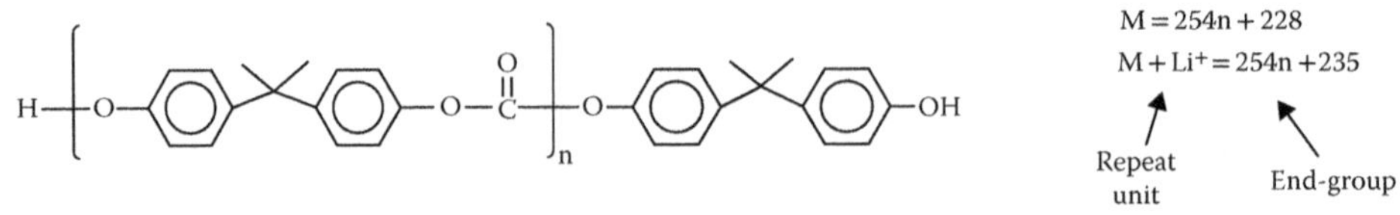

FIGURE 5.35 A pure oligocarbonate sample.

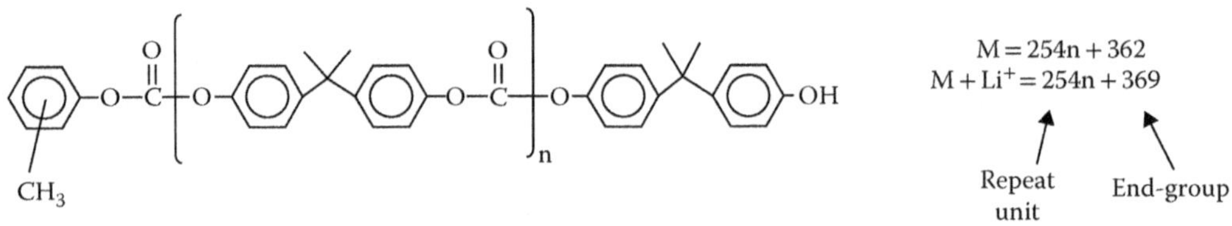

FIGURE 5.36 Determination of an impurity in an oligocarbonate.

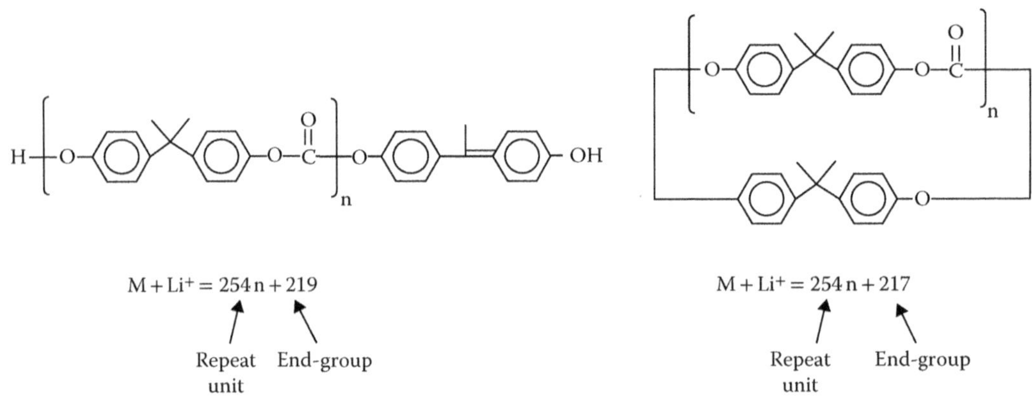

FIGURE 5.37 Speculation of the side reactions in the oligocarbonate.

5.8.5 Side Reaction in Oligocarbons

The peaks in the minor oligomer series denoted by the peaks with the + above them are again indicative of oligocarbonates due to the *m/z* 254 peak-to-peak mass increment. End-group calculations allow for the speculation of the two structures shown in Figure 5.37.

5.9 CIRCULAR DICHROISM (CD) SPECTROSCOPY

5.9.1 Principle

CD spectroscopy measures the differential absorption of right (R) and left (L) circularly polarized light as a function of wavelength. Light consists of EM waves vibrating in all directions perpendicular to the direction of the propagation of light. After passing through a Nicol prism or polaroid, light becomes plane polarized, that is, it consists of waves oscillating only in one plane. When two plane-polarized waves of equal amplitude and wavelength but differing in their planes of polarization by 90° are superimposed, circularly polarized light is obtained. It can be either right (R) circularly polarized or left (L) circularly polarized depending upon the relative positions of the peaks of the two component plane-polarized waves.

For an optically active compound, $\varepsilon_L \neq \varepsilon_R$, that is, the molar absorption coefficients of L and R circularly polarized light are unequal. It may absorb either the L or R circularly polarized light more than the other. Maximum absorption occurs when the electric field vector is parallel to the direction in which there is maximum electronic displacement within the absorbing molecules because of the differential absorption of the R and L circularly polarized light; the resultant light is elliptically polarized. It is this ellipticity (θ) rather than absorbance which is measured as a function of wavelength, and a CD spectrum is obtained.

CD spectroscopy is different from other forms of spectroscopy in the following ways:

- It requires the use of circularly polarized light as incident radiation.
- It gives information about the shape or 3D structure of an optically active molecule.
- Only optically active molecules can be analyzed. This is not a handicap for biological investigations since most biomolecules are optically active; for example, amino acids (generally of L-configuration), nucleotides (containing the sugars D-ribose and D-deoxyribose), and carbohydrates (D and L configurations) are the units from which polypeptides, proteins, nucleic acids, and polysaccharides are built up.

5.9.2 Instrumentation

Figure 5.38 diagrammatically shows the components of a CD spectrophotometer. R and L circularly polarized radiation is produced by passing plane-polarized light in an electro-optic modulator through which an alternating current is passed. Depending on the polarity of the electric field, the R or L component of light is transmitted.

The photomultiplier detector produces a voltage proportional to the ellipticity of polarization of the combined beam falling on it. This is recorded against the wavelength by the recorder.

$$\theta = 2.303\Delta E \frac{180}{4\pi}$$

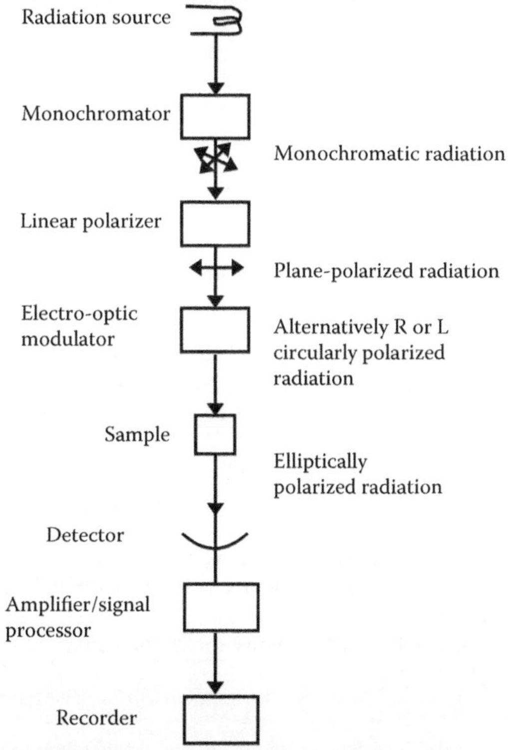

FIGURE 5.38 The main components of a CD spectrophotometer.

where
θ = ellipticity
ΔE = difference in the absorption of R and L waves

5.9.3 Applications

5.9.3.1 Protein Conformation

It is possible to study the 3D secondary structure of proteins. The CD spectra of α-helix, β-conformation, and random coil of poly-L-amino acids are known and can be used for calculating the amount of secondary structure present in a protein.

5.9.3.2 Nucleic Acid Structure

The CD spectrum of a single-stranded nucleic acid can be calculated from its nearest neighbor frequency. Any difference in the measured CD spectrum can be a result of conformational changes in the nucleic acid, that is, double-strandedness. These can therefore be studied in detail with the help of a CD spectrum.

5.9.3.3 Secondary Structures of Proteins

The technique is particularly useful for proteins which cannot be analyzed by x-ray diffraction studies.

Experiment: Isolation and Quantification of Pigments

A. Chlorophyll a (Chl a)

Principle

Cyanobacteria have only Chl *a*, and its quantitative estimation is important for the evaluation of growth and photosynthetic rates. Chl *a* is completely extractable in solvents like acetone/methanol and exhibits characteristic absorption at 663 nm.

Reagent

80% methanol.

Procedure

1. Centrifuge a known volume of homogenous cyanobacterial suspension at 5000 rpm for 10 minutes.
2. Wash the pellet twice in distilled water.
3. Suspend the pellet in 4 mL of methanol and vortex thoroughly.
4. Cover the mouth of the test tube with aluminum foil to prevent evaporation of the solvent.
5. Incubate the tube in a water bath at 60°C for 1 hour (preferably in the dark) with occasional shaking.
6. Cool the tube and centrifuge the contents at 5000 rpm for 5 minutes.
7. Transfer the supernatant to another tube and once again add 4 mL of the solvent and extract again like before.
8. To ensure complete extraction, add 2 mL of the solvent to the pellet and repeat the process.
9. Pool the supernatants and make up the volume to 10 mL with methanol (to compensate the solvent loss during heating).
10. Read at 663 nm in a spectrophotometer against methanol blank.

Comments

Chl *a* is not stable in light and easily gets oxidized. Care should be taken to extract Chl *a* in dim light or in the dark.

Calculations

$$\text{Chl } a\ (\mu\text{g}\cdot\text{mL}^{-1}) = \frac{A_{663} \times 12.63 \times \text{volume of sample}}{\text{Volume of methanol}}$$

where

A_{663} = absorbance at 663 nm
12.63 = correction factor

Using the above formula, calculate and express Chl *a* as $\mu g \cdot mL^{-1}$.

B. Carotenoids

Principle

Carotenoids include pigments like carotenes and xanthophylls. They are soluble in solvents like acetone and/or methanol, exhibiting a characteristic absorption at 450 nm.

Reagent

85% acetone.

Procedure

1. Centrifuge a known volume of homogenous cyanobacterial suspension at 5000 rpm for 10 minutes.
2. Wash the pellet twice in distilled water.
3. Homogenize the pellet with 3 mL of acetone.
4. Centrifuge the contents at 5000 rpm for 5 minutes and store the supernatant in the refrigerator.
5. Repeat the extractions until the acetone remains colorless.
6. Pool the supernatants and make up to a known volume with acetone.
7. Measure the absorbance at 450 nm against the acetone blank.

Calculation

$$\text{Caretonoids } (\text{mg}\cdot\text{mL}^{-1}) = \frac{D \times V \times f \times 10}{2500}$$

where

D = absorbance at 450 nm
V = volume of the sample
f = dilution factor
2500 = extinction coefficient

With the formula given above, calculate and express caretonoids as $mg \cdot mL^{-1}$

C. Phycobilins or phycobiliproteins

Principle

Phycobilins, which generally account for 24% of the dry weight of the soluble proteins of cyanobacterial cells, are important for physiological, biochemical, and ecological studies. These, namely phycocyanin, phycoerythrin, and allophycocyanin, are water-soluble pigments and are extracted in phosphate buffer (pH 6.8).

Reagent

Phosphate buffer (0.05 M), pH 6.8. Dissolve 1.72 g of dipotassium hydrogen phosphate (K_2HPO_4) and 1.36 g of potassium dihydrogen phosphate (KH_2PO_4) in 100 mL distilled water separately and set the pH to 6.8 by adding KH_2PO_4 to K_2HPO_4.

Procedure

1. Centrifuge a known volume of homogenous cyanobacterial suspension at 5000 rpm for 10 minutes.
2. Wash the pellet in distilled water (check for the loss of pigments).
3. Suspend the pellet in 3 mL of PO_4 buffer and homogenize/sonicate the contents.
4. Freeze–thaw the contents repeatedly and centrifuge at 5000 rpm for 5 minutes.
5. Store the supernatant in the refrigerator.
6. Repeat the same process until the pellet becomes colorless to ensure complete extraction.
7. Pool the supernatants and measure the absorbance at 565, 615, and 652 nm, respectively, against PO_4 buffer blank.

Calculation

$$\text{C-Phycocyanin (PC) mg mL}^{-1} = \frac{A615 - 0.474(A652)}{5.34}$$

$$\text{allo-Phycocyanin (APC) mg mL}^{-1} = \frac{A652 - 0.208(A615)}{5.09}$$

$$\text{C-Phycoerythrin (PE) mg mL}^{-1} = \frac{A562 - 2.41(PC) - 0.849(APC)}{9.62}$$

The values obtained from the calculation are for the extract (mg mL^{-1}). From this, the values for the original sample has to be calculated, based on the volume of the sample/culture used.

Experiment: Measurement of Absorption Spectra

A characteristic absorption spectrum is simply a plot of absorbance of light by a compound at different wavelengths. Absorption spectra of oxidized and reduced forms of cytochrome c are presented in Figure 5.39a. For instance, the presence of cytochromes, ham or flavins, as a prosthetic group of a number of enzymes was largely deduced from spectral studies. This technique has also been extremely useful in indicating the involvement of certain compounds in various complex processes. A notable example of this is in establishing the role of quinones, flavins, and various cytochromes in the mitochondrial electron transport chain as well as in the photosynthetic electron transfer chain. Special investigations revealed that, under anaerobic conditions, the addition of NADH to mitochondrial preparations results in the reduction of cytochrome c. It undergoes rapid oxidation on the introduction of oxygen, thereby suggesting that it acts as an intermediate carrier of electrons during their transport from NADH to oxygen via the mitochondrial electron transfer system. Such studies are generally carried out by examining

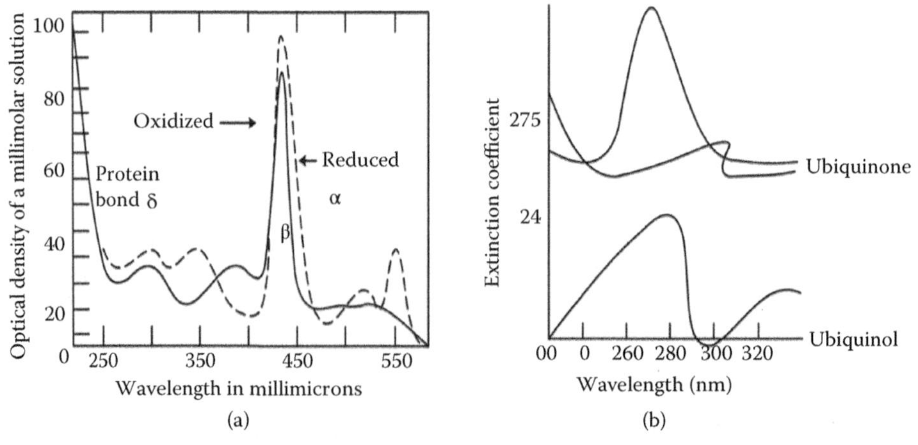

FIGURE 5.39 (a) Absorption spectra of oxidized and reduced forms of cytochrome c. (b) Absorption and difference spectra of ubiquinone and ubiquinol.

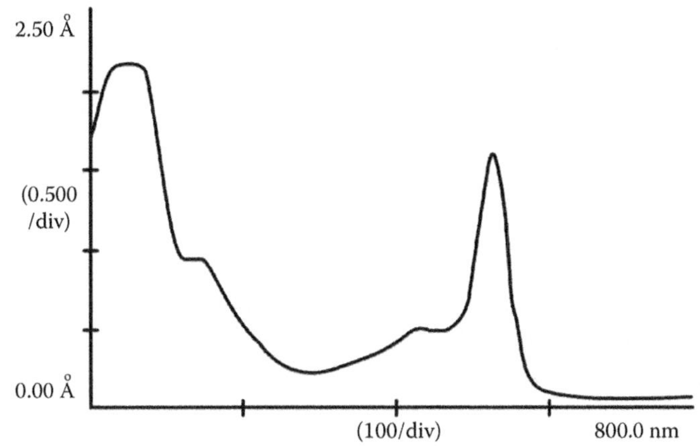

FIGURE 5.40 Absorption spectra of a cyanobacterium *Spirulina platensis*.

the difference spectra, because the various states of a compound exhibit qualitative alteration in their spectral characteristics. For example, the oxidized and reduced forms of quinones, flavins, and cytochromes have distinctive spectral difference. As the name denotes, the difference spectra is a graphical representation of the difference in the absorbance of light at different wavelengths by two forms of a compound. The absorption spectra and difference spectra of ubiquinone and ubiquinol are shown in Figure 5.39b.

Spectrophotometric analysis has been helpful in the identification of chromophores in light-mediated responses and processes. First, the rate of the light-dependent response or process is determined at varying wavelengths. A plot of the rate of the process (such as photosynthetic oxygen evolution) at various wavelengths is prepared. Such a plot is known as an action spectrum. An attempt is made to isolate the compound from the tissue that has an absorption spectrum superimposable on the action spectrum. This approach has successfully been employed in establishing the central role of chlorophylls as the primary light-harvesting pigments in photosynthesis and of phytochrome in red-/far-red-induced responses in plants.

Chl *a* was qualitatively measured by a spectrophotometer on samples filtered onto Whatman GF/C filters, extracted in 90% acetone, and scanned in a wavelength range of 400–700 nm through UV–VIS Spectrophotometer (Shimadzu; Kyoto, Japan) (Figure 5.40).

Experiment: Determination of the Molar Extinction Coefficient of NADH

Introduction

Quantitative colorimetric estimations are based on two laws, namely, Lambert's law and Beer's law. Although Lambert's law defines the relationship between the length of the light path through the solution, Beer's law states that the fraction of the length of the light transmitted is inversely proportional to the concentration of the light-absorbing compound in the solution. The amount of transmitted light is inversely related with the absorption of light by the media. Hence, as a corollary, the extent of light absorbed by a compound is directly proportional to its concentration. The experiment is designed to verify the validity of Beer's law.

Reduced forms of nictotinamide adenine nucleotides (NADH and NADPH) show a distinctive absorption peak at 340 nm. These nucleotides are required as cofactors by several oxidoreductases. The spectrophotometric determination of activities of such enzymes is based on the measurement of the rate of utilization or production of NADH by monitoring the change in absorbance at 340 nm. According to Beer's law, the absorbance of light that is often referred to as optical density (OD units) should be directly proportional to the concentration of NADH in the reaction mixture.

The molar extinction coefficient of NADH is defined as the absorption of light by 1 M concentration of a compound at an optimal wavelength with a fixed light path of 1 cm, and it is represented as ε with the wavelength shown as a subscript (for NADH at 340 nm, it is written as $\varepsilon_{340} = (6.2 \times 10^3)$. From this value, the concentration of the compound in a sample can directly be calculated without preparing a standard or reference curve.

Experimental Outline

1. Prepare an NADH standard solution of different molarities.
2. Record the absorbance of the sample.

Safety Guidelines

1. Prepare the NADH solution freshly as and when the absorption measurements are made.
2. Prepare NADH in a Tris–HCl (pH 7.5) buffer.

Materials

UV–visible spectrophotometer, standard quartz silica cuvettes with a light path of 1 cm, pH meter, NADH, Tris–hydroxylamine, and HCl.

Prelaboratory Precautions

NADH solution: 10 mL of 1 mM solution of NADH solution. Prepare 10 mL solution of NADH in 0.05 M Tris–HCl (pH 7.5).

Method

1. Switch "on" the spectrophotometer. Set the wavelength to 340 nm and after 1–2 minutes, switch "on" the UV (or deuterium) lamp and allow the instrument to warm up for 15–20 minutes (follow the instructions given in the operator's manual of the instrument supplied by the manufacturer).
2. Take a silica cuvette and add 3 mL of 0.05 M Tris–HCl. Place the cuvette in the cuvette holder and again adjust it to "zero" absorbance (or 100% transmission) using this blank (buffer alone) cuvette.
3. Check the "zero" and 100% transmission as described in step 2 to make sure that the instrument is properly adjusted.
4. In another matched cuvette, take 0.1 mL of 1 mM solution of NADH and 2.9 mL of 0.05 M Tris–HCl buffer so that the total volume in cuvette is 3 mL. Mix the contents thoroughly and record the absorbance of this sample solution (it corresponds to 33 μM NADH or 0.1 μmol in 3 mL).
5. Pour out the solution from the sample cuvette and rinse it with distilled water. Add 0.2 mL of NADH and increase the final volume 2.8 mL of 0.05 M Tris–HCl buffer. Record its absorbance.
6. Repeat step 5 by taking 0.40 mL of NADH solution and 2.6 mL of 0.05 M Tris–HCl buffer. Record its absorbance.
7. Repeat step 5 by taking 0.6 mL of NADH solution and increase the volume to 3 mL with 0.05 M Tris–HCl buffer, and record the absorbance of this solution.

Calculations

1. Present the result obtained in the form of a table as follows:

Sample No.	mL of NADH Solution	Conc. of NADH (μM)	Fold Conc. of NADH	A_{340}	Ratio
1	0.10	33.3	1.0	–	1.0
2	0.20	66.6	2.0	–	–
3	0.40	133.3	4.0	–	–
4	0.60	200.0	6.0	–	–

Calculate the ratios of *A*340 for samples 2, 3, and 4.

2. Check whether the calculated values of the ratios for samples 2, 3, and 4 correspond with the increase in the fold concentration of NADH in this sample as compared to sample 1.
3. Prepare a graph of A_{340} against the concentration of NADH in the sample. A straight line denotes the increasing concentration of NADH in the sample. This graph can be used as a standard curve for NADH. One can determine the concentration of NADH in the sample from its A_{340} by referring to this standard curve.
4. Calculate the molar extinction coefficient (ε_{340} nm) for NADH as below
   a. Suppose 133.3 μM solution of NADH gives A_{340} of x.
   b. 1 μM solution of NADH would correspond to (x/133.3) absorbance.
   c. 1 M solution of NADH would give A_{340} value of $(x/133.3) \times 10^6$.

   This is the ε_{340} for NADH. Thus, the concentration of NADH in an unknown sample can directly be calculated from its A_{340} by using the molar extinction coefficient of NADH.

Experiment: Preparation of Absorption Spectra of NADH

Introduction

The absorption spectrum refers to the absorbance of light of different wavelengths by a compound. Every compound or group of compounds has some characteristic features in its absorption spectra, which is used in analyzing the substance quality and also helps in estimating the quantity. For example, flavin adenine nucleotides (FAD) have broad absorption spectra with peaks at 370 and 450 nm. Enzymatic or chemical reduction of flavin nucleotides is accompanied by the disappearance of the peak at 450 nm. Thus, the presence of flavin nucleotides in a sample can be established by examining its absorption spectra before and after its reduction. Similarly, the oxidized form of cytochrome c shows a prominent peak at 415 nm and a broad but less intense peak at around 550 nm. Reduction with sodium dithionite leads to the appearance of a sharp distinct absorbance peak at 554 nm (α-band), another peak 524 nm (β-band), and an intense peak at 419 nm. The amount of cytochrome c can be quantitatively estimated from *A*554, which has a molar extinction coefficient value 27.7. Absorption spectra also helps in identifying wavelength which is characteristic and unique for a particular compound; for example, absorption peak at 340 nm is unique for NADH, and even though the peak at 260 nm is much more intense, it cannot be used for detection and quantitative estimation of NADH because NAD and other adenine nucleotides (like ATP, ADP, AMP) also show intense absorption at 260 nm but not at 340 nm.

The absorption spectra was obtained by examining the absorbance of light at different wavelengths by the compound. A dual beam spectrophotometer with a facility for scanning is required. In these spectrophotometers the wavelength changes in a predetermined range continuously at a constant rate (say, 25 nm per minute), and the difference in absorbance between the blank and the sample preparation is automatically recorded on chart paper. However, absorption spectra can also be obtained manually with simple single-beam spectrophotometer. At each wavelength, the instrument has to be set again to zero absorbance. The absorbance by the sample is then determined and is plotted in form of a graph of the absorbance value versus wavelength.

Experimental Outline

1. Preparation of 1 μm solution of NADH.
2. Measurement of absorption spectra at different wavelengths with UV–visible spectrophotometer.

Safety Guidelines

1. Prepare NADH solution as and when the experiment is carried out.
2. Prepare NADH in a Tris–HCl buffer.

Materials

Spectrophotometer, pH meter, NADH, Tris hydroxylamine, hydrochloric acid, NADH solution.

Prelaboratory Preparation

NADH solution: Prepare 1 mM solution of NADH in 0.05 M Tris–HCl.

Method

1. Switch "on" the spectrophotometer and after 1 minute, switch "on" the UV lamp.
2. Take two matched silica cuvettes. In the reagent blank cuvette, add 3 mL of 0.05 M Tris–HCl buffer (pH 7.5). In the sample cuvette, add 0.4 mL of 1 mM NADH solution and 2.6 mL of 0.05 Tris–HCl (pH 7.5).
3. Set the instrument at 200 nm using the reagent blank cuvette (Tris–HCl buffer alone) as described in steps 2 and 3 of the first experiment of this chapter and determine the absorption of the sample (containing NADH) against the blank.
4. Now reset the instrument at 210 nm with reagent blank and record the reading of the sample. Repeat this step at intervals of 10 nm each up to 40 nm. For taking readings beyond this wavelength, switch "off" the deuterium lamp and switch "on" the tungsten lamp.
5. Draw a graph of absorbance value versus the wavelength to obtain the absorption spectrum of NADH.

Experiment: Determination of Difference Spectra of Oxidized–Reduced NAD

Introduction

As the name indicates, the difference in the spectra represents or highlights the difference of the absorption of light by two forms of the same compound, for example, oxidized–reduced states of a compound like NAD and NADH, the oxidized form of FAD and its reduced form $FADH_2$, and so on. The differential spectra provides useful information as to which is the optimal wavelength to differentiate between the two states of the compound. In fact, the selection of the most appropriate wavelength to differentiate and specifically estimate the particular state of the compound from the other state is based on their difference spectra.

For obtaining difference spectra, an automatic, dual-beam scanning spectrophotometer is used, where NAD is placed in a blank position of the cuvette holder and that containing NADH in the sample position. The scanning range (say, 210–400 nm) and scanning speed (20 nm/min) are adjusted as per requirement of the investigations. The speed of recorder is synchronized (say, 1 cm–20 nm), scanning speed (20 nm/min) is adjusted, and the scanning operation is initiated. Such a spectrum will provide a graphical figure of the differences in the absorption spectra of the two states of the compounds, which is automatically recorded on scanning spectrophotometer. One can determine separately the absorption spectra of these two compounds by ordinary uv-vis spectrophotometer if scanning spectrophotometer is not available. Calculate the difference in absorption of light at different wavelengths and draw a graph of the difference in values of light absorbance versus wavelength. The value in the difference spectra will be zero and only peaks with positive values will appear at the wavelengths where the two compounds absorb light to the same extent; the blank (NAD solution) shows lesser absorption than the sample (reduced NAD). The absorption band of negative value will be obtained where the absorption of light by the sample is less.

Experimental Outline

1. Preparation of NAD solution.
2. Determination of absorption of light at different wavelengths.

Safety Guidelines

Same as the first two experiments of this chapter.

Materials

Same as the first two experiments of this chapter.

Prelaboratory Preparation

1 mM NAD solution in 0.05 M Tris–HCl (pH 7.5).

Method

Determine the absorption of light of different wavelengths by taking 0.1 mL of NAD manually as described in steps 1–4 of the second experiment of this chapter using 0.05 M Tris–HCl as the blank. Repeat the experiment using NADH in place of NAD in the sample cuvette.

Calculations

1. Subtract the absorbance values for NAD from those for NADH at each of the wavelengths.
2. Draw a graph of obtained differences in absorbance versus wavelength to obtain the difference spectra of oxidized–reduced NAD.

Experiment: Determination of Polyamines by Fluorescence Spectrophotometer

Introduction

Polyamines (PA) play an important role in biological process such as in proliferation of cells, growth and differentiation and in clinical disorders. They are highly conserved organic compound having two or more primary amino groups $-NH_2$. The most common PAs are putrescine (Put), cadavarine (Cad), spermidien (Spd), and spermine (Spm). Due to their cationic nature, PAs can associate with the amino components of biomembranes such as pectin. PAs may regulate DNA replication, transcription and translation, cell division, differentiation, and other morphogenetic processes. They are regarded as hormonal second messengers and as some of the reserves of carbon and nitrogen. In addition to free amines, PAs can bind covalently to protein and conjugate hydroxycinnamic acid and hydroxyferulic acid. The phenolic acids exhibit fluorescence when separated on TLC and viewed under UV light. They can be quantified with a spectroflurometer.

Experimental Outline

1. The extraction of free (soluble-free cations), soluble-conjugated (with phenolic acids and other low-molecular-weight compounds), and bound-conjugated (with macromolecules and cell walls) PAs.
2. Estimation of free, conjugated, and bound PAs.
3. Identification of PAs by UV fluorescence.
4. Quantitative estimation by using a spectrofluorimeter.

Safety Guidelines

Follow standard laboratory precautions.

Materials

Animal or plant tissue, proline, mortar and pestle, acetone, refrigerated centrifuge, dansyl chloride, perchloric acid (PCA), PA standards, hydrochloric acid, putrescine, benzene, spermidine, cyclohexane, spermine, ethyl acetate, agmatine, glass plates, chromatographic chambers, silica gel G, spectrofluorimeter.

Prelaboratory Preparations

10% PCA	Take 10 mL of PCA and increase the volume to 100 mL with distilled water
6 N HCl	Take 50 mL of conc. HCl and increase the volume to 100 mL with distilled water
PA standard	Prepare 1 mM solutions
Dansyl chloride	Weigh 50 mg of dansyl chloride and mix in 100 mL of water
Proline	Weigh 100 mg of proline and dissolve in 1 mL of distilled water

Method

Extraction and measurement: Homogenize the tissue in a prechilled mortar with 10% (v/v) PCA in the ratio of 100 mg fresh weight of tissue/PCA 10 mL. Incubate the homogenate at 4°C for 1 hour and then centrifuge at 16,000 × *g* for 20 minutes. After centrifugation, separate the supernatant and pellet. The supernatant contains free PAs, and PCA-soluble, conjugated, and bound PAs are released from 200 µL of original supernatant pellet and suspension, respectively, by acid hydrolysis with 200 µL 12 N HCl (1:1 v/v) for 16–18 hours at 110°C in flame-sealed glass ampoules. Filter the hydrolyzate through glass wool and wash with six drops of 6 N HCl, dry at 70°C on a dry bath in an exhaust hood, and resuspend in 20 µL of PCA. Dansylate aliquots of 200 µL nonhydrolyzed pellet containing bound PAs along with PA standards (1 mM) as control. Briefly, mix 100 µL of the samples and standard PA mixtures with 100 µL of saturated Na_2CO_3 and 200 µL of freshly prepared dansyl chloride (5 mg/mL of acetone). After vortexing, incubate the mixture overnight in the dark at 26 ± 1°C in reagent. Add 50 µL of proline (100 mg/mL H_2O) and incubate for a minute under dark conditions.

Extract dansylated PAs in 250 µL of benzene. Collect clear organic phase in glass vials loaded onto high-resolution silica gel TLC plates, and develop chromatogram developed for 1 hour with cyclohexane: ethyl acetate (5:4 v/v). After TLC, the bands can be detected and marked under a UV transilluminator and then scraped and eluted in 3 mL of ethyl acetate. Measure the fluorescence by using a Shimadzu RF 540 dual-wavelength spectrofluorimeter with excitation wavelength at 350 nm and emission wavelength at 495 nm.

Calculations

Calculate the PA content from standard graph made from any one of the PAs.

Experiment: Determination of Melting Temperature (T_M) and Base Composition of DNA from Thermal Denaturation Characteristics

Introduction

If the dilute aqueous solution of double-stranded DNA is heated, the two strands become separated due to the disruption of hydrogen bonds between the complementary bases. Such DNA is referred to as denatured DNA, and the process is called denaturation. Denaturation is a reversible process; if the heated DNA solution is slowly allowed to cool to room temperature, the complementary strands of DNA recombine to give duplex DNA. DNA is then said to be reannealed or renatured, and process is called reannealing or renaturation of DNA.

During denaturation, the absorbance of DNA at 260 nm is increased by 30–40% due to the exposure of the bases. The process of absorption increase is called the hyperchromic effect. Heating through the temperature range of 25°C to about 80°C results in only a minor increase in absorbance at 260 nm, followed by a sharp increase within a narrow temperature range (80–90°C), and then A_{260} becomes constant when DNA is fully denatured (Figure 5.41). A graph between absorbance at 260 nm and temperature gives a thermal denaturation curve, a temperature profile, or the melting curve of DNA. The temperature corresponding to the midpoint of the curve is defined as T_m, the transition temperature or melting temperature of DNA, and it denotes the temperature at which 50% of the DNA has undergone denaturation.

T_m is largely dependent on the (G+C) content of DNA; the higher the (G+C) content, the higher the T_m. Under standard conditions of ionic strength and pH, (G+C) content is related with T_m in accordance with the following equation:

$$\%(G+C) = T_m - 69.3 \times 2.44$$

Each species of DNA has a characteristic value that can be used for the purposes of identification and characterization.

Safety Guidelines

The extracted DNA should be pure. The glassware must be washed thoroughly with double-distilled water. Wear a lab coat.

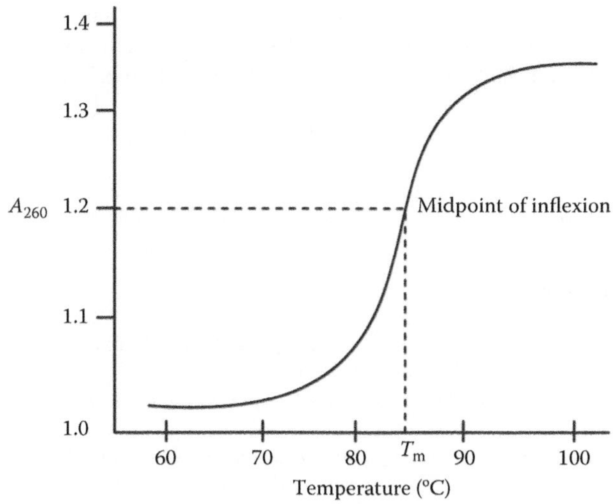

FIGURE 5.41 Melting curve of DNA sample.

Experimental Outline

1. The measurement of DNA samples with a spectrophotometer at different ranges of temperature from 25 to 100°C recording the absorbance at 260 nm.
2. Plotting a graph of absorbance against temperature.

Materials

UV spectrophotometer with thermoprogrammer, quartz cuvettes, sodium chloride, and sodium citrate.

Prelaboratory Preparations

1. Saline sodium citrate (SSC): Prepare 0.015 M sodium citrate solution (pH 7.0) and dissolve NaCl in it so that its final concentration in solution is 0.15 M.
2. DNA: Dissolve 50μg DNA/mL in SSC.

Method A

1. Switch "on" the spectrophotometer and after allowing a sufficient period for warming, set it to zero absorbance at 260 nm with SSC.
2. Measure A260 of the DNA samples at 25°C.
3. Heat the DNA solution at a rate of 1°C rise/min up to 100°C with the help of a thermoprogrammer. Record the absorbance values.
4. Calculate A260 (T°C)/A260 at 25°C for each of the following temperatures: 25, 35, 50, 70, 75, 80, 85, 90, 95, and 100°C, and plot the absorbance ratio against the temperature.
5. Determine the midpoint of increase in absorbance and by extrapolation find the corresponding temperature that represents Tm for the DNA sample.
6. Calculate % (G+C) content of the DNA using the following equation:

$$\%\,(\mathrm{G}+\mathrm{C}) = T_{\mathrm{m}} - 69.3 \times 2.44$$

Method B

If a UV spectrophotometer with a thermoprogrammer is not available, then the following procedure can be adopted:

1. Arrange a series of constant temperature water baths maintained at 25, 50, 75, 80, 85, 90, 95, and 100°C, respectively.
2. Record the absorbance at 260 nm of a DNA sample solution (50 mg/mL) kept at 25°C.

3. Distribute the above solution into eight test tubes and put one tube each in the water baths maintained at different temperatures. Allow the tubes to stand for 15 minutes. After incubation, quickly cool all the tubes (except one at 25°C) by placing them in a nice bath for 10 minutes.
   Note: Reannealing occurs only if the sample is cooled gradually, but reannealing is negligible if the cooling is instantaneous.
4. Record the A260 of these samples.
5. Then proceed as in steps 4–6 of the procedure described earlier.

SUGGESTED READING

Allen, D., C. Cooksey, and B. Tsai. 2010. Spectrophotometry. http://www.nist.gov/pml/div685/grp03/spectrophotometry.cfm (accessed October 5, 2010).

Bassler, G. C., and T. C. Morrill. 1981. *Spectrometric Identification of Organic Compounds*, 4th ed. New York: John Wiley and Sons.

Edisbury, J. R. 1967. *Practical Hints on Absorption Spectrometry*. New York: Plenum Press.

Hilton, W. A. 1974. *Experiments in Optical Physics*, 3rd ed. Liberty, MO: Department of Physics, William Jewell College.

Schwedt, G. 1997. *The Essential Guide to Analytical Chemistry*. Translated by Brooks Haderlie, 16–7. Chichester, NY: Wiley.

Sharma, B. K. 1991. *Instrumental Methods of Chemical Analysis*, 11th ed. New Delhi, India: Goyal Publishing House.

Snavely, B. B. 1969. "Flashlamp-Excited Organic Dye Lasers." *Proceedings of the IEEE* 57: 1374–90.

Westermeier, R., and T. Naven, 2002. *Proteomics in Practice: A Laboratory Manual of Proteome Analysis*, 3rd ed. Darmstadt Wiley-VCH Verlag Gmbh.

Wilson, K., and J. Walker. 2003. Practical Biochemistry: Principle and Techniques, 5th ed. Cambridge: Cambridge University Press.

IMPORTANT LINKS

1. UV–visible spectroscopy: http://www.uv-groebel.com/pms_spek1.php
2. Infrared spectroscopy: http://www.perkinelmer.com/Catalog/Product/ID/L1280002
3. Fluorescence spectroscopy: http://www.perkinelmer.com/Catalog/Category/ID/Fluorescence%20Spectroscopy
4. ESR spectroscopy: http://www.jeol.com/PRODUCTS/AnalyticalInstruments/ElectronSpinResonance/tabid/98/Default.aspx
5. NMR spectroscopy: http://www.magritek.com/kea.html
6. MALDI-TOFMS: http://www.thermoscientific.com/ecomm/servlet/productsdetail?productId=11962154&groupType=PRODUCT&searchType=0&storeId=11152&gclid=CLaVoIr-oKsCFYh_6wodyVEXfA
7. Circular Dichroism spectroscopy: http://www.biocompare.com/ProductDetails/665635/FVS-6000-Vibrational-CD-Spectrometer.html

6 Centrifugation

6.1 INTRODUCTION

Centrifugation is one of the most important and widely applied research techniques in biochemistry, cellular and molecular biology, and medicine. Current research and clinical applications rely on the isolation of cells, subcellular organelles, and macromolecules, often in high yields.

A centrifuge uses centrifugal force (*g*-force) to isolate suspended particles from their surrounding medium on either a batch or a continuous-flow basis. There are many applications for centrifugation, including the sedimentation of cells and viruses; separation of subcellular organelles; and isolation of macromolecules such as DNA, RNA, proteins, and lipids.

6.1.1 Increasing the Effect of Gravity: The Centrifuge

Many particles or cells in a liquid suspension, given time, eventually settle at the bottom of a container due to gravity (1 *g*). However, the length of time required for such separations is impractical. Other particles, which are extremely small in size, do not separate at all in a solution unless subjected to a high centrifugal force. When a suspension is rotated at a certain speed or revolutions per minute (RPM), centrifugal force causes the particles to move radially away from the axis of rotation. The force on the particles (compared to gravity) is called relative centrifugal force (RCF). For example, an RCF of 500 *g* indicates that the centrifugal force applied is 500 times greater than Earth's gravitational force. Table 6.1 illustrates common centrifuge classes and their applications.

6.2 PRINCIPLE OF CENTRIFUGATION

A centrifuge is a device for whirling an object with a high angular velocity. The consequent large acceleration $\varpi^2 r$ is equivalent to increasing the value of *g*. The acceleration due to gravity and such processes as sedimentation (settling of particles or precipitates out of a solution) can be greatly accelerated in this way. Hence, $\varpi^2 r$ is expressed as "number times *g*" or relative centrifugal field (RCF). This is the ratio of the weight of the particle in an applied centrifugal field to the weight of the same particle when acted upon by gravity alone. The force acting on a body moving in a circular path can be broken down into two components: (1) normal and (2) tangential to the path. Centrifugal force (*F*) is the force on a rotating object, which acts tangentially outward, therefore,

$$F = M \cdot \varpi^2 r$$

where

M = mass of the particle
ϖ^2 = angular acceleration of the particle moving in a circle of the square of angular velocity
r = the radial distance of the particle from the axis of rotation.

6.3 TYPES OF CENTRIFUGES

The different types of centrifuges are as follows:

1. Small benchtop centrifuges have a maximum rate of 3000 *g*–7000 *g*. They are used to precipitate coarse granules, yeast cells, and so on (Figure 6.1).
2. Large-capacity refrigerated centrifuges have a maximum rate of 6500 *g* with a capacity of 100 cc. They are refrigerated to control the temperature.

TABLE 6.1
Classes of Centrifuges and Their Applications

	Centrifuge Classes		
	Low Speed	**High Speed**	**Ultra/Micro-Ultra**
Maximum speed (rpm × 10^3)	10	28	100/150
Maximum RCF (×10^3)	7	100	800/900
Pelleting applications (bacteria)	Yes	Yes	(Yes)
Animal and plant cells	Yes	Yes	(Yes)
Nuclei	Yes	Yes	(Yes)
Precipitates	Some	Most	(Yes)
Membrane fractions	Some	Some	Yes
Ribosomes/polysomes	–	–	Yes
Macromolecules	–	–	Yes
Viruses	–	Most	Yes

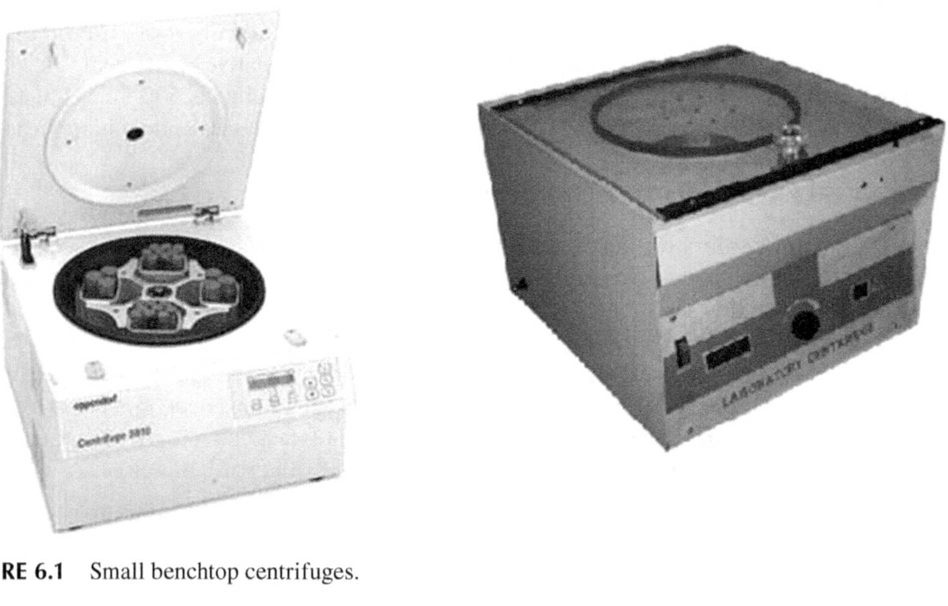

FIGURE 6.1 Small benchtop centrifuges.

3. High-speed refrigerated centrifuges have a maximum rate of 60,000 *g*; are refrigerated; and are used to sediment small microorganisms, cellular debris, and large cellular organelles (Figure 6.2).
4. Continuous-flow centrifuges are unlike other centrifuges where tubes are attached to the rotor; in continuous-flow centrifuges, the rotor itself is tubular. Particles are sedimented against the wall of the centrifuge and supernatant continuously flows out; they are used in the large-scale harvesting of bacteria.
5. A preparative ultracentrifuge has a maximum rate of 600,000 *g*. The rotor chamber is refrigerated, sealed, and evacuated to minimize heat production due to friction between air and the spinning rotor.

FIGURE 6.2 A large-capacity high centrifuge.

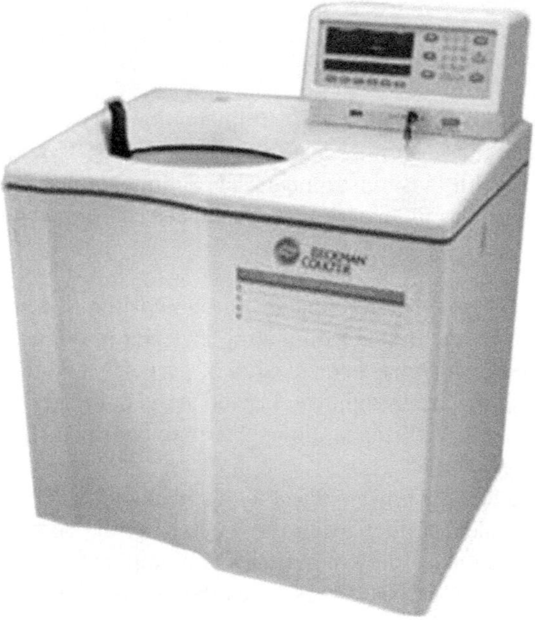

FIGURE 6.3 An ultracentrifuge.

6. In an analytical ultracentrifuge, a sample being spun can be monitored in real time through an optical detection system, using ultraviolet light absorption and/or interference optical refractive index sensitive system known as the (a) Schlieren optical system (Figure 6.3) or the Rayleigh interferometric system (Figure 6.4). This allows the operator to observe the evolution of the sample concentration versus the axis of rotation profile as a result of the

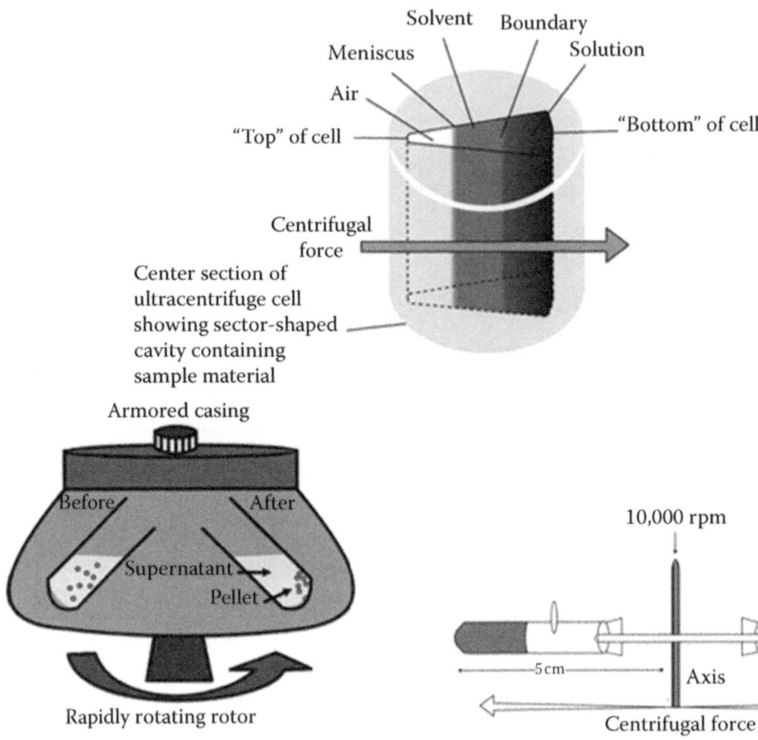

FIGURE 6.4 A rotor.

applied centrifugal field. With modern instrumentation, these observations are electronically digitized and stored for further mathematical analysis. Two kinds of experiments are commonly performed on these instruments: (1) sedimentation velocity experiments and (2) sedimentation equilibrium experiments. An analytical ultracentrifuge has a maximum rate of 500,000 *g*.

In the Schlieren optical system, when light passes through different density zones it is refracted at the boundary between the zones. In the sedimenting material in the analytical cell, a boundary is formed between the solvent, which has been cleared of particles, and the remainder of the solution containing the sedimenting material. Light is refracted at the boundary. The Schlieren system plots refractive index gradient against distance along the analytical cell. Concentration can be determined from the area of the peak.

A centrifuge is a device for whirling an object with a high angular velocity. The consequent large acceleration ($\varpi 2r$) is equivalent to increasing the value of *g*, the acceleration due to gravity, and processes such as sedimentation (settling of particles or precipitates out of a solution) can be greatly accelerated in this way. Hence, $\varpi 2r$ is expressed as the "number times *g*," or relative centrifugal force (RCF). This is the ratio of the weight of a particle in an applied centrifugal field to the weight of the same particle when acted on by gravity alone.

In an applied centrifugal field, the rate of sedimentation of a particle suspended in solution depends on the following factors:

1. Density and size of the particle
2. Density and viscosity of the medium in which the particle is suspended
3. Extent to which the particle's shape deviates from spherical

1. *Density and size of particle*: The net force on a particle suspended in a medium in a centrifugal field (F) is

$$F = M\,\varpi^2 r$$

$$\text{or } F = \frac{4}{3}\pi r^3 r_p(\rho_p - \rho_m)\varpi^2 r \quad (\text{since } M = \text{volume} \times \text{density})$$

where

ρ_p = density of particle
ρ_m = density of medium
r = radial distance of particle from axis of rotation
r_p = radius of particle
ϖ^2 = angular velocity

2. *Density and viscosity of the medium*: Due to viscosity of the medium, the frictional force (f_o) on a spherical particle, opposes motion through the medium.

$$f_o = 6\pi\,\eta\,r_p\,\nu$$

where

η = coefficient of viscosity
ν = velocity of the particle
r_p = radial distance from axis of rotation to the bottom of the tube
The particle continues to accelerate until

$$F = f_o$$

and in this condition

$$\text{or } F = \frac{4}{3}\pi r_p^3(\rho_p - \rho_m)\varpi^2 r = 6\pi\,\eta\,r_p\,\nu$$

$$\text{or } v = \frac{dr}{dt} = \frac{2r_p^3}{9\eta}(\rho_p - \rho_m)\omega^2 r \tag{6.1}$$

$$\text{or } t = \frac{9\eta}{2r_p^2(\rho_p - \rho_m)\omega^2 r} = \frac{\ln r_b}{r_t} \tag{6.2}$$

where

r_t = radial distance from axis of rotation to meniscus of liquid medium
r_b = radial distance from axis of rotation to the bottom of the tube
t = time in seconds

Extent to which the particle's shape deviates from spherical: The ratio of functional force f on a nonspherical particle to the functional force f_o on a spherical particle is equal to one ($f/f_o = 1$), whereas for a nonspherical particle the ratio is greater than one, therefore, the velocity of sedimentation v is

$$\text{or, } t = \frac{2r_p^2(\rho_p - \rho_m)\omega^2 r}{9\eta\,(f/f_o)} = \frac{\ln r_b}{r_t}$$

Thus, nonspherical particles sediment at a slower rate. Centrifugation is based on the principle that solution, occurs in an applied centrifugal field due to differences in shapes and masses of particle.

If the composition of the suspending medium is defined, the rate of sedimentation is proportional to $\varpi^2 r$ and Equation 6.1 simplifies to $v = s\varpi^2 r$, where s is the sedimentation coefficient. Sedimentation coefficients are usually very small for biological particles; hence, a basic unit of 10^{-13} seconds is taken for convenience. This is called the "Svedberg unit" (S). Thus, a sedimentation coefficient of 15×10^{-13} seconds is represented by 15S.

6.4 TYPES OF CENTRIFUGAL SEPARATIONS

6.4.1 Differential Centrifugation

Separation is achieved primarily based on the size of the particles in differential centrifugation. This type of separation is commonly used in simple pelleting and in obtaining partially pure preparations of subcellular organelles and macromolecules. For the study of subcellular organelles, sample tissue or cells are first disrupted to release their internal contents. This crude disrupted cell mixture is referred to as a homogenate. During centrifugation of a cell homogenate, larger particles sediment faster than smaller ones, which provides the basis for obtaining crude organelle fractions by differential centrifugation. A cell homogenate can be centrifuged at a series of progressively higher *g*-forces and times to generate pellets of partially purified organelles (Figure 6.4).

When a cell homogenate is centrifuged at 1000 *g* for 10 minutes, unbroken cells and heavy nuclei pellet to the bottom of the tube. The supernatant can be further centrifuged at 10,000 *g* for 20 minutes to pellet subcellular organelles of intermediate velocities such as mitochondria, lysosomes, and microbodies. Some of these sedimenting organelles can be obtained with partial purity and are typically contaminated with other particles. The repeated washing of the pellets by resuspending them in isotonic solvents and repelleting may result in the removal of contaminants that are small in size (Figure 6.5). Partial purification of organelles by differential centrifugation serves as the preliminary step for further purification using other types of centrifugal separation methods (e.g., density gradient separation).

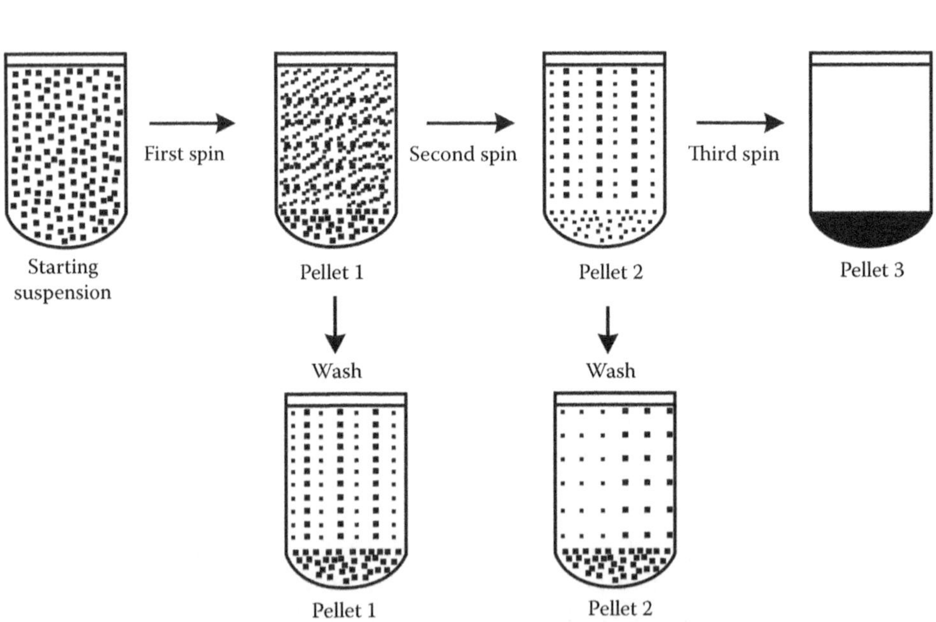

FIGURE 6.5 Demonstration of sedimentation and pelleting by centrifugation.

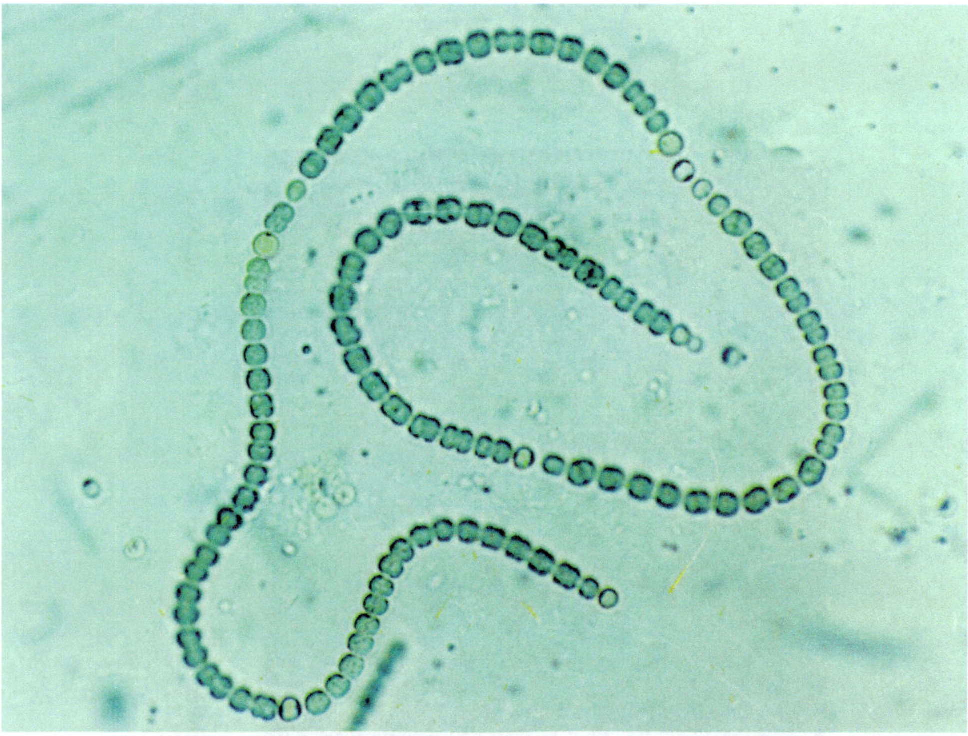

FIGURE 1.4 Microscopic view under a compound microscope of the cyanobacterium *Anabaena variabilis*, showing cell differentiation.

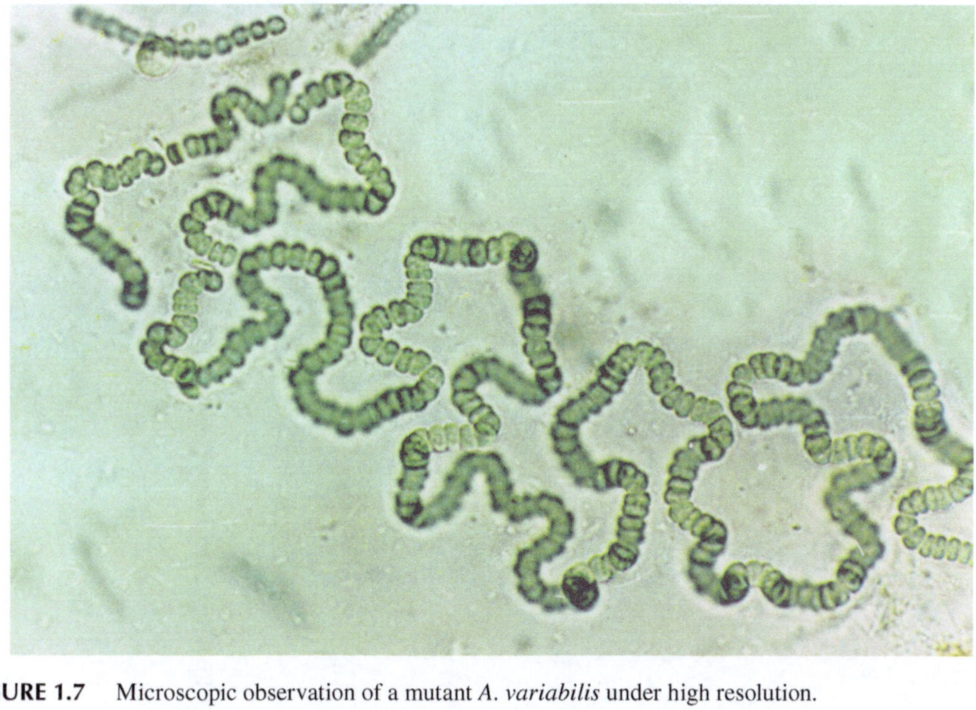

FIGURE 1.7 Microscopic observation of a mutant *A. variabilis* under high resolution.

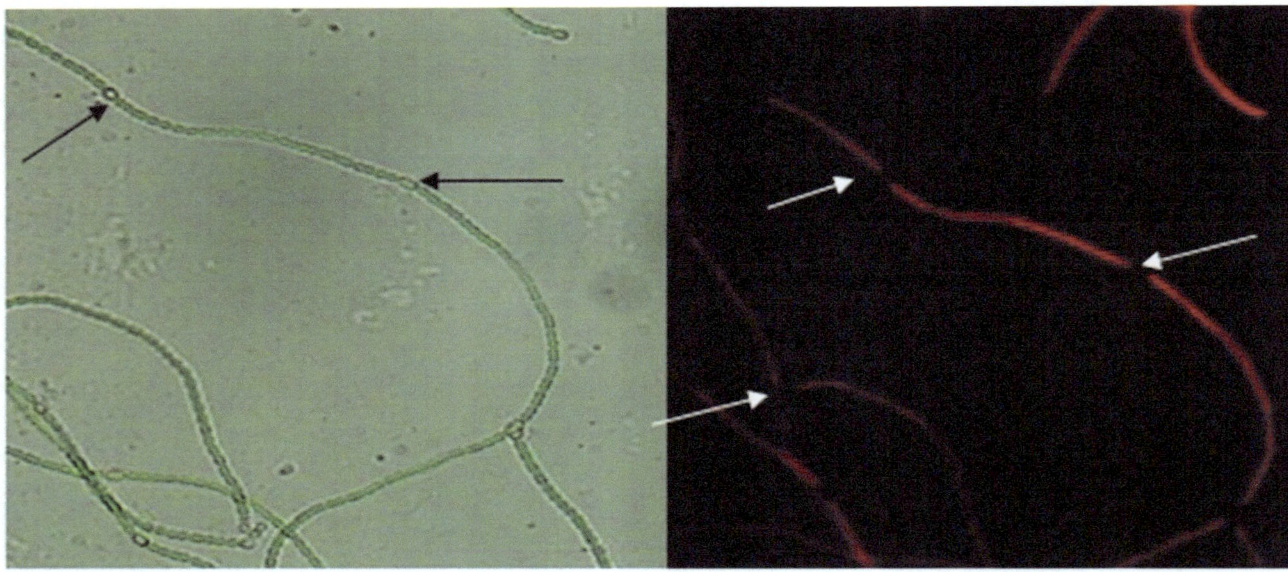

(a) (b)

FIGURE 1.10 Photomicrograph of *A. cylindrica* showing the position of heterocysts in between the vegetative cells: (a) bright field and (b) fluorescent photomicrographs.

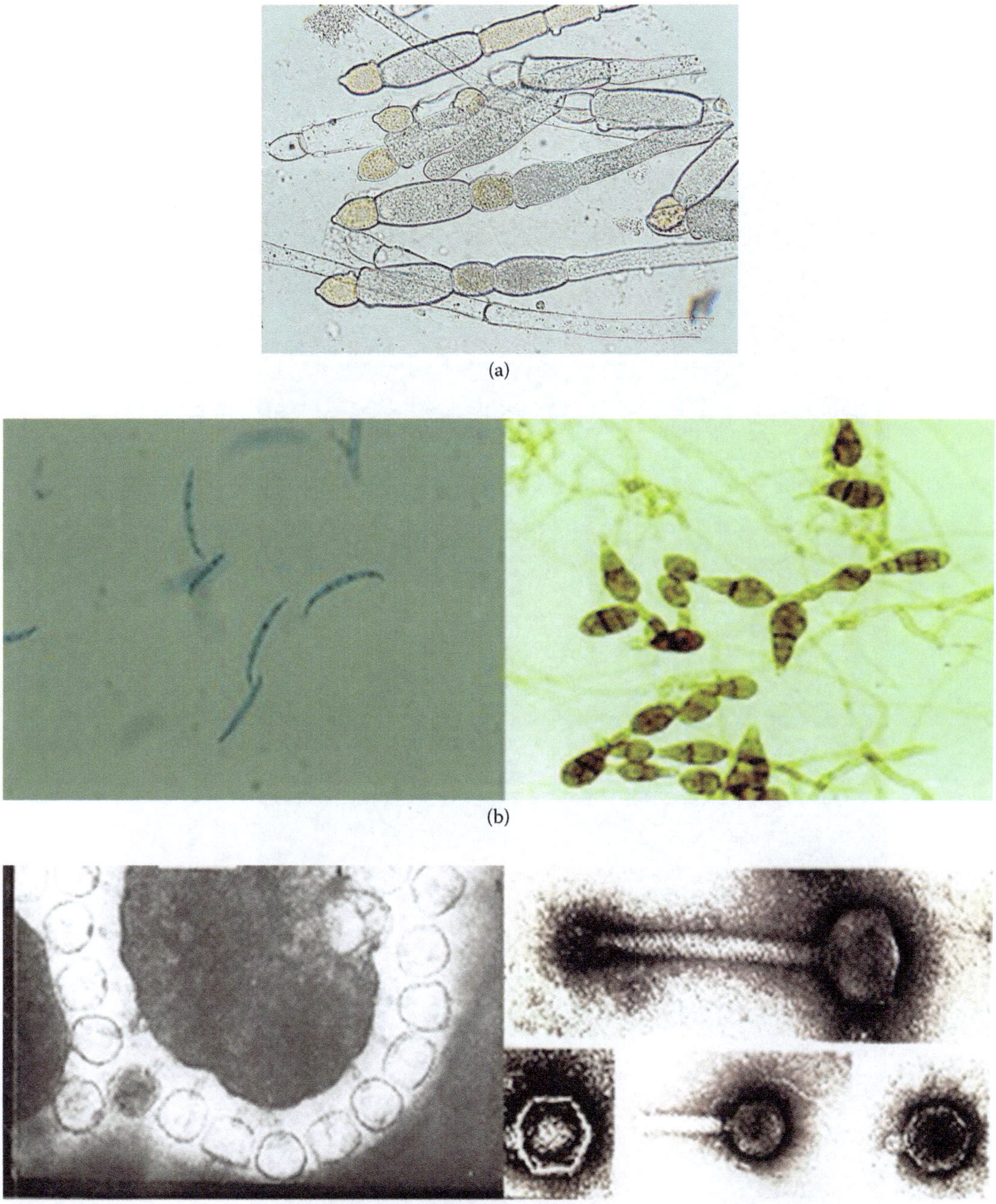

FIGURE 1.13 Figure showing (a) sporangium and sporangiospore in a slime mold viewed under a compound microscope, (b) fungal conidia of *Fusarium* and *Alternaria* under oil immersion, and (c) electron micrograph of phycovirus attached to cyanobacteria and different types of phycoviruses.

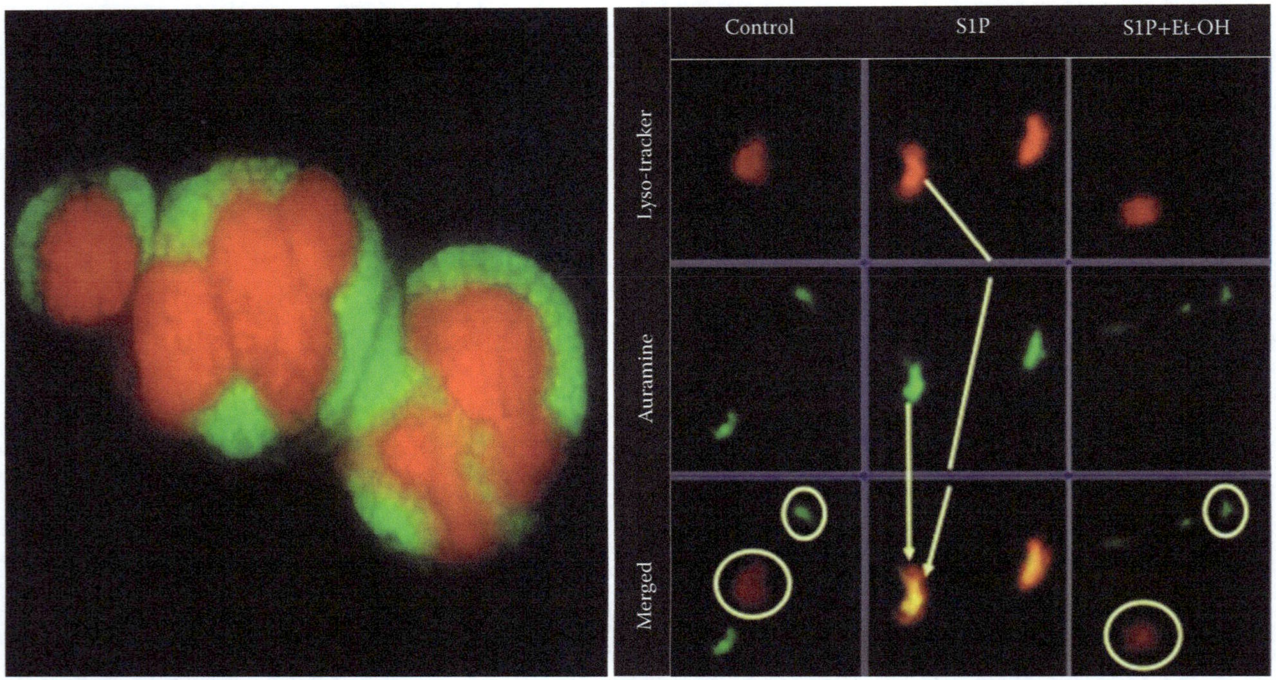

FIGURE 1.16 Figure showing (a) fluorescence in situ hybridization of *Betaproteobacteria* and (b) immunofluorescence detection of sphingosine 1-phosphate (S1P) in tuberculosis. (Courtesy of Pernthaler et al., 2008, *Proceedings of the National Academy of Sciences* 105, 7052–7057 10.1073/pnas.0711303105.)

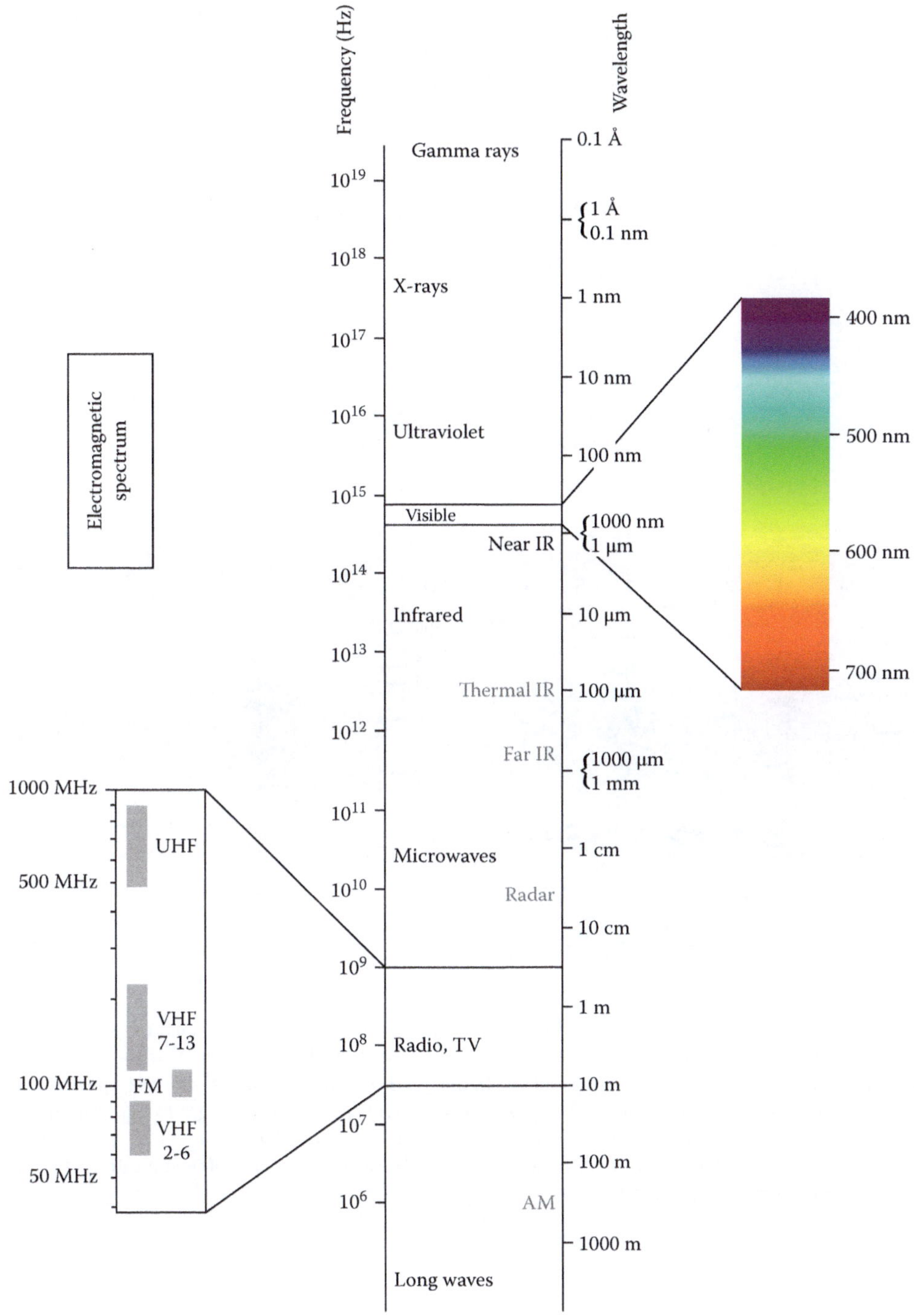

FIGURE 5.1 EM spectrum.

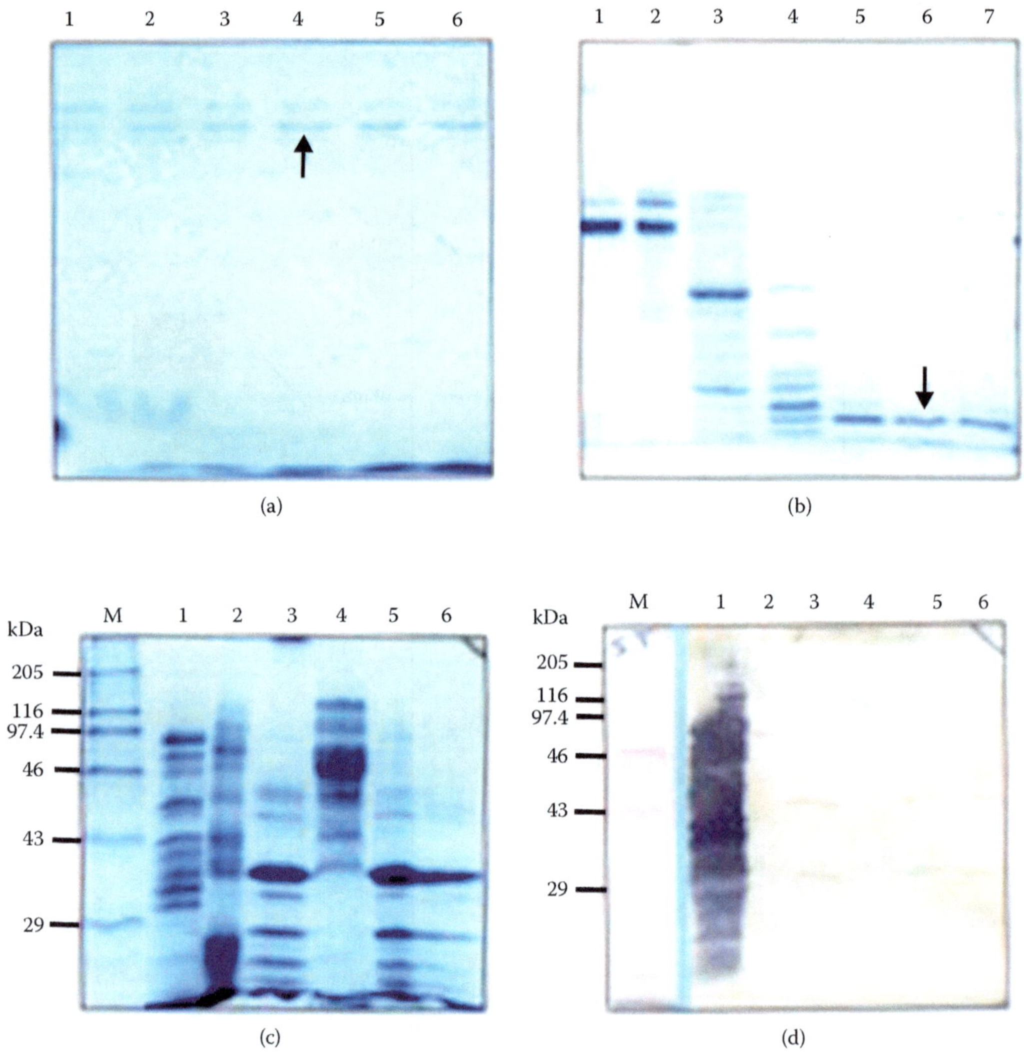

FIGURE 7.6 Top: Protein profile of different fractions of metabolic antigens of *Aspergillus fumigates* obtained by size-exclusion chromatography on 12% SDS-PAGE stained with Coomassie blue: (a) Arrow shows 70–72 kDa protein doublet bands and (b) arrow shows 18 kDa purified protein. (c) Protein profile of metabolic antigens of *Aspergillus fumigates* precipitated by graded saturation of ammonium sulfate. (d) Identification of immunogenic proteins by Western blotting. Lane M, molecular weight marker; 1-, 2-, 3-, 4-, and 5-proteins precipitated at 20%, 40%, 60%, 80%, and 100% saturation of ammonium sulfate in culture filtrate, respectively.

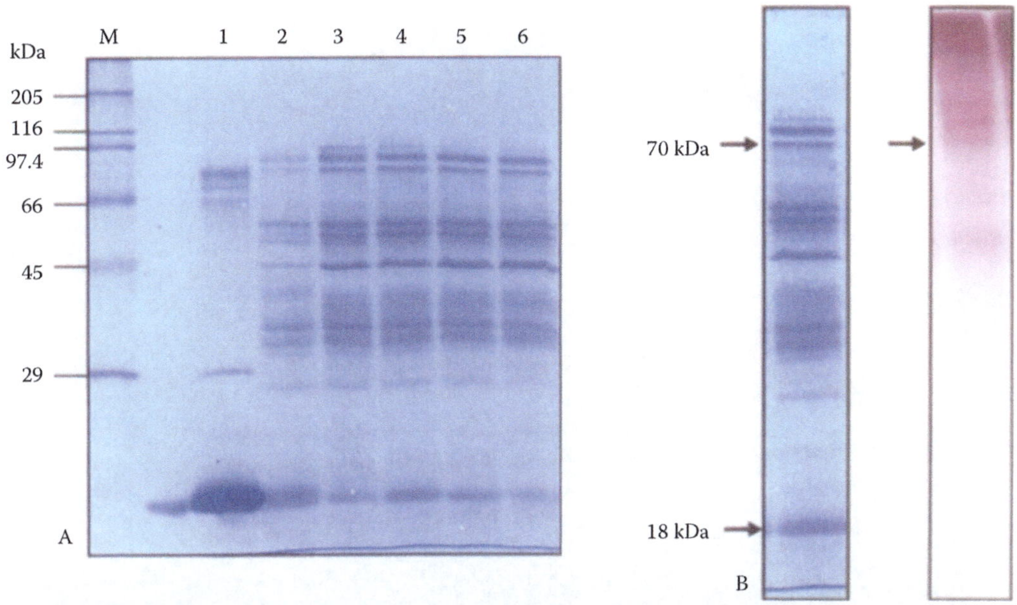

FIGURE 7.15 (b) Electrophoretic protein profile of a fungus *Aspergillus fumigates* with known-molecular-weight marker protein.

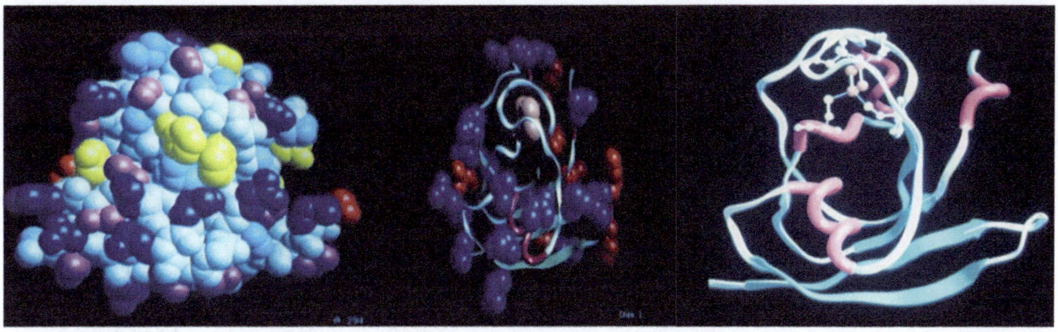

FIGURE 8.12 Computer-generated tertiary structures of crystallographic data of ferredoxin proteins FdxH1 and FDxH2 from the cyanobacterium *Anabaena variabilis.*

FIGURE 10.29 A laboratory-grade fermentor of capacity 2 L containing MRS broth.

FIGURE 10.30 A fermentor after 24 hours of incubation. The visible color difference in the media shows a high cell density of *S. thermophilus*.

6.4.2 Density Gradient Centrifugation

Density gradient centrifugation is the preferred method for purifying subcellular organelles and macromolecules. Density gradients can be generated by placing layer after layer of gradient media (Table 6.2) such as sucrose in a tube with the heaviest layer at the bottom and the lightest at the top in either a discontinuous or continuous mode. The cell fraction to be separated is placed on top of the layer and centrifuged. Density gradient separation can be classified into two categories: (1) rate-zonal (size) separation and (2) isopycnic (density) separation.

1. Rate-zonal (size) separation: Rate-zonal separation takes advantage of particle size and mass instead of particle density for sedimentation. Figure 6.5 illustrates a rate-zonal separation process and the criteria for successful rate-zonal separation. Examples of common applications include separation of cellular organelles, such as endosomes, and separation of proteins, such as antibodies. Antibody classes have very similar densities but different masses. Thus, separation based on mass separates the different classes, whereas separation based on density is not able to resolve these antibody classes. Certain types of rotors are more applicable for this type of separation than others (see Section 6.5 on rotor categories and Table 6.3).

TABLE 6.2
Application of Density Gradient Media for Isopycnic Separation

Gradient Media	Cells	Viruses	Organelles	Nucleoproteins	Macromolecules
Sugars (e.g., sucrose)	+	+ + +	+ + +	+	–
Polysaccharides (e.g., Ficoll)	+ +	+ +	+ +	–	–
Colloidal silica (e.g., Percoll)	+ + +	+	+ + +	–	–
Iodinated media (e.g., Nycodenz)	+ + + +	+ +	+ + + +	+ + +	+
Alkali metal salts (e.g., CsCl)	–	+ +	–	+ +	+ + + +

Notes: + + + + denotes excellent, + + + denotes good, + + denotes good for some applications, + denotes limited use, and – denotes unsatisfactory. Other rotors include continuous-flow and elutriation rotors.

TABLE 6.3
Types of Rotors and Their Applications

Type of Rotor	Pelleting	Rate-Zonal Sedimentation	Isopycnic
Fixed-angle	Excellent	Limited	Variable[a]
Swinging-bucket	Inefficient	Good	Good[b]
Vertical	NS	Good	Excellent
Zonal	NS	Excellent	Good

Note: NS = Not suitable.

[a] Good for macromolecules, and poor for cells and organelles.

[b] Good for cells and organelles, caution needed if used with CsCl.

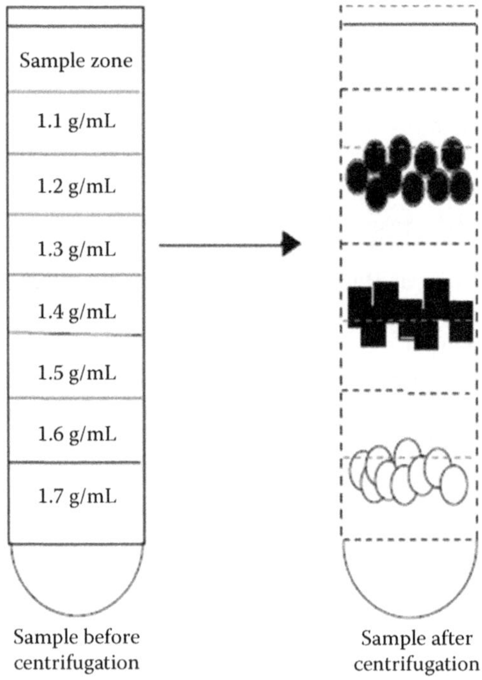

FIGURE 6.6 Isopycnic separation.

The criteria for successful rate-zonal centrifugation are as follows:

- Density of the sample solution must be less than that of the lowest-density portion of the gradient.
- Density of the sample particle must be greater than that of the highest-density portion of the gradient.
- Path length of the gradient must be sufficient for the separation to occur.
- Time is important. If you perform runs that are too long, particles may all pellet at the bottom of the tube.

2. Isopycnic separation: In this type of separation, a particle of a particular density will sink during centrifugation until a position is reached where the density of the surrounding solution is exactly the same as the density of the particle. Once this quasi-equilibrium is reached, the length of centrifugation does not have any influence on the migration of the particle. A common example for this method is the separation of nucleic acids in a CsCl gradient (Figure 6.6). A variety of gradient media can be used for isopycnic separations; their biological applications are listed in Table 6.2.

   The criteria for successful isopycnic separation are as follows:

   - Density of the sample particle must fall within the limits of the gradient densities (Figure 6.7).
   - Any gradient length is acceptable.
   - The run time must be sufficient for particles to band at their isopycnic point. Excessive run times have no adverse effect.

6.5 ROTOR CATEGORIES

Rotors can be broadly classified into three common categories: (1) swinging-bucket rotors, (2) fixed-angle rotors, and (3) vertical rotors (Figure 6.8; Table 6.3). Note that each type of rotor has strengths and limitations depending on the type of separation.

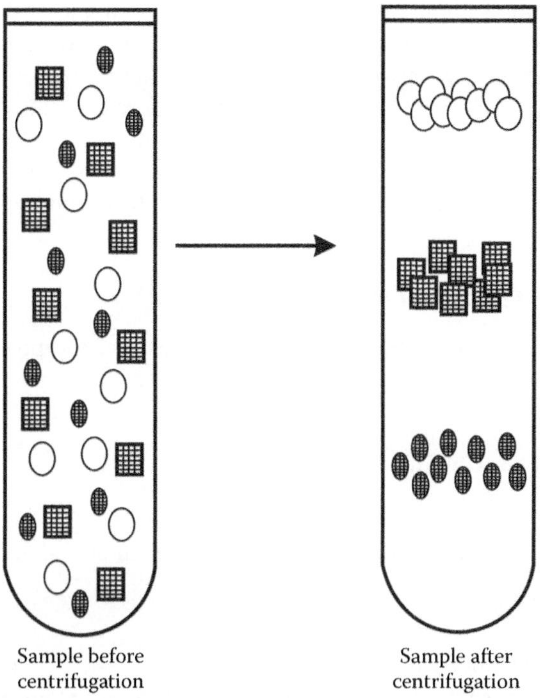

FIGURE 6.7 Rate-zonal (size) separation.

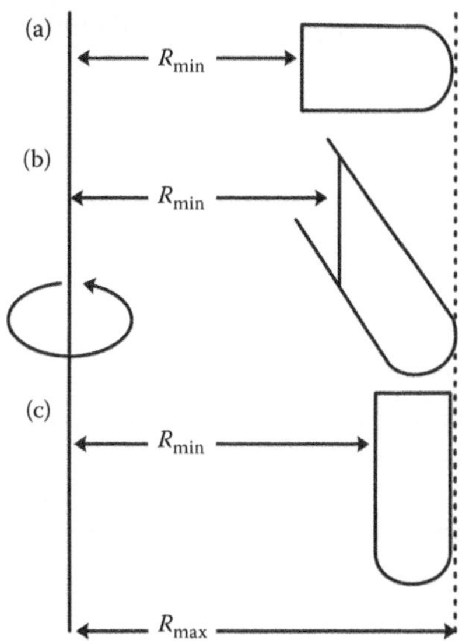

FIGURE 6.8 Figure showing rotor types: (a) swinging-bucket, (b) fixed-angle, and (c) vertical rotors.

In swinging-bucket rotors, the sample tubes are loaded into individual buckets that hang vertically while the rotor is at rest. When the rotor begins to rotate, the buckets swing out to a horizontal position (Figure 6.8). This rotor is particularly useful when samples are to be resolved in density gradients. The longer path length permits a better separation of individual particle types from a mixture. However, this rotor is relatively inefficient for pelleting. Also, care must be taken to avoid "point loads" caused by spinning CsCl or other dense gradient materials that can precipitate.

In fixed-angle rotors, the sample tubes are held fixed at the angle of the rotor cavity. When the rotor begins to rotate, the solution in the tubes reorients (Figure 6.8). This rotor type is most commonly used for pelleting applications. Examples include pelleting bacteria, yeast, and other mammalian cells. It is also useful for isopycnic separations of macromolecules such as nucleic acids.

In vertical rotors, sample tubes are held in the vertical position during rotation. This type of rotor is not suitable for pelleting applications but is most efficient for isopycnic (density) separations due to the short path length. Applications include plasmid DNA, RNA, and lipoprotein isolations.

6.6 SELECTION OF CENTRIFUGE TUBES

Tables 6.4 and 6.5 illustrate properties of centrifuge tubes and the proper rotors in which they should be used. Selection of the appropriate centrifuge tube

- Prevents sample leakage or loss
- Ensures chemical compatibility
- Allows easy sample recovery

Major factors considered in the selection of a tube (plastic) material are as follows:

- Clarity.
- Chemical resistance.
- Sealing mechanism (if needed).
- Check product guide pages or tube packaging for notes on recommended sample volume and maximum speed.
- Always run thin-walled, sealed tubes full in a fixed-angle or vertical rotor with open top tubes with multiple sealing assemblies and reseal tubes.
- Autoclave tubes only if it is absolutely necessary and only at 121°C for 15 minutes.

TABLE 6.4
Chemical Compatibility of Popular Tube Materials

Tube Plastic Type	Clarity	Chemical Resistance
Polypropylene (PP)	Opaque	Good
Polyallomer (PA)	Opaque	Good
Polycarbonate (PC)	Clear	Poor
Polyethylene terephthalate (PET)	Clear	Poor

TABLE 6.5
Tube Type and Rotor Compatibility

	Rotor Type		
Tube Type	Fixed Angle	Swinging Bucket	Vertical
Thin wall, open top	No	Yes	No
Thick wall, open top	Yes	Yes	No
Thin wall, sealed	Yes	Some tube types	Yes
Oak ridge	Yes	No	No

- Avoid cleaning plastic tubes in automated dishwashers or glassware washers, which may produce excessively hot temperatures.
- We recommend that you clean tubes with a mild laboratory detergent in warm water, rinse, and then air dry.
- Tube must be carefully matched with rotor type to prevent sample loss and/or failure, as illustrated in Table 6.5, and to prolong tube life and avoid tube breakage or collapse.

6.7 COMMON CENTRIFUGATION VOCABULARY AND FORMULAS

This section discusses common centrifugation vocabulary and formulas.

- *Pellet*: Hard-packed concentration of particles in a tube or rotor after centrifugation.
- *Supernatant*: The clarified liquid above the pellet.
- *Adapter*: A device used to fit smaller tubes or centrifugal devices in rotor cavities.
- *RPM*: Revolutions per minute (speed).
- R_{max}: Maximum radius from the axis of rotation in centimeters.
- R_{min}: Minimum radius from the axis of rotation in centimeters.
- *RCF*: Relative centrifugal force; 1 RCF = $11.17 \times R_{max}$ $(\text{RPM}/1000)^2$.
- *K-factor*: Pelleting efficiency of a rotor. The smaller the *K*-factor the better the pelleting efficiency:

$$K = \frac{2.53 \times 10^{11} \ln(R_{max}/R_{min})}{(\text{RPM})^2}$$

- *S-value*: The sedimentation coefficient is a number that gives information about the molecular weight and shape of a particle. The S-value is expressed in Svedberg units. The larger the S-value the faster the particle separates. (For more information on sedimentation coefficients, refer to the suggested readings section of this chapter.)
- *Pelleting time*: Time taken to pellet a given particle. $T = K/s$, where T is pellet time in hours, K is the *K*-factor of the rotor, and s is the sedimentation coefficient.
- *Rotor conversion formula*: If the time to pellet a sample in an "old" rotor is known, one can determine the time it takes to pellet the same sample in a "new" rotor. The formula for this determination is as follows:

$$\frac{T_1}{K_1} = \frac{T_2}{K_2} \rightarrow T_1 = T_2\left(\frac{K_1}{K_2}\right)$$

where

T_1 = time to pellet in the new rotor
T_2 = time to pellet in the old rotor
K_1 = *K*-factor of the new rotor
K_2 = *K*-factor of the old rotor

An example of a rotor conversion is as follows. The old rotor used is a Beckman® JA-10 and the new rotor used is a Sorvall® SLC-1500:

$$T_2 = 20 \text{ minutes}, \; K_\partial = 3610; \quad T_1 = 10 \text{ minutes}, \; K_1 = 1676$$

It is noted the old pelleting time is 20 minutes and new pelleting time is 9.2 minutes.

6.8 ANALYTICAL ULTRACENTRIFUGE

The ultracentrifuge is a centrifuge optimized for spinning a rotor at very high speeds; it is capable of generating accelerations as high as 1,000,000 *g* (9,800 km/s²). There are two kinds of ultracentrifuges: (1) preparative and (2) analytical. Both classes of instruments find important uses in molecular biology, biochemistry, and polymer science.

The analytical ultracentrifuge was first developed by Svedberg and associates in the 1920s. This tool proved invaluable in the study of macromolecules and provided some of the first evidence that proteins are indeed macromolecules comprising a huge number of atoms linked by covalent bonds. Since its creation, the ultracentrifuge has helped in further understanding the behavior of macromolecules.

6.8.1 Theory of Ultracentrifugation

A particle suspended in a solvent that is subjected to a gravitational field experiences several forces. We can draw the following diagram describing three forces acting on such a solute particle:

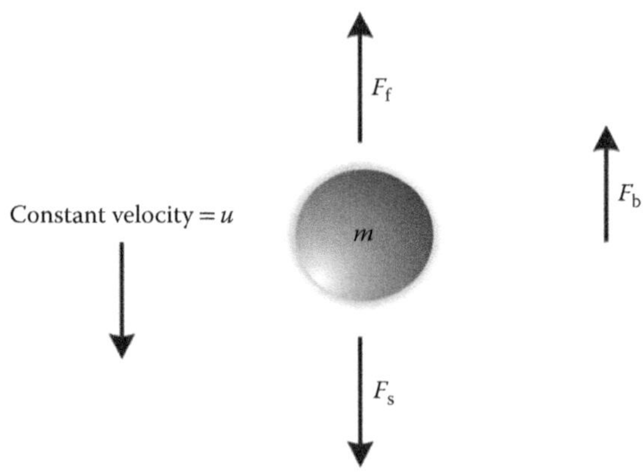

The sedimenting, or gravitational, force, F_s, is proportional to the mass of the particle and acceleration. For a particle subjected to a rotation, the acceleration is determined by the distance of the particle from the axis of rotation, *r*, and the square of the angular velocity, *w* (in radians per second), such that

$$F_s = m\omega^2 r = \frac{M}{N}\omega^2 r$$

where *m* is the mass of the single particle in grams, *M* is the molar weight of the solute in grams per mole, and *N* is Avogadro's number.

The buoyant force, F_b, from Archimedes' principle, is equal to the weight of the fluid displaced by the particle:

$$F_b = -m_0\,\omega^2 r$$

where m_0 is the mass of fluid displaced by the particle:

$$m_0 = m\bar{v}\rho = \frac{M}{N}\bar{v}\rho$$

Here, $v\rho$ is the volume in milliliters occupied by each gram of solute in the solution and *r* is the density of the solvent in grams per milliliter. If the density of a particle is greater than that of the solvent, the particle will sediment or sink. As the particle begins to move along a radial path toward the bottom of the cell, its velocity, u, will increase because of increasing radial resistance. Since the particle is moving through a viscous fluid, it will experience a frictional drag, F_f, which is proportional to the velocity:

$$F_f = -fu$$

where f is the coefficient of friction, which depends on the size and shape of the particle.

Within a very short time (on the order of 10^{-6} second), the forces come into balance. Then we can write the following force balance equation:

$$F_s + F_b + F_f = 0$$

$$\frac{M}{N}\omega^2 r - \frac{M}{N}\bar{v}\rho\omega^2 r - fu = 0$$

Rearranging, we get the following equation:

$$\frac{M}{N}(1-\bar{v}\rho)\omega^2 r - fu = 0$$

Collecting the terms that relate to particle size on one side and terms that relate to experimental conditions on the other side of the equation, we can write

$$\frac{M(1-\bar{v}\rho)}{Nf} = \frac{u}{\omega^2 r} \equiv s$$

The term u/w^2r is defined as, the velocity of the particle per unit gravitational acceleration, as the sedimentation coefficient. This coefficient depends on the properties of the particle and, in particular, it is proportional to the buoyant effective molecular weight of the particle. Also, it is independent of the operating conditions. Therefore, molecules with different molecular weights or different shapes and sizes, will, in general, move with different velocities in a given centrifugal field, that is, they will have different sedimentation coefficients.

6.8.2 Analytical Ultracentrifugation

An analytical ultracentrifuge spins a rotor at an accurately controlled speed and temperature. The concentration distribution of the sample is determined at known times using absorbance measurements. The concentration, c, is determined for solutes obeying the Beer–Lambert law:

$$A = \varepsilon \cdot c \cdot l$$

where the absorbance of the sample, A, is measured at a given wavelength, ε, knowing the fixed position in the cell, l.

Figure 6.9 displays a schematic diagram of the Beckman Optima XL-A absorbance system. A high-intensity xenon flask lamp allows the use of wavelengths between 190 and 800 nm. The lamp is fired briefly as a selected sector passes the detector.

Sedimentation velocity (SV) cells are cylindrical and have double-sector centerpieces (Figure 6.10 is a top-down view of the cylindrical centerpiece). One sector is for loading samples, and the other is the reference sector and contains the solvent. The reference sector is filled slightly more than the sample sector so that the reference meniscus does not obscure the sample profile (Figures 6.11 and 6.12).

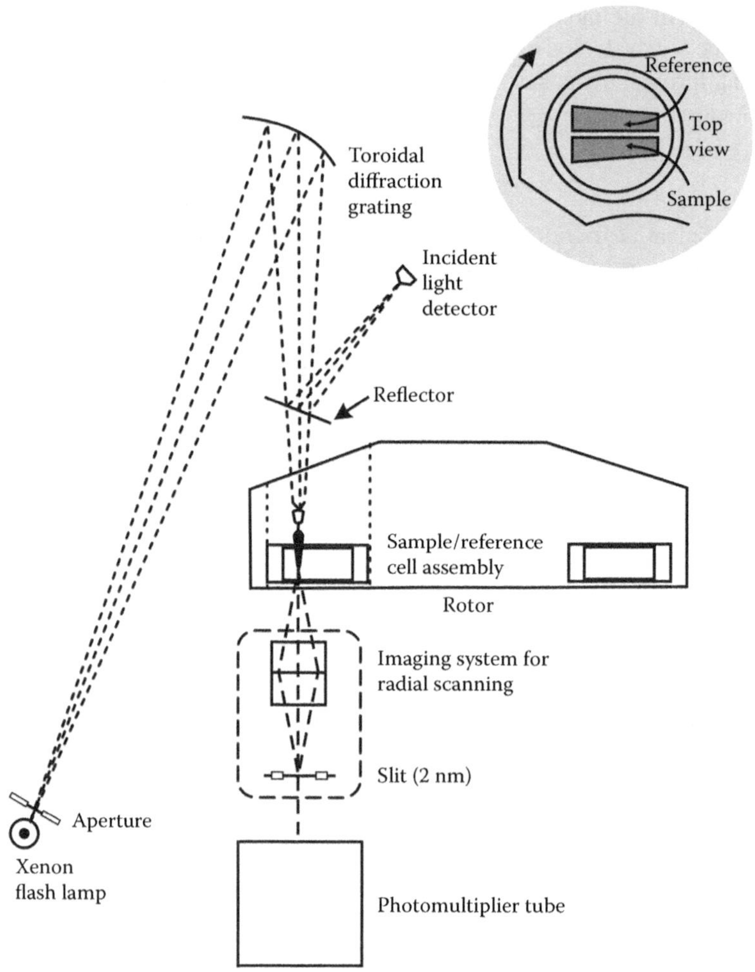

FIGURE 6.9 Schematic diagram of the Beckman Optima XL-A absorbance system.

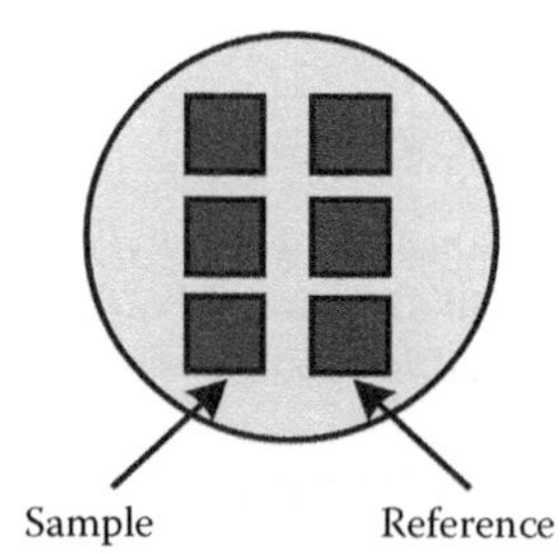

FIGURE 6.10 Top-down view of the cylindrical centerpiece.

For an SV experiment, an initially uniform solution is placed in a cell and a sufficiently high angular velocity is applied to cause rapid sedimentation of the solute toward the cell bottom. As a result, there is a depletion of solute near the meniscus, causing a characteristic spectrum as shown in Figure 6.13. A sharp boundary occurs between the depleted region and the sedimenting solute (the plateau).

Although the velocity of individual particles in SV experiments cannot be resolved, the rate of movement of the boundary region can be measured. From this, the sedimentation coefficient, *s*,

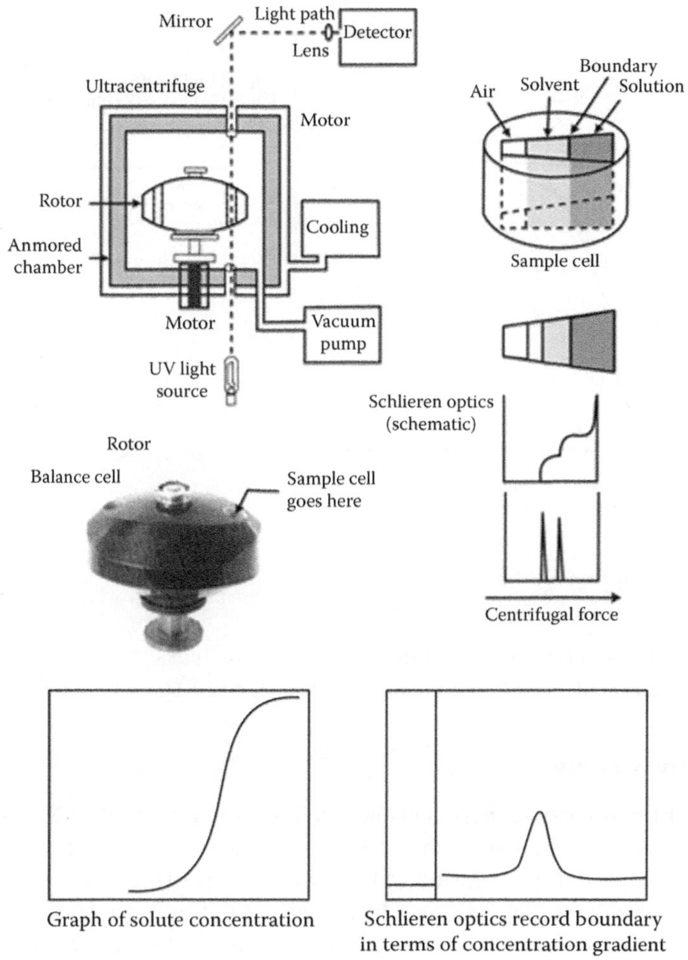

FIGURE 6.11 Rotor movement in an analytical centrifuge.

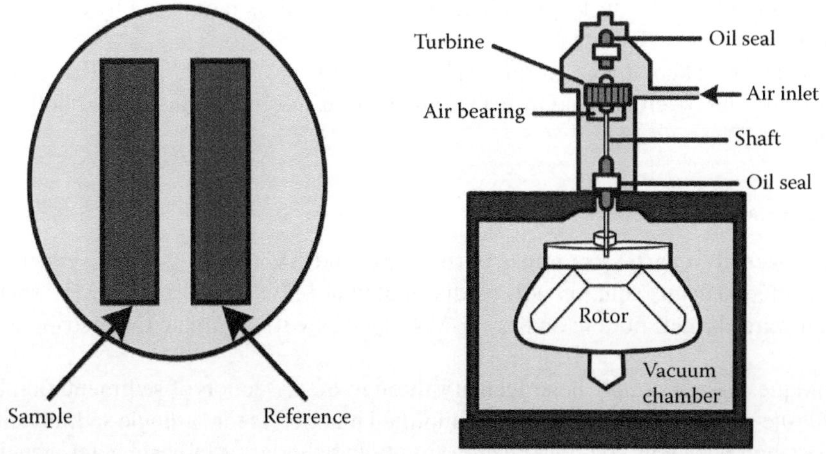

FIGURE 6.12 Examination of SV.

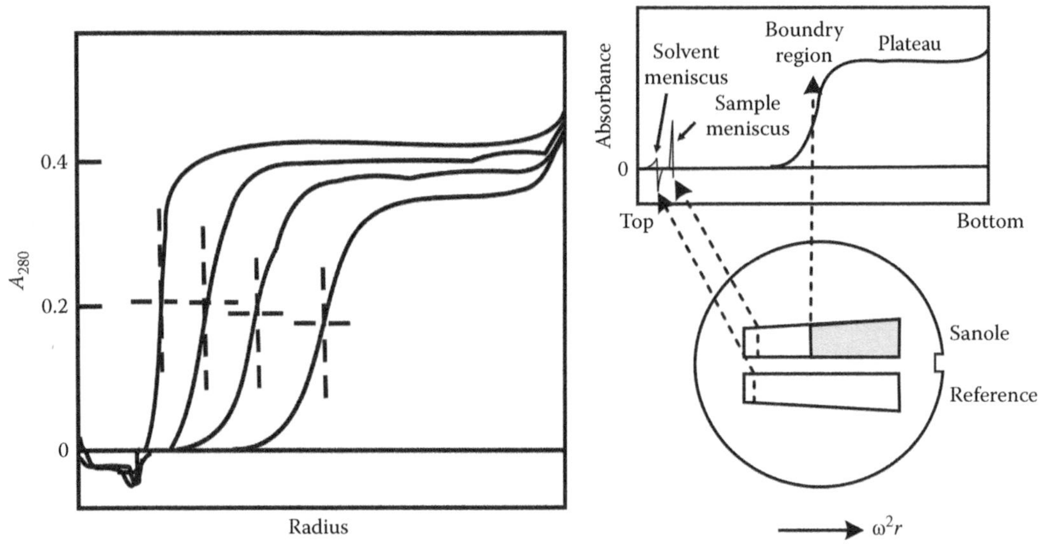

FIGURE 6.13 Determination of SV. "A" stands for absorbance at 280 nm.

can be determined. Remember that *s* depends directly on the mass of solute particles and inversely on the frictional coefficient, which is a measure of the size of the solute particles (Figure 6.10).

6.8.3 Examination of SV

Sedimentation equilibrium (SE) experiments have a lower rotor speed than SV experiments. Solute particles do not pellet at the bottom of the cell; instead, the process of diffusion opposes the process of sedimentation until, after a period of time, the two opposing forces reach equilibrium and the apparent concentration profile does not change. At equilibrium, the concentration of the solute increases exponentially toward the cell bottom. Each column displays a different absorbance profile, because the concentrations of sample are varied in each (Figure 6.14).

There are six columns in the SE experiment. One row is the sample row, and the other is the reference row containing only the solvent (specifically, the sample buffer). As in the SE experiment, the reference columns are filled more than the sample columns. In the solvent columns concentrations are varied, usually 0.25, 0.5, and 0.8 optical density (OD) (280), or the same fraction of absorbance is obtained at any selected wavelength where that wavelength is monitored by analytical ultracentrifugation (AU).

Several scans are taken at a given rotor speed to try to ascertain definitely that equilibrium has been reached (if it has been reached in scan 1 and you take scan 2 an hour later, both scans should look the same).

6.8.4 Examination of Sample Purity

Sample heterogeneity can be examined by both SE and SV methods. In SE experiments, each species of a heterogeneous solution will be distributed at SE such that higher-MW species will be distributed toward the cell bottom and lower-MW species will dominate the distribution at the top of the cell.

SV techniques assess sample heterogeneity through the detection of sedimentation boundaries. The general rule (although stated in an oversimplified manner) is that a single sedimentation boundary exists in a homogeneous solution, whereas multiple boundaries indicate heterogeneity.

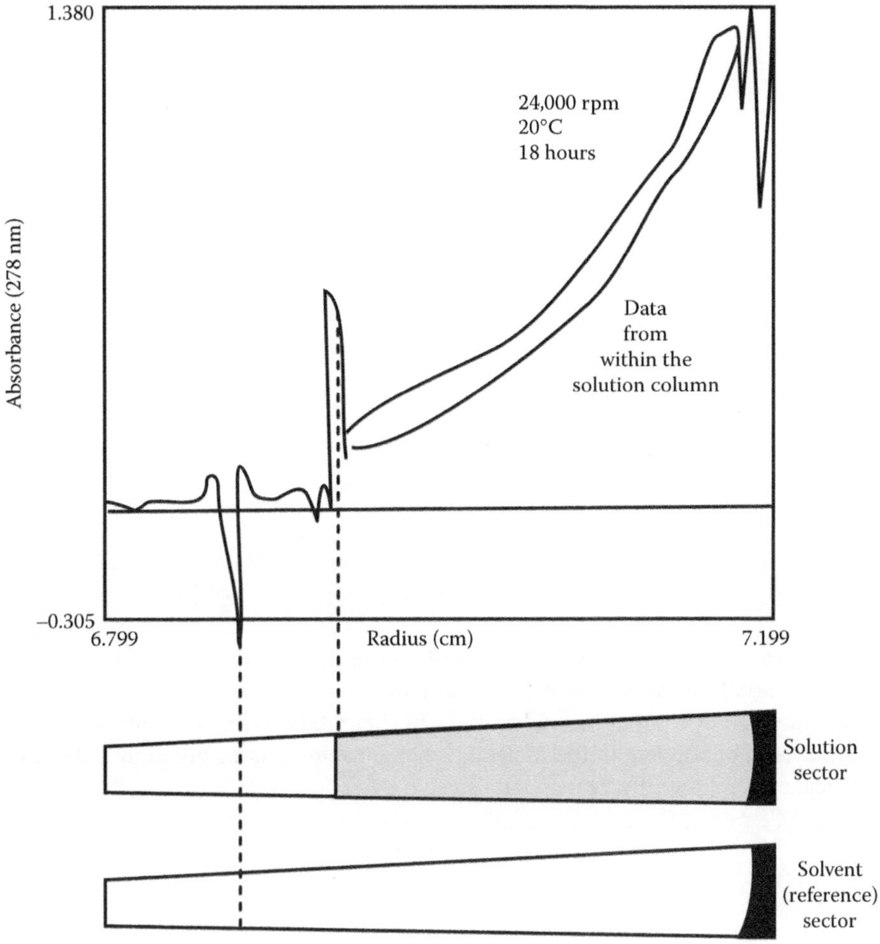

FIGURE 6.14 Autoscanning of absorbance during AU.

6.8.5 Determination of Molecular Weight

AU techniques for MW measurement are superior because they work with small sample sizes. The method is applicable to MW ranges from 100 to 1,000,000 (i.e., proteins, nucleic acids, and carbohydrates work with AU methods). AU techniques do not rely on assumptions or calibration; they work on any substance whose absorbance differs from that of the solvent. A final advantage is the fact that experimental design is simple in comparison with some other techniques of MW determination.

The molecular weights of biological materials can be determined using the analytical centrifuge. The molecular weight, *M*, is related to the sedimentation coefficient, *s*, of a molecule by the following equation:

$$M = \frac{RTs}{D\,(1 - v\rho)}$$

where

D = diffusion coefficient of the molecule

V = partial specific volume of the molecule (volume increases when 1 g of solute is added to an infinite volume of solution)

ρ = density of solvent at 20°C

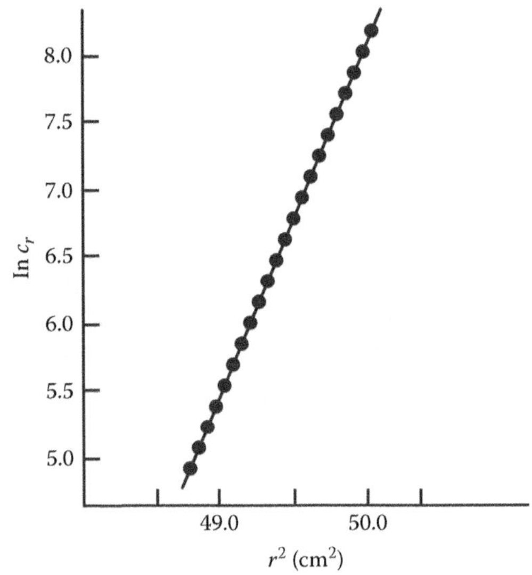

FIGURE 6.15 Sedimentation equilibrium for bovine serum albumin in 0.1 M NaCl at 27,700 rpm and 20°C.

The sample containing the molecule is centrifuged at high speeds in an ultracentrifuge. The molecules move radially outward, creating a distinct boundary between the solvent containing the molecules and the solvent without the molecules. This boundary moves radially outward depending on the sedimentation coefficient of the molecule. The movement is recorded by either Schlieren or Rayleigh optical system; thereby, D is found. Since

$$v = s\varpi 2r$$

$$s = \frac{v}{\varpi 2r}$$

Here, $\varpi 2r$ and v are known values; hence, s can be found and used to calculate M.

As a better alternative, equilibrium sedimentation can be used to determine the molecular weight. Centrifugation is continued until the solute particles acquire a static position in the tube. This happens when there is equilibrium between sedimentation due to diffusion. Using the concentration gradient, the molecular weight (M) can be calculated from the following formula:

$$M = \frac{2RT}{(1-\bar{v}\rho)\omega^2} \text{In} \frac{c_r}{c_\alpha} \cdot \frac{1}{r^2 - a^2}$$

where, RT is the gas constant at temperature T, $v\rho$ is patial specific volume, c_r is the concentration of the solute at a distance r from the axis of rotation and c_α is the distance of the meniscus from the axis of rotation. Hence, a plot of In (normal log) c_r versus r^2 is a straight line from whose slope M can be calculated (Figure 6.15).

6.8.6 Detection of Conformational Changes

Since the rate of sedimentation varies with not only the size but also the shape of a molecule, changes in conformation, for example, change of dsDNA to ssDNA, cause the molecule to band at different isodensity positions, thereby revealing the change in conformations.

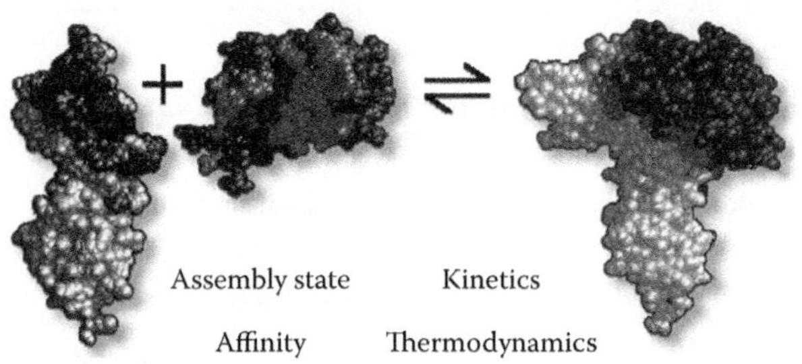

FIGURE 6.16 Demonstration of conformational changes of protein due to association and affinity binding.

6.8.7 Analysis of Associating Systems

Sedimentation analysis can yield valuable information relating to changes in molecular weight when a molecule associates to form more complex structures (Figure 6.16). Note that SE experiments can yield the following:

- The monomer MW
- The complex MW
- The stoichiometry of heterogeneous components
- The strength of interactions between components
- The thermodynamic nonideality of the solution

6.8.8 Ligand Binding

Absorbance measurement is well suited for the study of ligand binding because since the technique can distinguish ligands from acceptors. Ligands and acceptors can be labeled with a chromophore, provided that the modification does not alter the binding. When a ligand and an acceptor differ greatly in sedimentation coefficient, AU analysis of ligand/acceptor binding interactions is a simple matter.

6.8.9 Cell Fractionation and Metabolic Studies

The metabolic function of individual cell organelles and soluble components within the cell can be studied by separating the cell components from each other by centrifugation. When cells are subjected to high shear forces, the cell membrane ruptures and its contents are released into the medium. This is usually done by grinding animal or plant tissue in a pestle and mortar, and the process is called homogenization. Chloroplast material, blood cells, unicellular organisms, plant and animal tissue homogenates can also be ruptured in a pressure cell and sonication. Pressure cells, for example the French pressure cell, use hydraulic pressure controlled by a motor-driven pump to produce shear forces. The sonicator uses ultrasonic waves to produce cavitational forces within the suspending medium, which cause the cells to burst.

The homogenate is then suspended in a medium, which should be cheap, uncharged, and metabolically inert. Following this, either density gradient centrifugation or differential centrifugation can be used to separate the cell components. As described in Section 6.4.2, the cell components band at the isodensity positions. In differential centrifugation, the homogenate is separated into a number of fractions by centrifuging at various *g* values. The cell components sediment to form pellets at different rates according to their size and density. For example, rat liver homogenate can be separated into different components by subjecting it to differential centrifugation.

TABLE 6.6
Application of Centrifugation for Different Cell Organelles

Centrifugation Conditions *g* Values	Time (minutes)	Major Components in Fraction
500	5	Nuclei, whole cells
8,000	10	Mitochondria, some lysosomes
15,000	10	Mitochondria, lysosomes
100,000	60	Microsomes (membrane fragments, endoplasmic reticulum), ribosomes, soluble components of the cell
Final supernatant		

Certain tissues, for example pig brain tissues, are very heterogeneous in nature; thus, differential centrifugation of pig brain tissues gives rise to fractions with different compositions than those of rat liver (Table 6.6). Each fraction contains a very complex mixture of components, and density gradient centrifugation is required to purify them further.

Experiment: Isolation of Chloroplasts by Differential Centrifugation

Requirements

Fresh spinach; clean, sharp sand; 50 mL 0.5 M sucrose (17% w/v); cheesecloth; 12 × 12 in ice; ice bath; 25 mL graduated cylinder; mortar and pestle (or blender); table top clinical centrifuge; glass filter funnel; two 16 × 150 mm test tubes in rack; three 13 × 100 mm test tubes in rack; plastic-capped 15 mL centrifuge tube; double pan balance; glass stirring rods.

Procedure

1. Grind 8 g of deveined spinach with ½ teaspoon clean, sharp sand in a mortar and pestle to a paste.
2. Suspend in 0.5 M sucrose. Measure out 16 mL ice-cold 0.5 M sucrose solution in a 25 mL graduated cylinder. Add in 3–4 mL increments; grind to a smooth pulp with each addition (a blender may be used for volumes greater than 100 mL).
3. Filter the solution. Homogenate the solution through about eight layers of clean cheesecloth in a glass funnel into an ice-cold 16 × 150 mm test tube.
4. Pour the filtrate back into the 25 mL cylinder and record the volume. Save approximately 0.5 mL of the filtrate (F1) in a labeled 13 × 100 mm test tube to examine at 400× under a microscope to determine the composition, and illustrate the observations in a notebook. Note the appearance of components and degree of heterogeneity (label cells, ghosts, chloroplasts, mitochondria, and debris).
5. Centrifuge at low speed. Prepare a balance tube against the filtrate in a 16 × 150 tube and spin at 50 g for 10 minutes (speed 2 on the clinical centrifuge).
6. Decant the top 10 mL into a clean cold centrifuge tube, discard sediment, and record the volume. Save ~0.5 mL supernatant (S1) to examine under the microscope in order to determine composition; illustrate and label as in step 2.
7. Centrifuge the supernatant from step 3 opposite a carefully balanced tube at 1000 g for 10 minutes (speed 7) to precipitate chloroplasts. How does the supernatant appear? Does it appear as a precipitate? Carefully decant all the supernatant into the 16 × 150 mm tube, but save the pellet. Discard the supernatant if you have a significant pellet (you will lose some soft pellet, but there is no need to worry).
8. Resuspend the pellet from step 4 to one-tenth the volume of the step 2 filtrate in ice-cold 0.5 M sucrose with a clean, ice-cold stirring rod. Record the final volume. Keep on ice at all times. Examine the suspended organelles (SOs) under the microscope to determine composition; illustrate as in step 2.

Experiment: Isolation of Plasmid DNA by Centrifugation

The isolation of plasmid DNA from E. coli is a common routine in research laboratories. You will perform a widely practiced procedure that involves alkaline lysis of cells. This protocol, which is often referred to as plasmid "miniprep," yields fairly clean DNA quickly and easily.

Procedure

1. Fill a microcentrifuge tube with saturated bacterial culture grown in LB broth + antibiotic. Spin the tube in a microcentrifuge for 1 minute, and make sure the tubes are balanced in the microcentrifuge. Dump the supernatant and drain the tube briefly on a paper towel.
2. Repeat step 1 of the first experiment in the same tube, filling the tube again with more bacterial culture. The purpose of this step is to increase the starting volume of cells so that more plasmid DNA can be isolated per preparation. Spin the tube in a microcentrifuge for 1 minute. Pour off the supernatant and drain the tube on a paper towel.
3. Add 0.2 mL ice-cold solution 1 to a cell pellet and resuspend cells as much as possible using a disposable transfer pipette. Solution 1 contains glucose, Tris, and ethylene-diamine-tetraacetic acid (EDTA). Glucose is added to increase the osmotic pressure outside the cells. Tris is a buffering agent used to maintain a constant pH of 8.0. The EDTA protects the DNA from degradative enzymes (called DNAses); the EDTA binds divalent cations that are necessary for DNAse activity.
4. Add 0.4 mL solution 2, cap the tubes, and shake five time gently by inverting to mix uniformly. Let the tubes sit at room temperature for 5 minutes. Solution 2 contains NaOH and sodium dodecyl sulfate (SDS; a detergent). The alkaline mixture ruptures the cells, and the detergent breaks apart the lipid membrane and solubilizes cellular proteins. The NaOH also denatures the DNA into single strands.
5. Add 0.3 mL ice-cold solution 3, cap the tubes, and invert five times gently. Incubate tubes on ice for 10 minutes. Solution 3 contains a mixture of acetic acid and potassium acetate. The acetic acid neutralizes the pH, allowing the DNA strands to renature. The potassium acetate also precipitates the SDS from solution, along with the cellular debris. The E. coli chromosomal DNA, a partially renatured tangle at this step, is also trapped in the precipitate. The plasmid DNA remains in the solution.
6. Centrifuge the tubes for 5 minutes. Transfer the supernatant to a fresh microcentrifuge tube using a clean disposable transfer pipette. Try to avoid taking any white precipitate during the transfer. It is better to leave a little supernatant behind to avoid accidentally taking the precipitate. This fractionation step separates plasmid DNA from the cellular debris and chromosomal DNA in the pellet.
7. Fill the remainder of the centrifuge tube with isopropanol. Let the tube sit at room temperature for 2 minutes. Isopropanol effectively precipitates nucleic acids, but it is much less effective with proteins. A quick precipitation can therefore purify DNA from protein contaminants.
8. Centrifuge the tubes for 5 minutes. A milky pellet should form at the bottom of the tube. Pour off the supernatant without dumping out the pellet. Drain the tube on a paper towel. This fractionation step further purifies the plasmid DNA from contaminants. Cap the tubes and store them in a freezer.
9. Add 1 mL of ice-cold 70% ethanol. Cap the tube and mix by inverting it several times. Spin the tubes for 1 minute. Pour off the supernatant (one must be careful not to dump out the pellet) and drain the tube on a paper towel. Ethanol helps to remove the remaining salts and SDS from the preparation.
10. Allow the tube to dry for about 5 minutes. Add 50 μL Tris EDTA buffer (TE) to the tube. The DNA is ready for use and can be stored indefinitely in the freezer.

Solutions

Solution 1:

50 mM glucose	per 500 mL
25 mM Tris–HCl pH 8.0	9 mL 50% glucose

10 mM EDTA pH 8.0	12.5 mL 1 M Tris–HCl pH 8.0
Add H_2O to 500 mL	10 mL 0.5 M EDTA pH 8.0
Solution 2:	
1% SDS	per 500 mL
0.2 N NaOH	50 mL 10% SDS
Add H2O to 500 mL	100 mL 1N NaOH
Solution 3:	
3 M K +	per 500 mL
5 M acetate	300 mL 5 M potassium acetate
Add H2O to 500 mL TE	57.5 mL glacial acetic acid
10 mM Tris–HCl pH 8.0	per 100 mL
1 mM EDTA 0.5 mL	1 mL 1 M Tris–HCl pH 8.0
Add H2O to 100 mL	0.5 M EDTA pH 8.0

Optional: RNAse can be added to TE at a final concentration of 20 μg/mL.

SUGGESTED READING

Birnie, G. D., and D. Rickwood. 1978. *Centrifugation Separations in Molecular and Cell Biology*. London: Butterworth.

Ford, T. C., and J. M. Graham. 1991. *An Introduction to Centrifugation*. Oxford: Bios Scientific.

Griffith, O. M. 1983. *Techniques in Preparative, Zonal and Continuous flow Ultracentrifugation*, 4th ed. Palo Alto, CA: Beckman instruments Inc.

Ralston, G. 1993. *Introduction to Analytical Ultracentrifugation*. Palo Alto, CA: Beckman Instruments Inc.

Rickwood, D. 1984. *Centrifugation*, 2nd ed. Published in the Practical Approaches to Biochemistry series. Oxford/Washington, DC: IRL press.

Rickwood, T. C., and Steensgard, F. J. 1994. *Centrifugation Essential Data.* London: John Wiley & Sons Ltd.

Sharpe, P. T. 1998. "Laboratory Techniques in Biochemical and Molecular Biology." In *Methods of Cell Separation*, vol. 18, edited by R. H. Burden, and P. H. van Knippenberg. New York: Elsevier.

Van Holde, K. E., W. Johnson, and P. Shing Ho. 1998. *Principles of Physical Biochemistry*. Upper Saddle River, NJ: Prentice Hall.

Wilson, K., and J. Walker. 2003. *Practical Biochemistry-Principles and Techniques*, 5th ed. Cambridge: Cambridge University Press.

IMPORTANT LINKS

1. Centrifuge: https://www.beckmancoulter.com/wsrportal/wsr/research-and-discovery/products- and services/centrifugation/avanti-j-e/index.htm
2. Ultracentrifuge: https://www.beckmancoulter.com/wsrportal/wsr/research-and-discovery/products- and services/centrifugation/optima-tlx/index.htm

7 Electrophoresis

7.1 INTRODUCTION

Gel electrophoresis (GE) is a method that separates macromolecules on the basis of size, electric charge, and other physical properties. The term electrophoresis describes the migration of charged particles under the influence of an electric field. "Electro" refers to electricity and "phoresis," from the Greek word *phoros*, means "to carry across." Thus, GE refers to a technique in which molecules are forced across a span of gel, motivated by an electrical current. The driving force for electrophoresis is the voltage applied to electrodes at either end of the gel. The properties of a molecule determine how rapidly an electric field can move it through a gelatinous medium. The complete GE assembly along with the power pack is shown in Figure 7.1, whereas the various positions in which electrophoresis can be carried out, that is, upright, horizontal, and vertical, are shown in Figures 7.2 and 7.3.

Many important biological macromolecules (e.g., amino acids, peptides, proteins, nucleotides, and nucleic acids) possess ionizable groups and, at any given pH, exist in a solution as electrically charged species, either as cations (+) or anions (–). Depending on the nature of the net charge, the charged particles migrate either to the cathode or to the anode. For example, when an electric field is applied across a gel at a neutral pH, the negatively charged phosphate groups of DNA cause it to migrate toward the anode.

7.1.1 Structure of the Agarose GE Instrument

7.1.1.1 Components of the Agarose GE Instrument

The components of the agarose GE instrument are as follows (Figure 7.4):

Gel casting trays: Gel casting trays are available in a variety of sizes and comprise UV-transparent plastic. The open ends of the tray are closed with tape while the gel is being cast; the tape is removed prior to electrophoresis.

Sample combs: Molten agarose is poured around sample combs to form sample wells in the gel.

Agarose: A natural colloid extracted from seaweed, agarose is a linear polysaccharide (average molecular mass is ~12,000 Da) made up of the basic repeated unit agarobiose, which comprises alternating units of galactose and 3,6-anhydrogalactose. Agarose gels have large "pore" sizes and are used primarily to separate large molecules with molecular masses greater than 200 kDa. Table 7.1 gives the recommended agarose gel concentration for resolving linear DNA molecules.

Electrophoresis buffer: The electrophoretic mobility of DNA is affected by the composition and ionic strength of the electrophoresis buffer. Several buffers are available for electrophoresis of native double-stranded DNA. These contain EDTA (pH 8.0) and tris-acetate (TAE), tris-borate (TBE), or tris-phosphate (TPE) at a concentration of approximately 50 mM (pH 7.5–7.8). Electrophoresis buffers are usually prepared as concentrated solutions and stored at room temperature. TBE was originally used at a working strength of 1× for agarose GE.

Marker DNA: For a given voltage and agarose gel and buffer concentrations, migration distance depends on the molecular weight of the starting material. Therefore, marker DNA of a known size should be loaded into slots on both right and left sides of the gel. A marker generally contains a defined number of known DNA segments, which makes it easier to determine the size of the unknown DNAs if any systematic distortion of the gel occurs during electrophoresis.

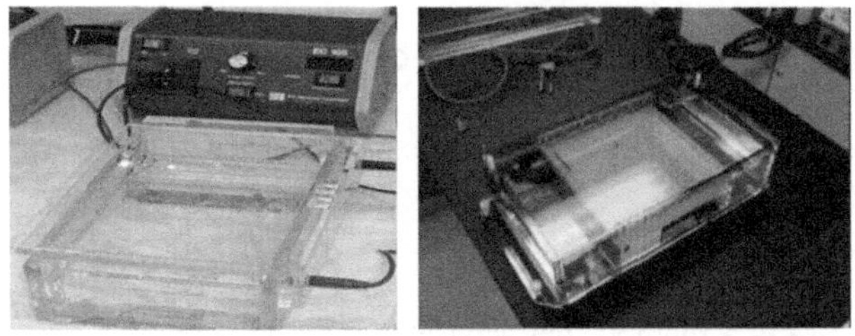

FIGURE 7.1 The GE apparatus.

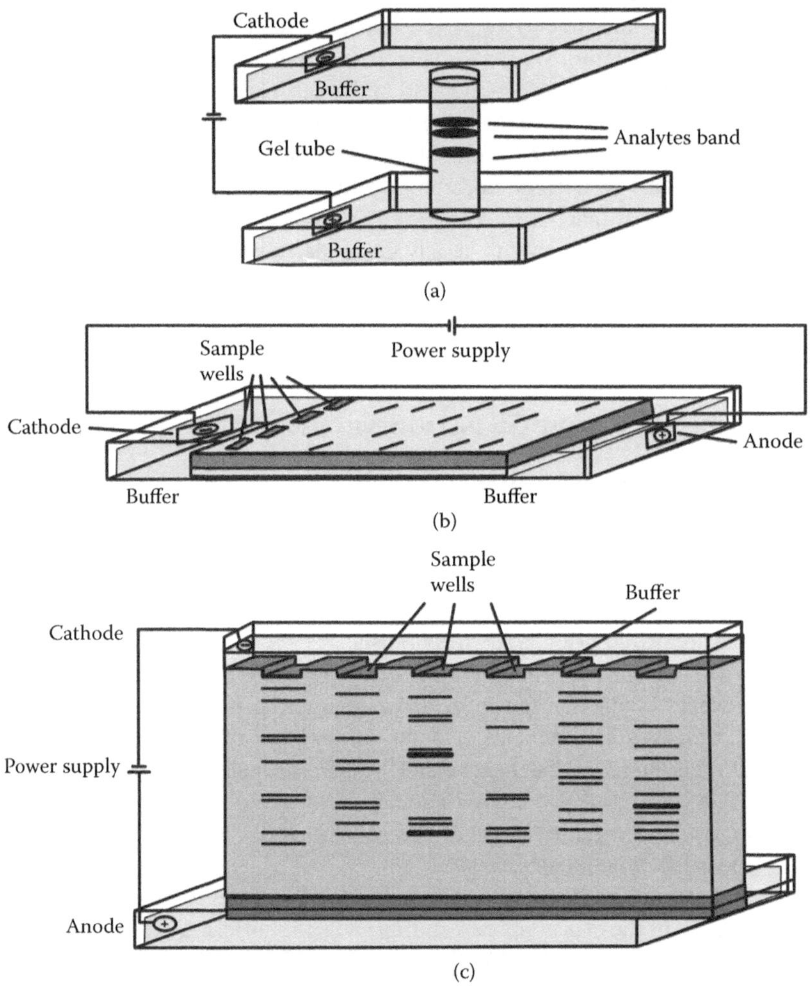

FIGURE 7.2 Schematically, GE can be performed in (a) an upright tube. Alternatively, flat rectangular slab gels can be used, which are positioned (b) horizontally or (c) vertically.

Loading buffer: The DNA samples to be loaded onto the agarose gel are first mixed with the loading buffer, which usually comprises water, sucrose, and a dye (e.g., xylene cyanole, bromophenol blue, bromocresol green). The loading buffer serves three purposes:

1. Increases the density of the sample, ensuring that the DNA drops evenly into the well.
2. Adds color to the sample, thereby simplifying the loading process.
3. Imparts a dye to the sample that moves toward the anode at a predictable rate in an electric field.

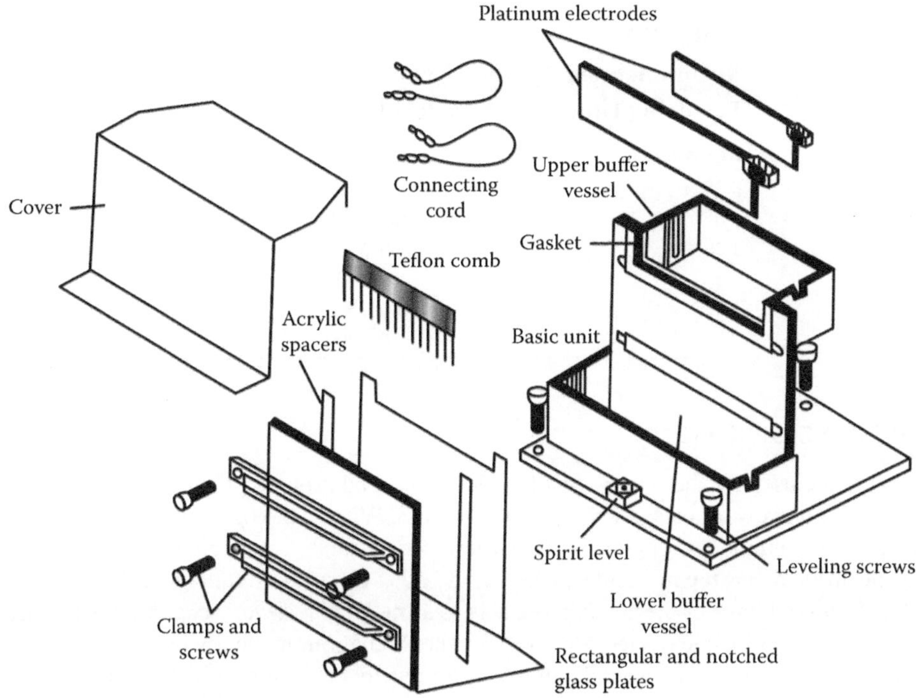

FIGURE 7.3 Vertical slab GE apparatus.

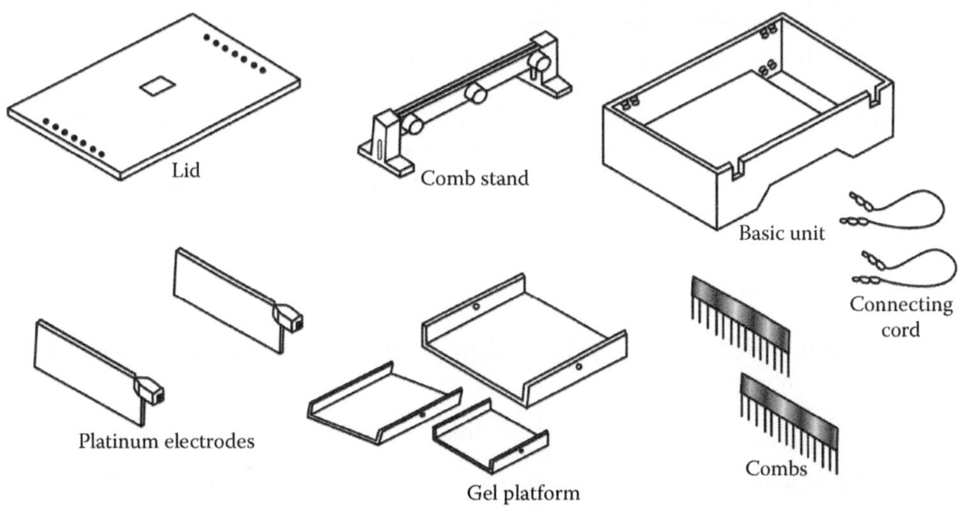

FIGURE 7.4 Parts of a submarine GE apparatus.

Ethidium bromide: Ethidium bromide is a fluorescent dye used for staining nucleic acids. This fluorescent dye intercalates between bases of DNA and RNA. It is often incorporated into the gel so that staining occurs during electrophoresis; the gel can also be stained after electrophoresis by soaking in a dilute solution of ethidium bromide. Ethidium bromide is a known mutagen and should be handled as a hazardous chemical; wear gloves while handling it.

Transilluminator: A transilluminator is a UV light box, which is used to visualize ethidium bromide–stained DNA in gels.

TABLE 7.1
Recommended Agarose Gel Concentrations for Resolving Linear DNA Molecules

Agarose (%)	DNA Size Range Base Pair [bp]
0.75	10,000–15,000
1.0	500–10,000
1.25	300–5000
1.5	200–4000
2.0	100–2500
2.5	50–1000

7.2 PRINCIPLES OF GE

GE is a technique used for the separation of nucleic acids and proteins. The separation of macromolecules depends on two variables: (1) charge and (2) mass. When a biological sample, such as DNA, is mixed in a buffer solution and applied to a gel, these two variables act together. The electrical current from one electrode repels the molecules, whereas the other electrode simultaneously attracts the molecules. The frictional force of the gel material acts as a "molecular sieve," separating the molecules by size. The separation principle of electrophoresis is shown diagrammatically in Figure 7.5.

During electrophoresis, macromolecules are forced to move through the pores; their rate of migration through the electric field depends on the following:

- Strength of the field
- Size and shape of the molecules
- Relative hydrophobicity of the samples
- Ionic strength and temperature of the buffer in which the molecules are moving

To completely understand the separation of charged particles in GE, it is important to look at the simple equations relating to electrophoresis. When a voltage is applied across the electrodes, a potential gradient, E, is generated; it can be expressed by the following equation:

$$E = V/d \tag{7.1}$$

where V, measured in volts, is the applied voltage and d is the distance in centimeters between the electrodes.

When the potential gradient, E, is applied, a force, F, on a charged molecule is generated, which is expressed by the following equation:

$$F = Eq \tag{7.2}$$

where q is the charge in coulombs bearing on the molecule. It is this force, measured in newtons, that drives a charged molecule toward an electrode.

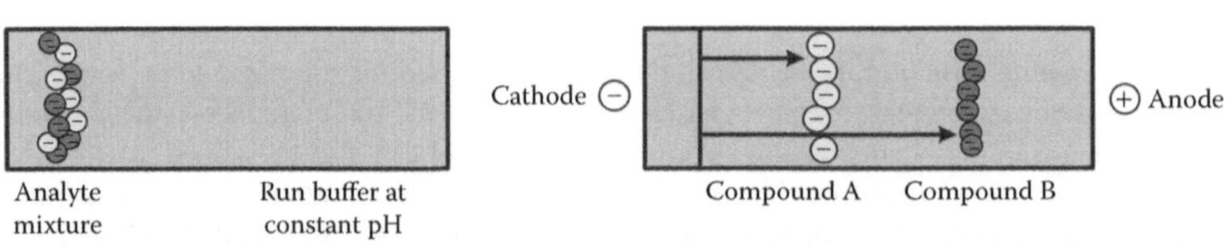

FIGURE 7.5 The separation principle of electrophoresis. Particles with different charges, in this case negative charges, and different sizes migrate toward the electrodes at different velocities in an applied electric field.

7.3 WORKING WITH THE ELECTROPHORESIS APPARATUS

A direct current (DC) power source is connected to the electrophoresis apparatus and an electrical current is applied. Charged molecules in the sample enter the gel through the walls of the wells. Molecules with a net negative charge migrate toward the positive electrode (anode), whereas net positively charged molecules migrate toward the negative electrode (cathode). Within a range, the higher the applied voltage the faster the samples migrate (EDVOTEK Manual). At the end of a run, the separated molecules can be detected in position in the gel by staining or autoradiography and quantified by scanning with a densitometer, and the gel can be dried for permanent storage.

7.3.1 Applications of GE

Major applications of GE are as follows:

- GE is used in forensics, molecular biology, genetics, microbiology, and biochemistry.
- The results can be analyzed quantitatively by visualizing the gel with UV light and a gel imaging device. The image is recorded with a computer-operated camera, and the intensity of the band or spot of interest is measured and compared against standard or markers loaded on the same gel. The measurement and analysis are mostly done with specialized software.
- Depending on the type of analysis being performed, other techniques are often implemented in conjunction with the results of GE, providing a wide range of field-specific applications (Figure 7.6).

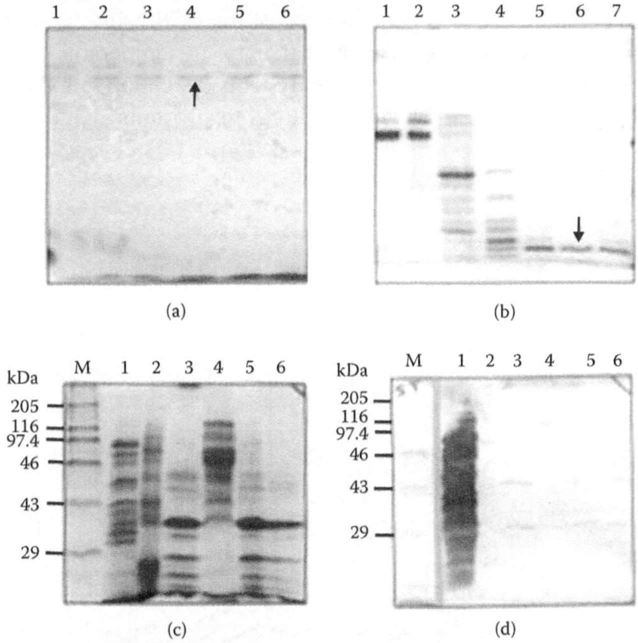

FIGURE 7.6 (See color insert.) Top: Protein profile of different fractions of metabolic antigens of *Aspergillus fumigates* obtained by size-exclusion chromatography on 12% SDS-PAGE stained with Coomassie blue: (a) Arrow shows 70–72 kDa protein doublet bands and (b) arrow shows 18 kDa purified protein. (c) Protein profile of metabolic antigens of *Aspergillus fumigates* precipitated by graded saturation of ammonium sulfate. (d) Identification of immunogenic proteins by Western blotting. Lane M, molecular weight marker; 1-, 2-, 3-, 4-, and 5-proteins precipitated at 20%, 40%, 60%, 80%, and 100% saturation of ammonium sulfate in culture filtrate, respectively.

7.4 DENATURING GRADIENT GE

Denaturing gradient GE (DGGE) is a robust method for point mutation detection that has been widely used for many years. It is a polymerase chain reaction (PCR)-based method, the principle being the altered denaturing temperature of a PCR product with a mutation compared to the wild-type product. Figure 7.7 shows different components of the DGGE instrument. DGGE is used to detect changes (mutations) in the genetic code within a sample and can detect as little as one base pair difference between strands of DNA.

7.4.1 Structure of the DGGE Instrument

7.4.2 Principle of DGGE

The principle of this technique is to separate DNA strands based on their actual base composition or the ratio of GC to AT base pairs that make up a particular segment of DNA. This is accomplished by exposing the DNA to a gradient of denaturant at elevated temperatures within a polyacrylamide gel. The products are run on an acrylamide gel with a gradient of denaturing agents, urea and formamide. The working principle of DGGE is diagrammatically shown in Figure 7.8. These denaturing agents alone are not sufficient. In addition to their action, the gel is run at a high temperature, usually 60°C. During electrophoresis, the PCR products will run through the gel as double-stranded DNA until they reach the point where they start to denature.

Once denatured, the PCR products continue running through the gel as single-stranded DNA, but the fragments have to remain precisely where they denatured. To achieve this, a so-called GC clamp is attached to prevent complete denaturing. This GC clamp is a string of 40–60 nucleotides comprising only guanine and cytosine and is attached to one of the PCR primers. A PCR with a GC clamp results in a product with one end having a very high denaturing temperature. A PCR product running through a DGGE gel will therefore denature partially. The GC clamp remains double stranded. The fragment will form a Y-shaped piece of DNA that sticks firmly at its position on the gel.

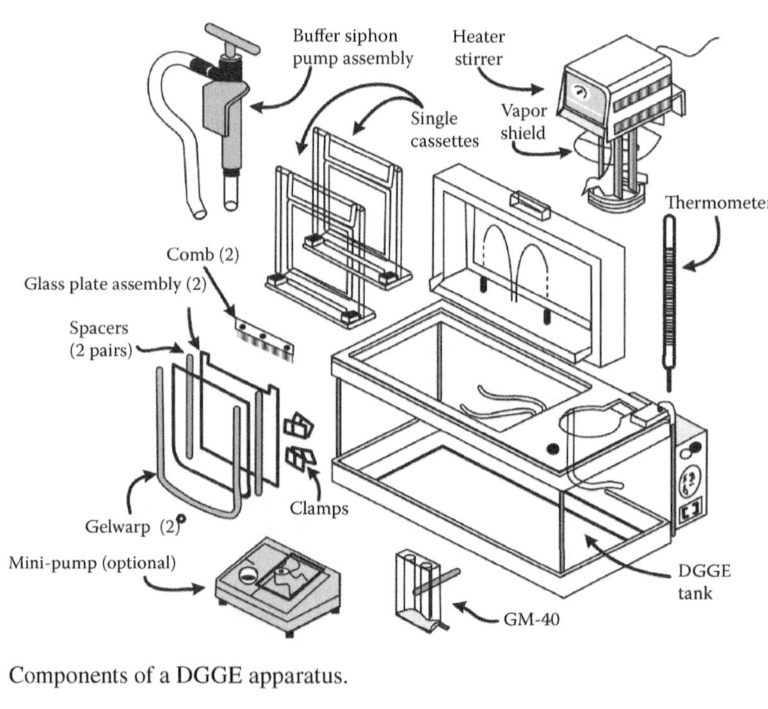

FIGURE 7.7 Components of a DGGE apparatus.

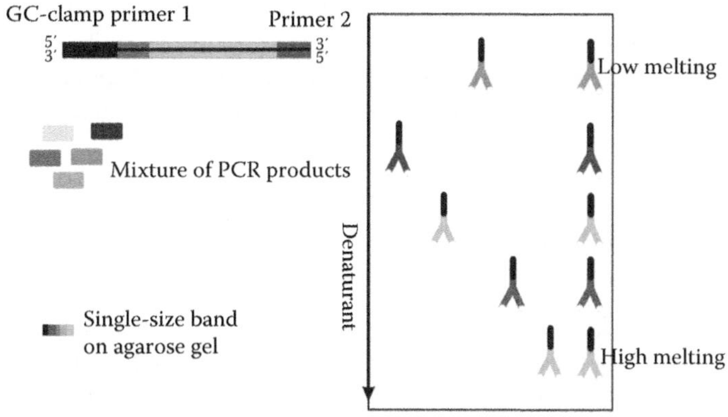

FIGURE 7.8 Working principle of DGGE.

7.4.3 Working with the DGGE Instrument

DGGE gels are poured and run to separate similarly sized PCR products. Gels are created by combining two solutions containing acrylamide (structural material) and differing amounts of denaturants (urea and formamide) to form a gradient of denaturant in which double-stranded DNA fragments of differing sequences are denatured during electrophoresis. The gel is stained and visualized to reveal band patterns, which can be used to determine the similarity of sampled microbial communities.

7.4.4 Applications of DGGE

Applications of DGGE include the following:

- DGGE is a useful tool for tracking a contamination to its source. For example, if a certain microbial contaminant is appearing in a particular product, DGGE can be used to track this organism to its source. This can be particularly useful in production facilities in food, pharmaceutical, and various manufacturing systems industries.
- Using DGGE, the amplified segments of DNA can be separated into individual species bands, recovered, and then used for sequencing to identify each organism within the sample.
- DGGE is a rapid and economical way of comparing large numbers of samples to one another without having to culture, isolate, and analyze each sample individually.
- The ability of DGGE to separate individual species within a sample also enables one to follow the progression of communities over a period of time. This application is extremely useful for remediation studies. In such cases, sample sites require sampling over extended periods of time to follow the degradation of contaminants and the organisms degrading them.
- DGGE eliminates several problems by eliminating the need for media, minimizing sample volumes, and being able to detect organisms that either are unculturable or may have perished during transit from the sample site to the lab.
- DGGE also allows the researcher to examine which organisms are forced out of a community over time, which organisms are stable, and what new organisms may be appearing.
- DGGE is a method to identify small mutations (e.g., point mutations). However, there are more types of small mutations, such as deletions or insertions of one or more nucleotides, which can be identified by DGGE as well.

- With the breadth of PCR primers available, DGGE can also be used to investigate broad phylogenies or specific target organisms such as pathogens or xenobiotics degraders.
- DGGE has been used to elucidate and characterize microbial populations in many bioleaching environments, including bioreactors, acidic mining-impacted environments, and bioleaching heaps.

7.5 TEMPERATURE GRADIENT GE

Temperature gradient GE (TGGE) is a powerful technique for the separation of nucleic acids or proteins. The TGGE method, which is covered by patents, uses the temperature-dependent changes of conformation for separating molecules. Figure 7.9 shows the complete apparatus of TGGE along with a power pack system. Since the introduction of the first commercially available TGGE apparatus in 1989, TGGE has gained much interest in scientific and clinical research laboratories due to its unprecedented resolution capability and ease of analysis. The range of scientific publications using the TGGE method is broad and covers all disciplines using molecular biology methods, such as oncology, virology, immunology, RNA viroid research, prion research, and population analysis. The TGGE method has also been used for quantitative analysis in the industry and for the conformational analysis of proteins.

7.5.1 Structure of the TGGE Instrument

7.5.1.1 Components of the TGGE Instrument

A typical TGGE instrument consists of different parts, which are diagrammatically shown in Figure 7.10.

7.5.1.2 Electrophoresis Unit

The electrophoresis unit consists of four parts:

1. A safety lid with two electric plugs (anode and cathode)
2. Two removable electrophoresis chambers, each with platinum wires and electric connectors (maximum volume: 250 mL)
3. Housing with Peltier-element-powered gradient block
4. Thirty-seven pins connecting the cable to the control unit

7.5.1.3 Controller Unit

The controller is a highly integrated microprocessor-driven unit for controlling the temperature, ramping time, and ramping rate of the gradient block, as well as supplying the power for

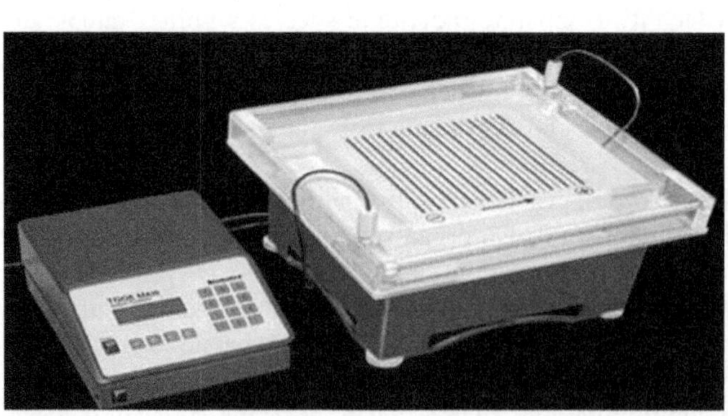

FIGURE 7.9 The TGGE unit.

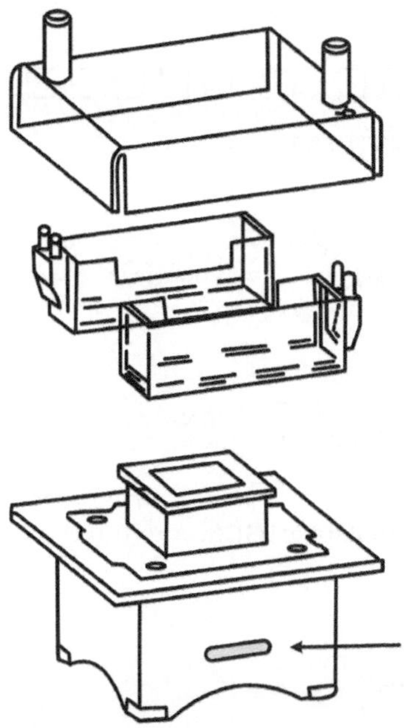

FIGURE 7.10 Different components of a TGGE instrument.

the electrophoresis unit. During the run, the display of the controller continuously shows the current parameters.

7.5.2 Principles

Conventional protein or nucleic acid electrophoresis separates molecules according to their size or charge. TGGE adds a new parameter for separation: the melting behavior of a molecule. Figure 7.11 describes the basic principles of TGGE. The melting behavior is determined by primary sequence and the secondary and tertiary structures of the molecule and can be changed by external influences such as temperature, salt concentration, and pH.

During electrophoresis, the sample migrates along a temperature gradient. As the temperature rises, the molecules start to denature. Working with PCR fragments, for example; during electrophoresis it starts separating from double stranded molecules. At a certain temperature, the DNA starts to melt, resulting in a forklike structure (partial single strand). In this conformation, the migration is slowed down compared to a DNA fragment of same size that is completely double stranded. Since the melting temperature strongly depends on the base sequence, DNA fragments of the same size but different sequences can be separated. This is used in mutation detection, where PCR fragments of identical size but different sequences are separated. Thus, TGGE not only separates molecules but also gives additional information about the melting behavior and stability of molecules.

7.5.3 Working with the TGGE Instrument

The sample is loaded directly to the gel. The gel is connected with the buffer chambers by paper wicks soaked with the running buffer. To protect the gel from drying, it is covered with a gel cover film. The complete setup, consisting of gel with cover film and buffer wicks, is covered with the gel cover plate. The gel cover plate has two sealings and fits precisely onto the thermoblock. It holds the buffer wicks in place and helps to build a humidity chamber around the gel. This is important to prevent evaporation during a run.

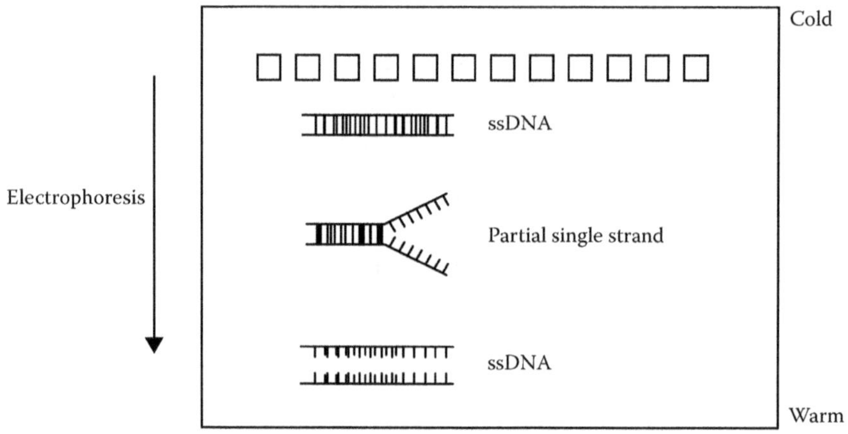

FIGURE 7.11 Different conformations of DNA during TGGE.

7.5.4 Applications of TGGE

Applications of TGGE include the following:

- Study community complexity.
- Monitor population shifts.
- Analyze enrichment cultures and the isolation of bacteria.
- Detect sequence heterogeneities of 16S rRNA genes in single genomes.
- According to a recent investigation, TGGE can be utilized to examine mutations in the mitochondrial DNA of an individual.
- Compare DNA extraction methods.
- Screen clone libraries.
- Determine PCR and cloning biases.

7.6 PULSED-FIELD GE

DNA molecules larger than 15–20 kb migrating through a gel will essentially move together in a size-independent manner. In 1984, Schwartz and Cantor described pulsed-field GE (PFGE), thereby introducing a new way to separate DNA. In particular, PFGE resolved extremely large DNA for the first time, raising the upper size limit of DNA separation in agarose from 30–50 kb to well over 10 Mb (10,000 kb). The complete instrument of PFGE is shown in Figure 7.12. The development of PFGE expanded the range of resolution of DNA fragments by as much as two orders of magnitude.

7.6.1 Structure

7.6.2 Principles

During electrophoresis, DNA above 30–50 kb migrate with the same mobility regardless of size. This is seen in a gel as a single large diffuse band. If, however, the DNA is forced to change direction during electrophoresis, fragments of different sizes within this diffuse band begin to separate from each other. With each reorientation of the electric field relative to the gel, smaller DNA begin moving in the new direction more quickly than larger DNA. Larger DNA lags behind, providing a separation from smaller DNA. The basic principle underlying the PFGE process is described in Figure 7.13.

FIGURE 7.12 The PFGE unit.

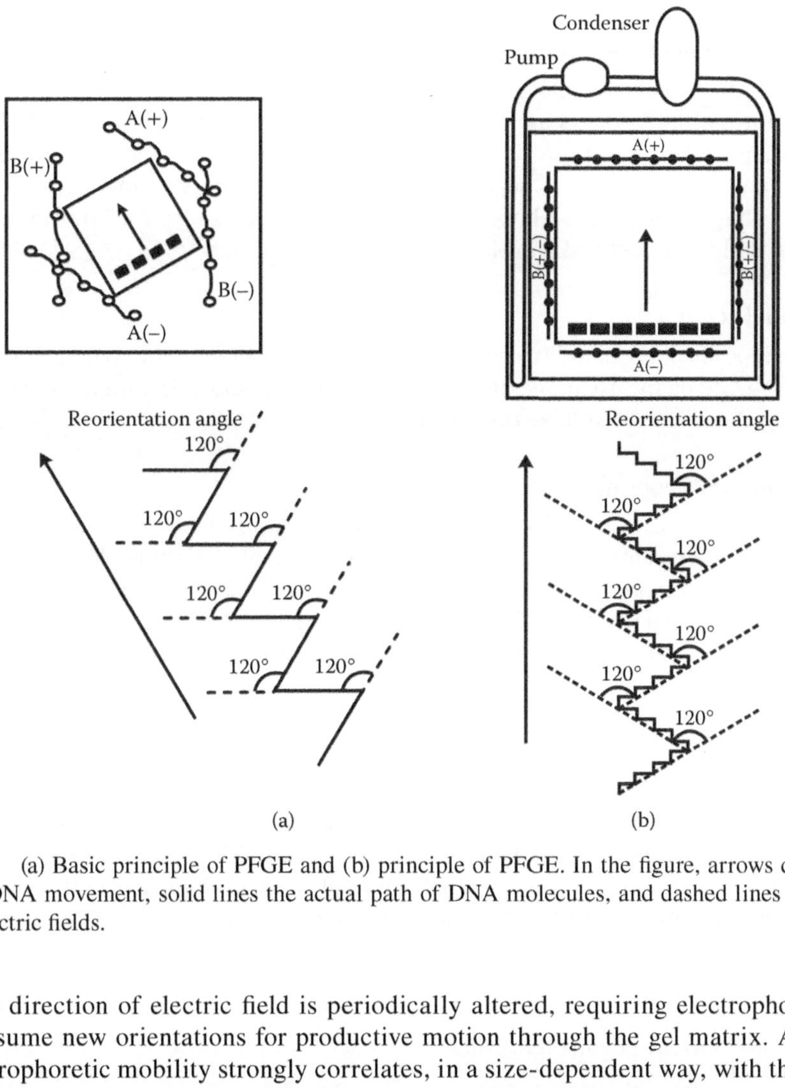

FIGURE 7.13 (a) Basic principle of PFGE and (b) principle of PFGE. In the figure, arrows denote the net direction of DNA movement, solid lines the actual path of DNA molecules, and dashed lines the trajectory created by electric fields.

Here, the direction of electric field is periodically altered, requiring electrophoresing molecules to assume new orientations for productive motion through the gel matrix. Accordingly, the net electrophoretic mobility strongly correlates, in a size-dependent way, with the frequency of the applied electrical fields. Consequently, as molecules approach complete orientation they move more effectively through the gel matrix with vanishing electrophoretic size dependencies. As such, PFGE pulsing routines use switching frequencies that optimize the periods in which molecules undergo size-dependent reorientation, causing zigzag traversals and separations up to 12 Mb.

7.6.3 Working with the PFGE Instrument

The basic components of a PFGE system are a gel box with some means of temperature regulation, a switching unit for controlling the electric fields, a cooler, and a power supply.

7.6.3.1 Gel Box

The basic design of PFGE boxes consists of an immobilized gel within an array of electrodes and a means for circulating the electrophoresis buffer. Voltage gradients of 10 V/cm are commonly used in PFGE. Voltage gradients as high as 15 V/cm have been used in field inversion separations of cosmid clones. The temperature of the buffer is controlled by a heat-exchange mechanism. Generally, the buffer is recirculated throughout the gel box using inlet and outlet ports.

7.6.3.2 High-Voltage Power Supply

Precise control of the electric field gradient is necessary to obtain consistent PFGE separations. The output ratings of the power supply should therefore be high enough to meet both voltage and current requirements of the gel box. To achieve the commonly used range of voltage gradients of 1.5–15 V/cm, one requires a power supply with a maximum voltage rating of 750 V. The current drawn at this voltage in most PFGE boxes is about 0.5 A at 14°C using 0.5× TBE (1× TBE is 89 mM Tris pH: 7, 89 mM boric acid and 2 mM EDTA) as running buffer.

7.6.3.3 Switch Unit

The ability to reproducibly control the switch interval is critical for separation. Such switching units are commonly based on the use of metal oxide semiconductor field-effect transistors (MOSFETs) in both switching and electrode voltage control circuits.

7.6.3.4 Computer Program

Careful control of the switch interval is crucial in controlling the resolution in PFGE. A versatile switching unit should have software with the same characteristics. The algorithm should be fast enough so that switch times at least as short as 1 millisecond can be achieved, and switch interval increments should have at least 1 millisecond resolution. Linear switch interval ramping is the most commonly used procedure because of its simple implementation. The maximum run time should be about two weeks to allow for the separation of very large DNA molecules. This is controlled by a computer program.

7.6.3.5 Cooler

DNA molecule migration is sensitive to temperature, and thus a uniform temperature across the gel is needed to ensure even migration in each of the lanes. Buffer is recirculated through the gel chamber by a reciprocating solenoid pump at a rate of about 450 mL/min. The buffer is chilled in its reservoir tank by circulating cold water (5°C) through a glass tubing heat exchanger. Buffer temperature is thus maintained at 13°C–15°C throughout a typical run.

7.6.4 Applications

Applications of PFGE include the following:

- The advent of PFGE techniques for the resolution of large DNA molecules provides a new analytical approach for bacterial genomes. The PFGE of DNA fragments obtained using different enzymes is a powerful technique for quick resolution of the bacterial genome into a small number of large fragments.
- PFGE is an amazing and useful tool for microbial strain identification and epidemiological tracking and tracing. PFGE techniques are precise, reliable, and reproducible.

- PFGE techniques may be used for genotyping or genetic fingerprinting.
- Yeast artificial chromosome (YAC) libraries have been constructed by PFGE.
- PFGE helps in identifying restriction fragment length polymorphisms (RFLPs) and construction of physical maps.
- PFGE is commonly considered a gold standard in epidemiological studies of pathogenic organisms.
- PFGE helps in detecting *in vivo* chromosome breakage and degradation.

7.7 CAPILLARY ELECTROPHORESIS

Capillary electrophoresis (CE) was developed in the 1980s by James Jorgenson and Krynn Lukas. They separated derivatized amino acids in a tube of 75 μm inner diameter. The technique has variously been referred to as high-performance CE (HPCE), capillary zone electrophoresis (CZE), free solution CE (FSCE), and simply CE. The CE method (Figure 7.14) can be used to separate a wide spectrum of biological molecules including amino acids, peptides, proteins, DNA fragments (e.g., synthetic oligonucleotides), and nucleic acids as well as many small organic molecules such as drugs or even metal molecules.

7.7.1 Principles

CE is based on the same principle as GE. Charged analytes can be separated in an applied electric field based on their mobility. In contrast to GE, however, separations are carried out in a capillary with a small diameter containing a free solution of electrolyte rather than on a slab gel. Moreover, convective flows due to Joule heating occur more easily in a free solution than in a gel. In contrast to GE, electroosmotic flow is often part of the separation process.

7.7.2 Structure of the CE Instrument

The instrumentation required for CE is relatively simple, consisting of vials with samples and buffer, a high-voltage power supply, a capillary enclosed in a thermostatically controlled compartment, an on-column detector, and a data output system as well as a vacuum system for sample injection. The parts of a typical CE instrument are discussed in Sections 7.7.2.1 through 7.7.2.4.

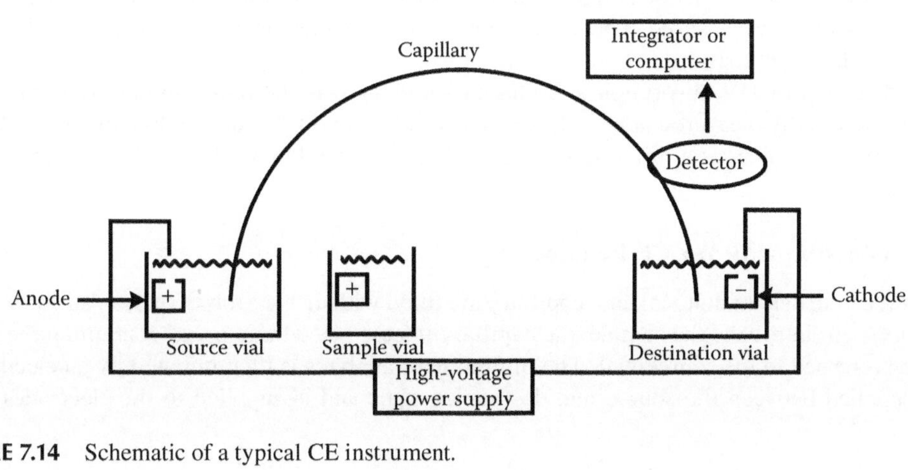

FIGURE 7.14 Schematic of a typical CE instrument.

7.7.2.1 Capillaries

The capillaries used in CE have internal diameters of 20–100 nm and outer diameters of about 400 nm. They are typically between 10 and 100 cm long. The most popular capillary material used is fused silica, that is, amorphous quartz, which is transparent to UV and visible light. These capillaries are externally coated with a polyimide layer of about 10 nm thickness to increase flexibility.

The capillary is usually enclosed in a thermostatically controlled environment for temperature control. This is because the viscosity of the buffer varies with temperature and Joule heating must be dissipated effectively to avoid temperature fluctuations, which can have dramatic effects on the efficiency and reproducibility of CE separations.

7.7.2.2 Buffer

A buffer is also referred to as a carrier electrolyte or background electrolyte. The purpose of a buffer is to maintain the pH as well as the conductivity during an electrophoretic separation. A controlled pH is crucial for maintaining a constant net charge on the biomolecules and thus maintaining their electrophoretic mobility. Buffer concentrations in CE are typically on the order of 10–100 mM.

7.7.2.3 Injection System

The injection system must be capable of reproducibly introducing very small sample volumes into the capillary. Two injection methods are commonly used:

1. In electrokinetic injection, voltages are used to introduce the sample into the capillary. The source ends of the capillary together with the source end of the electrode are placed into the sample solution. A high voltage is applied over the capillary between the sample vial and the destination vial for a given period of time. This causes the sample to move into the capillary according to its apparent mobility.
2. Hydrodynamic injection can be performed in three different ways: (1) In pressure injection, a precisely controlled external pressure is used to force a controlled amount of sample into the capillary. (2) In vacuum injection, a vacuum is applied to the buffer reservoir at the detector end of the capillary for a controlled period of time at a regulated reduced pressure. (3) For gravity flow injection, the sample vial with one end of the capillary is elevated to a certain height above the other end of the capillary for a given period of time. Gravity forces a sample plug into the capillary.

7.7.2.4 Detectors

Detection schemes used for CE include the measurement of UV absorption, fluorescence, and refractive index. Electrochemical signals and conductivity as well as radioactivity from radioisotopes have also been measured. The signals obtained are plotted against the migration time in the form of an electropherogram.

The detection of UV absorption at a chosen wavelength is the most commonly used scheme. Peptides are usually measured at $\lambda = 210$ nm, and proteins and DNA at $\lambda = 260$ nm or $\lambda = 280$ nm. The absorbance is measured directly through a detection window in the capillary, approximately 1 mm long.

7.7.3 Working with the CE Instrument

The source vial, destination vial, and capillary are filled with an electrolyte such as an aqueous buffer solution. To introduce the sample, the capillary inlet is placed into a vial containing the sample and then returned to the source vial. The migration of analytes is then initiated by an electric field that is applied between the source and destination vials and is supplied to the electrodes by the high-voltage power supply. The output of the detector is sent to a data output and handling device such as an integrator or a computer. The data is then displayed as an electropherogram, which reports detector response as a function of time. Separated chemical compounds appear as peaks with different retention times in an electropherogram.

7.7.4 Applications

- CE can be used to quantify DNA. For example, CE analysis of PCR products from Human Immunodeficiency Virus-I [HIV-I] allowed the identification of between 200,000 and 500,000 viral particles per cubic centimeter of serum.
- A range of small molecules, drugs, and metabolites can be measured in physiological solutions such as urine and serum.
- Point mutations in DNA, such as those occurring in a range of human diseases, can be identified by CE.
- Chiral compounds can be resolved using CE. Most of the efforts to this end have been carried out in a free solution using cyclodextrins as chiral selectors.

Experiment: GE of Proteins

Principles

Any charged ion or group will migrate toward the electrodes when placed in an electric field. Negatively charged particles in solution move toward the positive electrode and vice versa. These particles move at different speeds in solution depending on their net charge and size of the molecule.

SDS–Polyacrylamide GE (SDS-PAGE)

SDS, in the presence of the reducing agent beta-mercaptoethanol, dissociates proteins into their subunits and binds in large quantities with the protein. This completely masks the natural charge of the protein, giving a constant charge to mass ratio. Therefore, the larger the molecule the greater the charge. The electrophoretic mobility of the complex depends on the size (molecular weight) of the protein.

Protein Sample Preparation

A. For Bacteria

1. Raise 50 mL of the culture.
2. Centrifuge at 6000*g* for 10 minutes. Wash the pellet three times with 0.1 M Tris buffer (pH 6.8).
3. Resuspend the pellet in 2 mL of 0.1 M Tris buffer (pH 6.8) containing 15% glycerol and SDS to a final concentration of 1% (w/v).
4. Heat it in a boiling water bath for 10 minutes.
5. Centrifuge and store the supernatant containing proteins.
6. Mix 100 µL of protein sample with 100 µL of sample solubilizing buffer. Heat it at 100°C for 3 minutes.

B. For Cyanobacteria

1. Take 500 mg of cyanobacterial pellet and add 5 mL of 10% cold trichloroacetic acid (TCA).
2. Incubate in ice for 30 minutes.
3. Centrifuge at 12,000*g* for 5 minutes.
4. Wash the pellet with ethanol–ether (1:1 v/v) repeatedly to remove the TCA.
5. Dissolve the TCA precipitate to 0.0625M Tris–HCl (pH 6.8).
6. Subject the sample to SDS-PAGE.

Reagents

1. Acrylamide stock: Dissolve in 100 mL of distilled water. Store at 4°C in a dark bottle.

Acrylamide	29.2 g
NN′-methylene bisacrylamide	0.8 g

2. Resolving gel buffer (pH 8.8): Dissolve the salt in 50 mL of distilled water and adjust the pH with 6N HCl. Make up to 100 mL with distilled water. Store at 4°C.

Tris	18.15 g
SDS	0.4 g

3. Stacking gel buffer (pH 6.8): Dissolve the salt in 50 mL of distilled water and adjust the pH with 6N HCl. Make up to 100 mL with distilled water. Store at 4°C in a dark bottle.

Tris	6.05 g
DSD	0.4 g

4. Reservoir buffer: Dissolve and make up to 1000 mL with distilled water.

Tris	3.0 g
Glycine	14.4 g
SDS	1.0 g

5. Protein stain: Dissolve and make up to 500 mL with distilled water.

Coomassie brilliant blue-R250	1.25 g
Methanol	200 mL
Acetic acid	35 mL

6. Destaining solution: Mix and make up to 1000 mL with distilled water.

Methanol	500 mL
Acetic acid	75 mL

7. Storing solution:

Methanol	5 mL
Acetic acid	7 mL
Distilled water	87.5 mL

8. Sample buffer (stock):

0.0625 M Tris–HCl, pH 6.8	1.2 mL
SDS	1 g
Glycerol	3 mL
Beta-mercaptoethanol	200 μL
Bromophenol blue (1mg/mL)	2 μL
Distilled water	5 mL

Procedure

1. Clean the glass plates by soaking them in chromic acid overnight; rinse them with water and then with ethanol. Keep the plates down on a clean tissue paper, with the side that is to be in contact with the gel facing upward, and swab them with an acetone-soaked tissue held in a gloved hand. After a final rinse with ethanol, allow the plates to air dry.
2. Fix the gel plate with appropriate spacers on the gel maker. To avoid leakage, apply vacuum grease on both sides of the spacers.
3. Prepare the resolving gel solution. Degas the mixture and then add

Acrylamide (A)	10 mL
Distilled water	12.5 mL
Resolving gel buffer	7.5 mL

Pour the solution into a glass plate up to a level such that 1–2 cm is allowed for stacking the gel. Remove air bubbles, if any.

TEMED	45 μL
Ammonium persulfate (APS)	
(APS 1.5%)	150 μL

4. Add an even layer of isobutanol on the top of the separating gel solution to get a flat surface on top of the gel.
5. Allow it to polymerize for 30 minutes, remove the isobutanol layer, and wash it with water.
6. Prepare the stacking gel solution by mixing

Acrylamide (A)	1 mL
Stacking gel buffer (C)	2.4 mL

Degas the mixture and then add

TEMED	20 μL
1.5% APS	100 μL

7. Insert the comb between the plates. Pour the solution carefully on the top of the separating gel and remove the bubbles, if any.
8. After 20 minutes, remove the bottom spacer and fix the gel plate with the slab-gel unit. Remove the comb and fill up the wells with reservoir buffer (D).
9. Add equal volume of sample buffer (I) to the sample solution and boil in a water bath for 3 minutes.
10. Load about 100–200 μL of the aforementioned sample mixture into each well and, instead of the sample, load an equal amount of standard protein mixture to any one of the wells to compare the molecular weight of the sample proteins.
11. Add a reservoir buffer to the anode and cathode chambers until the buffer touches the gel.
12. Connect the power supply and apply 60 V until the marker dye enters the separating gel. Then increase the voltage to 100–200 V.
13. Continue the power supply until the marker dye reaches the bottom of the gel. Disconnect the power supply and remove the slab-gel setup.
14. Remove the glass plate and place the gel in Coomassie brilliant blue R-250 stain (E) for 2–4 hours.
15. Destain the gel in the destaining solution mixture (F) until a clear background is obtained.
16. Store the gel in the storing solution (G; Figure 7.15).

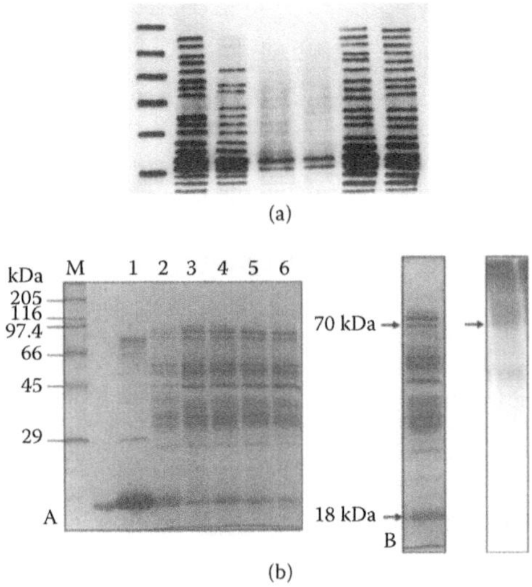

FIGURE 7.15 Protein profiles: (a) whole cell protein profile of *Anabaena variabilis* under different physiological conditions with lane 1 as marker, and (b) **(See color insert)** electrophoretic protein profile of a fungus *Aspergillus fumigates* with known-molecular-weight marker protein.

Experiment: Determination of the Molecular Weight of Plasmid DNA by Agarose GE

Introduction

DNA molecules are negatively charged at neutral or alkaline pH and migrate toward the anode when an electric field is applied. The charge/mass ratio in nucleic acids is unity; thus migration occurs largely on the basis of the molecular size of the DNA molecules (Figure 7.16).

Experimental Outline

1. Preparation of agarose gels
2. Loading plasmid DNA preparation along with the standard
3. Conducting electrophoresis
4. Examining the gels under UV light

Safety Guidelines

High-purity chemicals and water should be used. Other guidelines are as follows:

1. All glassware, plasticware, and solutions should be sterile.
2. Ethidium bromide is a carcinogen. Gloves must be worn while handling gels in ethidium bromide solution.
3. Wear safety glasses while viewing the gel on a UV transilluminator.
4. Do not touch the electrophoresis chamber or the electrical leads while electrophoresis is in progress.
5. At the time of casting the gel, ensure that no air bubbles are entrapped in the gel.

Materials

Plasmid DNA (isolated and standard); minigel horizontal agarose gel electrophoretic unit, which comprises

- Gel casting plate
- Electrophoretic tank
- Comb

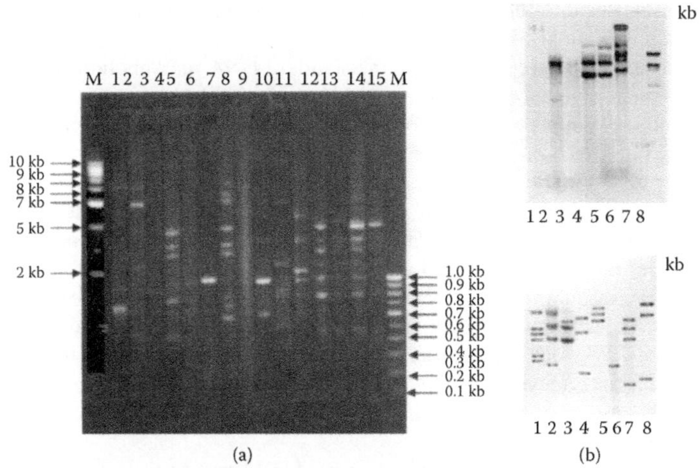

FIGURE 7.16 Figure showing (a) agarose GE of plasmid DNA with known ladder, and (b) genomic library of a bacterium with chromosomal DNA from wild and mutant strains after DNA digestion with restriction analysis.

- Electrophoretic leads
- Adhesive tape
- Power pack (0–500 V), UV transilluminator with camera, micropipette or syringe
- Gloves

Prelaboratory Preparations

1. TAE stock solution (5×): A fivefold concentrated TAE buffer stock solution. Adjust the pH of the aforementioned solution to pH 9 and add water to make 1 L. Dilute five times before use to obtain the working buffer (1× buffer).

Tris–base	24.2 g
Glacial acetic acid	7.71 mL
0.5 M EDTA	10 mL

2. 1% agarose in 1× TAE buffer: Dissolve 0.75 g agarose in 75 mL of 1× TAE buffer (working TAE buffer), by boiling and maintaining it at 50°C until it is to be used.
3. Gel loading solution: Dissolve 10% glycerol and 0.025% bromophenol blue in water.
4. Ethidium bromide: Dissolve 10 mg of ethidium bromide per milliliter of the 1× TAE buffer. Gloves must be worn while preparing this solution.
5. Plasmid DNA preparation: To 20 μL of plasmid DNA preparation, add 10 μL of gel loading solution and mix properly.
6. Standard DNA marker: Take 20 μL of the λ DNA–Hind III Digest; add 10 μL of gel loading solution and mix well.

Method

1. Take a clean, dry gel casting plate and make a gel mold using adhesive tape along the sides of the plate to prevent the material to be poured on the plate from running off.
2. Pour 50 mL of 1% agarose solution kept at 50°C onto the casting plate. Immediately place the comb about 1 cm from one end of the plate, ensuring that the teeth of the comb do not touch the glass plate. Wait until a firm layer of gel is formed.
3. Remove the comb and the tape surrounding the plate carefully and transfer the gel plate to the electrophoresis tank such that the wells are toward the cathode.

4. Pour lx TAE buffer into the tank until the gel is completely submerged. Connect the electrodes to the power supply.
5. Load the plasmid DNA preparation and the standard DNA markers into separate wells with the help of a micropipette or a syringe.
6. Turn on the power supply and run at 100 V (10–15 mA). Monitor the progress of the fast-running tracking dye (bromophenol blue) during electrophoresis.
7. Turn off the power supply when the tracking dye reaches near the opposite edge of the gel.
8. Transfer the gel from the casting plate onto a UV-transparent thick plastic sheet, place it in a staining tray containing ethidium bromide solution, and stain for 20–30 minutes.
9. Place the gel for destaining in water for 15–20 minutes.
10. Observe the gel on a UV-transparent sheet with a UV transilluminator for the presence of orange-colored bands. The gel may be photographed for a permanent record (Figure 7.16). (Caution: UV light is extremely injurious to eyes; wear UV protection glasses while viewing the gel.)
11. Measure the distance moved by each band from the edge of the loading well. Draw a graph of logarithm to base 10 of the molecular weight of the standard DNA markers versus the distance traveled by each of them.
12. From the distance traveled by the supplied plasmid DNA preparation, determine its molecular weight using the calibration curve prepared in step 11.

Experiment: Electroblotting (Western Blotting of Proteins from SDS Polyacrylamide Gel)

Introduction

When an electric field is applied, proteins migrate from cathode to anode. Nitrocellulose sheets are able to bind proteins. The proteins get immobilized onto the sheet, forming an exact replica of protein bands on the nitrocellulose sheet as they come in contact.

Experimental Outline

Transfer the proteins separated on polyacrylamide gels onto the nitrocellulose sheet. Examine the blue bands of the transferred proteins (Figure 7.17).

Safety Guidelines

1. Do not touch the nitrocellulose sheet using your hands.
2. Wear gloves while handling the gel and the nitrocellulose sheet.
3. Avoid the entrapment of bubbles in the assembly.
4. The proteins get immobilized onto the sheet and form the exact replica of protein bands on the nitrocellulose sheet.

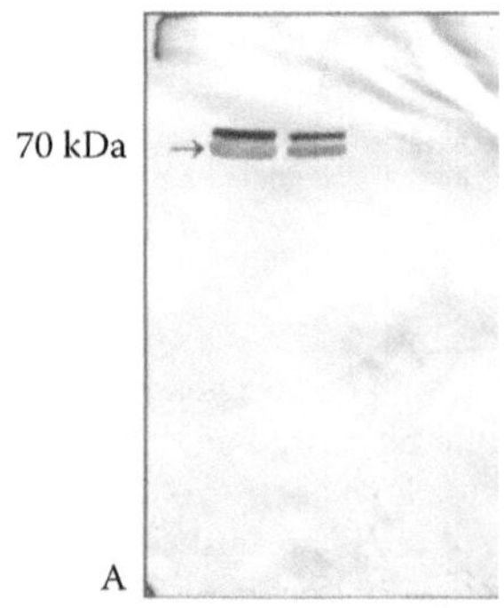

FIGURE 7.17 Western blot on a Whatman no. 3 nitrocellulose sheet.

Materials

1. Western blot apparatus consisting of gel holder, sponge, and transfer tank
2. Power pack and electrical leads
3. Slab gel containing separated proteins (use the gel obtained after the electrophoresis of proteins on SDS-PAGE in the first experiment of this chapter)
4. Nitrocellulose sheet cut to size
5. Whatman no. 3 MM paper cut to size of the gel

Prelaboratory Precautions

Transfer buffer consists of

Tris	3 g
Glycine	14.7 g
Methanol	200 mL

After mixing the aforementioned components, adjust the pH of the solution to 8.3 and make the final volume 1 L with distilled water.

Method

1. Take the gel obtained after electrophoresis in the first experiment of this chapter. Do not stain it, and mark it by cutting off one corner. Place the gel for 30 minutes in the transfer buffer for equilibration.
2. Take a nitrocellulose sheet, cut it to the size of the gel, and dip it in the transfer buffer by carefully wetting one edge and then slowly lowering the sheet into the buffer. Leave it in the buffer for 30 minutes.
3. Soak the sponge in transfer buffer and place the wet sponge on the gel holder. Now keep a sheet of Whatman no. 3 MM paper (presoaked in transfer buffer) on the sponge.
4. Place the equilibrated gel carefully on the filter paper, avoiding trapping any air bubbles.
5. Carefully lay down the nitrocellulose membrane with its shining side toward the gel on top of the gel. Gently roll a sterile 10 mL pipette over the membrane to remove air bubbles to ensure good contact between the membrane and the gel.
6. Complete the sandwich by placing a wet Whatman no. 3 MM filter paper over the membrane and a second sponge on the filter paper. Close the gel holder and place it in the transfer tank containing sufficient quantity of the transfer buffer to completely cover the blot.
7. Connect the setup to the power supply and run for 5 hours at 60 V or at 30 V overnight.
8. When the transfer is complete, lift the membrane from the gel. Stain and destain it using the Coomassie brilliant blue R-250 stain. Examine the nitrocellulose sheet for the presence of blue bands of the transferred proteins.

SUGGESTED READING

Amersham. 1999. *Protein Electrophoresis: Technical Manual.* USA: Amersham Biosciences Inc.

Andreas, M., P. Nicole, and L. Dimitri. 2004. *Bioanalytical Chemistry.* London: Imperial College Press.

Arakawa, T. et al. 1993. "Analysis of the Heat-Induced Denaturation of Proteins Using Temperature Gradient Gel Electrophoresis." *Analytical Biochemistry* 208:255–9.

Baughman, K. 2005. *Principles and Applications of DGGE.* Microbac Laboratories, Inc.

Baumstark, T., and D. Riesner. 1995. "Only One of Four Possible Secondary Structures of the Central Conserved Region of Potato Spindle Tuber Viroid Is a Substrate for Processing in a Potato Nuclear Extract." *Nucleic Acids Research* 23:4246–54.

Biometra a Whatman company. 1999. *TGGE System Manual.* Göttingen: Biometra, biomedizinische, Analytik GmbH.

Biometra a Whatman company. 2002. *TGGE System Manual*, Version 6.0. Goettingen: Biometra, biomedizinische, Analytik GmbH.

Birmes, A. et al. 1990. "Analysis of the Conformational Transition of Proteins by Temperature-Gradient Gel Electrophoresis." *Electrophoresis* 11:795–801.

Birren, B. W., E. Lai, S. M., Clark, L. Houd, and M. I. Simon. 1988. "Optimized Conditions for Pulsed-Field Gel Electrophoretic Separations of DNA." *Nucleic Acids Research* 16:7563–81.

Carle, G. F., and M. V. Olson. 1984. "Separation of Chromosomal DNA Molecules from Yeast by Orthogonal-Field-Alternation Gel Electrophoresis." *Nucleic Acids Research* 14:5647–63.

Chrambach, A., M. J. Dunn, and B. J. Radola. 1983. *Advances in Electrophoresis*, vol. 4, 189–95. Weinheim: VCH Verlagsgesellschaft.

Dempsey, J. A. F., W. Livaker, A. Madhure, T. L. Snodgrass, and J. G. Cannon. 1991. "Physical Map of the Chromosome of Neisseria Gonorrhoea FA1090 with Locations of Genetic Markers, Including Opa and Pil Genes." *Journal of Bacteriology* 173:5476–86.

EDVOTEK Manual. 2003. EDVO-Kit # 101 Principles and Practice of Agarose Gel Electrophoresis, 1–31. Washington, DC: The Biotechnology Education Company.

Gardiner, K. 1991. "Pulsed-Field Gel Electrophoresis." *Analytical Chemistry* 63:658–65.

Hecker, R. et al. 1988. "Analysis of RNA Structure by Temperature-Gradient Gel Electrophoresis: Viroid Replication and Processing." *Gene* 72:59–74.

Henco, K., and M. Heibey. 1990. "Quantitative PCR—The Determination of Template Copy Numbers by Temperature Gradient Gel Electrophoresis." *Nucleic Acids Research* 18:6733–4.

Herschleb, J., G. Ananiev, and D. C. Schwartz. 2007. "Protocol: Pulsed-Field Gel Electrophoresis." *Nature Protocols* 2:677–84.

Horn, D. et al. 1996. "Three Novel Mutations of the NF1 Gene Detected by Temperature Gradient Gel Electrophoresis of Exons 5 and 8." *Electrophoresis* 17:1559–63.

Instruction Manual. 2011. *Denaturing Gradient Gel Electrophoresis Systems*. Del Mar, CA: C.B.S. Scientific Company, Inc.

Jeroen, H. R., and J. M. P. Dorien. 1998. "Denaturing Gradient Gel Electrophoresis (DGGE)." In *Medical Biomethods Handbook*, edited by J. M. Walker and R. Rapley. Totowa, NJ: Humana Press, Inc.

Joppa, B., S. Li, S. Cole, and S. Gallagher. 1992. HIS Laboratories, Hoefer Scientific Instruments. Fall San Francisco: Probe Volume 2(3).

Kappes, S. et al. 1995. "p53 Mutations in Ovarian Tumors, Detected by Temperature-Gradient Gel Electrophoresis, Direct Sequencing and Immunohistochemistry." *International Journal of Cancer* 64: 52–9.

Kuhn, J. E. et al. 1995. "Quantitation of Human Cytomegalovirus Genomes in the Brain of AIDS Patients." *Journal of Medical Virology* 47:70–82.

Lai, E., B. W. Birren, S. M. Clark, M. I. Simon, and L. Hood. 1989. "Pulsed-Field Gel Electrophoresis." *Biotechniques* 7:34–42.

Lessa, E. P., and G. Applebaum. 1993. "Screening Techniques for Detecting Allelic Variation in DNA Sequences." *Molecular Ecology* 2:119–29.

Linke, B. et al. 1995. "Identification and Structural Analysis of Rearranged Immunoglobulin Heavy Chain Genes in Lymphomas and Leukemia." *Leukemia* 9:840–7.

Loss, P., M. Schmitz, G. Steger, and D. Riesner. 1991. "Formation of a Thermodynamically Metastable Structure Containing Hairpin II is Critical for the Potato Spindle Tuber Viroid." *EMBO Journal* 10:719–28.

Menke, M. A., Tiemann, M., Vogelsang, D. et al. 1995. Temperature gradient gel electrophoresis for analysis of a polymerase chain reaction-based diagnostic clonality assay in the early stages of cutaneous T-cell lymphomas. *Electrophoresis* 16: 733–8.

Milde-Langosch, K. et al.1995. "Presence and Persistence of HPV and p53 Mutation in Cancer of the Cervix Uteri and the Vulva." *International Journal of Cancer* 63:639–45.

Nubel, U. et al. 1996. "Sequence Heterogeneities of Genes Encoding 16S rRNAs in *Paenibacillus Polymyxa* Detected by Temperature Gradient Gel Electrophoresis." *Journal of Bacteriology* 178:5636–43.

Orbach, M. J., D. Vollrath, R. W. Davis, and C. Yanofsky. 1998. "An Electrophoretic Karyotype of *Neurospora Crassa*." *Molecular Cell Biology* 8:1469–73.

Prischmann, J. 2011. Basics and Theory of Electrophoresis (ppt). Diagnostic Lab, North Dakota State Seed Department.

Richter, A., L. Plobner, and J. Schumacher. 1997. "Quantitatives PCR-VerfahrenzurBestimmung der Plasmidkopienzahl in rekombinanten Expressionssystemen." *BIOforum* 20:545–7.

Riesner, D. 1998. "Nucleic Acid Structures." In *Antisense Technology*. Practical Approach Series, 1–24. Oxford, UK: Oxford University Press.

Riesner, D., K. Henco, and G. Steger. 1991. Temperature gradient gel electrophoresis: A method for the analysis of conformational transitions and mutations in nucleic acids and proteins. *Adv Electrophoresis* 4:169–250.

Sambrook, J., E. F. Fritsch, and T. Maniatis. 1989. "Gel Electrophoresis of DNA." In *Molecular Cloning: A Laboratory Manual*, edited by J. Sambrook, E. F. Fritsch, and T. Maniatis. Cold Spring Harbor, NY: Cold Spring Harbor Laboratory Press. Chapter 6.

Schwartz, D. C., and C. R. Cantor. 1984. "Separation of Yeast Chromosome-Sized DNAs by Pulsed Field Gradient Gel Electrophoresis." *Cell* 37:67–75.

Sheffield, V. C., D. R. Cox, L. S. Lerman, and R. M. Meyers. 1989. "Attachment of a 40-Base-Pair G+C-Rich Sequence (GC-Clamp) to Genomic DNA Fragments by the Polymerase Chain Reaction Results in Improved Detection of Single-Base Changes." *Proceedings of the National Academy of Sciences of the United States of America* 86:232–6.

Skoog, D. A., F. J. Holler, and S. R. Crouch. 2007. *Principles of Instrumental Analysis*, 6th ed. Belmont, CA: Thomson Brooks/Cole Publishing.

Westermeier, R. 1997. *Electrophoresis in Practice: A Guide to Methods and Applications of DNA and Protein Separation*. Weinheim: VCH.

Wieland, U. et al. 1996. "Quantification of HIV-1 Proviral DNA and Analysis of Genomic Diversity by Polymerase Chain Reaction and Temperature Gradient Gel Electrophoresis." *Journal Virology Methods* 57:127–39.

Wiese, U. et al. 1995. "Scanning for Mutations in the Human Prion Protein Open Reading Frame by Temporal Temperature Gradient Gel Electrophoresis." *Electrophoresis* 16:1851–60.

Wilson, K., J. Walker. 2000. *Practical Biochemistry: Principles and Techniques*, 5th ed. Cambridge: Cambridge University Press.

IMPORTANT LINKS

1. DGGE: http://www.cbsscientific.com/dgge.aspx
2. TGGE: http://www.cbsscientific.com/ttge.aspx
3. PFGE: http://www.biolabo.com/Electrophoresis/Pulsed-Field-Gel-Electrophoresis-PFGE-system
4. CE: http://www.biocompare.com/ProductDetails/3177514/Agilent-7100-Capillary-Electrophoresis-System.html?fi=3177514

8 X-Ray Microanalysis

8.1 INTRODUCTION

X-ray microanalysis makes it possible for chemical analysis to be performed on biological tissue within very small and well-defined regions of specimens. All elements from Na to U can be detected while observing a specimen through an electron microscope. Using x-ray microanalysis, as little as 10^{-17}–10^{-18} g of an element can be detected. It is used in such diverse scientific areas as metallurgy, physics, electronics, mineralogy, environmental pollution, geology and, lately, in pathology, zoology, biochemistry, and other biological fields. The excess energy of an electron that migrates to an inner shell to fill a newly created hole can do more than emit an x-ray. Often, instead of x-ray emission, the excess energy is transferred to a third electron from a further outer shell, which prompts its ejection. This ejected species is called an Auger electron, and the method for its analysis is known as Auger electron spectroscopy (AES).

8.2 PRINCIPLES

Atoms, when struck by electrons from an external source, yield x-rays, which are characteristic of those atoms and are used to identify and quantify the elements present in a molecule. X-ray microanalysis combines electron microscopy and x-ray spectroscopy. When a specimen is observed through the electron microscope, the sample is bombarded with high-energy electrons, which generate characteristic and continuous background x-rays in the irradiated area of the sample. If the exciting electron beam is focused to a diameter of 100 nm on an ultrathin section of same thickness, a lightly conical electron probe of this size is formed within the specimen and x-ray emission is restricted to this probed area. Thus, we can observe the specimen in the electron microscope, select the feature of interest, focus the electron beam onto this feature, and record the generated x-ray spectrum.

In Figure 8.1, a nucleus comprising neutrons and protons is surrounded by orbital electrons distributed in different energy levels: K, L, M, and so on. X-ray microanalysis is based on the excitation of these electrons to produce an emitted x-ray spectrum that is characteristic of the element concerned. If one of the orbital electrons is removed from its energy level by an incident electron, then the atom is said to be in an excited state. When this occurs, an electron from a higher energy state will fall down into the gap to stabilize the atom. Because of the difference in potential energy levels, the excess energy is emitted during this electron transition as an x-ray photon. Thus, if an electron in the K shell is removed a second electron from the L shell may instantaneously replace it, giving off its excess energy as a photon of energy (E_L–E_K), which is generally called "$E\alpha$ radiation." Now, the filling of the vacancy in the K shell by an electron from the L shell will also produce a vacancy in the L shell, which in turn may be filled with an electron from the M shell and so on; each electron transition is associated with the production of x-ray photons of energies determined by the orbital energies. Thus, a single ionization can give rise to a whole spectrum of characteristic x-rays, and this energy spectrum identifies the atom.

A primary electron beam may, instead of interfering with the orbital electrons to produce characteristic x-rays, interact with the nucleus. As the incoming electron beam is decelerated by the field of nuclear charge, it radiates energy, which can be anything from the maximum energy originally carried by the electrons to a small fraction of it. Figure 8.2 shows the energy spectrum produced by this

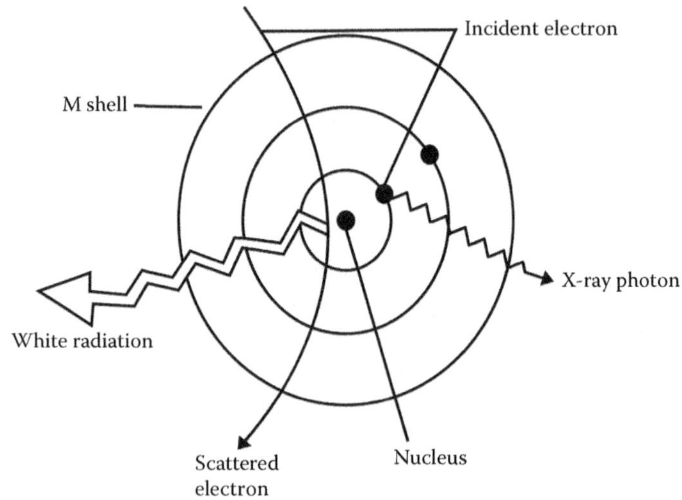

FIGURE 8.1 Simple schematic representation of an atom showing x-ray production.

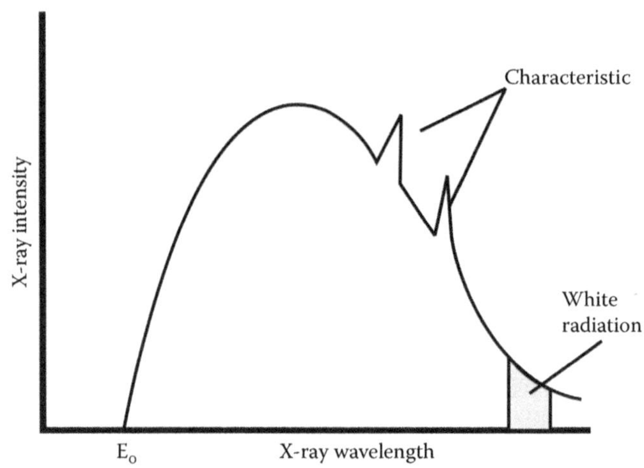

FIGURE 8.2 X-ray continuum.

effect, with the maximum energy being that of the primary electron beam E_0. This general spectrum is called “x-ray continuum,” “continuous radiation,” “white radiation,” or “Bremsstrahlung.” It forms the background on which the characteristic x-ray lines are superimposed at specific wavelengths or energies.

White radiation forms a basic limitation to the ability to detect a characteristic “line” and it is considered x-ray noise. However, it is useful in calculating elemental concentrations.

8.3 INSTRUMENTATION

For viewing biological material, there are two methods: (1) thin specimens are viewed in transmission and (2) bulk specimens are viewed by reflection. Accordingly, instrumental arrangements are of two types:

1. Scanning electron microscope and x-ray detection system
2. Transmission electron microscope and x-ray detection system

X-ray detection systems used generally are also of two types: (1) wavelength-dispersive crystal spectrometers and (2) energy-dispersive solid-state detectors. Wavelength-dispersive crystal spectrometers work on the principle that of the x-rays leaving the specimen, a narrow cone falls on a curved crystal and a fraction of this signal is reflected into a detector, which is usually a gas flow or a sealed, proportional counter. In the detector, the x-rays are converted to electrical signals that are amplified and transferred to a multichannel analyzer. The individual channels are calibrated by energy and the channel number or energy position gives qualitative analytical information. The fraction of signal that is reflected depends on the crystal's ability to "diffract" a particular wavelength maximally and is governed by Bragg's law, which states that for crystals,

$$2d\sin\theta = n\lambda$$

where n is an integer, λ is the x-ray wavelength maximally diffracted, d is the lattice spacing of the crystal, and θ is the angle of incidence (and of reflection) of the x-ray beam at the crystal. Every element in the periodic table has x-ray lines corresponding to characteristic wavelength (λ) values. A crystal having a value of maximally diffracted wavelength the same as the λ value of x-rays emitted by an element not only detects this element but also quantifies it. In order to extend the range of elements detected, a number of crystals are incorporated in the x-ray differection system. All the crystals have different d values or lattice spacings, so that for the same θ range the wavelength (λ) range is large, thereby covering a number of elements.

The energy-dispersive solid-state detector can detect all x-ray energies leaving the specimen at once, unlike the crystal spectrometer, which can detect only one x-ray energy at a time. The solid-state detector provides an energy spectrum of all elements analyzed simultaneously. Here, the incoming x-rays are sorted and analyzed by their energies and the apparatus consists of a liquid nitrogen–cooled Si/Li semiconductor.

Figure 8.3 illustrates the incorporation of crystal detectors or solid-state detectors, or both, at appropriate positions with respect to the electron microscope either in the scanning mode or the transmission mode. In a scanning electron microscope, the detectors can be easily brought close to the specimen for greater sensitivity.

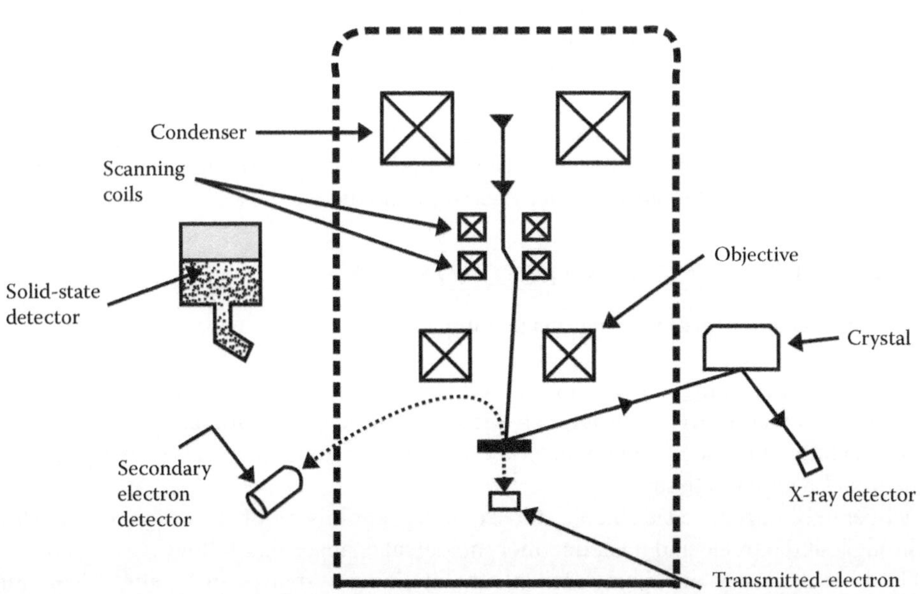

FIGURE 8.3 Schematic representation of a scanning electron microscope incorporating x-ray microanalysis.

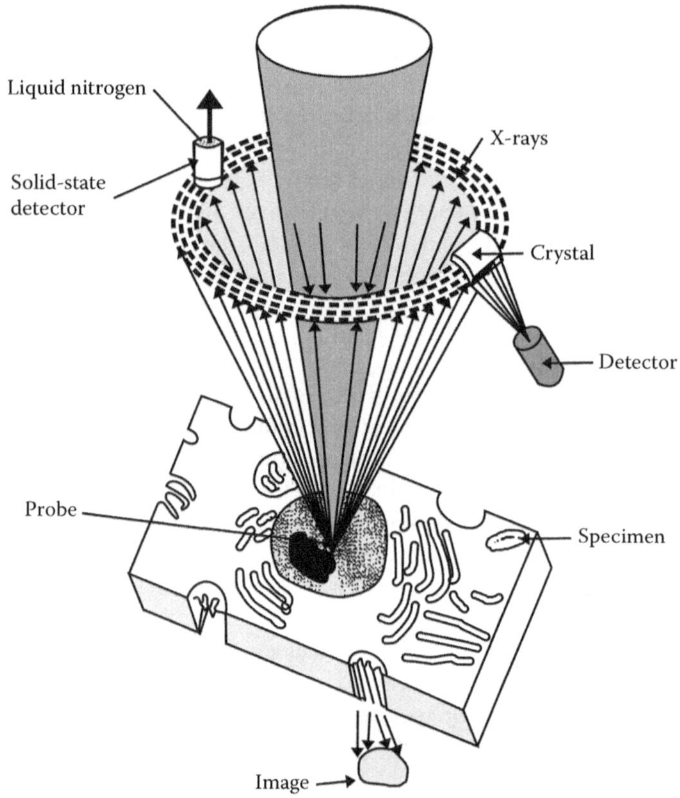

FIGURE 8.4 Schematic representation of x-ray microanalysis of thin sections with TEM.

Both 1D and 2D analysis can be performed with a scanning electron microscope. For 2D analysis, the detector is set so as to collect x-rays from only a particular element, and then the focused electron beam is made to scan the specimen. The x-ray signal from the detector is displayed on a cathode ray oscilloscope by synchronizing it with the electron beam raster. For one-dimensional analysis, that is, a line trace, the output is fed to a chart recorder so that changes in elemental content can be monitored across that line in the specimen. To analyze a particular area of interest, a static probe is directed over that area and x-ray emission is analyzed over a period of time (Figure 8.4).

Almost any electron microscope can thus be converted to an x-ray analyzer by attaching a suitable detector to the specimen region. This idea was first given by Castaign in 1949, and in 1968 the first electron microscope with an x-ray analyzer was constructed.

8.4 APPLICATIONS OF X-RAY MICROANALYSIS

Several problems can be tackled using this technique:

- The natural elemental composition of tissues can be demonstrated by analysis of normal physiological levels after appropriately preparing the specimen for analysis.
- Accidentally introduced foreign material, for example, toxic chemicals, can be located and identified within the tissue.
- Deliberately introduced elements, for example, administration of drugs, can be traced throughout the tissue and related to morphological changes that follow.
- The histochemistry and immunochemistry of living systems can be studied in situ by analysis of biochemical events.

8.5 TECHNIQUES FOR THE ANALYSIS OF SECONDARY, TERTIARY, AND QUATERNARY PROTEINS BY X-RAY CRYSTALLOGRAPHY

8.5.1 Introduction

Protein x-ray crystallography is a technique used to obtain the 3D structure of a particular protein by x-ray diffraction of the crystallized form of the protein. This 3D structure is crucial in determining a protein's functionality. X-ray crystallography can reveal the precise 3D positions of most atoms in a protein molecule because x-rays and covalent bonds have similar wavelengths and, therefore, this method currently provides the best visualization of protein structure. The specificity of the protein's active sites and binding sites is completely dependent on the protein's precise conformation. It was x-ray crystallography that made it possible for J. D. Watson and F. H. C. Crick to figure out the double-helix structure of DNA.

The technique imparts knowledge of the cellular mechanism and the 3D structure of enzymes and other macromolecules. It is critical that we understand better how each chemical reaction that occurs in a cell needs a specific enzyme for it to happen. Two common techniques used for the analysis of protein structure are nuclear magnetic resonance (NMR) and x-ray crystallography. X-ray crystallography can be used to analyze different compounds up to a molecular weight of 10^6 (g/mol), for instance, whereas NMR is restricted to the analysis of biopolymers (polymers produced by a living organism such as starch, peptides, and sugars) of molecular weights not more than 30,000 (g/mol). It can also measure compounds that are very small because the appropriate size to measure the distance between atoms in a molecule is 0.5–1.5 Å. X-rays are used as the form of radiation to this end because their wavelengths are on the same order of magnitude as the bond length of a covalent bond (~1 Å or 1×10^{-10} m) and this is necessary to obtain a diffraction pattern that reveals information about the structure of the molecule. If the radiation had a wavelength much bigger or much smaller than the bond length of a covalent bond, the light would not diffract and no new knowledge of molecular structure would be obtained.

8.5.2 Principles of X-Ray Crystallography

Crystals, including those of globular proteins, consist of repetitions of a basic structural component called a "unit cell," which may be a single molecule or a symmetrical arrangement of several molecules. Thus, each atom in a crystal must lie in a specific position with respect to all other atoms in the crystal, enabling the structure to be determined by x-ray diffraction analysis. This consists of directing a beam of x-rays of a single wavelength on a crystal and studying the characteristics of the emerging rays. Most rays pass straight through the crystal without being affected, but the rays that come into contact with an atom in the crystal are scattered by the clouds of electrons surrounding it. More precisely, these electrons act as secondary sources of x-rays, which then radiate out from the atom in all directions. The intensity of the x-rays leaving an atom of high electron density, such as a heavy metal, is much greater than that of x-rays leaving an atom of low electron density, such as hydrogen; thus, areas of high electron density can be said to scatter x-rays more strongly than areas of low electron density.

X-rays, similar to other forms of electromagnetic radiation, are best regarded as waves of characteristic lengths and amplitudes (Figure 8.5); the intensity of a ray is proportional to the square of its amplitude. If two rays of identical wavelength are traveling on a common path so that they are exactly in phase, that is, the crests and troughs of the waves correspond exactly, then they will combine to give a ray of the same wavelength and phase but greater amplitude. The amplitude, and hence intensity, obtained under these conditions will be the maximum that can be obtained by any combination of these two rays. If the two rays are one quarter of a cycle out of phase, the intensity of the combined ray will be about one-quarter of the maximum possible value, and the phase of the combined ray will

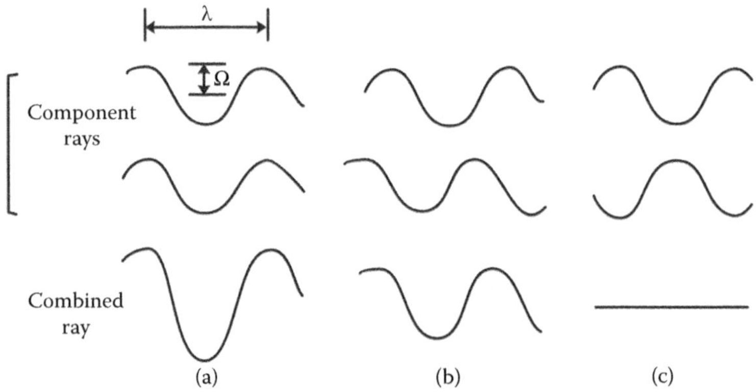

FIGURE 8.5 The combination of rays of identical wavelength (λ) and amplitude (Ω) when directed along a common path. The component rays are (a) exactly in phase, (b) one-quarter of a cycle out of phase, and (c) half a cycle out of phase.

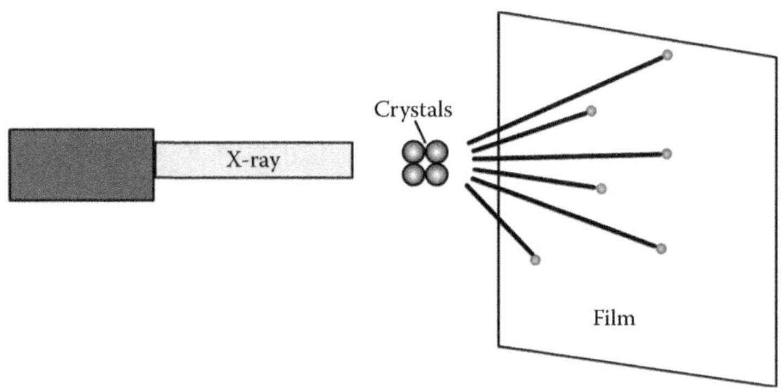

FIGURE 8.6 Diffracted x-rays.

be a combination of the phases of its component rays. If the two rays are exactly half a cycle out of phase, the waves will cancel each other out and the intensity of the combined ray will be zero.

The scattered x-rays emerging from a crystal can combine in the following way: Rays emerging at certain angles to the incident ray will combine to give rays of maximum intensity, whereas those emerging at other angles will cancel each other out. The results can be observed by placing a photographic plate behind the crystal to register the impact of emerging rays. In general, development of the plate will show a spot at the center caused by the undeflected x-rays, which will be in phase; this is surrounded by a pattern of other spots corresponding to the angles where emerging rays combine to give intensity maxima. The overall effect is known as a "diffraction pattern" (Figure 8.6).

Before discussing x-ray crystallography in more detail, let us consider why we cannot observe the atoms in a molecule by the use of optical or electron microscopy. Vision consists of two processes: The beam of light (another form of electromagnetic radiation) that strikes an object is scattered by the atoms, and these scattered rays are brought back together (focused) by the lens in the eye to produce an image of the object on the retina. A magnified image may be produced by the use of further lenses (optical microscopy), enabling one to clearly distinguish (resolve) features that are too close to be seen separately by the unaided eye. The limit of resolution in microscopy depends on the wavelength of the type of electromagnetic radiation used and the focusing properties of the instrument. With optical microscopy, the limit of resolutions is about half the wavelength of the light used. Hence, individual atoms separated in a molecule by distances on the order of 1–2 Å (1 Å = 0.1 nm) cannot be resolved by an optical microscope, since the wavelength of visible light is in excess of 4000 Å.

Electron microscopes give a much greater resolving power than optical microscopes, but despite the very low wavelengths of electron beams it is still not possible to visualize individual atoms using an electron microscope because of the generally poor performance of the electromagnetic lenses used in electron microscopy. Similarly, x-rays have wavelengths much smaller than those of light rays; in fact they are of the same order of magnitude as inter-atomic distances. However, no procedure has yet been devised for focusing x-rays, therefore no image can be produced using them. Nevertheless, the detailed structure of a crystal scattering x-rays can be deduced from the diffraction patterns obtained.

Although each unit cell in a crystal may contain many atoms in a complex arrangement, let us for the moment consider a unit cell simply as a region of high electron density, which can act as a scattering center for x-rays. Thus the crystal consists of a regular arrangement of major scattering centers, each corresponding to a unit cell, as shown in Figure 8.7.

First of all, let us look at the plane of scattering centers containing A, B, and C, which is inclined at an angle (θ) to the incident beam of x-rays (Figure 8.7). Some rays will be scattered by the electron-dense regions in the plane, whereas most of the rays will pass straight through the plane. Each scattered ray has the same wavelength and phase as the incident ray and regarded as being deflected. Rays will be scattered in all directions, so some will emerge at an angle ϕ to the plane. Those leaving A and B at the same angle to the incident beam will reach a given point after traveling exactly the same distance through space (so that distance QAU = distance RBT in Figure 8.7) only when $\phi = \theta$. This is known as the "reflection condition" since the phenomenon of reflection at a planar surface is also characterized by these angles being equal. Therefore, all rays reflected by the scattering centers on the plane ABC, that is, all rays emerging at an angle such that $\phi = \theta$, will be exactly in phase and will combine to give a ray of maximum intensity.

Now, let us consider a second plane of scattering centers containing D, E, and F, which is parallel to the plane ABC and is separated from it by a distance d. Rays leaving B and E in the same direction can never reach a given point after traveling the same distance through space since distance PES must be greater than distance RBT. However, if the difference in distance ($d_1 + d_2 = d\sin\theta + d\sin\phi$) is exactly a whole number of wavelengths, the emerging waves will still be exactly in phase. Therefore, rays that emerge at the same angle to the incident beam from scattering centers on different but parallel planes will combine to give a ray of maximum intensity if $n\lambda = d\sin\theta + d\sin\phi$, where n is a whole number and λ is the wavelength of the rays. In summary, all the scattering centers in a single plane will combine to give a diffracted ray of maximum intensity at the angle where the reflection condition is met, whereas centers in different planes will combine to give rays of maximum intensity at angles where $n\lambda = d\sin\theta + d\sin\phi$. If these two conditions are put together, rays

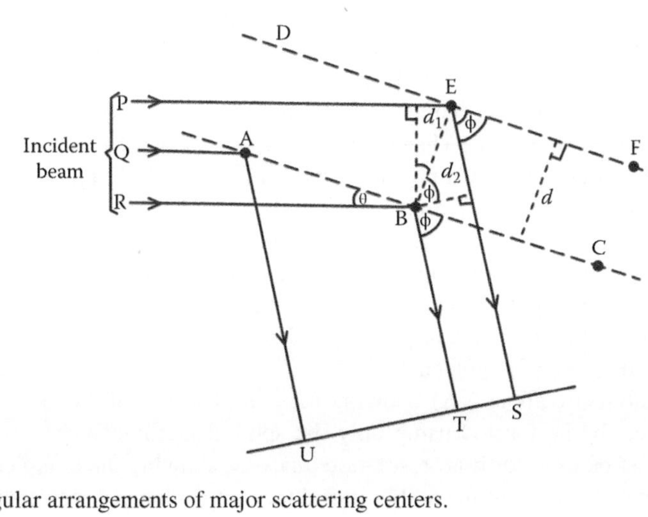

FIGURE 8.7 Regular arrangements of major scattering centers.

emerging from all the scattering centers on any number of parallel planes that are at a distance d apart will combine to give an intensity maximum where $n\lambda = 2d\sin\theta$. This was first stated by Bragg and Bragg (father and son) in 1913 and is known as the "Bragg condition."

Thus, it can be seen that regular repeating units are essential for the establishment of clear diffraction patterns, since patterns from different scattering centers may reinforce each other under these conditions. Crystals are rotated in a beam of x-rays, allowing time in each position for the investigation of the diffraction pattern, until the pattern obtained indicates that planes of scattering centers are inclined at a suitable angle to the incident beam for reinforcement of patterns to take place. A clear diffraction pattern may be obtained without rotating the specimen if the crystal in question is not a single crystal but comprises separate regions of repeating units set at random angles to each other and, thus, to the incident beam; this is often the case where the specimen is a powder of fine crystals or a natural fiber.

In general, the greater the repeating distances within a specimen the closer the intensity maxima on the photographic plate. If clear diffraction patterns can be obtained with a crystal in three different orientations, then the dimensions of the unit cell and arrangement of unit cells within the crystal can be deduced from the maxima values nearest to the center, which correspond to the largest repeat distances.

If we now turn our attention to the structure of the molecule or molecules making up a unit cell, we immediately realize that not all the atoms in a complex molecule can lie on the same plane. Hence, although we have hitherto considered a unit cell to be a single scattering center lying on a specific plane, a unit cell may in fact consist of a large number of component scattering centers (atoms) of which some lie at various distances in front of the plane and some at various distances behind the plane. This inevitably affects the diffraction pattern obtained. If intensity (and hence amplitude) and phase is known for each x-ray causing a spot in the diffraction pattern, then a 3D contour map showing the distribution of electron density within the unit cell can be drawn up using a mathematical procedure called Fourier synthesis. From this map, the structure of the molecule can be deduced. The intensity of an x-ray can be determined from a photograph of a diffraction pattern or measured directly using a Geiger counter; however, there is no direct method for determining the phase of the x-ray. This is called the phase problem.

The phase problem can be overcome in one of two ways: (1) A model may be having a built of possible data on a molecular structure and the theoretical diffraction patterns. This may be useful in explaining certain repeating features, but otherwise the number of possible structures of a complex molecule is too immense to enable the successful implementation of this method when used alone. (2) The alternative method is that of isomorphous replacement, which was introduced by Perutz in 1954. A heavy metal atom, such as mercury or uranium, is attached to a specific site of each molecule in the crystal without altering the 3D structure of the molecule. The heavy metal atoms, being in regions of very high electron density, will cause appreciable changes to the amplitude and phase of the rays producing a diffraction pattern. If the intensities of the spots in this case are compared to those of the spots in the original diffraction pattern, it is possible to deduce the location of the substituted atoms within the unit cells. The contribution of rays from the heavy metal atoms to each spot in the diffraction pattern may then be calculated, in terms of both phase and amplitude. This enables two possible solutions of the phase problem to be obtained for each spot, one in which the phase of the original ray is in advance of that of the ray from the substituted atom, and one in which it is an equal distance behind. It is possible to determine which of the alternative solutions is correct by substituting between the amplitude and the phase of the rays producing a diffraction pattern with a heavy metal atom in a different place, or possibly several different places. It is thereby possible to deduce the complete structure of a molecule.

Low-resolution analysis (to about 5 Å), showing the main features of the molecular structure but not the fine detail, may be performed using only the spots near the center of the diffraction pattern. All the spots must be used for high-resolution analysis, showing the complete structure of the molecule. Isomorphous replacement inevitably introduces some changes, however slight, in the 3D structure, and the higher the resolution attempted the more significant these changes. Hydrogen atoms are extremely weak scatterers of x-rays, so they are very difficult to pinpoint using these techniques.

It has often been found advantageous to use model-building and isomorphous replacement techniques to complement each other. Low-resolution studies indicate the general shape of the molecule, and from this information models can be built to elucidate the fine structure. X-rays are commonly obtained in the x-ray tube by accelerating electrons to a high velocity released from an indecent tungsten filament against a copper target. This produces rays of approximate wavelength 1.5 Å.

A more recent development is that x-rays selected from the electromagnetic radiation emitted by highly expensive devices called "synchrotrons" or electrons to rings are of higher intensity than those obtained from a conventional source. This is advantageous for the determination of the structure of proteins of high molecular weight where the crystalline structure is unstable over the exposure period to weak radiation. Another possibility is the use of neutron beams, which are scattered by atomic nuclei rather than electrons, for high-resolution analysis, since they are scattered strongly by hydrogen atoms. Also, they cause very little radiation damage to macromolecules, enabling irradiation to be carried out for far longer periods than is possible with x-rays.

8.5.3 Some Results of X-Ray Crystallography

In 1939, Pawling and Corey and their coworkers began a systematic investigation of the 3D structures of amino acids, dipeptides, and other molecules to provide data with a view to the eventual elucidation of protein structure. X-ray diffraction analysis soon showed that the C–N bond length on a peptide bond is shorter than what is expected for a single covalent bond. Therefore, some double-bond character must be present; the actual structure deduced between two extremes is shown as follows:

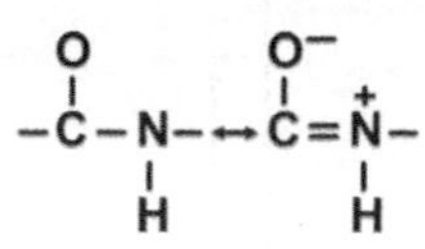

A consequence of the presence of this partial double-bond character is that rotation about the axis is restricted and all the atoms involved lie in the same plane. Two isomer arrangements are possible in the *trans* form, with the oxygen and hydrogen atoms lying diametrically opposite to each other, and the *cis* form, with these atoms lying adjacent to each another. In fact, only the *trans* isomer is found.

The most significant factor accounting for the stability of the *trans* form is the spacing between the carbon atom and the oxygen atom (2.8 Å), which is marginally less than the van der Waals contact distance between these atoms (3.4 Å), so repulsion is slight between the atoms. In the unstable *cis* form, the two (3.4 Å) atoms lie adjacent to each other and are separated by a distance (2.8 Å) more than the van der Waals contact distance between two carbon atoms (4.0 Å), so repulsive forces would be high between the atoms (Figure 8.8).

Pawling and Corey and their colleagues also noted that, in crystals, there is a high degree of hydrogen bonding between oxygen atoms in one peptide bond and nitrogen atoms in another. The distance between such atoms is often about 2.9 Å, which is much less than the van der Waals contact distance between nonbonded oxygen and nitrogen atoms, thus indicating the presence of the hydrogen bond. Furthermore, the N–H…O linkage is usually approximately linear. On the basis of these findings, they suggested various theoretical types of secondary structures that might be found in proteins. In particular, both right- and left-handed α-helices seemed consistent with the available data, but the arrangement of London dispersion forces (LDFs) was more favorable in the former structure.

Astbury had already used x-ray scattering to demonstrate regular features in the structures of several fibrous proteins, and in some cases these were found to be consistent with the postulated right-handed α-helix. The main features of the x-ray diffraction pattern of α-keratin (Figure 8.9a)

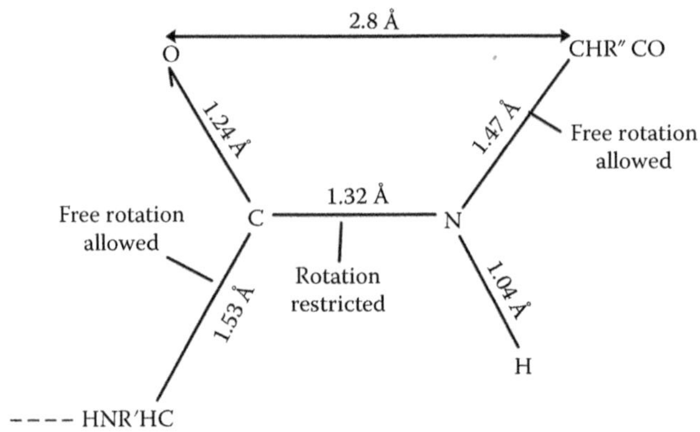

FIGURE 8.8 Dimensions of the peptide bond.

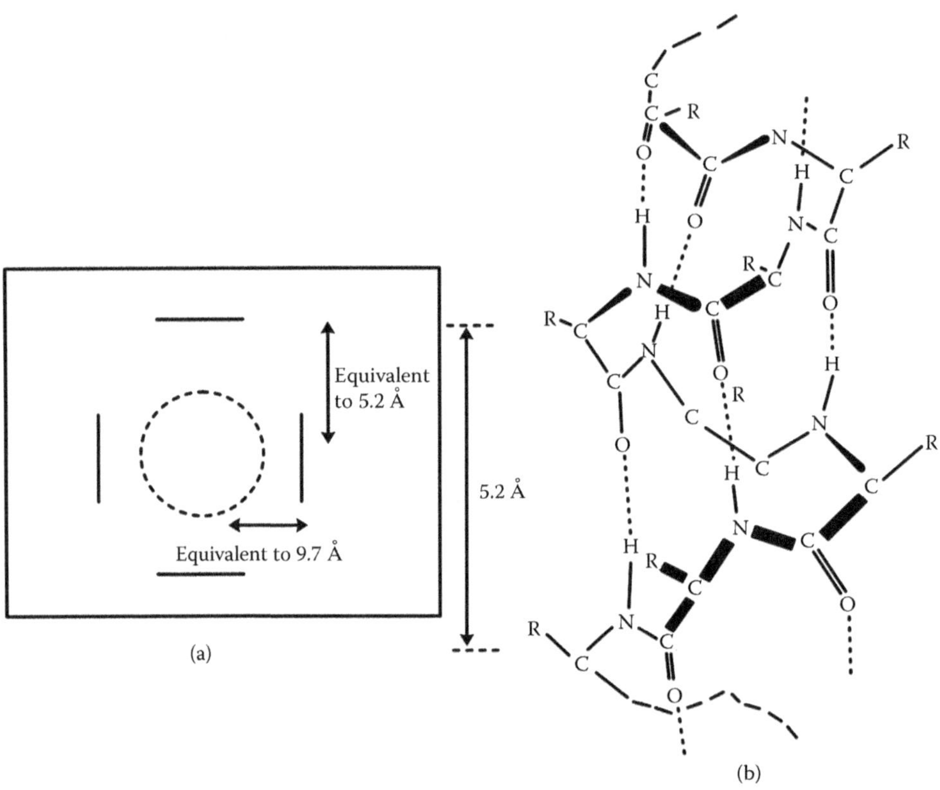

FIGURE 8.9 Figure showing (a) the main features of the x-ray diffraction pattern of α-keratin and (b) a polypeptide chain in the form of a right-handed α-helix (note the α-hydrogen atom has been omitted from the diagram to minimize congestion).

are a periodicity of 5.2 Å along the axis of the fibers, which is the distance between the turns of the α-helix (Figure 8.9b), and a periodicity of 9.7 Å at right angles to this axis, which is presumably the distance between adjacent α-helices.

An α-helix contains approximately 3.6 amino acid residues per turn. This results in each pair of peptide oxygen and nitrogen atoms being in a suitable position to form hydrogen bonds with the corresponding atoms in the next turn of the helix (Figure 8.9b); these hydrogen bonds are all approximately parallel to the axis of the helix.

Other types of secondary structures have also been demonstrated. If α-keratin is stretched under humid conditions, it is converted to β-keratin, which has characteristic periodicities of 3.3 Å along the axis of the fiber and 4 and 9.7 Å perpendicular to the axis, which gives evidence of a more extended form of secondary structure called a β-pleated sheet. Again, hydrogen bonds can be formed between oxygen and nitrogen atoms in different peptide bonds, but in this instance they are perpendicular to the axis of the fiber. The structure is unstable in keratin but not in other fibrous proteins such as silk fibroin. Keratin is said to form parallel β-pleated sheets, since the N-termini of adjacent polypeptide chains lie in the same direction, whereas silk fibroin forms antiparallel β-pleated sheets, the N-termini of adjacent chains being in opposite directions. Antiparallel β-pleated sheets may also be formed by the doubling back of a single polypeptide chain (Figure 8.9a and b).

The stability of these α-helix and β-pleated sheet structures depends on the nature of the amino acid side chains present; large or charged side chains tend to be disruptive. Hence, fibrous proteins usually consist only of amino acids with small and uncharged side chains and have well-developed secondary structures. Proline, because of the restricted rotation resulting from its ring structure, is another amino acid that cannot form part of an α-helix or a β-pleated sheet, but it is incorporated into the unique triple-helix structure of the fibrous protein collagen. The first globular protein to have its 3D structure elucidated by x-ray crystallography was sperm whale myoglobin. Myoglobin is an iron- and oxygen-binding protein found in the muscle tissue of vertebrates in general and in almost all mammals. It is related to hemoglobin, which is the iron- and oxygen-binding protein in blood, specifically in the red blood cells. This close relative of hemoglobin is a single polypeptide chain of 153 amino acid residues; despite its relatively small size, over 10,000 diffraction spots had to be accurately analyzed to give a resolution to 2 Å. Shortly afterward, the structural analysis of hemoglobin itself was completed. Each of the four component polypeptide chains was found to have a tertiary structure almost identical to that of myoglobin. Kendrew for his work on myoglobin structure and Perutz for his studies on hemoglobin were awarded the Nobel Prize in 1962. The structure of lysozyme, an enzyme from egg white consisting of 129 amino acid residues in a single chain, was given by Blake and Phillips and their colleagues in 1965. These and other studies on globular proteins have shown that a limited degree of secondary structure is usually present in them. Lysozyme has about 25% of its amino acids in α-helical zones and some in sections of β-pleated sheet (Figures 8.10 and 8.11).

The amino acids at the positions where the secondary structure is disrupted are, as expected, those with large side chains, for example, proline and leucine, or those with charged side chains where two or more with like charges are close together. The molecules are very compact, with space for very few water molecules within the interior. Most of the amino acids with nonpolar side chains

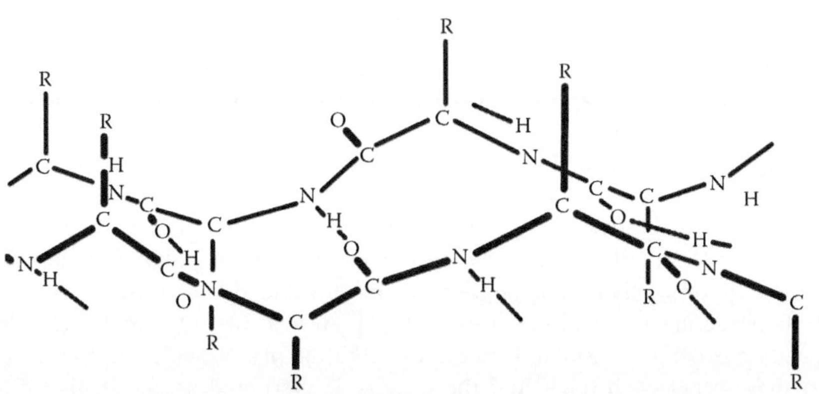

FIGURE 8.10 A section of antiparallel β-pleated sheet (note the α-hydrogen atom has been omitted from the diagram to minimize congestion).

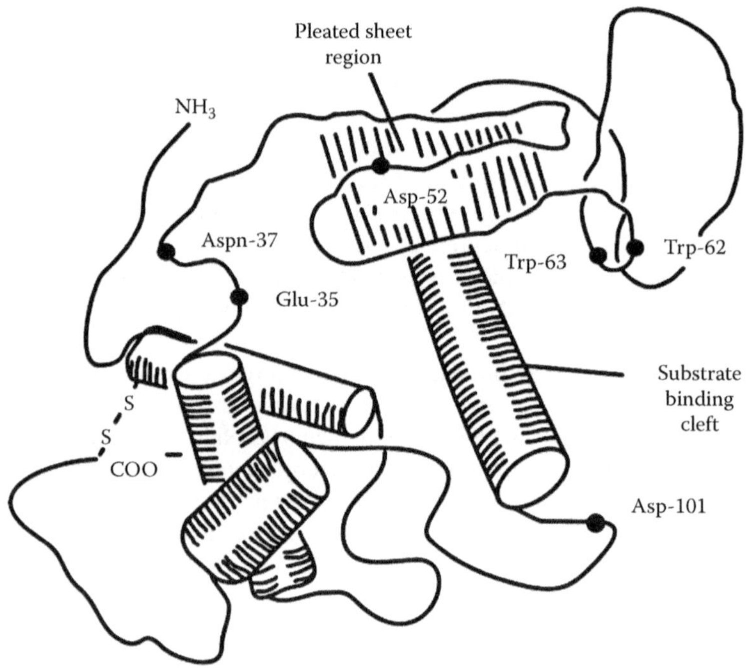

FIGURE 8.11 Simplified representation of the 3D structure of egg white lysozyme as revealed by x-ray diffraction studies. Only the backbone of the polypeptide chain is shown, and α-helical regions are represented by cylinders. The amino acid side chains fill up most of the available space within a molecule, but a clearly defined cleft for the binding of substrate is apparent. The positions of certain important amino acid residues are indicated.

are found within the interior of the molecule, where they are unlikely to come into contact with water, whereas those with polar side chains are usually exposed to the solvent.

8.5.4 Investigation of Protein Structure in Solution

X-ray diffraction analysis is not suitable for investigating proteins in solution, since the molecules are not fixed in a regular arrangement. However, other techniques may be used to give information on the structures of proteins in solution, particularly on the degree of secondary structure present.

There will be differences in both IR and UV spectra between a polypeptide chain in an α-helix conformation and one existing as a random coil (i.e., one without regular, repeating 3D features); these are due to the presence or absence of hydrogen bonding between atoms in different peptide bonds. Also, since a right-handed α-helix is an asymmetric structure, there will be differences in optical rotation between a polypeptide in such a conformation and one consisting of the same amino acid residues in a random coil.

Such investigations have helped to demonstrate that polypeptide chains can exist as α-helices in solution. Such a structure is most readily formed if all amino acid side chains present are small and uncharged, as is the case with polyalanine, a synthetic polypeptide consisting entirely of L-alanine residues. If all the side chains are large, as with polyisoleucine, no α-helix is formed. In the case of synthetic polypeptides with ionizable side chains, for example, polyglutamic acid, the structure in solution varies with pH. At acid pH, glutamic acid side chains are uncharged and an α-helix is formed. However, at alkaline pH, the side chains will have a negative charge; these repel each other and the α-helix is disrupted as shown by optical rotation measurements. The reverse effect is found with polylysine, whose side chains are uncharged at alkaline pH but have a positive charge at acid pH. Today, direct measurements of optical rotation are rarely carried out as similar information can be obtained more easily by the use of circular dichroism (CD) spectroscopy, which measures the differential absorption of right and left circularly polarized light over a range of wavelengths.

Spectrophotometry may give useful information about protein structure since peptide bonds, aromatic and imidazole side chains, and disulfides all give absorbance bands in the ultraviolet (UV) range, which may vary according to the conformation of the protein and the micro-environment of the absorbing group. Spectrofluorimetry too may be of value, for example, in investigations of the fluorescence of tryptophan side chains. The electron spin resonance (ESR) technique, which detects unpaired electrons, is useful for the investigation of metal ions in enzymes, whereas NMR spectrometry currently gives the best structural information about proteins in solution. The NMR technique detects atoms with an odd number of protons in their nuclei by giving them a magnetic moment. If a kilogauss magnetic field is applied, such nuclei will move around with a frequency depending on the magnetic moment (m) and the magnitude of the applied field (H_0); if a radiofrequency around 100 MHz field is then generated so that its magnetic vector rotates perpendicular to the kilogauss field and the conditions are adjusted, resonance will occur when the frequency of oscillation of the field corresponds to the precession of a nuclear dipole, which can enable the nuclei concerned to move to higher energy levels. For the single proton in the hydrogen nucleus, for example, there are two possible orientations: One, aligned with the kilogauss field, has an energy level given by $-\mu H_0$, whereas the other, aligned against the field, has an energy level of $+nH_0$. Therefore, moving from the orientation of lower energy to that of higher energy requires an energy input of $2\mu H_0$, and the resonance frequency (ν) is given by $h\nu = 2\mu H_0$.

In this expression, h is Planck's constant. This resonance is detected as the absorption of energy ($= 2\mu H_0$) by the proton from the radiofrequency field. In general, the intensity of a resonance absorption line is directly proportional to the number of nuclei in an identical environment. However, chemical shifts in resonance frequency for identical nuclei in different electronic environments can be detected, as can the splitting of resonance peaks into multiple fine structures because of the interaction between neighboring nuclear spins. Switching off the radiofrequency field and studying the characteristics of the change back (termed relaxation) to the original distribution of nuclei at different energy levels can also give information about interactions between neighboring nuclei. Among the nuclei that can be investigated by NMR techniques, there are ^{1}H, ^{13}C, and ^{15}N. Isotopes such as ^{13}C and ^{15}N, which do not occur naturally, can be used in the investigation of protein structure. They must first be incorporated into the protein, for example, by utilizing bacteria to synthesize the protein in a medium rich in the isotope, depending on the type of investigation consistent with the assumption that 3D structures found in protein crystals may also occur in solution.

Experiment: Analysis of Proteins by X-Ray Crystallography

Principle

The technique imparts knowledge of the cellular mechanism and the 3D structure of enzymes and other macromolecules (Figure 8.12). It is critical that we understand better how each chemical reaction that occurs in a cell needs a specific enzyme for it to happen. Two common techniques used for the analysis of protein structure are NMR and x-ray crystallography. X-ray crystallography can be used to analyze any different compounds up to a molecular weight of 10^6 (g/mol), for instance, whereas NMR is restricted to the analysis of biopolymers (polymers produced by a living organism such as starch, peptides, and sugars) of molecular weights no more than 30,000 (g/mol). It can also measure compounds that are very small because the appropriate size to measure the distance between atoms in a molecule is 0.5–1.5 Å. X-rays are used as the form of radiation because their wavelengths are on the same order of magnitude as the bond length of a covalent bond (~1 Å or 1×10^{-10} m), which is necessary to obtain a diffraction pattern that reveals information about the structure of the molecule. If the radiation had a wavelength much bigger or much smaller than the bond length of a covalent bond, the light would not diffract and no new knowledge of the molecular structure would be obtained.

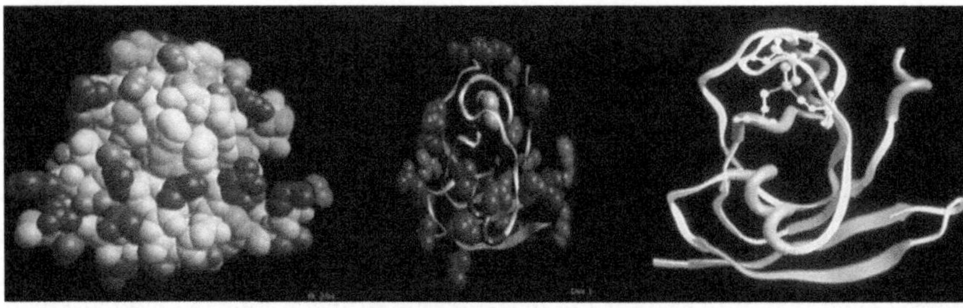

FIGURE 8.12 **(See color insert.)** Computer-generated tertiary structures of crystallographic data of ferredoxin proteins FdxH1 and FDxH2 from the cyanobacterium *Anabaena variabilis.*

Prelaboratory Preparation

The three components needed to complete an x-ray crystallography analysis are a protein crystal, a source of x-rays, and a detector.

Procedure

A. First Step

The process begins by crystallizing a protein of interest. Crystallization of protein causes all the protein atoms to be orientated in a fixed way with respect to one another while maintaining their biologically active conformations, which is a requirement for x-ray diffraction. A protein must be precipitated out or extracted from a solution. The rule of thumb here is to get as pure a protein as possible to grow lots of crystals (this allows the crystals to have charged properties and surface-charged distribution for better scattering results). Four critical steps are performed to achieve protein crystallization, as follows:

1. Purify the protein. Determine the purity of the protein; if it is not pure (usually greater than 99%), then it must undergo further purification.
2. Precipitate the protein. This is usually done by dissolving the protein in an appropriate solvent (water–buffer solution with an organic salt such as 2-methyl-2,4-pentanediol; if the protein is insoluble in water–buffer solution or water–organic buffer solution, then a detergent such as sodium dodecyl sulfate [SDS] must be added).
3. The solution must be brought to supersaturation (condensing the protein from the rest of the solvent-forming condensation nuclei). This is done by adding a salt to the concentrated solution of the protein, thereby reducing its solubility and allowing the protein to form a highly organized crystal (this process is referred to as salting out). Other methods include batch crystallization, liquid–liquid crystallization, vapor diffusion, and dialysis.
4. Let the actual crystals grow. Since nuclei crystals are formed, this will lead to actual crystal growth.

B. Second Step

For the next step, x-rays are generated and directed toward the crystallized protein. X-rays can be generated in four different ways:

1. By bombarding a metal source with a beam of high-energy electrons
2. By exposing a substance to a primary beam of x-rays to create a secondary beam of x-ray fluorescence
3. From a radioactive decay process that generates x-rays (gamma rays are indistinguishable from x-rays)
4. From a synchrotron (a cyclotron with an electric field at constant frequency) radiation source

The first and the last methods utilize the phenomenon that an accelerating charge gives off radiation. Then the x-rays are shot at the protein crystal, which results in some of the x-rays going through the crystal and the rest being scattered in various directions. The scattering of x-rays is also known as x-ray diffraction. Such scattering results from the interaction of electric and magnetic field of the radiation with electrons in the atoms of the crystal.

The patterns are a result of the interference between the diffracted x-rays, which is governed by Bragg's law. It gives the expression $2d\sin\theta = n \times \lambda$, where d is the distance between two regions of electron density, θ is the angle of diffraction, λ is the wavelength of the diffracted x-ray, and n is an integer. If the angle of reflection satisfies the condition

$$\sin\theta = \frac{(n \times \lambda)}{2d}$$

the diffracted x-rays will interfere constructively. Otherwise, destructive interference occurs. Figure 8.13 shows an example of constructive interference. Here is an example of destructive interference:

Constructive interference indicates that the diffracted x-rays are in phase or lined up with each other, whereas destructive interference indicates that the x-rays are not exactly in phase with each other. The result is that the measured intensity of the x-rays increases and decreases as a function of angle and distance between the detector and the crystal.

The x-rays that have been scattered in various directions are then caught on an x-ray film, which shows a blackening of the emulsion in proportion to the intensity of the scattered x-rays hitting the film, or by a solid-state detector, similar to the ones found in digital cameras. The crystal is rotated so that the x-rays are able to hit the protein from all sides and angles. The pattern on the emulsion reveals much information about the structure of the protein in question. The three basic physical principles underlying this technique are as follows:

1. Atoms scatter x-rays. The amplitude of the diffracted x-ray is directly proportional to the number of electrons in the atom.
2. Scattered waves recombine. The beams reinforce one another at the film if they are in phase and cancel one another out if they are out of phase. Every atom contributes to a scattered beam.
3. The 3D atomic arrangement determines how the beams recombine.

The intensities of the spots and their positions are thus the basic experimental data of the analysis.

C. Final Step

The final step involves creating an electron density map based on the measured intensities of the diffraction pattern on the film. A Fourier transform can be applied to the intensities on the film to reconstruct the electron density distribution of the crystal. In this case, the Fourier transform takes the spatial arrangement of electron density and gives out the spatial frequency (how closely spaced the atoms are) in the form of the diffraction pattern on the x-ray film. An everyday example of Fourier transform is the music equalizer on a music player. Instead of displaying the actual music waveform, which is difficult to visualize, the equalizer displays the intensity of various bands of frequencies. Through the Fourier transform, electron density distribution is illustrated as a series of parallel shapes and lines stacked on top of each other (contour lines), similar to a terrain map. The mapping gives a 3D representation of the electron densities observed in x-ray crystallography. When interpreting the electron density map, the resolution needs to be taken into account. A resolution of 5–10 Å can reveal the structure of polypeptide chains, 3–4 Å the structure of groups of atoms, and 1–1.5 Å the structure of individual atoms. The resolution is limited by the structure of the crystal, and for proteins it is about 2 Å.

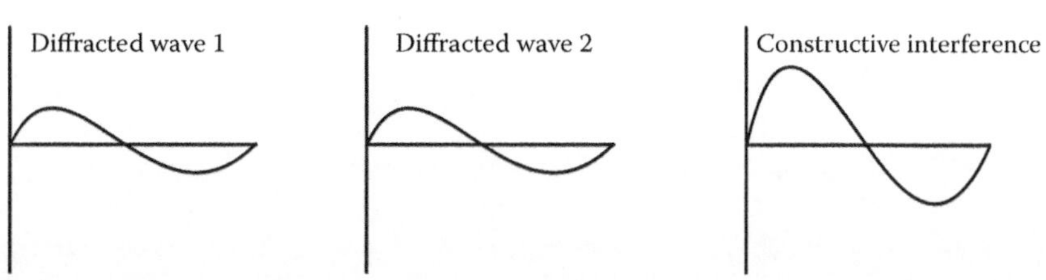

FIGURE 8.13 Diffracted wave and constructive interference.

SUGGESTED READING

Goldstein J., D. Newbury, D. Joy, C. Lyman, P. Echlin, E. Lifshin, L. Sawyer, and M. Joseph. 2003. *Scanning Electron Microscopy and X-Ray Microanalysis*. New York: Springer.

Kraut, J. "How Do Enzymes Work?" *Science*. 242:534.

National Institutes of Health. 2006a. NIH Publication No. 06-474 July 2006 *Medicine by Design*. 25–27. http://www.nigms.nih.gov.

National Institutes of Health. 2006b. NIH Publication No. 06-474 July 2006 *Medicine by Design*. 31. http://www.nigms.nih.gov.

National Institutes of Health. 2007. NIH Publication No. 07-5423 September, 2007 *Structure of Life*. 37–38, 40, 44. http://www.nigms.nih.gov.

Orr, P. J., and S. L. Kearns. 2011. "X-Ray Microanalysis of Burgess Shale and Similarly Preserved Fossils." *Quantifying the Evolution of Early Life. Topics in Geobiology* 36:271–99.

Shafranovskii, I. I., and N. V. Belov. 1962. "E. S. Fedorov." In *50 Years of X-Ray Diffraction*, edited by Paul Ewald, 351. New York: Springer.

Viadiu, H. November 2011. "Why Do We Need Crystals?" UCSD Lecture.

IMPORTANT LINK

1. X-ray microanalysis: http://www.thermoscientific.com/ecomm/servlet/newsdetail?contentId=52420&storied=11152&gclid=CNS785KeoasCFYEa6wodZWwYFA

9 Techniques with Radioisotopes

9.1 INTRODUCTION

A radioisotope is a version of a chemical element that has an unstable nucleus and emits radiation during its decay to a stable form. Radioisotopes have important uses in medical diagnosis, treatment, and research. A radioisotope is so named because it is a radioactive isotope, an isotope being an alternate version of a chemical element that has a different atomic mass.

Most radioisotopes are made by bombarding a stable element with neutrons in the core of a nuclear reactor (see fission). The radiations given off by radioisotopes are easy to detect (hence their use as tracers), can in some instances penetrate substantial thicknesses of materials, and have profound effects (such as genetic mutation) on living matter. Most natural isotopes of relative atomic mass less than 208 are not radioactive. Those from 210 and up are all radioactive.

Radioisotopes have many uses in medicine, for example, in radiotherapy and radioisotope scanning. The use of radioactive isotopes in the diagnosis, investigation, and treatment of disease is called nuclear medicine. The nucleus of a radioisotope is unstable and undergoes changes by breaking down into a more stable form. A radioisotope decays over a period of time into a new element. As it decays, it emits radiation energy in the form of alpha and beta particles and gamma radiation. The time taken for half of the original atoms to decay is known as the half-life. The product of radioactive decay is called a "daughter" atom.

9.2 ISOTOPES AND RADIOACTIVITY

There are many different types of radiation. Some types have a particulate character; others, called electromagnetic (EM) radiation, have wave-like properties. *Alpha particles* result from the decay of relatively heavy radioisotopes. They consist of two neutrons and two protons, bound together and identical to helium nuclei, He:

$$^{210}_{84}\text{Po} \rightarrow {}^{206}_{82}\text{Pb} + {}^{4}_{2}\text{He} \quad (\alpha\text{-particle})$$

The particles move relatively slowly, but due to their mass, they have a high-momentum travel in straight lines not deflected from their path. Normally, they are only deflected by direct collision with a nucleus (Figure 9.1). Americium 241 (Am-241) is a familiar example, commonly found in household smoke detectors.

In *beta* decay, a neutron converts to a proton, emitting a *beta particle* in the process. The beta particle is identical to an ordinary electron. Beta particles are electrons released from the nucleus by the decay of a neutron into a proton and an electron (Figure 9.2). Their energy spectrum depends on the speed with which the electron leaves the nucleus. Beta particles are emitted from a given radioisotope over a continuous range of energy up to a maximum value (E_{max}), which is the characteristic of each radioisotope. They have little mass—about 1/7400 the mass of alpha particles. As a beta particle traverses material, it causes the ionization and excitation of orbital electrons. Beta particles produced by ^{3}H and ^{14}C are weak, with little penetrating power ("soft" beta particle), but those produced by ^{32}P are more energetic ("hard" beta particles) and have greater penetrating power.

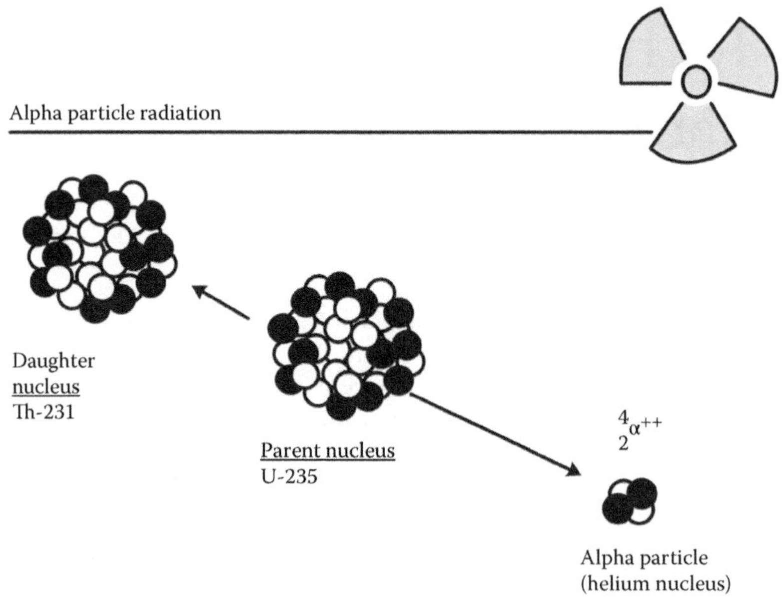

FIGURE 9.1 Alpha radiation and decay process.

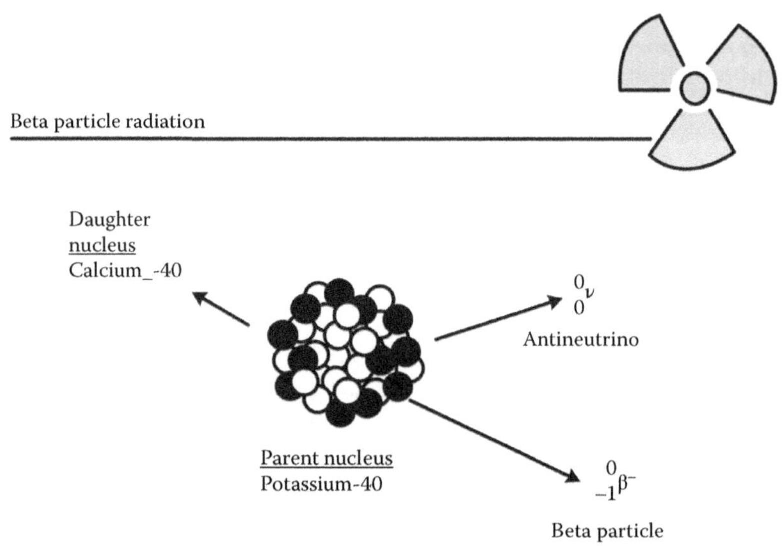

FIGURE 9.2 Beta radiation and decay process.

Carbon 14 (^{14}C) is a radioisotope of carbon, which undergoes beta decay and may be familiar for its use to establish the age of ancient artifacts ("carbon dating"):

$$^{14}_{6}C \rightarrow ^{14}_{7}N + \textit{beta} \text{ particle}$$

Gamma rays are emitted if a nucleus still has excess energy following the decay and emission of other particles. They are electromagnetic in nature (called photons), with a discrete, unique energy (this is used to identify different radioisotopes). *Gamma rays* are not physical particles, but their interactions with matter are described by assigning them particle-like properties (Figure 9.3).

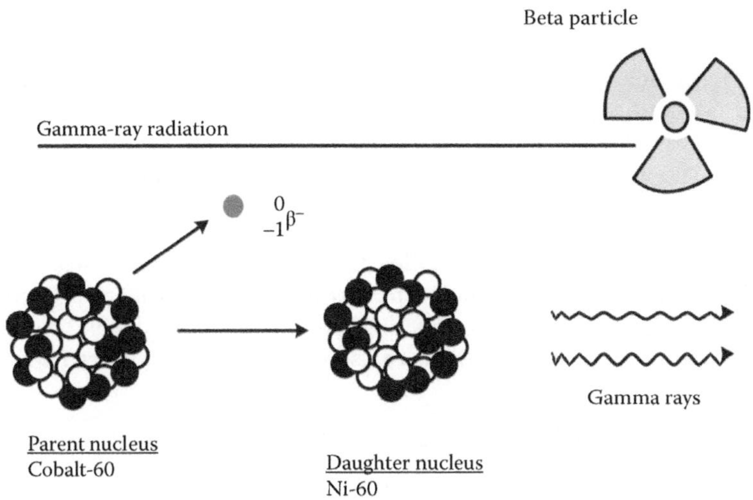

FIGURE 9.3 Gamma radiation and decay process.

Radioisotopes or unstable versions of an element that emit radiation as they try to reach more stable forms are critical in modern medicine. The nucleus of an atom of a radioactive isotope has a certain probability, characteristic for that isotope, that it will decay at any instant. The probability remains constant independent of chemical reactions the atom may undergo. It is also independent of temperature or other physical conditions. The probability is constant for all atoms of a given isotope in a specialized period of time; a constant fraction of the nuclei will have decayed. The half-life ($T_{1/2}$) of an isotope is a period of time during which half the radioactive atoms originally present will have decayed. It is mathematically related to the radioactivity decay constant, which is as follows:

$$T_{1/2} = \frac{0.693}{\lambda}$$

^{99}Mo is used to produce ^{99}Mo/^{99m}Tc generators (a generator technology developed at Brookhaven National Laboratory) for use in nuclear medicine. ^{99m}Tc is the most widely used radionuclide in nuclear medicine, both for the detection of disease and for the study of organ structure and function. More than 15 million procedures are performed each year in the United States using ^{99m}Tc.

9.3 IONIZATION EFFECTS

Because alpha particles are relatively massive and doubly charged, when they pass close to another atom, they may strip some orbital electrons from that atom, producing positive ions. An alpha particle with energy of 1.5 MeV can produce about 2×10^5 positive ions in the air before its energy is expended. Beta particles also lead to the ionization of materials through which they pass. Being less massive, they travel faster and hence spend less time in the vicinity of other atoms and have less time to remove electrons from them. The ionization effects caused by alpha particles are therefore much less intense than those produced by beta particles. Gamma rays being uncharged have no appreciable force fields; they do not lead directly to the ionization of materials. However, they interact with matter in three unique ways, indirectly producing energetic electrons:

1. Low-energy gamma radiation (<0.5 MeV) can transfer all their energy to the orbital electrons of atoms of the absorbing medium. The energy is thus transformed to the kinetic energy of the electrons, which ejects out as the "photoelectron."
2. Medium-energy gamma radiation (0.5–1.0 MeV) may interact with electrons by inelastic collision and transfer only a portion of their energy in the process. The electron ejected

(now termed the "compton electron") has lesser energy than the energy of the incident gamma radiation.

3. High-energy gamma rays having energy more than 1.02 MeV react directly with the nucleus producing an electron–positron pair. This is a unique process in which high-energy gamma rays are transformed into matter. The resulting positron reacts with the surrounding matter by colliding with an electron. The mass of both the positron and electron are annihilated to become two 0.51 MeV photons.

Several devices for the measurement of radioactivity are based on the measurement of this ionization in gas-filled chambers, gas ionization detectors such as ion chambers, proportional counters, and Geiger–Mueller counters. Before considering any of the measurement techniques, it makes sense to first make ourselves familiar with the measurement units of radioactivity.

9.4 MEASUREMENT UNITS

Measurement units can be classified into three types:

1. *Unit of exposure of radioactivity*, that is, the amount of radiation energy directed at a material. The most common unit is Rontgen (R), which is defined as the quantity of X radiation or gamma radiation in the air, an ion of either sign carrying a charge of 2.58×10^4 C/Kg of air. Rontgen was restricted to radiations below 3 MeV. Because of the difficulty of measuring the ionizations in air of the very energetic secondary electrons, radiation absorbed dose (RAD) was introduced.
2. *Radiation absorbed dose or RAD* is the unit of absorbed radiation energy. It is defined as the absorption of 10^{-2} J of radiation energy per kilogram of material. The SI unit of absorbed dose is Gray (Gy) defined as 1 J/kg and therefore one hundred times larger than RAD. Since the energy absorbed in tissue corresponding to an exposure to 1 Rontgen is 0.0095 J/Kg, 1 Rontgen gives all absorbed dose of 0.95 RAD.
3. *Unit of activity*: This was until recently Curie (Ci), defined as the number of disintegrations per second (dps) occurring in 1 g of pure radium (^{236}Ra). It was arbitrarily fixed at 3.7×10^{10} dps. That is,

$$1 \text{ Ci} = 3.7 \times 10^{10} \text{ dps}$$

In SI, the Curie has been replaced by the Becquerel (Bq). The two are related as

$$2.703 \times 10^{-11} \text{ Ci} = 1 \text{ Bq}$$

$$3.7 \times 10^{10} \text{ Bq} = 1 \text{ Ci}$$

Specific activity is defined as the amount of radioactivity in a given weight of material and is usually expressed as Curies, disintegrations rate, or count rate, per unit mass of element; for example, the specific activity of C is 19 counts per minute per gram.

9.5 MEASUREMENT TECHNIQUES

Radioactivity is detected and calculated by measurements made by observing physical or chemical changes in appropriate devices. The most common method is by measuring the amount of ionization caused in air or other gases. Though this is the most important dosimetric method, there are many other systems of detection, some of which are discussed next.

Scintillation detectors convert the radiation energy of ionizing particles into pulses of light. The size of the pulse is proportional to the energy deposited in the crystalline or liquid scintillant. Some semiconductors show an altered conductivity during exposure to radiation. This method is similar to the ionization chamber measurement, except that the current is flowing in the solid semiconductor crystals and not in the gas of the ionization chamber. In the ferrous sulfate dosimeter, $FeSO_4$ is converted to $Fe_2(SO_4)_3$ when irradiated. Chemical titrations with $K_2Cr_2O_7$ may be used to determine the amount of ferric ion and therefore the radioactivity present.

Calorimetry involves the measurement of the rise in temperature produced by radiation in an insulated mass of unknown thermal capacity. When LiF and many other crystals are heated after being exposed to radiation, they emit light; this is called thermoluminescence. The absorption of radiation energy causes free electrons to be trapped in the lattice imperfections of the crystalline structure. These trapping levels lie between the valence band and the conduction band; the electrons can remain trapped for a considerable time. If the temperature of the crystal is raised, the electrons are excited from the trapping levels to the conduction band and then return to the valence level with the emission of light. Thermoluminescence devices may be used as powders in capsules or sachets. Thermoluminescent dosimeters are widespread in laboratories for personnel monitoring.

Photographic emulsions consist of AgX crystals or grains dispersed in gelatin. Radiation absorbed in an individual grain forms a latent image, and the chemical action of development reduces the grain to silver. The blacker the film is, as measured by densitometry, the more doses it has received. This again is used for personnel protection, in the form of film badges. Certain plastics and glasses become increasingly opaque with increasing radiation dose.

9.5.1 Scintillation Counting Systems

Scintillation counting is a means of detecting radiation through the production of light flashes (scintillation). The energy of radiation is used to produce pulses of light, which are counted. Substances called "fluors" or scintillators fluoresce upon irradiation. The phenomenon of fluorescence has been discussed elsewhere, and here it should suffice to say that fluorescence is the emission of radiation of longer wavelengths as compared to the wavelengths of incident radiation. This may appear in the visible region of the spectrum, and energy is emitted as photons, which are counted by associated electronic equipment.

Figure 9.4 shows the mechanism of scintillation counting. The photomultiplier tube transforms the emitted photons into an electrical signal; the magnitude of this signal is increased by more than a million times. An extremely stable high-voltage power must be supplied in order to maintain the reproducibility in multiplication. The preamplifier further multiplies the photomultiplier output. This allows high counting rates to be achieved using scintillation counting.

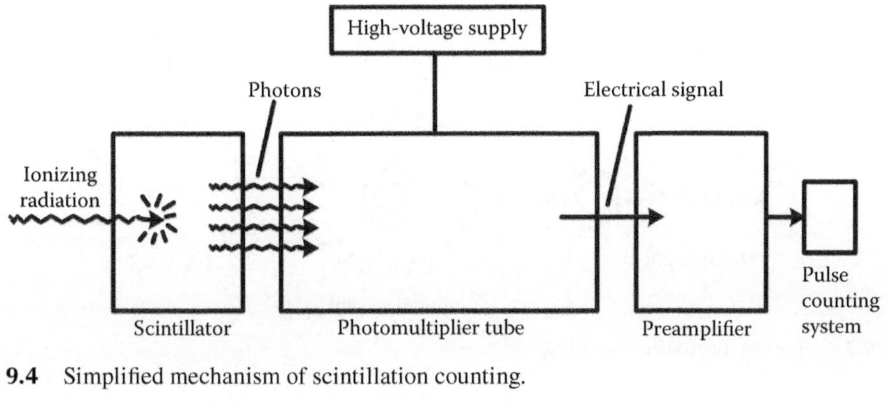

FIGURE 9.4 Simplified mechanism of scintillation counting.

The pulse height analyzer differentiates among the energies of the incident gamma rays. This analyzer classifies pulses according to their height or amplitude. A single channel pulse height analyzer consists of two variable discriminators, which allow for the selection of lower and upper levels of detection. The lower discriminator setting is termed the base. The upper discriminator setting is selected by adding a voltage increment to the base. These two discriminators, together with an anticoincidence circuit, allow only those energies between the two discriminator levels to pass to the scaler. The scaler displays the counts accumulated during the counting period, that is, it registers the number of pulses received.

Depending on whether the fluor is a solid or liquid, there are two systems, namely solid scintillation counting and liquid scintillation counting. The fluor generally used for solid scintillation counting is a single large crystal of sodium iodide containing thallium as an activator. The crystal surrounds the sample to increase counting efficiency. The sodium iodide crystals, being hygroscopic (absorb water) in nature, are kept in light-reflecting aluminum sachets, which are completely closed except where attached to the photomultiplier tube through a transparent window.

Low-energy beta emitters like ^{3}H and ^{14}C cannot be counted by solid scintillation counting because the weak emissions cannot traverse the aluminum envelope of the scintillation crystals. This necessitates an intimate mixture of the isotope dissolved in liquid fluor so that there is no self absorption of beta or alpha particles and efficient energy transfer occurs. Liquid scintillation derives its name from the liquid mixture composed of isotope sample, the fluor dissolved in an organic solvent. The mixture of an organic solvent (usually toluene) and the fluor is often called a "cocktail." For samples containing water, dioxane can be used as a solvent. The fluors are generally complex heterocyclic, organic compounds that when excited emit photons in the near ultraviolet and visible regions. The fluors generally comprise less than 1% of the cocktail. Some organic fluors used are PPO and POPOP. PPO is the primary solute. POPOP is the secondary solute and is used to shift the fluorescence of PPO to a longer wavelength for better matching to the spectral response of photomultiplier tubes. This procedure increases the counting yield and is generally used in the measurement of H.

The emissions from the sample interact with the material surrounding them and cause excitation first of the solvent molecules. This excitation energy of the solvent is transferred to the solute and the fluor, causing the excitation of fluor electrons. The excited electrons in the fluor emit photons of light as they fall back to the ground state. These photons of light are detected by the paired photomultiplier tube (PMT) and result in electrical pulses as in solid-state counters (Figure 9.5).

Since the low-energy ranges of beta emissions result in a photomultiplier output of the same magnitude as the thermionic emission, coincidence circuitry has been used. This ensures that only pulses seen simultaneously (usually within less than 20 nanoseconds) are registered. Pulses from

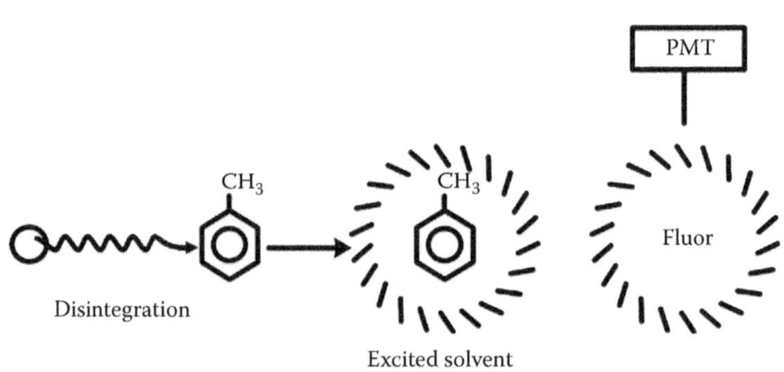

FIGURE 9.5 Liquid scintillation counting.

the two detection circuits are passed through a coincidence circuit and electronically added up. Pulses then pass to two or more channels, each consisting of a linear amplifier, pulse height analyzer, and scaler, as in solid-state systems (Figure 9.6)

Liquid scintillation counting, though ideal for counting weak emissions, poses the problem of "quenching." This is especially true of biological samples, which may contain a great variety of chemical compounds. Even oxygen in the air and chemicals in the sample result in quenching. Quenching is of three types: chemical, chromatic, and optical. Chemical quenching is caused by various polar absorbing energies from excited solvent molecules partly preventing the excitation of the fluor. Color quenching can be expected whenever the fluor solution does not have its characteristic light blue color due to impurities. Yellow and red solutions especially may absorb the blue light emitted by the fluors. Optical quenching results when the mixture of the scintillation cocktail and sample results not in true solutions but in suspensions. Quenching decreases the detectable energy released by isotopes, hence it should be minimized. Because quenching is often unavoidable and the

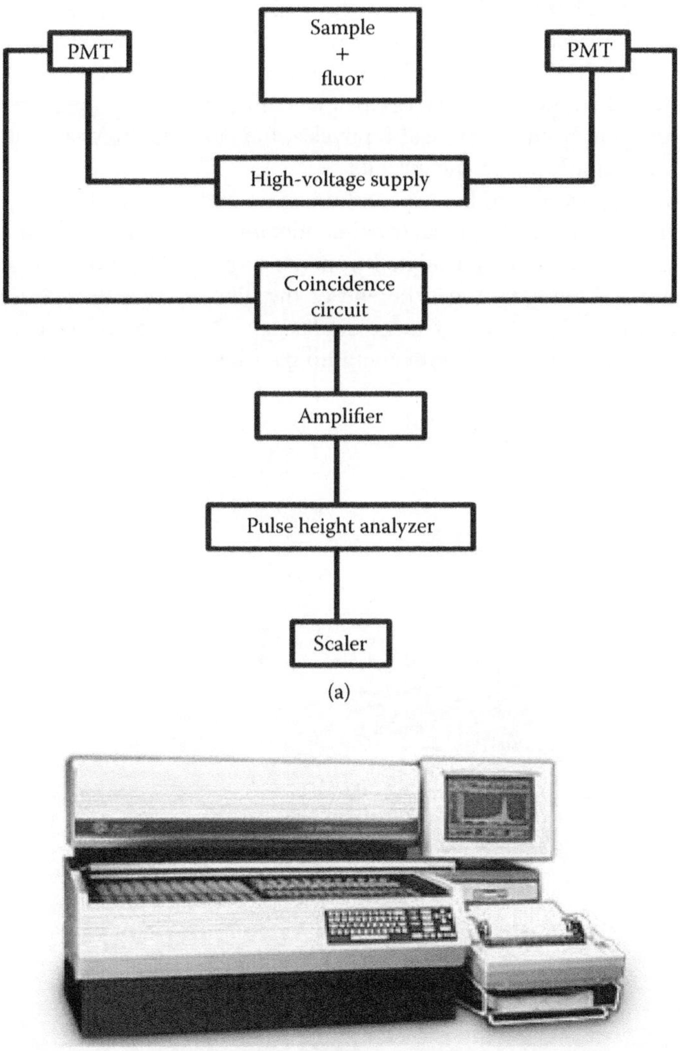

FIGURE 9.6 (a) Block diagram of a liquid scintillation counter; (b) a liquid scintillation counter.

degree of quenching is unpredictable, therefore, it should be corrected. Correction for quenching, that is, the determination of efficiency (E), can be brought about by using the following relationship:

$$\%\,\mathrm{E} = \frac{\mathrm{cpm}}{\mathrm{dpm}} \times 100$$

Disintegrations per minute (dpm) is an absolute number proportional to the isotopic content of a sample, while counts per minute (cpm) refers to the detected counts, which is always less than dpm due to quenching.

9.5.2 Geiger–Mueller Counter

The Geiger–Mueller counter utilizes the ionizations produced in air or gas by radioactive disintegrations. A *Geiger counter*, also called a *Geiger–Müller counter*, is a type of particle detector that measures ionizing radiation. They detect the emission of nuclear radiation: alpha particles, beta particles, or gamma rays. A Geiger counter detects radiation by ionization produced in a low-pressure gas in a Geiger–Müller tube. Each particle detected produces a pulse of current, but the Geiger counter cannot distinguish the energy of the source particles. Geiger counters are popular instruments used for measurements in health physics, industry, geology, and other fields, because they can be made with simple electronic circuits.

The electrical signal is produced by a special tube (Figure 9.7). The tube consists of a chamber with the inner surface coated with an electrical conductor and it acts as a cathode. At the central axis of the chamber is a wire (anode) insulated from the cathode. The tubes are of different designs. In the most common end–window tube, the end of the tube is covered with a thin membrane or window that is permeable to particles of sufficient energy. The inside of the chamber is filled with a monoatomic gas, usually argon or helium containing a small amount (0.1%) of halogen or organic

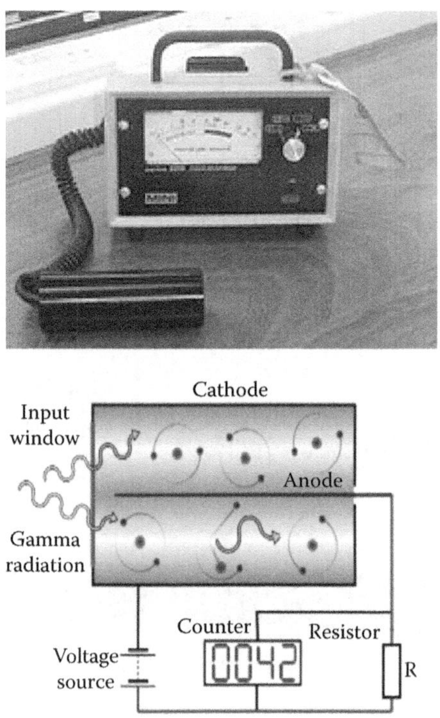

FIGURE 9.7 Geiger–Mueller counter.

gas. This type is used mostly for quantitative estimations on solid materials. Other designs include the thin wall type used mainly inside contamination monitors. There are others not discussed here, which are useful for measuring activity in liquids. These include the dipping, liquid flow, annular well, and tissue probe types.

For accurate quantitative work, Geiger–Mueller tubes (Figure 9.8) are contained in a lead block, which also surrounds the sample chamber to shield the tube and chamber from outside radiations giving rise to counts in the apparatus when no sample is present. When a radioactive emitter is brought close to the window of the tube, some of the ionizing radiation (gamma) or particles (alpha and beta) penetrate the window and pass into the gas inside the tube, leading to the formation of pairs of ions—positive ions and electrons. If a high, potential difference is applied across the electrodes, these ions are accelerated toward the electrode of the opposite charge. The accelerated ions also react with gas atoms of the tube to produce more ions, and this chain reaction continues to produce an avalanche of ions. An amplification of about 10^6–10^8 is normally obtained; when the ion avalanche reaches the electrodes, it is neutralized, producing a flow of electrons in the external circuit and giving a measurable potential of 1–10 V. Thus, the reaction is terminated and another particle can be detected. The halogen gas inside the tube absorbs some of the energy of the accelerated ions, thereby quenching them and helping to terminate the reaction in a very short time. Therefore, sufficient quenching gas must be added to return the tube to its unionized state in less than a millisecond from the particle's entry.

The electronic circuits associated with the tube can be designed to indicate the average current flow (a rate meter) or to total the number of electrical pulses (a scaler). The count rate is dependent upon the potential applied across the electrodes. This can be shown by varying the applied voltage and determining the count rate on a suitable sample. On plotting the electrical potential against the count rate, a characteristic curve is obtained (Figure 9.9). At low voltages, the curve is exponential, and slight changes in voltage cause considerable changes in the count rate. At higher voltages, the curve becomes linear over this region; the voltage across electrodes is a great enough to cause particles to move sufficiently fast to produce maximum ionization of the tube gas.

At even higher voltages, the count rate increases exponentially as the Geiger tube goes into continuous discharge, the voltage being great enough to break down the insulating properties of the gas inside the tube. For maximum stability, the tube is operated in the plateau region where

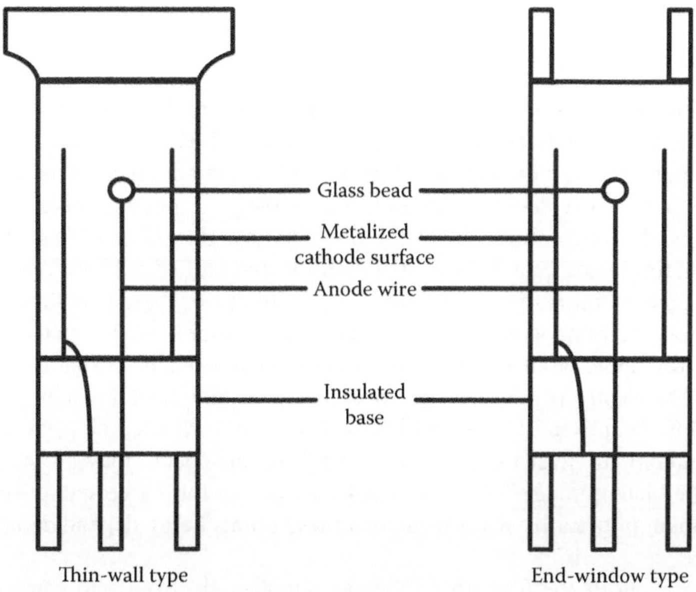

FIGURE 9.8 Types of Geiger–Muller tubes.

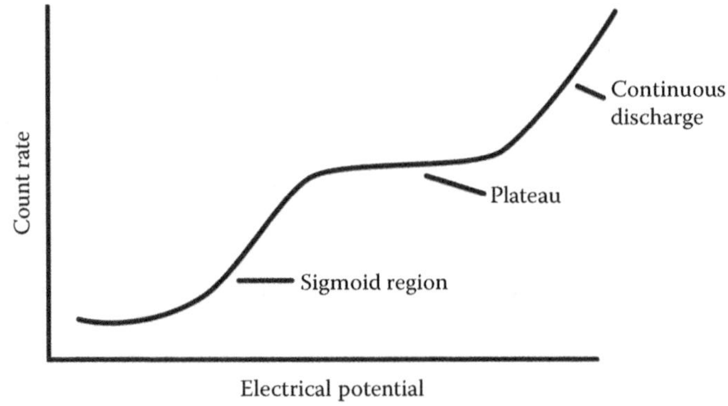

FIGURE 9.9 Characteristic sigmoid curve.

changes in voltages have little effect upon the count rate. In the plateau region, the tube is operating at near its maximum efficiency; therefore, no distinction can be made between particles of differing energy content (that the particles have sufficient energy to penetrate the tube window). This method of counting is therefore an all-or-nothing process. As in the case of scintillation counters, the efficiency of the tube is cpm/dpm × 100.

When the energy of radiation is low, as from ^{3}H particles, most of the particles are absorbed either before they reach the Geiger tube of the counter or by the window material. Hence, tritium cannot be counted on in this type of system unless special windowless gas-flow tubes are used.

The condition under which a Geiger–Mueller tube is usually operated is such that the tube gas is almost totally ionized, and hence the electrical pulse size is not influenced by the nature of the incident radiation. However, at low voltages, alpha and beta particles produce different degrees of ionization and different intensities of signal. Consequently, by the use of suitable discriminator circuits either in a normal Geiger system or in a low-voltage ionization chamber, alpha and beta particles from same or different sources can be detected. This type of counting is termed "proportional counting" since the size of the signal is proportional to the amount of ionization produced by the incident radiation and therefore varies with the nature of this radiation.

9.6 AUTORADIOGRAPHY

Autoradiography is the detection of radioactivity using photographic emulsions. This is a technique by which radioactivity can only be detected and not measured. However, this factor does not in any way lessen its importance as a unique technique in biology. In conjunction with several other techniques, autoradiography is widely used in molecular biology. A photographic emulsion consists of very small crystals of silver bromide in gelatin mounted on a glass or flexible support, and the silver bromide also contains small quantities of silver sulfide and colloidal silver. Also, the lattice structure is not perfect but contains spaces or "holes" in the structure where ions should be, and there are silver ions out of place and in between the other ions (interstitial ions). When a radioactive emission passes into a photographic emulsion, it causes ionization and sets electrons free in the ionic silver bromide lattice. The electrons migrate over short distances to areas known as "sensitivity specks," which are probably composed of silver sulfide, and these in turn become negatively charged. This charge attracts interstitial silver ions, and upon reaching the speck, they are neutralized to silver, forming a so-called latent image. Thus, the specks act as loci for the growth of the nuclei of metallic silver, the amount of metallic silver formed at these points being dependent upon the amount of radiation received.

During development of the film, the developer supplies electrons and causes the reduction of silver ions to metallic silver atoms, which are responsible for the visible image. However, because of the catalytic effect of the metallic atoms already produced by the radiation, the silver ions in the region of the latent image are reduced by the developer much faster than those regions where there is no latent image. Hence, if development is correctly timed, only the former produces a visible image, which will reflect the sites of exposure of the emulsion to radiation. Thiosulfate present in the fixer solution removes unexpected silver bromide and thereby stabilizes the image.

9.7 COUNTING STATISTICS

If a single radioactive sample is counted several times under identical conditions and the count rate is corrected for radioactive decay or the decay correction is negligible, the individual count rates will deviate about a mean value. These deviations are due to the random nature of radioactive decay. Understanding these statistical effects is necessary in the consideration of experimental design and in the interpretation of experimental results. Counting statistics closely follow the Poisson probability distribution. By a special property of the Poisson distribution, the standard deviation (D) of a registered number of counts, C, is equal to the square root of that number for c registered counts:

$$D = \sqrt{c}$$

The standard deviation increases as the square root of the number of counts but decreases as a percentage of the counts. If in the equation directly above both sides are divided by the period of counting, t, the result is the standard deviation of the count rate, r:

$$\frac{D}{t} = \frac{\sqrt{c}}{t} = Dr$$

$$\text{since } c = rt$$

$$Dr = \frac{\sqrt{rt}}{t}$$

The number of counts, C, collected in any counting interval is due to the true counts of the sample plus those due to background. There is significant background radiation in almost any location. This is due to cosmic rays, cosmic ray-induced activity such as ^{14}C, and naturally occurring radioactive materials in the Earth's crust, for example, ^{226}Ra, ^{232}Th, and ^{40}K. The latter all have associated gamma rays. The cosmic ray contribution varies with altitude, and the composition of the Earth's crust varies with location. All radiation detector counter systems have an associated background counting rate due to the above-mentioned sources and also electronic noise. The background count rate is commonly reduced by shielding or by special electronic circuitry. The background radiation will be a function of the type of detector shielding, location, discriminator settings, and so on.

The variations of background are independent of the variations of the sample activity and the appropriate error terms added as the sum of the squares. Therefore, the variance (D^2) due to the sample activity alone is

$$D_s^2 = D_{a+b}^2 + D_b^2$$

where

D_s^2 = Variance due to sample alone
D^2_{a+b} = Variance due to count of sample plus background
D_b^2 = Variance due to background

For a total of C accumulated counts due to sample and background counts, the standard deviation of the sample count is given by

$$D_s = \sqrt{C}_{a+b} + C_b$$

where

C_{a+b} = Total counts due to sample plus background
C_b = Total counts due to background counted alone

If t_s is the counting time for a sample in the presence of background and t_b is the counting time for background, then the standard deviation of the sample counting rate Dr_s is as follows:

$$Dr_s = \frac{r_{a+b}}{t_s} + \frac{r_b}{t_b}$$

In tracer experiments, discussed in Section 9.8, the net counting rate of the samples very commonly approaches, or is even less than, the background counting rates. In order to choose the best division of time for counting the sample and for counting the background to obtain the minimum error, the following equation may be used:

$$\frac{t_b}{t_s} = \frac{r_b}{r_{a+b}}$$

9.8 BIOLOGICAL USES OF RADIOISOTOPES

Radioisotopes are widely used to study the physiological processes of biology. A small quantity of radioisotope called a tracer or label is introduced into the system to be studied, and its behavior is observed by tracing the position of the radioisotope within the system. The substance to be traced is called a tracee. The label is incorporated into the tracee through biological growth, chemical synthesis, or exchange processes.

Radioisotopes have many uses in medicine, for example, in radiotherapy and radioisotope scanning. The use of radioactive isotopes in the diagnosis, investigation, and treatment of disease is called nuclear medicine. The nucleus of a radioisotope is unstable and undergoes changes by breaking down into a more stable form. A radioisotope decays over a period of time into a new element. As it decays, it emits radiation energy in the form of alpha and beta particles and gamma radiation. The time taken for half of the original atoms to decay is known as the half-life. The product of radioactive decay is called a "daughter" atom.

9.9 TRACER DILUTION TECHNIQUE

The tracer dilution technique introduced by Hevesey is particularly useful when quantitative separations are not possible or are too tedious for the systems under study. If a radioactive specimen of one of the compounds present in the mixture is added to the mixture, the label becomes uniformly distributed throughout the mixture. When the compound is reisolated, some of the label will be present and hence can be counted, and the specific activity of the sample can be calculated. The concentration of the nonradioactive compound in the mixture can be determined from the extent to which it has diluted the radioactive additive, that is,

$$\text{Weight of unlabeled material} = \text{weight labeled}\ \frac{\text{specific activity additive}}{\text{specific activity isolate}} - 1$$

It is not necessary to isolate all material under study from the mixture because the label will be uniformly diluted in it. It is, however, necessary to purify the extract rigorously to remove any radioactive contaminants. A variation of the tracer dilution technique called inverse tracer dilution enables the determination of the amount of tracer in a system by the addition of a known amount of tracee.

9.10 RADIOIMMUNOASSAY

Radioimmunoassay (RIA) is one of the most important techniques in the clinical and biochemical field for quantitative analysis of steroids, hormones, and drugs and is especially useful for testing the sera from immunized animals for antibody response and for screening hybridomas for secretion of a specific antibody. Immunoassays are very sensitive and specific and therefore are commonly used for a great variety of measurements in both research and analytical laboratories. Several refinements of the basic technology have been developed. The most common techniques use a radioactively labeled antigen or antibody and involve competition for antibody binding between labeled and unlabeled antigens. The technique is based on the competition between an unlabeled antigen and a finite amount of the corresponding radiolabeled antigen for a limited number of antibody-binding sites in a fixed amount of antiserum. At equilibrium (with excess antigen), there will be both free antigen and antigen bound to the antibody. Under standard conditions, the amount of labeled antigen bound to the antibody will decrease as the amount of unlabeled antigen in the sample increases:

$$4\ Ag^* + 4\ Ab \qquad 4\ Ag^*\ Ab$$
$$4\ Ag + 4\ Ag^* + 4Ab \qquad 2Ag^*\ Ab + 2Ag\ Ab + 2Ag + 2\ Ag^*$$
$$12\ Ag + 4\ Ag^* + 4Ab\ Ag \qquad Ab + 3Ag^*\ Ab + 3Ag^* + 9\ Ag$$

where Ab, Ag, Ag*, and AgAb are antibody, unlabeled antigen, labeled antigen, and antibody–antigen complex, respectively.

By using known amounts of unlabeled antigen and a fixed amount of antibody and labeled antigen, the amount of labeled antigen bound as a function of the total antigen added is measured, and a calibration curve is constructed (Figure 9.10). This calibration curve may then be used to determine the amount of antigen in samples treated similarly.

RIAs offer the following potential advantages: (1) sensitivity, (2) simplicity, (3) specificity, and (4) the ability to measure any compound that is immunogenic, which means the technique can be universally applied. In spite of these advantages, RIA is not free from drawbacks, including the high cost of instruments used for measuring radioactivity and also the dangers involved in handling

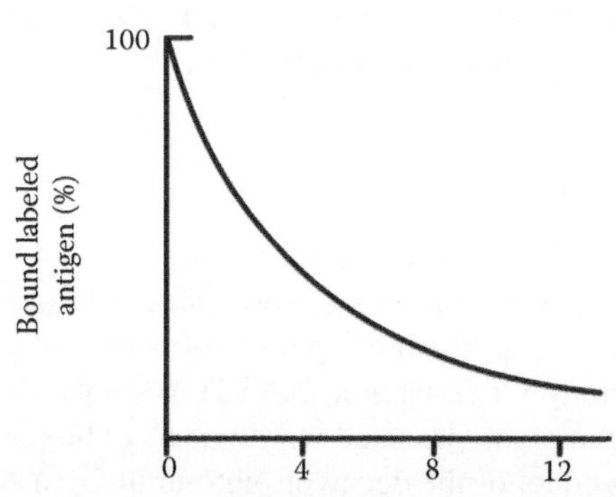

FIGURE 9.10 Calibration curve.

radioactive compounds. Hence, alternative labels such as enzymes (ELISA—enzyme linked immunosorbent assay) or fluorochromes (fluorescence immune assay) have been used.

9.11 MISCELLANEOUS USES OF RADIOISOTOPES

It is often left to the researcher's imagination to determine the number of ways in which he or she can use a particular technique. One such enterprising scientist has been Southern, after whom the technique of Southern blotting is named. Molecular biology involves the analysis of DNA in many ways. Electrophoresis of DNA results in the separation of DNA according to size. Now, it is often required to detect whether the required DNA piece is present or not. This can be achieved by transferring the DNA from the intact gel onto a piece of nitrocellulose paper placed in contact with it, using denaturation conditions, so that the DNA becomes bound to the paper in exactly the same pattern as that originally on the gel. This transfer, called the Southern blot, can be achieved electrophoretically or by drawing large volumes of buffer through both gel and paper. DNA thus transferred can now be treated with a radiolabeled DNA molecule acting as a "probe" to discover which bands of DNA contain sequences complementary to the probe after autoradiography of the paper. The use of radiolabeled DNA probes together with subsequent autoradiography has been put to practice in many molecular biology techniques, such as colony hybridization to detect and locate mutant colonies of microorganisms growing on a petri plate.

9.11.1 A Short Guide to Isotopes

The simplest way to envision an atom is to imagine a nucleus, which consists of protons and neutrons, surrounded by a cloud of electrons. The number of protons in the nuclei of any element's atoms determines the chemical character of that element.

Different isotopes of an element all contain the same number of protons, but different numbers of neutrons, and hence have different atomic masses. While some isotopes are stable, others are not; for the 82 stable elements, there are about 275 associated stable isotopes and 800 unstable isotopes or radioisotopes. (The U.S. Nuclear Science Advisory Committee [NSAC], in the report quoted above, estimates the number of natural and artificial radioactive isotopes as exceeding 3200 and "growing every year.")

The nuclei of unstable isotopes usually stabilize through radioactive decay (hence the name "radioisotope"). That is, as the isotope tries to reach a more stable form, it kicks off alpha particles, which consist of two protons and two neutrons, and/or beta particles, which are electrons or positrons. These beta particles are often (but not always) accompanied by the emission of gamma rays or EM radiation.

Unstable isotopes rarely occur in nature and are nearly always produced artificially. For example, ^{99}Tc is produced as the artificially produced radioisotope. ^{99}Mo decays, through gamma ray emission (and a rearrangement of its nucleus), to become ^{99}Tc.

9.11.2 Using Radioisotopes

Naturally occurring or artificially produced radioisotopes are useful because they emit radiation, whether alpha, beta, or gamma. Carbon dating uses a naturally occurring isotope of carbon, ^{14}C (carbon-14); ^{14}C can measure the age of water up to 50,000 years old. According to the Australian Nuclear Science and Technology Organisation (ANSTO), smoke detectors represent "the largest number of devices based on radioisotopes used worldwide." (The devices use a minute quantity of Americium-241, which is a product of the decay of plutonium-241.) Around 200 different radioisotopes are currently used in a variety of applications, such as in food and agriculture, industry, and medicine.

9.11.3 Food and Agriculture

In food and agriculture, the radiation emitted by radioisotopes is used in a variety of different ways.

9.11.3.1 Insect Eradication

Each year, insects probably destroy around 10% of global harvests. In developing countries, this figure has been estimated as high as 25–35%. In what is called the sterile insect technique (SIT), huge numbers of insects are bred, irradiated with gamma rays to sterilize them, and then released en masse in infested areas. Since they are sterile, they cannot reproduce, and subsequently, after a series of such releases, the insect populations in the areas in which they are released can be decimated. SIT has been used successfully against such insects as the Mediterranean fruit fly (in Argentina, Chile, and Mexico), the screwworm (in Mexico and the southern United States), and the tsetse fly (in Africa).

9.11.3.2 Food Preservation

Irradiation, using gamma rays, is now an acceptable method of destroying bacteria, insects, and various harmful organisms in a number of different foodstuffs, for example, cereals, fresh fruit, seafood, and vegetables. Indeed, it is the only way to kill bacterial pathogens in frozen and raw food.

9.11.3.3 Fertilizer Labeling

When used as labels, the two radioisotopes, nitrogen-15 (^{15}N) and phosphorus-32 (^{32}P), can indicate how much fertilizer is taken up or lost by plants.

9.11.3.4 Genetic Alteration

Either gamma or neutron irradiation can be used to help with genetic modification and the production of strains of plants that are either more productive or resistant to pests.

9.11.3.5 Sterilization

Gamma rays are also used to sterilize such things as archival documents, wood, and raw wool (before it is exported from Australia).

9.11.4 Industry

Radioisotopes also have a range of industrial uses, including nuclear gauging, gamma radiography, and tracing.

9.11.4.1 Nuclear Gauging

Beta radiation can be used to gauge (and control) the thickness of anything from continuous sheets of paper to metal, glass, and plastic film as they speed through the machinery that makes them. In the coal industry, gamma radiation is used to measure not only the ash content of coal as it passes on a line but also its sulfur and moisture contents. When the coal is in its hoppers, gamma radiation is used to gauge exactly how full the hoppers actually are. In addition, radioisotopes are used both in borehole logging and, more simply, to measure the water content of soil and its density.

9.11.4.2 Gamma Radiography

Gamma rays can be used to produce images much in the same way as x-rays. However, gamma rays can not only produce an image of the object through which they are passing but also help provide a degree of analysis of the object itself. This is one reason why gamma radiography is often used to screen luggage at airports. More mainstream uses of gamma radiography include weld inspections in gas and oil pipelines using "pipe crawlers" and stress testing, for example, of the structural integrity of jet turbine blades. Autoradiography and neutron radiography can also be used to locate things that might otherwise be difficult to see, for example, detecting corrosion and moisture entrapment, especially in aircraft structures.

9.11.4.3 Tracing

Since the tiniest amount of a radioactive material can be easily traced, radioisotopes are ideal when conditions are particularly harsh or complex. Measuring liquid flows through, for instance, a blast furnace is an example of the former. Using a radioisotope to detect leaks in, say, a power station is an instance of the latter. One other fascinating example is provided by the World Nuclear Association in their article "Radioisotopes in Industry": "The extent of termite infestation in a structure can be found by feeding the insects radioactive wood substitute, then measuring the extent of the radioactivity spread by the insects. This measurement can be made without damaging any structure as the radiation is easily detected through building materials" (see Table 9.1).

9.11.5 Medicine

Radioisotopes are key in modern nuclear medicine. Three of their most important uses are sterilization, medical diagnostics, and medical treatments.

TABLE 9.1
Some Common Industrial Radioisotopes

Radioisotope	Half-life	Decay	Some Uses
Carbon-14[a]	5,730 years	Beta	Dating. Tracer in photosynthesis studies
Chlorine-36[a]	301,000 years	Beta	Measuring age of water—up to 2 million years
Hydrogen-3[a] (tritium)	12.32 years	Beta	Measuring age of "young" (up to 30 years) groundwater. In red building-exit signs
Lead-210[a]	22.3 years	Beta	Dating layers of sand and soil up to 80 years old
Americium-241	432 years	Alpha	Smoke alarms and neutron gauging
Caesium-137	30.2 years	Beta	Tracing and thickness gauging
Cobalt-60	5.3 years	Beta and gamma	Gamma radiography, gauging, and sterilization
Gold-198	2.69 days	Beta	Tracing factory waste pollution in oceans
Iridium-192	73.83 days	Beta	Gamma radiography
Krypton-85	10.76 years	Beta and gamma	Industrial gauging and locating leaks
Magnesium-27	9.5 minutes	Beta and gamma	Locating leaks
Manganese-54	312.2 days	Gamma	Predicting the behavior of heavy metals in effluents from mining wastewater
Nickel-63	100.1 years	Beta	Light sensors in cameras and thickness gauging
Selenium-75	119.78 days	Gamma	Gamma radiography and nondestructive testing
Sodium-24	15 hours	Beta and gamma	Locating leaks
Strontium-90	28.8 years	$Beta^-$	Thickness gauging
Ytterbium-169	32 days	Gamma	Gamma radiography
Zinc-65	244.26 days	Gamma	Predicting the behavior of heavy metals in effluents from mining wastewater

Note: $Beta^-$ is positron emission.
[a] Naturally occurring.

9.11.5.1 Sterilization

Many medical devices that cannot be sterilized in any other way are sterilized using gamma radiation. This applies, in particular, to single-use devices such as gloves, syringes (and needles), alcohol wipes, and scalpel blades. In addition, irradiation is used to sterilize plasma, serum, and tissue that is going to be transplanted.

9.11.5.2 Medical Diagnostics

Some 10,000 hospitals around the world use radioisotopes for diagnostic tests. Whether taken orally, inhaled, or injected, medical isotopes are used in a wide variety of different procedures. They are used, for example, to assess bone damage; to know how the heart, kidneys, liver, and lungs are functioning; and, extensively, for testing for cancer, heart, and thyroid diseases. On a more mundane level, they are commonly used in biochemical analyses in vitro (known as radio immunoassays) to test blood, urine, hormones, and serum, among other biological samples. Some 15 million such analyses are undertaken each year in Europe alone.

9.11.5.3 Medical Treatments

Irradiation can control or even eliminate some cancerous growths. Some radioisotopes, for example, strontium-89 and samarium-153, are excellent for reducing the pain induced by certain cancers. Tiny seeds, or sometimes wires, of radioisotopes are used in the treatment of various cancers found in the breast, head, prostate, and thyroid (see Table 9.2). (Interestingly, however, the frequency of use of radioisotopes in therapeutics is only about a tenth of their use in diagnosis.)

9.12 RADIOISOTOPE PRODUCTION

Although a number of radioisotopes occur naturally, nearly 200 are being used on a regular basis and are produced artificially. This mainly occurs via three methods: (1) nuclear reactors, (2) research reactors, and (3) cyclotrons and linear accelerators.

9.12.1 Nuclear Reactors

While not the most common source of radioisotopes, a number of power reactors do produce radioisotopes. Put at its simplest, isotopes are made in a nuclear reactor through neutron gain; that is, some of the million errant neutrons winging around inside the core of the reactor are absorbed by the nuclei of particular materials placed in the reactor's core. So, for example, if one of cobalt's stable isotopes, cobalt-59, is stuck in the core of a nuclear reactor and exposed to a high flux of

TABLE 9.2
Some Commonly Used Isotopes in Medicine

Radioisotope	Half-life	Decay	Some Uses
Bismuth-213	46 minutes	Alpha	Cancer therapy
Cobalt-60	5.27 years	Gamma	Sterilization
Erbium-169	9.4 days	Beta$^-$	Relief of pain from arthritis
Iodine-125	60 days	Gamma	Treatment of brain and prostate cancer
Phosphorous-32	14 days	Beta	Treatment of excess red blood cells
Technetium-99m	6 hours	Gamma	Many different imaging applications
Thallium-201	73 hours	Gamma	Diagnosis of coronary artery disease
Xenon-133	5 days	Beta$^-$	Pulmonary ventilation studies

Note: Beta$^-$ is positron emission.

neutrons, it can absorb some of these neutrons (adding one to each of its nuclei) and become the radioisotope cobalt-60.

9.12.2 Research Reactors

Currently, radioisotopes are most commonly produced in research reactors through fission, or the splitting of the nuclei in atoms. Rather than being absorbed by the nuclei of other atoms, they are instead hurled at those other atoms to break up their nuclei and create by-products in the form of radioisotopes. The speed of these neutrons and the temperature will determine exactly which radioisotopes are produced.

9.12.3 Cyclotrons and Linear Accelerators

Radioisotopes can be produced in either cyclotrons or linear accelerators. In both of these, a beam of protons is accelerated, using magnets, and smashed into a target to produce the required radioisotope. One major difference between using a cyclotron and using a nuclear reactor is that a cyclotron needs neither fissionable material nor uranium to produce radioisotopes. On the other hand, however, it can only accelerate charged particles, and only protons can be added in this way to atomic nuclei.

In fact, while reactors produce neutron-rich nuclei, cyclotrons and linear accelerators, by adding protons, produce neutron-*deficient* nuclei. Because neutron-deficient and neutron-rich radioisotopes decay by different means, they will have different properties and are better suited for different purposes.

9.13 MOST WIDELY USED RADIOISOTOPES

By far the most widely used radioisotopes, excluding those used in power production, are three that are used for medical purposes: cobalt-60, technetium-99m (molybdenum-99), and thallium-201 in that order. A 2008 article by Lawrence Kidd in *Nuclear Engineering International* estimated total yearly global production of cobalt-60 at 60 million Curies, while production of molybdenum-99/technetium-99m was some 468,000 6-day Curies. While these figures may reflect how things were back in July 2008, the scene has changed radically since Kidd published his groundbreaking article.

Both cobalt-60 and molybdenum-99 are produced in reactors. But perhaps surprisingly, both radioisotopes are produced only in a handful of power and research reactors, a couple of which have now not been operating for quite some time. This has recently led to a major supply crisis for molybdenum-99 and, hence, technetium-99m.

Before it closed for repairs, the National Research Universal (NRU) reactor probably produced more than 80% of the world's cobalt-90 and nearly 40% of its molybdenum-99. Today it produces nothing. The latest update indicates that only 56% of the repair work necessary to return the reactor to service has been completed. Then, in late February, the high flux reactor (HFR) at Petten in the Netherlands, which is estimated to have produced 25% or more of the world's molybdenum-99, was also shut down for repairs. This reactor is not expected to start operating again for at least several years (see Table 9.3).

Experiment: Distribution of $^{14}CO_2$ in Different Plant Parts

Introduction

The objectives of this activity are (1) feeding plants with $^{14}CO_2$ and (2) determining the radioactivity in different plant parts and fractions. ^{14}C (radioactive carbon) and ^{12}C (naturally occurring nonradioactive isotope of carbon) have the same atomic number with the same chemical properties. Living organisms, including plants, are unable to differentiate between these two isotopes of carbon. Thus, the plant

TABLE 9.3
Some of the World's Major Radioisotope-Producing Reactors

Radioisotope	Source	Country
Cobalt-60	Atucha 1	Argentina
	Embalse	Argentina
	Bruce B 1 and 2	Canada
	NRU	Canada
	Qinshan 1 & 2	China
	Leningrad 1	Russia
	Wolsong 1 & 2	South Korea
	Clinton BWR	United States
Molybdenum-99	BR-2	Belgium
	HFR	Netherlands
	NRU	Canada
	Osiris	France
	Maria	Poland
	Safari-1	South Africa

leaves utilize ^{12}C and ^{14}C carbon dioxide at the same rate. The photosynthetically fixed carbon is then partitioned between various plant organs, namely roots, shoots, seeds, fruits, and so on. The partitioning of assimilates is influenced by a variety of factors such as plant species, age, stage of development, and environmental factors. The partitioning pattern can be quantitatively determined by estimating the radioactivity in different plant parts following exposure of the seedlings to $^{14}CO_2$ in light. The percent distribution of ^{14}C in various parts is then calculated.

Safety Guidelines

Radioisotopes emit ionizing radiations and affect all life forms. They are dangerous if used indiscriminately. The use of radioactive materials in laboratories is controlled under a license with stringent restrictions on use and waste disposal due to the health hazards involved. One must follow the simple precautions listed below

1. The place where the isotopes are to be used must be kept very clean.
2. Never pipette by mouth in a radioactive laboratory, even if a particular liquid is not radioactive. It may be contaminated.
3. Wear a laboratory coat or apron.
4. Always spread some disposable sheets of cellophane or absorbent paper, or at least an old newspaper, on your work tables.
5. Wipe all spillage several times using tissue paper.
6. Discard the wiped papers, used containers, or the soluble waste in their respective containers.
7. Clearly mark all glassware used as "radioactive."
8. Wash all the glassware personally first with "carrier" solution. (This is a solution that contains the same nonradioactive compound that was used in the experiment. When we use ^{14}C glucose in the experiment, normal glucose solution will be the carrier solution.)
9. It is advisable to use a pair of rubber gloves.
10. Wash your hands well after the completion of the experiment, before using the counters. If you handle planchets or planchet holders with contaminated hands, they may increase the counting rate and mess up the background counting.
11. Never prepare the samples in the counting room where the counters are placed.
12. Subscribe for film badge service and get the dosage level reported once every 15 days.
13. Using an ordinary rate meter, check the background activity as often as possible.

Disposal of Radioactive Waste

Keep all glassware, paper, towels, and so on used in experiments in a separate container, and do not mix them with other materials in the laboratory. It is advisable to use a foot-operated bin for radioactive waste material. It should be periodically emptied and buried. Liquid waste is best kept diluted and stored in carboys, which can be disposed of from time to time.

Where to Buy the Isotopes

Compounds labeled with radioactive isotopes are manufactured by regulatory authorities in respective countries. Permission should be obtained for the recognition of the department or laboratory, enclosing details of the space, and other facilities available. For film badge service, apply to the Radiation Protection Service. Isotopes can be obtained from Amersham Inc., England.

Experiment: Determination of Radioactivity in Different Parts of Plants

Experimental Outline

1. Feeding plants with $^{14}CO_2$.
2. Determination of radioactivity in different parts of the plants.

Materials

Syringes, double-walled (water-jacketed) Perspex chamber, airtight circulation pump, Geiger–Muller counter, gas proportional counter or liquid scintillation counter, planchets or scintillation vials, HCl and NAOH scintillation fluid, ^{14}C sodium bicarbonate, fully grown plants of wheat, pea or tomato, PPO (*p*-terphenyl, 2,5-diphenyloxazole) and POPOP (1,4-bis-2-(5-phenyloxazolyl)-benzene) and toluene, reagent bottles, balance, clinical centrifuge, water bath.

Prelaboratory Preparations

1. 0.1 N HCl: Take 8.3 mL of hydrochloric acid, and increase the volume to a liter with distilled water.
2. 0.2 N HCl: Take 16.6 mL of HCl and dilute to 1 L with distilled water, ^{14}C-labeled $NaHCO_3$ (specific activity of 50 mCi/mmol).
3. 0.1 N NaOH: Weigh 4.12 g of sodium hydroxide, and dissolve in a liter of distilled water.
4. Scintillation fluid: Dissolve 4 g of POP (2,5 diphenyloxazole) and 100 g of POPOP (1,4 bis 2) (phenyl oxazole) benzene in 1 L of toluene.

Method

A. Feeding of Plants with $^{14}CO_2$

1. Enclose the plant (e.g., wheat, tomato or pea plant) in a double-walled Perspex chamber of an appropriate size and place it under natural sunlight.
2. Circulate water between the chamber walls through the water jacket to prevent any rise in temperature in the chamber during the feeding period.
3. Connect the outlet of an airtight CO_2-generating tube with a flexible propylene or rubber tubing to the inlet of the airtight circulating pump and the outlet of the pump to the inlet of the chamber. With different tubing, connect the outlet of the chamber with the inlet of the CO_2-generating tube.
4. Place 1–2 µmol of $NaH^{14}CO_3$ (specific activity 50 mCi/mmol) at the bottom of the CO_2-generating tube.
5. Release 0.1 N HCl from the stopper into the $NaH^{14}CO_3$-containing CO_2-generating tube so that 50–100 µCi of labeled ^{14}C is liberated.
6. Circulate the evolved $^{14}CO_2$ through the Perspex chamber (in which the plant has been enclosed) using a small airtight electrical pump.
7. Allow the plant to fix $^{14}CO_2$ for an appropriate time (5 minutes to 1 hour).

8. Remove excess $^{14}CO_2$ diverting the circulation of gas through NaOH trap, which is attached via a three-way stop clock in between the CO_2-generating tube and the pump.
9. Return the ^{14}C-fed plants to natural growth conditions.
10. After 24 hours of feeding, harvest the plants and separate the parts, for example, leaf, stem, roots, seeds, fruits, and so on.
11. Dry different parts separately in an oven at 65°C to a constant weight and record their weight.

B. Extraction and Measurement of Radioactivity

Radioactivity can be measured by any one of the following methods after suitably processing the plant sample.

Method 1

1. Grind the oven-dried sample of different plant parts to fine powder.
2. Place a known weight of the sample in a planchet with some adhesive material on its surface so that the sample gels stuck to the planchet.
3. Count the dpm either by a gas proportional counter or by Geiger–Muller counter or liquid scintillation counter.

Method 2

Prepare the ethanol-soluble and ethanol-insoluble factions of the powder sample, and then determine the radioactivity in each of them individually.

1. Take an appropriate amount of the dry sample in a centrifuge tube.
2. Add 5 mL of 80% aqueous ethanol (v/v), and place in a water bath at 40°C for 1 hour. Shake the tube occasionally.
3. Centrifuge at 5000×g for 15 minutes, and decant the supernatant.
4. Extract the residue in the similar manner with 50% (v/v) aqueous ethanol, centrifuge, and collect the supernatant.
5. Finally extract the residue with water, centrifuge, and collect the supernatant.
6. Pool all the supernatants obtained in steps 3–5, which will constitute the ethanol-soluble fraction.
7. Suspend the residue in 2 N HCl, and autoclave at 15 kg/cm^2 for 1 hour.
8. Centrifuge the hydrolyzed material, and wash the pellet twice with 2 N.
9. Pool all the acid fractions representing the ethanol-insoluble fraction.
10. Evaporate both the fractions (ethanol soluble and insoluble) separately to dryness, and redissolve each of them in 2 mL distilled water.
11. Pipette 0.2 mL of the above extract on a planchet, dry under an infra-red lamp, and then record the counts in a gas-flow proportional counter.

Method 3

Measurement of radioactivity by a liquid scintillation counter:

1. Take a suitable aliquot (1 mL) of the above fractions in scintillation vials and dry it.
2. Add 10 mL of scintillation fluid to each vial.
3. Record the counts on a liquid scintillation counter with an appropriate correction for quenching.

Calculations

Determine the total amount of ^{14}C assimilates in the plant by adding the radioactivity (dpm) in both the fractions of all plants. While doing this, take into account the weight of the part used for sample preparation, dilution factors (if any in case of fractions), and the total weight of that plant or organ in the plant. For example, if radioactivity is determined in a powdered nonfractionated leaf sample, then:

$$\text{Total } ^{14}\text{C-labeled material} = \frac{\text{dpm} \times \text{total weight of leaves on the plant}}{\text{weight of powdered sample taken}}$$

For the fractionated sample, take into account the dilution factor as well, that is, the volume of the fraction used for determination of radioactivity and the total volume of the fraction.

1. Calculate the total ^{14}C-labeled assimilates in each plant part, a, b, c, and d for leaves, stem, seeds, and roots, respectively, from the sum of the counts in both the fractions of an organ.
2. Compute the percent distribution of ^{14}C in different parts as follows:

   % ^{14}C in leaves = a/X× 100 dpm
   % ^{14}C in stem = b/X× 100 dpm
   % ^{14}C in seeds = c/X× 100 dpm
   % ^{14}C in roots = d/X× 100 dpm

   where X is the total ^{14}C assimilated by the entire plant and a, b, c, and d represent total ^{14}C-labeled material in each of these plant parts.

Result

Tabulate the results of ^{14}C labeling in each plant part in terms of percent incorporation.

Experiment: Photosynthetic $^{14}CO_2$ Reduction to the Primary Metabolic Product

Introduction

The mechanism of photosynthetic CO_2 assimilation in plants varies with species and may occur via either the C_3 or C_4 pathway. The C_3 or C_4 mechanism in a particular plant can be established by identifying the first product of a CO_2 fixation reaction. In case of the C_3 pathway, 3-phosphoglycerate (3-PGA) is the first product, whereas in the C_4 type the initial product of carboxylation reaction is a C_4 organic acid such as oxaloacetate, which is rapidly converted to either malate or aspartate. When plants are exposed to $^{14}CO_2$, the label is incorporated into the primary products. Determination of these primary products (namely, 3-PGA or malate and aspartate) for radioactivity provides the confirmatory evidence for operation of a particular photosynthetic pathway in a plant species.

Safety Guidelines

Follow all the safety guidelines given in the experiment.

Experimental Outline

1. Feed $^{14}CO_2$ to the actively photosynthesizing leaves.
2. Trace $^{14}CO_2$ photo assimilates by the paper chromatographic method.
3. Identify the chemical nature of the assimilate by using specific spray reagents.

Materials

Experimental plants (select one C_3 and one C_4 plant), chamber made of Perspex or glass for feeding the plants with $^{14}CO_2$, gas mixture pump, ^{14}C-labeled sodium bicarbonate ($NaH^{14}CO_3$), chromatographic chamber for descending paper chromatography, Whatman no. 1 filter paper, PGA, glucose-6 phosphate, malic acid, POP (2,5-diphynyl oxazole), POPOP (1,4 bis (-2,4,5-phynyl oxazole)) benzene, tricine, KOH, sorbitol, liquid scintillation counter, 50 mM tricine-KOH buffer (pH 7.5) containing 0.33 M sorbitol buffer, *n*-butanol, propionic acid.

Prelaboratory Preparations

1. Prepare the scintillation fluid by dissolving 8 g 2,5-diphenyl oxazole (POP) and 100 mg of 1,4-bis (-2,4,5-phenyl oxazole) benzene (POPOP) in 1 L of toluene.
2. Authentic standards: Prepare a solution containing 1 mg/mL of authentic sample of 3-PGA, glucose-6-phosphate, aspartic acid, and malic acid.

3. Solvent system for paper chromatography: Mix *n*-butanol:propionicacid:water in the ratio of 10:5:7 (v/v), and use as the solvent system.
4. Detection reagents: Prepare various spraying reagents as given in step 2 in the following procedure.
5. 50 mM Tris-KOH buffer.

Method

1. Soak the glass fiber discs in 50 mM tricine-KOH buffer (pH 7.5) containing 0.33 M sorbitol.
2. Quickly transfer 0.1 g plant leaf segments on these glass discs to prevent dehydration of plant tissues. Place these glass fiber discs in steel planchets. Prepare five such planchets and place them on a strip with a well in its center. Add 2 N HCl to this central well and keep the strip in the Perspex chamber (19 × 12 × 37 cm) with an internal volume of 50 mL.
3. Expose the chamber to direct sunlight for 5 minutes. Then inject $NaH^{14}CO_3$ through a rubber septum into the acid-containing well on the strip such that the concentration of $^{14}CO_2$ in the chamber is 0.1%, and circulate the air through the chamber with the help of an airtight gas mixture pump.
4. Remove the samples from the chamber after 20, 40, 60, 120, and 300 seconds of exposure to $^{14}CO_2$.
5. Immediately kill the tissue by dropping it in boiling ethanol.
6. Extract the plant material with 80% (v/v) ethanol, centrifuge the homogenate at 8000×*g*, and collect the supernatant.
7. Extract the residue with distilled water, centrifuge again, and carefully decant the supernatant. Pool the obtained supernatants and increase the volume to a known final volume of 10 mL with 80% ethanol.
8. Take 1 mL of the extract in a scintillation vial containing 1.5 mL scintillation fluid and evaporate it to dryness. Record the counts in a liquid scintillation counter.
9. Take another aliquot of the extract, evaporate to dryness, and extract chlorophyll from the residue with chloroform. Evaporate off the excess chloroform from this preparation, and dissolve the residue in water.
10. Apply a suitable aliquot of the above water extract onto Whatman no. 1 filter paper sheet in the form of a spot. Perform descending paper chromatography using *n*-butanol:propionicacid:water in the ratio of 10:5:7 as the mobile phase.
11. Remove the chromatogram and dry it at room temperature. Spray the chromatogram for developing color using the different spraying reagents detailed as follows:

Compound Detected	Spray Reagent
Malate sucrose	0.08% (w/v) bromocroesol green in ethanol 1.5 g benzidine + 10 mL glacial acetic acid + 10 mL 40% (w/v) TCA and ethanol to increase the volume to 100 mL
3-PGA and glucose-6-phosphate	5 mL 60% (w/v) $HClO_4$ + 25 mL 4% (w/v) ammonium molybdate + 10 mL N HCl + water to increase the final volume to 100 mL
Aspartate	0.4% (w/v) ninhydrin in 95% (v/v) acetone

After spraying, dry it at room temperature, and then keep it at 80–100°C in an oven for 5 minutes. Colored spots appearing on the chromatogram are due to malate, aspartate, and sucrose.

12. Autoclave the paper for 2 minutes at 8–10 psi for detecting 3-PGA and glucose-6-phosphate. The paper becomes blue. Now expose the paper to ammonia vapors. The blue color of the phosphomolybdate is due to the reaction with phosphorylated compounds, while the blue color of the background disappears.
13. Identify the compounds by comparing their R_f values with those of authentic standards run parallel. Punch the area of paper containing labeled compounds from the developed and sprayed chromatograms, and count in the scintillation counter. Make sure that necessary corrections for quenching are made for any colored segments of the chromatogram.

Conclusions

By examining the kinetics of distribution of ^{14}C label in 3-PGA and C_4 dicarboxylic organic acids, the mode of photosynthetic CO_2 assimilation can be deduced. If at the initial stages the ^{14}C label is initially present predominantly in C_4 dicarboxylic acids (oxaloacetate, malate, or aspartate) but little in 3-PGA, and this is followed by decrease in ^{14}C label in C_4 acids accompanied by concomitant increase of ^{14}C in 3-PGA, then this is an indication of a C_4 pathway of CO_2 fixation. If, in contrast, at the early stages the ^{14}C is largely in 3-PGA and ^{14}C labeled dicarboxylic acids appear at a much later stage, this denotes a C_3 mechanism of photosynthetic CO_2 assimilation.

Result

Express your results as a percentage incorporation of radio activity in each fraction.

Autoradiography

Radioactive isotopes of various elements can be used to follow the specific molecules biochemically; one of the earliest uses of radioactivity was to trace the chemical pathway of carbon during photosynthesis. The soluble contents can be separated by paper chromatography (either by ascending, descending, or two-dimensional); small molecules containing ^{14}C atoms derived from CO_2 can be detected by a sheet of x-ray film placed over the dried chromatogram enclosed in a x-ray cassette. The cassette is kept in the dark room for 48–72 hours. Remove the x-ray film, and develop the film in the dark room in the manner that is followed for the development of photographic film. The metabolites formed can be identified. The location of the individual molecules blotted on the x-ray film can be compared with the compounds identified on the chromatographic paper with the help of spray reagents. After identification of the compounds, each area on the chromatogram is cut and eluted in ethanol and added to the scintillation vial containing scintillation fluid, and the radioactivity of each biomolecule is measured.

SUGGESTED READING

Biilington, D., G. G. Jayson, and P. J. Maltby. 1992. *Radioisotopes*. Oxford: Bios Scientific.

Chapman, J. M., and G. Avery. 1981. *The Use of Radioisotopes in the Life Sciences*. London: George Allen & Unwin Ltd.

Connor, K. J., and I. S. Mclintock. 1994. *Radiation Protection Handbook for Laboratory Workers*. Leeds, UK: H and H Scientific Consultants.

Slater, R. J. 1995. "Radioisotopes in Molecular Biology." In *Molecular Biology and Biotechnology: A Comprehensive Desk Reference*, edited by R. A Myers, 209–20. New York: VCH.

Slater, R. J. 2000. *Radioisotopes in Biology—A Practical Approach*, 2nd ed. Oxford: IRL Press.

IMPORTANT LINKS

1. Scintillation counter: http://www.hidex.com/products/hidex-300-sl.aspx
2. Autoradiography: http://www.gelifesciences.com/aptrix/upp01077.nsf/Content/na_blotting_site~na_detection~na_autoradiography_cassettes?opendocument&cmpid=ppcaw100

10 Fermentation

10.1 INTRODUCTION

Fermentation is typically the conversion of sugar to alcohol under anaerobic conditions in the presence of yeast. Chemical conversion of carbohydrates into alcohols is a general example of fermentation. When fermentation stops prior to the complete conversion of sugar to alcohol, it is called stuck fermentation. The science of fermentation is known as zymology. The French chemist Louis Pasteur was the first known zymologist; he defined fermentation as "respiration without air." The German scientist Eduard Buchner, winner of the 1907 Nobel Prize for chemistry, found that the secretion of zymase in yeast is responsible for fermentation.

Fermented foods have been with us since the dawn of civilization. They will be with us far into the future, as they are the source of alcoholic foods, beverages, vinegar, pickled vegetables, sausages, cheeses, yogurts, vegetable protein amino acid/peptide sauces and pastes with meat-like flavors, and leavened and sourdough breads. All consumers today have a considerable portion of their nutritional needs met through the consumption of fermented foods and beverages.

10.2 HISTORY

Fermentation is an ancient process by means of which bread, wine, beer, and cheese are made. Egyptians found that uncooked dough left standing became lighter and softer. The Egyptians and the Romans discovered that the use of yeast produced lighter and leavened bread. Around 4000 BC, wine was made from grape juice through a fermentation process. Beer-making by the ancients involved soaking barley in water; at the time, beer was probably a serendipitous by-product of bread making.

For the production of yogurt, cheese, wine, vinegar, and different types of sauces, microorganisms were used by the Chinese. Earlier, cheese was made by storing milk in animal skins or bladders made from animal stomachs. The bacteria and enzymes present in these urns would cause the separation of casein (milk protein) to form curd.

Louis Pasteur in the late nineteenth century was the first individual who understood the process of fermentation; his understanding was based on his experience and implicit knowledge. Antony van Leeuwenhoek, a Dutch biologist, was the first to see microorganisms in samples of fermenting beer through a microscope in 1680. Pasteur, a French chemist, discovered that yeasts convert sugars to alcohol and carbon dioxide during fermentation.

For the first time, large-scale aerobic fermentors were used in Central Europe in the 1930s for the production of compressed yeast. The fermentor consisted of a large cylindrical tank with air introduced at the base via a network of perforated pipes. In later modifications, mechanical impellers were used to increase the rate of mixing and to break up and disperse air bubbles, which led to the requirement of compressed air. Baffles on the walls of the vessels prevented the formation of a complex vortex in the liquid and later a system was patented in which aeration tubes were introduced with water and steam for cleaning and sterilization.

Fermentation is the process by which alcoholic beverages or acidic dairy products (cheeses, yogurt) are manufactured. It is a way for a cell to obtain energy without using oxygen. During the process, complex organic substances are broken down into simpler ones. The cell (microbial or animal) obtains energy through glycolysis, that is, the splitting of a sugar molecule to release electrons. The by-product of this process is excreted from the cell in the forms of substances such as alcohol, lactic acid, and acetone. With the development of microbiology and technologies such as biotechnology, microorganisms are subjugated to produce an extensive assortment of products using fermentation. These include

Dairy products: Cheese, yogurt.
Beverages: Beer, wine.
Single-cell proteins (SCPs): SCPs are cell monocultures of bacteria, fungi, and algae. Since the cells contain large amounts of protein, SCPs are used as food or food supplements for humans and cattle. They are regarded as cheap sources of dietary protein and are produced from methanol and the by-products of cheese production and papermaking.
Antibiotics: Antibiotics are one of the most important compounds produced by fermentation. In 1929, Alexander Fleming was the first to discover penicillin, an antibiotic. Large numbers of antibiotics are produced today by fermentation using various bacteria and fungi.
Chemicals: Citric and acetic acids, amino acids, enzymes, vitamins.
Fuels: Ethanol, methanol, methane.

10.3 PRINCIPLES OF FERMENTATION

Fermentation is a metabolic pathway by which cells catabolize, or break down, a carbon source to produce energy. The cells consume carbohydrates, leading to an increase in cell mass by replication and the production of carbon dioxide and other organic compounds. To track the consumption and production of various carbon products, we analyzed the following carbon balance:

$$C_{\text{glucose consumed}} = C_{CO_2} + C_{\text{cell mass}} - C_{\text{initial cell mass}} + C_{\text{ethanol}}$$

The calculation of carbon for the components in the aforementioned equation is relatively simple and self-explanatory, except for the cell mass. To obtain these experimental values, we estimated that 70% of the dry cell mass is water by weight and approximately half of the remaining mass is carbon. Thus, the experimental dry cell mass is multiplied by 0.15 to calculate the mass of carbon in the cell mass.

Microorganisms undergoing fermentation go through four distinct phases of growth, that is, lag, exponential, stationary, and death (Figure 10.1). They are described as follows:

1. Lag phase: The period of apparent inactivity in which cells are adapting to a new environment and preparing for reproductive growth. Cells are usually synthesizing new components. In practice, bacteria from one medium to another, where there are chemical differences between the two media, typically results in a lag in cell division. This lag in division is associated with a physiological adaptation to the new environment. Cells may increase in size during this time, but they simply do not divide (by binary fission). The lag phase varies considerably in length depending on the condition of the microorganisms and the nature of the medium.
2. Log (exponential) phase: The period in which organisms are growing at the maximal rate possible given their genetic potential, the nature of the medium, and the conditions under which they are growing. In this stage, yeast consumes food and multiplies to form new cells at a high rate. Here, the production of carbon dioxide rises rapidly and the concentration of glucose decreases rapidly.
3. Stationary phase: Eventually, population growth ceases and the growth curve becomes horizontal. The increase in cell number due to cell divisions is exactly balanced by a

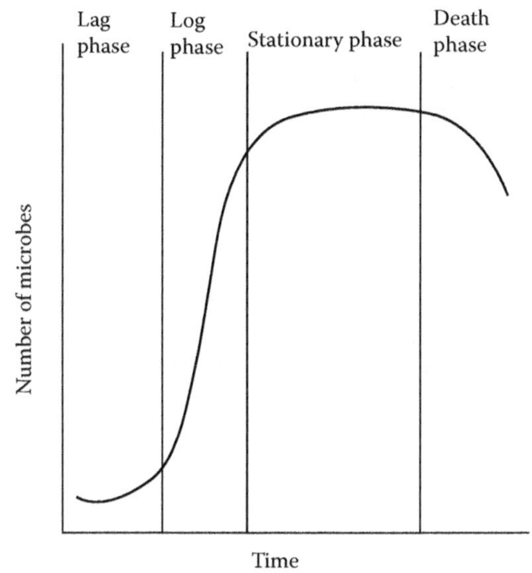

FIGURE 10.1 Bacterial growth pattern.

decrease in cell number due to cell death. Carbon dioxide production levels off and glucose consumption also slows down.

4. Death phase: The stationary phase in a standard bacterial growth curve is followed by a die-off of cells, called the death phase. It is the period in which cells die at an exponential rate. Some of the reasons for cell death are continued accumulation of wastes and the loss of a cell's ability to detoxify toxins. During this period, both sugar concentration and carbon dioxide production approach zero.

In addition, stirring rates may also affect the rate of fermentation reactions as yeast cells need to be in contact with sugar and oxygen (in aerobic fermentation) in order to react. The yeast cells are incapable of swimming; if they are just suspended in the fermentor broth, they quickly exhaust the surrounding resources and smother in their own waste products.

10.4 INTRODUCTION TO MICROORGANISMS

Microorganisms are unicellular, that is, they contain only a single cell. Cellular organisms are broadly classified into prokaryotes and eukaryotes, aerobic and anaerobic, and according to the type of metabolism.

10.4.1 Prokaryotes

These include simple bacteria and archaea (Figure 10.2). They are molecules surrounded by a membrane and a cell wall. The cells of these organisms do not contain membrane structures such as mitochondria or a nucleus. Chromosomes are present as circular DNA distributed in the cytoplasm; they can be classified into archaebacteria and eubacteria with further classifications under those types.

Archaebacteria include:

- Methanogens: Anaerobic methane-producing bacteria
- Extreme halophiles: Bacteria that require high salt concentrations to survive
- Extreme thermophiles: Bacteria that require high temperatures and sulfur to survive

Eubacteria include:

- Gram-positive bacteria: Bacteria whose cell walls retain crystal violet dye during iodine treatment, for example, lactic acid bacteria
- Gram-negative bacteria: Bacteria that vary in cell-wall structure and do not retain crystal violet dye during iodine treatment

10.4.2 Eukaryotes

The cells of these organisms contain mitochondria, nucleus, plasma membrane, cytoplasm, and other membranes. In these organisms, bound structures and chromosomes are present in the nucleus, for example, multicellular organisms and fungi (Figure 10.3).

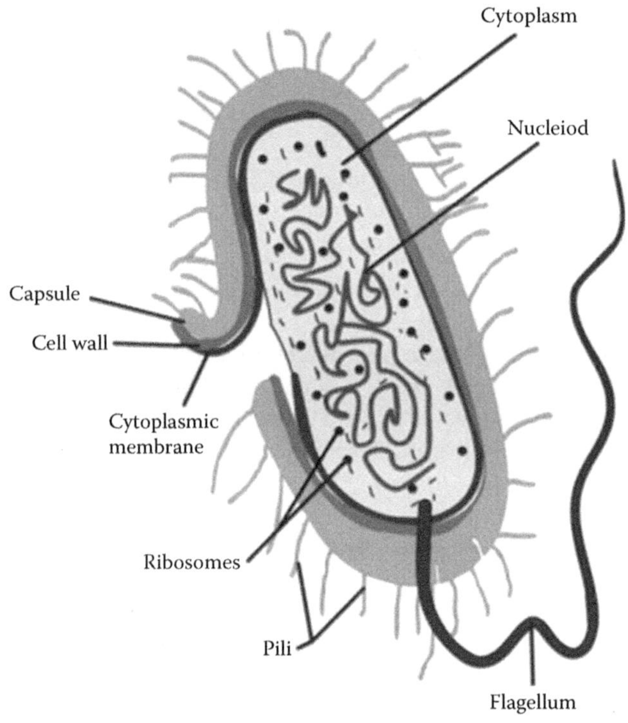

FIGURE 10.2 Prokaryotic cell structure.

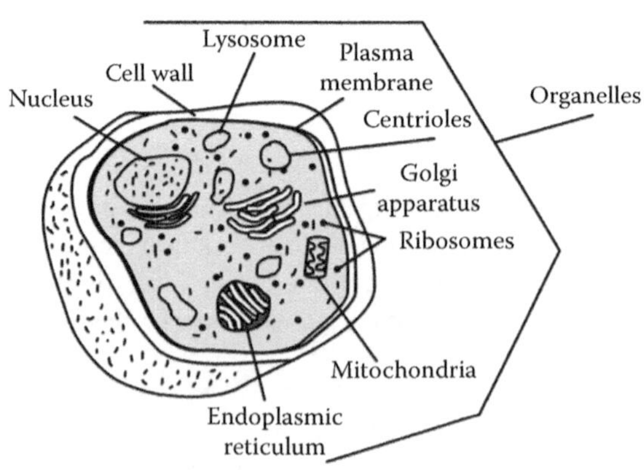

FIGURE 10.3 Eukaryotic cell structure.

10.5 TYPES OF FERMENTATION METABOLISM

Fermentation is commonly defined as the process in which energy is formed by the oxidation of organic compounds such as carbohydrates and sugars. This leads to the conversion of these organic compounds into an acid or an alcohol, which releases energy. It can be carried out by microorganisms with and without the help of oxygen.

10.5.1 Aerobic Fermentation

The fermentation process that includes fresh air and is similar to vat, crock, tank, and polyethylene pail is referred to as aerobic fermentation. The presence of air is most important for the growth of yeast, and it reproduces and increases in density and alcohol is produced in a conductive manner. Aerobic fermentation needs adequate aeration. The quantity of air that is needed per hour is almost 60 times the fermentation medium volume. An adequate supply of clean air is more essential in aerobic fermentation in a bioreactor and clean air is usually sparged inside the fermentation medium. Aerobic fermentation has a mechanism for the complete mixing and stirring of the fermentation medium and biological cells (Figure 10.4).

When oxygen is present (i.e., under aerobic conditions), most organisms undergo two more steps, Kreb's cycle and electron transport, to produce adenosine triphosphate (ATP). In eukaryotes, these processes occur in the mitochondria, whereas in prokaryotes they occur in the cytoplasm. These are microorganisms that can grow and live in the presence of oxygen. In compost heaps, bacteria of this type generate heat when they convert carbon and nitrogen, reducing the volume of waste. As they exhaust the oxygen in the compost pile, they die, leaving behind a germ-free product.

10.5.1.1 Acetyl Co-A: The Transition Reaction

Pyruvic acid is first altered in the transition reaction by removing a carbon atom and two oxygen atoms (which forms carbon dioxide). When the carbon dioxide is removed, energy is released and nicotinamide adenine dinucleotide (NAD^+) is converted into the higher energy form NADH. Coenzyme A attaches to the remaining 2-C (acetyl) unit, forming acetyl Co-A. This process is a prelude to the Kreb's cycle.

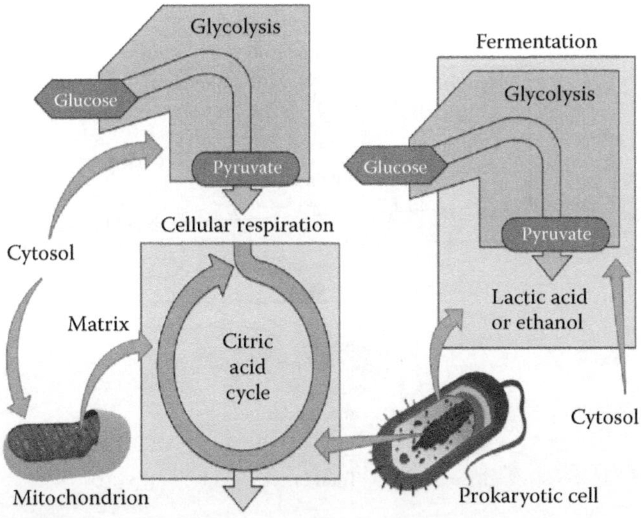

FIGURE 10.4 Biochemical reactions in prokaryotic and eukaryotic cells.

10.5.1.2 Kreb's Cycle (the Citric Acid Cycle)

The acetyl Co-A (2-C) is attached to a 4-C chemical (oxaloacetic acid [OA]). The Co-A is released and returns to await another pyruvic acid. The 2-C and 4-C make another chemical known as citric acid, which is a 6-C. Thus, Kreb's cycle is also known as the citric acid cycle. The process that occurs after citric acid is formed essentially involves the removal of carbon dioxide; release of energy in the form of ATP, guanine triphosphate (GTP), NADH, and $FADH_2$; and, finally, regeneration of the cycle. Between isocitric acid and α-ketoglutaric acid, carbon dioxide is given off and NAD^+ is converted into NADH. Between α-ketoglutaric acid and succinic acid, a release of carbon dioxide and reduction of NAD^+ into NADH happens again, resulting in a 4-C chemical, that is, succinic acid. The GTP (which transfers its energy to ATP) is also formed here (GTP is formed by attaching a phosphate to GDP).

The remaining energy carrier–generating steps involve the shifting of atomic arrangements within the 4-C molecules. Between succinic and fumaric acids, such molecular shifting does not release enough energy to make ATP or NADH outright; instead, this energy is captured by a new energy carrier, flavin adenine dinucleotide (FAD). FAD is reduced by the addition of two Hs, resulting in $FADH_2$. $FADH_2$ is not as rich an energy carrier as NADH, the former yielding less ATP than the latter.

The last step, between malic acid and OA, reforms OA to complete the cycle. Energy is given off and trapped by the reduction of NAD^+ to NADH. The carbon dioxide released by cells is generated by the Kreb's cycle, as are the energy carriers NADH and $FADH_2$, which play a role in the next step (Figure 10.5).

10.5.1.3 Electron Transport Phosphorylation

Whereas Kreb's cycle occurs in the matrix of the mitochondrion, electron transport system (ETgS) chemicals are embedded in the membranes known as cristae. Kreb's cycle completely oxidizes the carbons in the pyruvic acids, producing a small amount of ATP and reducing NAD and FAD into higher-energy forms. In ETS these higher-energy forms are cashed in, producing ATP. Cytochromes are molecules that pass the "hot potatoes" (electrons) along the ETS chain. Energy released by the "downhill" passage of electrons is captured as ATP by adenosine diphosphate (ADP) molecules. The ADP is reduced by the gain of electrons. ATP formed in this way is made by the process of oxidative phosphorylation. The mechanism for the oxidative phosphorylation process is the gradient of H^+ ions discovered across the inner mitochondrial membrane. This mechanism is known as chemiosmotic coupling. This involves both chemical and transport processes. Drops in the potential

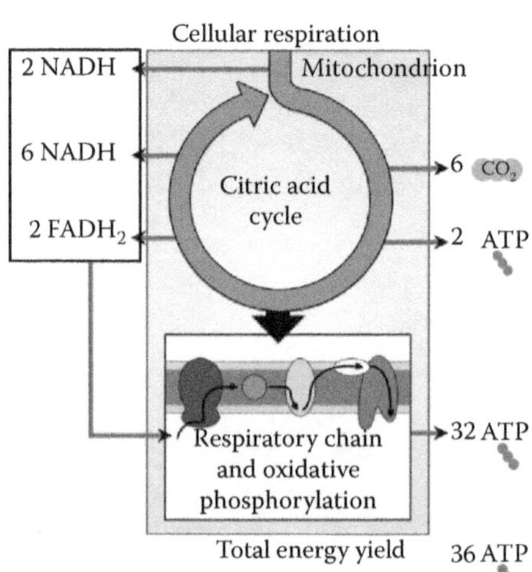

FIGURE 10.5 Biochemical reactions depicting energy yield in complete oxidation.

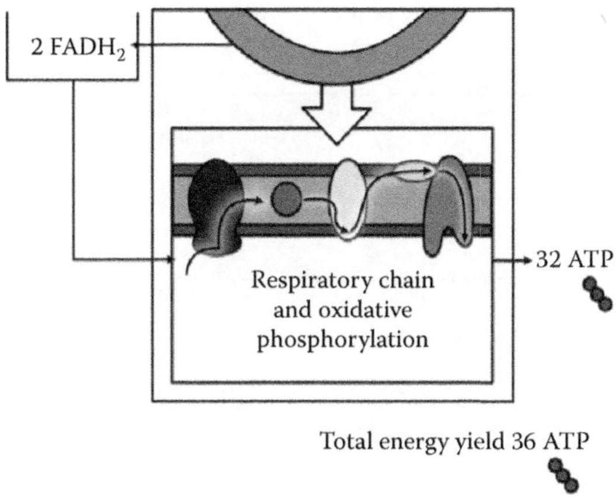

FIGURE 10.6 Depiction of total energy yield in oxidative phosphorylation.

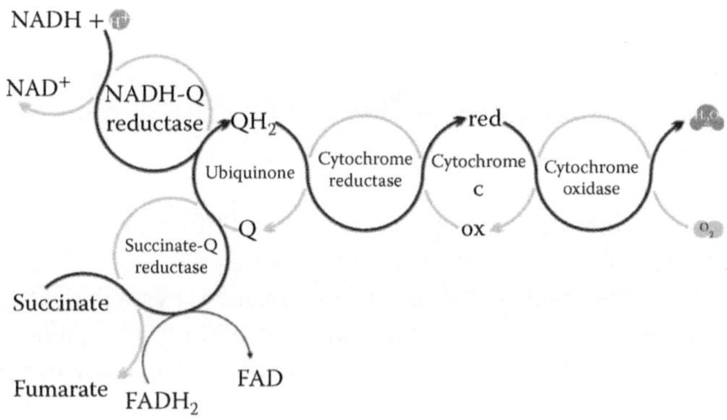

FIGURE 10.7 The ETS.

energy of electrons moving down the ETS chain occur at three points. These points turn out to be where ADP + P are converted into ATP. Potential energy is captured by ADP and stored in the pyrophosphate bond. NADH enters the ETS chain at the beginning, yielding three ATPs per NADH. $FADH_2$ enters at Co-Q, producing only two ATPs per $FADH_2$ (Figures 10.6 and 10.7).

10.5.2 Anaerobic Fermentation

When fermentation is carried out in the absence of oxygen, it is called anaerobic fermentation. Anaerobic fermentation can be supported with the help of an equation that describes the substrates used in the fermentation process to give the desired products. One of the examples of anaerobic fermentation is alcoholic fermentation, whose equation is given here to illustrate how the process of anaerobic fermentation occurs:

$$\text{Acetaldehyde} + \text{NADH} + \text{H}^+ \xrightarrow{\text{dehydrogenase}} \text{Alcohol} + \text{NAD}^+$$

In this reaction, anaerobic respiration occurs in the sugars to cause fermentation, with the help of the fungus yeast, which is not in contact with the atmosphere or oxygen. Acetaldehyde and NADH are substrates of the reaction, which are fermented along with one hydrogen ion to form the product, that is, alcohol. This reaction takes place in presence of the active enzyme alcohol dehydrogenase to give the alcohol and a cofactor, which is one ion of NAD. This is what anaerobic fermentation in yeast produces and this reaction follows the decarboxylation of pyruvate with the help of the

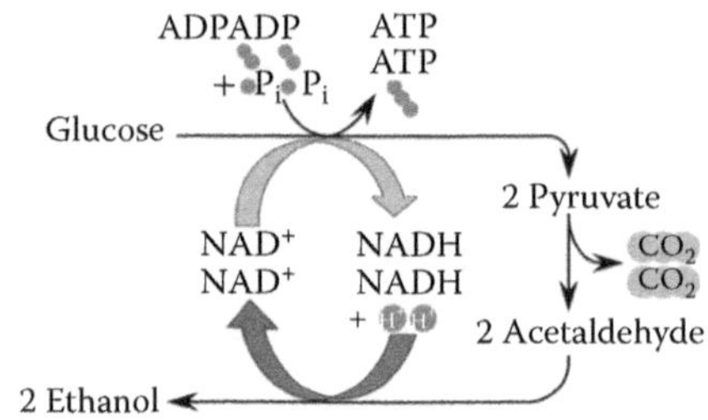

FIGURE 10.8 Alcoholic fermentation reaction.

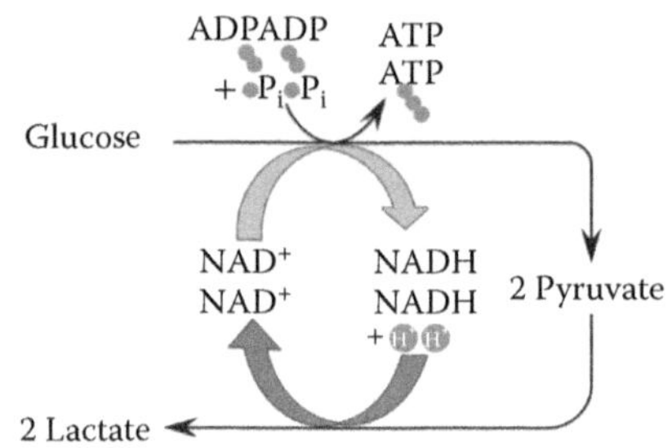

FIGURE 10.9 Lactic acid fermentation reaction.

enzyme pyruvate decarboxylase, thiamine pyrophosphate (TPP), and two Mg ions. After this acetaldehyde is produced, CO is formed to give the aforementioned reaction.

Under anaerobic conditions, that is, the absence of oxygen, pyruvic acid can be routed by the organism into one of three pathways: (1) lactic acid fermentation, (2) alcohol fermentation, or (3) cellular (anaerobic) respiration. Humans cannot ferment alcohol in their own bodies; we lack the genetic information to do so. These biochemical pathways with their myriad reactions catalyzed by reaction-specific enzymes, which are all under genetic control, are extremely complex. Alcohol fermentation is the formation of alcohol from sugar. Yeast, under anaerobic conditions, converts glucose to pyruvic acid through the glycolysis pathway and then goes one step further to convert pyruvic acid into ethanol, a C-2 compound (Figure 10.8).

Many organisms also ferment pyruvic acid into other chemicals, such as lactic acid. Humans ferment lactic acid in the muscles where oxygen becomes depleted, resulting in localized anaerobic conditions. This lactic acid causes the muscle stiffness that couch potatoes feel after beginning exercise programs. The stiffness goes away after a few days since the cessation of strenuous activity allows aerobic conditions to return to the muscle, and the lactic acid can be converted into ATP through normal aerobic respiration pathways (Figure 10.9).

10.5.2.1 Anaerobic Fermentation and Glycolysis

Anaerobic fermentation is defined as a form of metabolism in which an organic compound serves as both the electron donor and electron acceptor and ATP is produced by substrate level phosphorylation. Fermentation products can be grouped into two general categories: alcoholic fermentation products and acidic fermentation products.

Glycolysis, which is defined as the process of converting glucose to pyruvate, can happen anaerobically. This method of transforming glucose into pyruvate with the help of a very limited amount of oxygen is known as anaerobic glycolysis. The word "lysis" means breakdown and "glyco" means glucose; so from this pathway, there is fermentation or breakdown of glucose to pyruvate, which is a by-product of the same. This process produces energy (38 ATP molecules) in about 10 seconds to 2 minutes and is carried out in the cytoplasm of muscle cells. This energy is stored in the Golgi apparatus inside these cells and then transported when required throughout the body.

Anaerobic glycolysis starts when glucose is phosphorylated into glucose-6-phosphate in the first step. This product is further converted into fructose-6-phosphate, which is then converted into fructose-1,6-bisphosphate. This product gives two further products, phosphoglyceraldehyde (PGAL) and dihydroxyacetone phosphate (DHAP), which are interconvertible by a reversible reaction.

The PGAL then produces 1,3-bisphosphoglycerate by reducing one ion of NAD and an inorganic phosphate to NADH. This is followed by the production of 3-phosphoglycerate and 2-phosphoglycerate one after the other with the formation of one ATP molecule. The 2-phosphoglycerate then gives phosphoenolpyruvate (PEP) by the dehydrogenation of one water molecule. This PEP is converted to pyruvate with the generation of another ATP molecule, and thus energy is produced by the anaerobic fermentation of glucose.

Anaerobic microorganisms are useful in biodegradation, breaking down organic chemicals into smaller compounds, and producing methane and carbon dioxide. Some anaerobic organisms can break down organic chemicals by fermentation. Such organisms are useful in hazardous-waste sites.

10.5.3 Pathways/Fermentation

10.5.3.1 Fatty Acid and Hydrocarbon Pathway

Microbes utilize fatty acids and hydrocarbons for their metabolism and are usually added to various fermentation reactions, such as the manufacture of antibiotics. Fats and hydrocarbons are converted to fatty acids, which are finally converted to acetyl-CoA by a series of reactions. This is also called β-oxidation.

10.5.3.2 Oxidation of Methane and Methanol

Microorganisms using methane as the carbon source are called methanotrophs, and those using methanol are called methylotrophs. Methane is oxidized initially to methanol and finally to carbon dioxide in a reaction involving a series of steps. The overall reaction can be summarized as follows:

$$\text{Methane} \rightarrow \text{methanol} \rightarrow \text{formaldehyde} \rightarrow \text{formic acid} \rightarrow \text{carbon dioxide}$$

10.5.3.3 Oxidation of Amino Acids

Amino acids are metabolized by microorganisms through oxidation and are utilized as excellent sources of carbon and nitrogen.

10.5.3.4 Oxidation of Polymers

A polymer is a compound with a large number of monomeric units linked together. As an example, polysaccharides (starch, cellulose, etc.) contain a large number of monomeric sugar (carbohydrate) molecules linked together. Proteins contain a large number of amino acids linked together to form large chains. Microorganisms metabolize polymers by breaking the chains into individual monomeric units in a process called hydrolysis, a process that requires the presence of enzymes called hydrolytic enzymes or hydrolases.

10.6 FERMENTATION PROCESS

10.6.1 Selection of Substrates in a Culture Medium

Although specific media requirements depend on the type of microorganism being used in the fermentation process, the basic requirements remain the same, that is, all media should contain sources of energy, water, carbon sources, nitrogen sources, vitamins, and minerals. Designing media for small-scale laboratory purposes is relatively easy, but media for industrial purposes are difficult to prepare.

The culture medium should allow for a high yield of the desired product at a fast rate; allow for a low yield of undesired products; be sterilized easily; yield consistent products, that is, have minimum batch variation; be cheap and readily available; be compatible with the fermentation process; and not pose environmental problems before, during, or after the fermentation process.

The culture medium will affect the design of the fermentor. For example, hydrocarbons in some media require high oxygen content, so an air-lift fermentor (ALF) should be used in such cases. Natural media ingredients are cheap, but they have high batch variation. On the other hand, pure ingredients (also called defined media or formulated media) have very little batch variation but are expensive. The media should support the metabolic process of the microorganisms under study and allow biosynthesis of the desired products. Media are designed based on the following equation using minimum components required to produce maximum product yield:

$$\text{Carbon and energy source} + \text{nitrogen source} + O_2 \\ \rightarrow \text{biomass} + \text{products} + CO_2 + H_2O + \text{heat}$$

Knowledge of the elemental composition of process microorganisms is required for solving the elemental balance equation. Some microorganisms cannot synthesize specific nutrients, for example, amino acids, vitamins, and nucleotides. Once a specific growth factor is identified, it can be incorporated into a medium in adequate amounts as a pure compound or as the component of a complex mixture. The carbon substrate plays a dual role in biosynthesis and energy generation; the carbon requirement for biomass production under aerobic conditions may be estimated from the "cellular yield coefficient," which is defined as follows:

$$\text{Cellular yield coefficient} = \frac{\text{quantity of cell dry matter produced}}{\text{quantity of carbon substrate utilized}}$$

The other nutrient required is oxygen, which is provided by aerating the culture; this aspect is considered in detail in Section 10.8. The design of a medium and selection of substrates will influence the oxygen demand of a culture in that more reduced carbon sources result in a higher oxygen demand. Energy for growth comes from either the oxidation of medium components or light. Most industrial microorganisms are chemoorganotrophs; therefore, the most common sources of energy will be carbon sources such as carbohydrates, lipids, and proteins. Important components of the medium are carbon sources, nitrogen sources, minerals, growth factors, chelating agents, buffers, antifoaming agents, air, steam, and fermentation vessels.

10.6.1.1 Carbon Sources

Product formation is directly dependent on the rate at which the carbon source is metabolized; also, the main product of fermentation determines the type of carbon source to be used. Carbon sources include carbohydrates, oils and fats, and hydrocarbons.

10.6.1.2 Carbohydrates

Starch is an easily available carbohydrate obtained from maize, cereals, and potatoes. It is widely used in alcohol fermentation. Grains such as maize are used directly in the form of ground powder as carbohydrates. Malt and beer made from barley grains contain high concentrations of different carbohydrates such as starch, sucrose, cellulose, and other sugars. Sucrose is obtained from sugar cane and molasses. Molasses is one of the cheapest sources of carbohydrate. It contains a high sugar concentration and other components such as nitrogenous substances and vitamins; it is used in alcohol, SCPs, amino acids, and organic acid fermentations. Extraction and purification of the products is expensive. Sulfite waste liquor is the by-product of the paper industry; it contains carbohydrates and is used in yeast cultivation. Whey is the by-product of the dairy industry. It is used in alcohol, SCP, gum, vitamins, and lactic acid fermentation.

10.6.1.3 Oils and Fats

Vegetable oils are used as carbon sources. Oils provide more energy per weight compared to sugars. They also have antifoaming properties, but they are generally used as additives rather than as the sole carbon source. Examples are olive oil, cottonseed oil, soybean oil, linseed oil, and lard (animal fat).

10.6.1.4 Hydrocarbons

The C12–C18 alkanes can be used as carbon sources. They are cheap and have more carbon and energy content per weight than sugars. They can be used in organic acids, amino acids, antibiotics, enzymes, and proteins fermentations.

10.6.1.5 Nitrogen Sources

Ammonia, ammonium salts, and urea are the most commonly used nitrogen sources in the fermentation process. Ammonia also serves the purpose of pH control. Other substances used as nitrogen sources are corn steep liquor, soya meal, peanut meal, cottonseed meal, amino acids, and proteins.

10.6.1.6 Minerals

All microorganisms have certain mineral requirements for growth and metabolism. Calcium, chlorine, magnesium, phosphorous, potassium, and sulfur are the essential minerals for all media. Other minerals such as copper, cobalt, iron, manganese, molybdenum, and zinc are needed in trace amounts and are generally present as impurities in other components. The specific concentrations of these elements required depend on the microorganism being used.

10.6.1.7 Growth Factors

Some microorganisms cannot synthesize the full complement of cell components and therefore require preformed compounds called growth factors. Vitamins, amino acids, and fatty acids are used as growth factors in the fermentation process to complement the cell components of microorganisms.

10.6.1.8 Chelating Agents

Chelating agents prevent the formation of insoluble metal precipitates. They form complexes with metal ions present in the medium and can be utilized by microorganisms. Chelating agents are not required in large-scale fermentation processes since some of the other ingredients in such processes, such as yeast extract, perform the function of forming complexes with metal ions. One example of a chelating agent is EDTA, which is a versatile chelating agent and is able to form six bonds with a metal ion. It is frequently used in soaps and detergents because it forms a complex with calcium and magnesium ions. These ions are present in hard water and interfere with the cleaning action of soaps and detergents. Other chelating agents are citric acid and pyrophosphates.

10.6.2 Controlling pH

Buffers are used to maintain the pH of a medium as microbial growth is affected by changes in pH. The optimum pH for most microorganisms is 7.0. Commonly used buffers are calcium carbonate, ammonia, and sodium hydroxide. The balanced use of carbon and nitrogen sources will also form a basis for pH control as the buffering capacity can be provided by proteins, peptides, and amino acids, such as in corn steep liquor. The pH may also be controlled by adding ammonia or sodium hydroxide and sulfuric acid to the medium.

10.6.3 Antifoaming Agents

Microbial processes produce a large amount of foam in the fermentation vessel. This is caused by microbial proteins or other components of the media. Foaming causes the removal of cells from the media and their autolysis; thus, releasing more microbial foam-producing proteins aggravates the problem. Foam will reduce the working volume of the fermentation vessel, decrease the rate of heat transfer, and deposit cells on the top of the fermentor. The air filter in the fermentor becomes wet, allowing the growth of contaminating microorganisms. Antifoaming agents are also called surfactants, that is, they reduce the surface tension in the foam and destabilize foam-producing proteins. Commonly used antifoaming agents are stearyl alcohol, cottonseed oil, linseed oil, olive oil, castor oil, soybean oil, cod liver oil, silicones, and sulfonates.

An ideal antifoam should be highly dispersible in foam, fast acting, nontoxic to humans and microorganisms, long acting in preventing new foam formation, cheap, heat sterilizable, and have no effect on oxygen transfer. Compounds that meet most of these requirements are alcohols, stearyl, octyldecanol, esters, and fatty acids and their derivatives, particularly glycerides, which include cottonseed oil, linseed oil, soybean oil, olive oil, castor oil, sunflower oil, rapeseed oil, cod liver oil, silicones, sulfonates, alkaterge C, oxazaline, and polypropylene glycol. These antifoams are generally added when foaming occurs during fermentation. Because many antifoams have low solubility, they need a carrier such as lard oil, liquid paraffin, or castor oil, which may be metabolized and affect the fermentation process. Unfortunately, the concentrations of many antifoams required to control fermentations will reduce the oxygen-transfer rate by as much as 50%; therefore, antifoam additions must be kept to an absolute minimum. There are also other antifoams that increase the oxygen-transfer rate. If the oxygen-transfer rate is severely affected by antifoams additions, then mechanical foam breakers may have to be considered as a possible alternative.

10.6.4 Air

Air is required for aeration and is supplied to the fermentor by means of pumps or compressors. It is sterilized by passing through filters before being introduced. The amount of air and the extent of purity required depend on the fermentation process being carried out. The majority of fermentation processes are aerobic and, therefore, require the provision of oxygen. If stoichiometry of respiration is considered, then the oxidation of glucose may be represented as follows:

$$C_6H_{12}O_6 + 6O_2 = 6H_2O + 6CO_2$$

Oxygen is normally supplied to microbial cultures in the form of air, this being the cheapest available source of the gas. The method for provision of a culture with a supply of air varies with the scale of the process. Laboratory-scale cultures may be aerated by means of the shake flask technique, where the culture is grown in a conical flask shaken on a platform contained in a controlled environment chamber, whereas pilot and industry-scale fermentations are normally carried out in stirred, aerated vessels. It is often advantageous to culture relatively small volumes (1 dm^3) in stirred, aerated vessels as this enables the cultural conditions to be better monitored and controlled and facilitates addition of supplements and removal of samples. Some fermentors are so designed that adequate oxygen transfer is obtained without agitation.

10.6.5 Steam

Steam is used to sterilize fermentors and other equipment and to control temperature. Continuous dry steam supply is required for the fermentation process and care should be taken to prevent steam condensation.

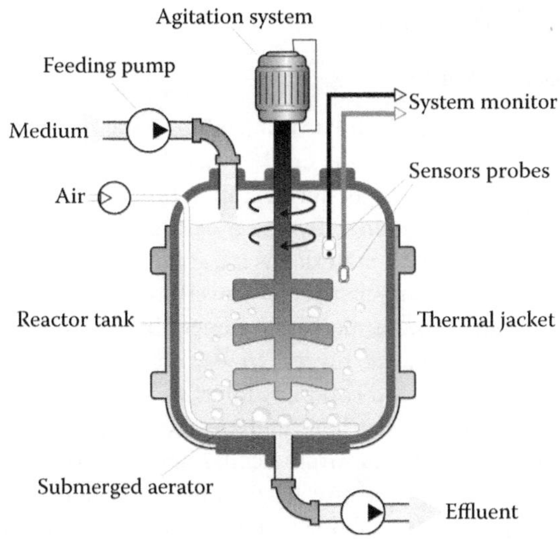

FIGURE 10.10 External view of an ideal fermentor.

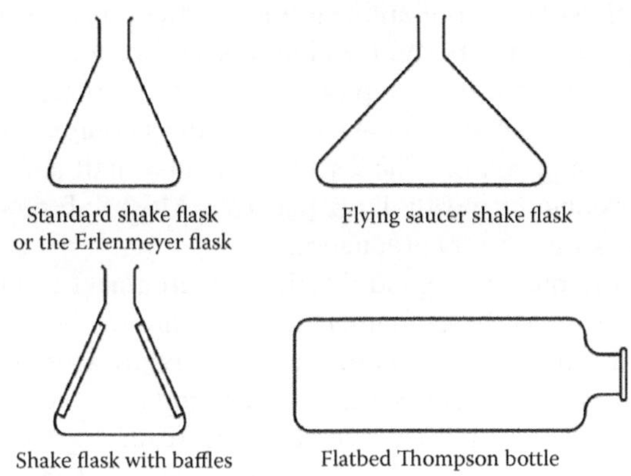

FIGURE 10.11 Different types of shaker flasks.

10.6.6 Fermentation Vessels

Laboratory-scale fermentations are carried out in shaker flasks and flatbed bottles. Large-scale fermentations are carried out in glass or stainless steel tank fermentors. A fermentation vessel should be cheap, nontoxic to the microorganism used for the process, easy to sterilize, easy to operate, robust and reliable, and leak-proof; allow visual monitoring of the fermentation process and sampling; and not allow the contamination of its contents (Figure 10.10).

10.6.7 Shaker Flasks

Shaker flasks are conical vessels made of glass that are available in different sizes. The typical volume of these flasks is 250 ml. There are different types of shaker flasks, such as baffled, unbaffled, Erlenmeyer, and flying saucer. Shaker flasks are used for screening microorganisms and cultivating them for inoculation. Shaker beds or shaker tables are used to allow oxygen transfer by their continuous rotary motion. Baffled flasks are used to increase oxygen transfer. Shaker flasks need to be plugged to prevent contamination with other microorganisms. Cotton wool, polyurethane foam, glass, and synthetic plugs are commonly used. Fernwald shaker flasks and flatbed Thompson bottles are expensive and are not commonly used (Figure 10.11).

10.7 TYPES OF FERMENTATION

10.7.1 Solid-State Fermentation

Solid-state (substrate) fermentation (SSF) holds tremendous potential in the production of enzymes. It is of special interest in processes where the crude fermented product may be used directly as the enzyme source. SSF has been defined as the fermentation process occurring in the absence or near absence of free water. SSF processes generally use a natural raw material such as carbon and an energy source. SSF can also use an inert material as the solid matrix, which requires supplementing a nutrient solution containing necessary nutrients as well as a carbon source. The solid substrate (matrix), however, must contain enough moisture. Depending on the nature of the substrate, the amount of water absorbed can be one or several times more than its dry weight, which leads to relatively high water activity (a_w) on the solid/gas interface in order to allow a higher rate of the biochemical process. Low diffusion of nutrients and metabolites takes place in lower-water-activity conditions, whereas compaction of the substrate occurs under conditions of higher water activity. Hence, maintenance of an adequate moisture level in the solid matrix along with suitable water activity is essential for SSF processes. Solid substrates should generally have a large surface area per unit volume (say, in the range of 10^3–10^6 m^2/cm^3 for the ready growth on the solid/gas interface). Smaller substrate particles provide larger surface areas for microbial attacks but pose difficulty in aeration/respiration due to limitations in interparticle space availability. Larger particles provide better aeration/respiration opportunities but less surface area. In bioprocess optimization, sometimes it may be necessary to use a compromised size of particles (usually a mixed range) for cost-effectiveness. For example, wheat bran, which is the most commonly used substrate in SSF, is obtained in two forms: fine and coarse. The former contains small particles (mostly smaller than 500–600 μ) and the latter contains mostly large particles. Most SSF processes use a mix of these two forms at different ratios for optimal production.

Solid substrates generally provide a good dwelling environment to microbial flora comprising bacteria, yeast, and fungi. Among these, filamentous fungi are the best-studied organisms for SSF due to their hyphal growth; they have the ability to not only grow on the surface of the substrate particles but also penetrate through them. Several agricultural crops such as cassava and barley and agro-industrial residues such as wheat bran, rice bran, sugarcane bagasse, cassava bagasse, various oil cakes (e.g., coconut oil cake, palm kernel cake, soybean cake, and groundnut oil cake), fruit pulps (e.g., apple pomace), corn cobs, sawdust, seeds (e.g., tamarind, jack fruit), coffee husk and coffee pulp, tea waste, and spent brewing grains are the most commonly used substrates for SSF processes. During growth on such substrates, hydrolytic exoenzymes are synthesized by the microorganisms and excreted outside the cells, which create and help in accessing simple products (carbon sources and nutrients) by the cells. This in turn promotes biosynthesis and microbial activities.

Based on the type of microorganisms involved, SSF processes can be classified into two main groups, natural (indigenous) SSF and pure-culture SSF, using individual strains or a mixed culture. Composting and ensiling are the two best examples of SSF processes, which involve natural microflora. Bioprocesses involving pure-culture SSF have also been known for a long time.

10.7.2 Semicontinuous Fermentation

The semicontinuous fermentation process is based on a continuous feeding rate and discontinuous medium withdrawal. A semicontinuous culture, once established, can be continued almost indefinitely, requiring relatively brief interruptions for harvesting and resupplying the fermentor with a fresh medium. To overcome some of the limitations of batch processing, continuous operation in a plug flow system must be considered. As the contaminated solids travel through the system, specific conditions develop in the successive reactor. One practical way of achieving a plug flow system is using a sequence of batch reactors. In an experimental cascade of three stirred batch slurry reactors, the contaminated solids (oil contamination) are periodically transferred to the next step.

10.7.3 Continuous Fermentation

Continuous fermentation process is the process in which cells are maintained in a continuous culture in the exponential growth phase by the frequent addition of a fresh medium that is exactly balanced by the removal of cell suspension from the bioreactor for the recovery of cells or fermentation products. On the one hand, there is a continuous addition of nutrients, and on the other hand, there is a continuous removal of fermented broth. This is in contrast to a batch fermentation process in which a large volume of nutrient medium is inoculated, and growth and biochemical synthesis are allowed to proceed only until the maximum yield is obtained. At this point, batch fermentation is stopped for product recovery, the fermentor is cleaned and resterilized, and a new fermentation is started. At first glance, continuous fermentation appears to be the better of the two procedures because the fermentation equipment is in constant use with little shutdown time and, theoretically at least, after the initial inoculation further production of inoculums is not required. However, the inherent problems associated with the continuous fermentation process often do not allow the achievement of this goal.

Continuous fermentation can be conducted in various ways. It can be carried out as a "single stage" in which a single fermentor is inoculated and then kept in continuous operation by balancing the input and output of the nutrient solution and harvested culture, respectively. In "recycle" continuous fermentation, a portion of the withdrawn culture, or of the residual unused substrate plus the withdrawn culture, is recycled to the fermentation vessel. For example, the immiscible hydrocarbon substrate of hydrocarbon fermentation can be recycled for further microbial attack. A portion of the organisms being produced during a continuous fermentation process also can be recycled in certain instances in which the actual available substrate level in the nutrient solution for microbial growth is quite low. An example of this type of substrate is sulfite waste liquor with its low available carbohydrate content; in this instance, the recycling of cells provides a higher population of cells in the fermentor and, hence, greater productivity. Multistage ethanol production with the recycling of yeast is another such example. A third type of continuous fermentation is "multistage" continuous fermentation. It involves two or more stages with the fermentor being operated in sequence. The multistage continuous fermentation process is particularly applicable to fermentation in which growth and synthetic activities of the cells are not simultaneous, that is, synthesis is not growth-related but occurs after the cell multiplication rate slows down.

There are two commonly used, possible means by which microbial activity in a continuous culture can be controlled: (1) the "turbidostat" and (2) the "chemostat." In the turbidostat, the total cell population is held constant by using a device that measures culture turbidity so as to regulate both the nutrient feed rate to the fermentor and the culture withdrawal rate from the fermentor. If the population numbers rise above a predetermined level, a greater amount of fresh medium is added to the fermentor so as to dilute the cell concentration; thus, there is no limiting nutrient consciously imposed with this process so that the cell growth rate should always be maximal. However, the growth must be maintained in the logarithmic growth phase or very close to it. This factor is a disadvantage in that fermentation must be performed at a lower maximum cell population than is possible with a chemostat, and this causes a greater residue of unused nutrient to be lost from the fermentation vessel with the withdrawal of the harvested culture.

In contrast to the turbidostat, a chemostat maintains the nutrient feed and culture withdrawal rates at constant values that are always less than the values that allow a maximum growth rate. The growth rate is controlled by supplying only a limiting amount of a critical growth nutrient in the feed solution. Thus, cell multiplication cannot proceed at a rate greater than that allowed by the availability of this critical nutrient. The controlling factor for growth, however, does not necessarily have to be a limiting nutrient; it can also be a relatively high concentration of a toxic product of the fermentation process. The chemostat is used more often than the turbidostat, because fewer mechanical problems are encountered in this process and because of the occurrence of less residual unused nutrient in the harvested culture. Table 10.1 presents a list of representative microorganisms investigated for their possible use in continuous fermentations.

TABLE 10.1
Representative Genera of Organisms Grown in Continuous Cultures

Organization	Genera
Algae	*Chlorella, Euglena, Scenedesmus, Spirulina, Anacystis, Aerobacter*
Bacteria	*Bacillus, Brucella, Clostridium, Salmonella*
Fungi	*Ophiostoma, Penicillium, Aspergillus*
Protozoa	*Tetrahynema*
Yeast	*Saccharomyces*

TABLE 10.2
Chemical Products from Continuous Fermentation

Growth Associated	Not Growth Associated
Acetic acid	Acetone
Butanediol	Butanol
Ethanol	Glycogen
Gluconic acid	Subtilin
Hydrogen sulfide	Chloramphenicol
Lactic acid	Penicillin, streptomycin
	Vitamin B_{12}

From among the aforementioned potential continuous fermentation processes, only the production of beer, fodder yeast (from sulfite paper mill waste), vinegar, and baker's yeast (from molasses) have found commercial applications. However, the activated sludge system for the processing of wastewaters also may be considered a commercialized continuous fermentation process, which differs from the more conventional fermentation only in that it deals with a mixed microbial population acting on a heterogeneous substrate and that the products of this fermentation do not have as much commercial importance (Table 10.2).

10.8 TYPES OF FERMENTORS

The main function of a fermentor is to provide a controlled environment for the growth of microorganisms or animal cells to obtain a desired product. Some of the bioreactor types are discussed in Sections 10.8.1 through 10.8.7.

10.8.1 Stirred-Tank Fermentors

Stirred-tank fermentors are the most commonly used fermentors. They are cylindrical vessels with a motor-driven agitator to stir the contents in the tank. The top-entry stirrer (agitator) model is most commonly used because it has many advantages, including ease of operation, reliability, and robustness (Figure 10.12).

The bottom-entry stirrer (agitator) model is rarely used. Stirred-tank reactors have the following functions: homogenization, suspension of solids, dispersion of gas–liquid mixtures, aeration of liquid, and heat exchange. The stirred-tank reactor is provided with a baffle and a rotating stirrer is attached either at the top or at the bottom of the bioreactor. The typical decision variables are the type, size, location, and number of impellers, and sparger size and location. These determine the hydrodynamic pattern in the reactor, which in turn influences mixing times, mass and heat transfer coefficients, shear rates, and so on.

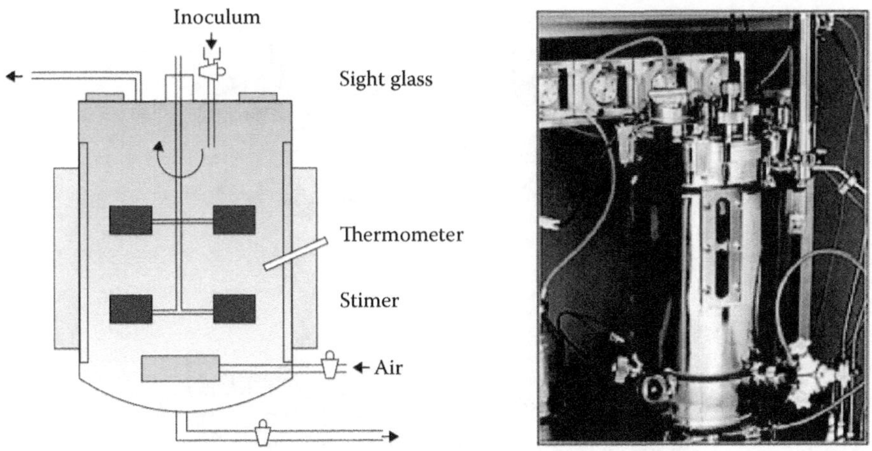

FIGURE 10.12 Stirred-tank fermentor.

Conventional fermentation is carried out in a batch mode. Since stirred-tank reactors are commonly used for batch processes with slight modifications, these reactors are simple in design and easy to operate. Even today, many of the industrial bioprocesses are carried out in batch reactors; although significant developments have taken place in recent years in reactor design, the industry still prefers stirred tanks because in case of contamination or substandard product formation the loss is minimal. The batch stirred tanks generally suffer due to their low volumetric productivity. The downtimes are quite large and unsteady-state fermentation imposes stress on microbial cultures due to nutritional limitations. The fed batch mode adopted in recent years eliminates this limitation. The stirred-tank reactor offers excellent mixing and reasonably good mass-transfer rates. The cost of operation is low and the reactors can be used with a variety of microbial species. Since the stirred-tank reactor is commonly used in the chemical industry, mixing concepts are well developed. Stirred-tank reactors with immobilized cells are not favored generally due to attrition problems; however, by separating the zone of mixing from the zone of cell culturing one can successfully operate such systems.

Laboratory-scale stirred-tank fermentors are made of borosilicate glass with a stainless steel lid and a top-entry stirrer. The typical volume of such fermentors is 1–100 L. Stirrers consist of a motor attached to a shaft. The shaft contains impellers. Stainless steel fermentors are also used in laboratories having special requirements. They are made of high-grade stainless steel with an internal surface that should be polished to reduce the adhesion of contents to the walls of fermentor joints; the surface must be smooth and free from any pinholes.

10.8.1.1 Fermentation Vessel Additional Equipment

10.8.1.1.1 Agitator

The agitator is required to achieve a number of mixing objectives, for example, bulk fluid and gas phase mixing, air dispersion, oxygen transfer, heat transfer, suspension of solid particles, and maintaining a uniform environment throughout the vessel contents. The agitator consists of a shaft, impellers with 4–6 blades, and a motor to drive it. Shafts should have double seals to prevent the leakage of contents. The main function of the agitator is mixing the contents, aeration, and removal of carbon dioxide produced during fermentation process by mixing action. Agitators may be classified as follows (Figure 10.13):

- Rushton blade or disk turbine: This is the most commonly used impeller because of its simple design, robustness, and ease of operation. It has 4–6 blades (Figure 10.13).
- Open turbine impellers.
- Marine impellers.

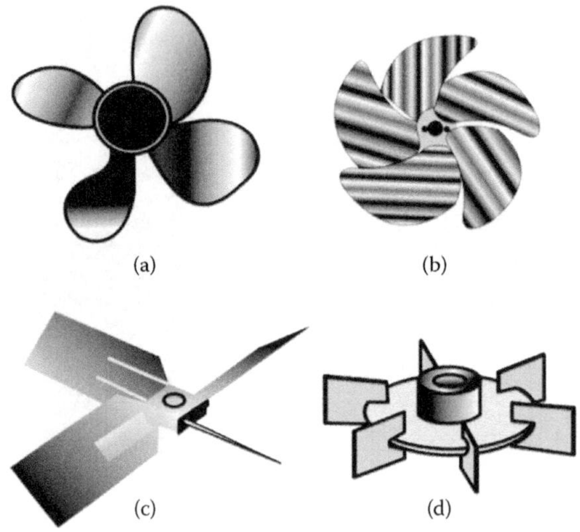

FIGURE 10.13 Different types of agitators: (a) and (b) Rushton blade disk turbine, and (c) and (d) marine impeller.

The disk turbine consists of a disk with a series of rectangular vanes set in a vertical plane around the circumference, and the vaned disk has a series of rectangular vanes attached vertically to its under side. Air from the sparger hits the underside of the disk and is displaced toward the vanes, where the air bubbles are broken up into smaller bubbles. The vanes of a variable pitch open turbine and the blades of a marine propeller are attached directly to a boss on the agitator shaft. In this case, the air bubbles do not initially hit any surface before getting dispersed by the vanes or blades.

Since the 1940s, a Rushton disk turbine of one-third the fermentor diameter has been considered the optimum design for use in many fermentation processes. It was established experimentally that the disk turbine is most suitable in a fermentor since it can break up a fast air stream without itself becoming flooded with air bubbles.

10.8.1.1.2 Baffles

Four baffles are fixed on the walls of the fermentor and are used to prevent the formation of a vortex. Impellers and baffles produce axial or radial flow patterns of the contents in the fermentor. Baffles are metal strips of roughly one-tenth the vessel diameter and are attached radially to the wall. The agitation effect is only slightly increased with wider baffles and drops sharply with narrower baffles. Extra cooling coils may be attached to baffles to improve the cooling capacity of a fermentor without unduly affecting the geometry (Figure 10.14).

10.8.1.1.3 Compressor

Compressors are used to pump air under pressure into a fermentor from the bottom to promote aeration and mixing. They are used mainly in large fermentors. For laboratory-scale, small fermentors, air pumps are used. Compressors should be oil-free in order to provide food-grade clean air to the fermentor.

10.8.1.1.4 Filters

A filter is used to filter the air supplied by the compressor before it enters the fermentor to ensure the air supply is clean. Types of filters are as follows:

- Membrane filters: They are made of cellulose acetate and nitrate with a constant pore size of 0.2 mm to ensure filtering of particulate matter. They are not cheap, but they are easy to maintain.

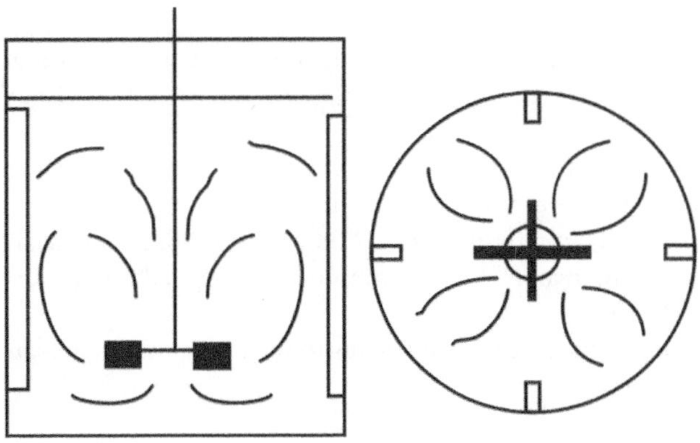

FIGURE 10.14 Baffles.

- Packed-bed filters: They contain glass wool or cotton wool packed in a container through which air is passed. Pore size is not uniform and these filters are susceptible to compaction and wetting, thereby causing channeling.
- Cartridge filters: The filter element is present in a stainless steel or polycarbonate cartridge in these filters. They are reliable and durable, but they are also expensive.

10.8.1.1.5 Aeration System (the Sparger)

A sparger may be defined as a device for introducing air into the liquid in a fermentor. Three basic types of spargers are used: (1) porous sparger, (2) orifice sparger (a perforated pipe), and (3) combined sparger-agitator. A combined sparger-agitator may be used in laboratory fermentors.

10.8.1.1.5.1 Porous Sparger A porous sparger is made of sintered glass of ceramics or metal. It is used only in laboratory-scale nonagitated vessels. The size of the bubble formed is 10–100 times larger than the pore size. There is a pressure drop across the sparger and the holes tend to be blocked by the organism's growth, which is the limitation of a porous sparger.

10.8.1.1.5.2 Orifice Sparger An orifice sparger is used in small stirred fermentors. It is a perforated pipe kept below the impeller in the form of crosses or rings. Its size should be approximately three-quarters of the impeller diameter. Air holes drilled to the undersurfaces of the tubes and the holes should be at least 6 mm in diameter. This type of sparger is used mostly with agitation. It is also used without agitation in some cases such as in yeast manufacture, effluent treatment, and production of SCP.

10.8.1.1.5.3 Nozzle Sparger A nozzle sparger is mostly used in large-scale applications. It is a single open/partially closed pipe positioned centrally below the impeller. When air is passed through this pipe, there is lower pressure loss than in a porous sparger and the holes do not get blocked.

10.8.1.1.5.4 Combined Sparger-Agitator This is air supply through the hollow agitator shaft. Air is emitted through holes in the disk or blades of the agitator.

10.8.1.1.6 Sensors

Sensors are used to monitor and control various fermentation parameters such as pH, temperature, and oxygen content.

10.8.1.1.7 Pumps

Liquid/medium is introduced to the fermentor using a pump. Various types of pumps are available, including

- Peristaltic pumps: They provide a constant flow rate and mild pumping. As liquids do not backflow, check valves are not required (Figure 10.15).
- Mini or delta pumps: They are used for adding pH control agents such as acid and alkali and for adding antifoam agents. Mini pumps are fixed-speed pumps and can pump against pressure (Figure 10.16).
- Larger pumps: They are used to add nutrients to the medium. Speed can be varied by altering the bore size of tubes.

10.8.1.1.8 Additional Equipment

10.8.1.1.8.1 Feed Ports The addition of nutrients and acid/alkali to small fermentors is normally made through silicone tubes, which are autoclaved separately and pumped by a peristaltic

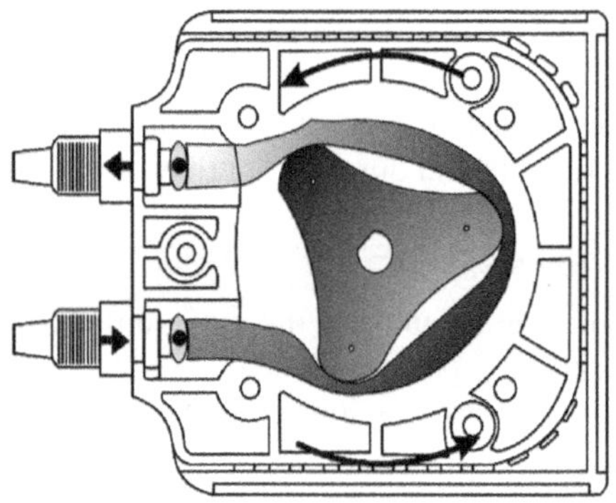

FIGURE 10.15 A peristaltic pump.

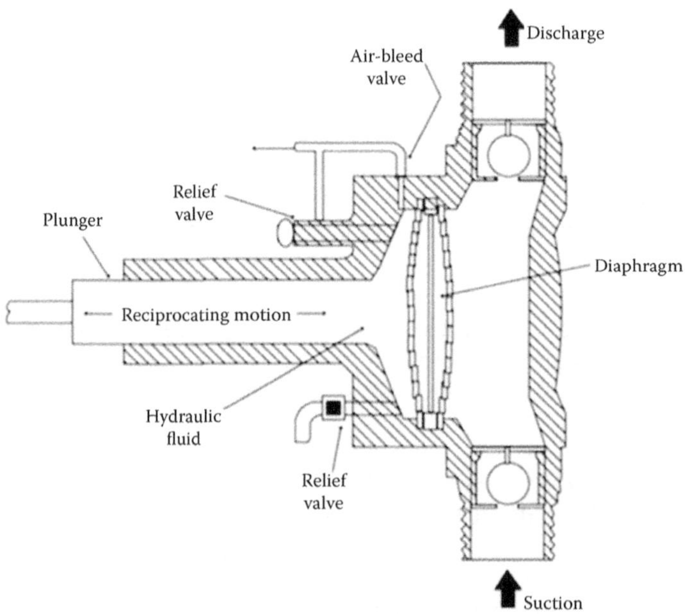

FIGURE 10.16 A diaphragm pump.

pump after aseptic connection. In large fermentor units, nutrient reservoirs and associated piping are usually an integrated part which can be sterilized with the vessel. However, there may also be ports that are used intermittently. These can be sterilized in situ with steam after the connection is completed and before any conditions are made.

10.8.1.1.8.2 Valves and Steam Traps Valves attached to fermentors and ancillary equipments are used for controlling the flow of liquids and gases in a variety of ways:

- Simple on/off valves, which are either fully open or fully closed
- Valves that provide coarse control of flow rates
- Valves that may be adjusted very precisely so that flow rates may be accurately controlled
- Safety valves, which are constructed in such a way that liquids or gases flow in only one direction

There are different models of valves:

1. Opening and closing valves, raising or lowering the blocking unit
    - Gate valve: A sliding disk moving in/out of the flow path by a turn of the stem
    - Globe valve: A horizontal disk/plug, which is raised or lowered
    - Piston valve: Similar to a globe valve except that a piston controls the flow
    - Needle valve: Similar to a globe valve except that the disk is replaced with a tapered plug/needle
2. Drilled sphere/plug
    - Plug valve: Parallel/tapered plug with an orifice on a 90° turn closes/opens the flow path
    - Ball valve: Similar to a plug valve except that balls with orifices replace the plug
3. Disk rotating between bearings
    - Butterfly valve: A disk rotates about a shaft and closes against the seal to stop the flow
4. Rubber diaphragm/tube pinching
    - Diaphragm valve: Similar to a pinch valve except that there is no pinching but pushing from one side against a diaphragm
    - Pinch valve: A flexible sleeve closed by a pair of pinch bars (rubber, neoprene, etc.)

Valves are chosen based on four types of applications:

1. On/off application: Globe valve, butterfly valve
2. Crude flow control: Gate valve
3. Accurate control: Needle valve
4. Very sterile operation: Pinch valve, diaphragm valve

10.8.1.1.8.3 Check Valves Valves used to prevent the accidental reversal of the flow of liquid or gas due to a break down are called check valves. There are three types of check valves: (1) swing check, (2) lift check, and (3) combined stop and check valves.

10.8.1.1.8.4 Pressure-Control Valves Pressure-control valves are used for two purposes:

1. Pressure reduction
2. Pressure retaining

10.8.1.1.8.5 Safety Valves There are various types of safety valves by which the increase in pressure is released. They are

- A spindle lifted from its seating against the pressure, which releases pressure
- Bursting/rupturing of disks to release pressure

In case of releasing a gas, the escaping gas must be treated before release.

10.8.1.1.8.6 Steam Traps A steam trap is important in removing any steam condensate. There are two components to a steam trap: (1) a valve and seat assembly and (2) an open/close device. The operation of the components is based on the following:

- Density of fluid: A float (ball/bucket) floats in water but sinks in steam. When it floats the valve closes, and when it sinks the valve opens.
- Temperature of fluid: The steam trap has a water/alcohol mixture, which senses any change in temperature. This mixture expands in hot steam and closes the valve. When it contracts in cool water, the valve is opened.
- Kinetic effect of fluid in motion: Low-density steam flows with a high velocity. Similarly, high-density steam flows with a low velocity. The conversion of pressure energy into kinetic energy controls the opening and closing of the valve.

10.8.2 ALF

ALFs do not have mechanical agitation systems (e.g., motor, shaft, impeller blades); the contents are agitated by injecting air from the bottom. Sterile atmospheric air is used if the microorganisms are aerobic and inert gas is used if the microorganisms are anaerobic. This is a gentle method of mixing the contents and is most suitable for fermentation of animal and plant cell cultures, since mechanical agitation produces high shearing stress that may damage the cells. ALFs are most widely used for large-scale production of monoclonal antibodies (Figure 10.17).

An ALF is generally classified as a pneumatic reactor without any mechanical stirring arrangements for mixing. The turbulence caused by the fluid flow ensures adequate mixing of the liquid. A draft tube is provided in the central section of the reactor. The introduction of the fluid (air/liquid) causes an upward motion and results in circulatory flow in the entire reactor. The air/liquid velocities will be low and, hence, energy consumption is also low. ALFs can be used for both free and immobilized cells. There are very few reports on ALFs for metabolite production. The advantage of air-lift reactors is the elimination of attrition effects generally encountered in mechanically agitated reactors. It is ideally suited for aerobic cultures since its oxygen mass transfer coefficient is quite

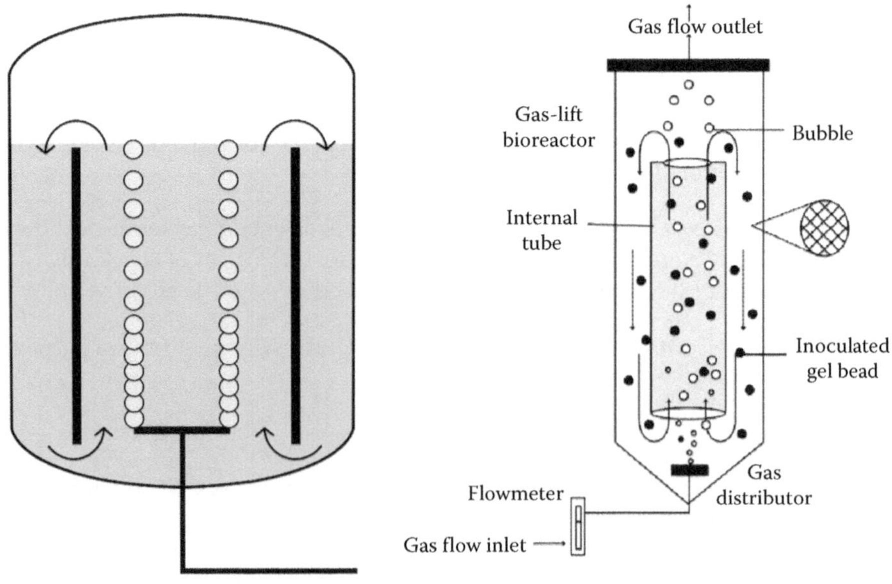

FIGURE 10.17 Air-lift fermentor.

high in comparison with stirred-tank reactors. This is ideal for SCP production from methanol as carbon substrate. This is used mainly to avoid the production of excess heat during mechanical agitation.

10.8.3 Fixed-Bed Fermentors

Packed-bed or fixed-bed bioreactors are commonly used with attached biofilms, especially in wastewater engineering. The use of packed-bed reactors gained importance after the potential of the whole-cell immobilization technique was demonstrated. The immobilized biocatalyst is packed in the column and fed with nutrients from either the top or bottom.

One of the disadvantages of packed beds is the change in flow characteristics resulting from alterations in bed porosity during the operation. While working with soft gels such as alginates and carrageenan, the bed compaction that generally occurs during fermentation results in a high pressure drop across the bed. In many cases, the bed compaction is so severe that gel integrity is severely hampered. In addition, channeling may occur due to turbulence in the bed. Although packed beds belong to the class of plug flow reactors in which back mixing is absent, in many packed beds a slight amount of back mixing occurs, which changes the characteristics of fermentation.

Packed beds are generally used in cases where substrate inhibition governs the rate of reaction. Packed-bed reactors are widely used with immobilized cells. Several modifications such as tapered beds to reduce the pressure drop across the length of the reactor, inclined bed, horizontal bed, and rotary horizontal reactors have been tried with limited success (Figure 10.18).

10.8.4 Tower Fermentors

Tower fermentors are simple in design and easy to construct. They consist of a long cylindrical vessel with an inlet at the bottom, an exhaust at the top, and a jacket to control temperature. They do not require agitation; hence, they do not have shafts, impellers, or blades. Tower fermentors are used for the continuous fermentation of beer, yeast, and SCPs (Figure 10.19).

A bubble column fermentor is the simplest type of tower fermentor; it consists of a tube that is air sparged at the base. It is an elongated, nonmechanically stirred fermentor with an aspect ratio of 6:1. This type of fermentor is used in citric acid production.

10.8.5 Batch Culture Fermentation

Batch culture is a closed culture system, which contains an initial limited amount of a medium. As the growth of the microorganism proceeds, the medium availability changes and the organism goes through a number of phases as described in Section 10.3.

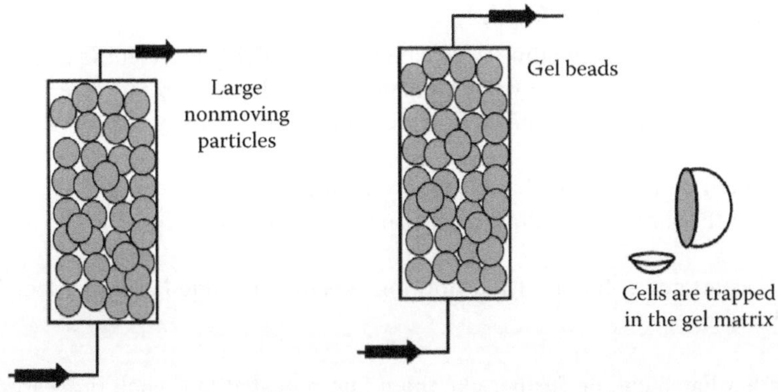

FIGURE 10.18 Fixed-bed fermentor.

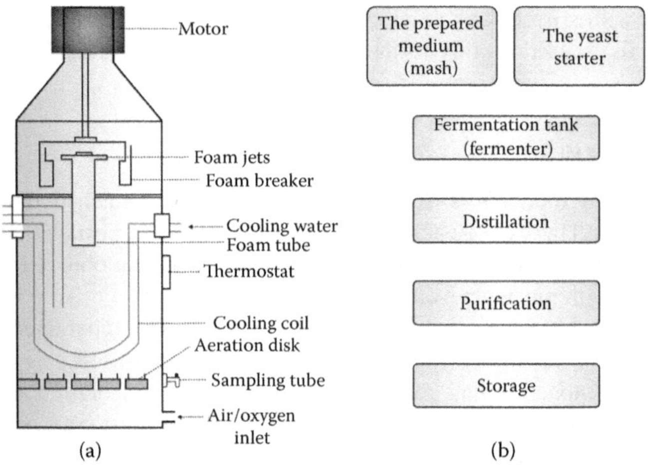

FIGURE 10.19 Tower fermentor: (a) Diagrammatic view of tower fermentation, and (b) a flow chart of tower fermentation.

10.8.6 Fed-Batch Culture

Two basic approaches to fed-batch fermentation are possible: (1) constant-volume fed-batch culture or fixed-volume fed-batch and (2) variable-volume fed-batch.

10.8.6.1 Fixed-Volume Fed-Batch

In this type of fed-batch fermentation, the limiting substrate is fed without diluting the culture. The culture volume can be maintained practically constant by feeding the growth-limiting substrate in its undiluted form, for example, as a very concentrated liquid or gas (e.g., oxygen). Alternatively, the substrate can be added by dialysis or, in a photosynthetic culture, radiation can be the growth-limiting factor without affecting the culture volume. A certain type of extended fed-batch culture, that is, the cyclic fed-batch culture for fixed-volume systems, refers to periodic withdrawal of a portion of the culture and use of the residual culture as the starting point for a further fed-batch process. Basically, once the fermentation reaches a certain stage (e.g., when aerobic conditions cannot be maintained anymore), the culture is removed and the biomass is diluted to the original volume using sterile water or a medium containing the feed substrate. Dilution decreases the biomass concentration and results in an increase in the specific growth rate. Subsequently, as feeding continues the growth rate will decline gradually as biomass once more increases and approaches the maximum amount sustainable in the vessel, at which point the culture may be diluted again.

10.8.6.2 Variable-Volume Fed-Batch

As the name implies, a variable-volume fed-batch is one in which the volume changes with the fermentation time. The way in which the volume changes depends on the requirements, limitations, and objectives of the operator. The feed can be provided according to one of the following three options:

1. The same medium used in the batch mode is added.
2. A solution of the limiting substrate at the same concentration as that in the initial medium is added.
3. A very concentrated solution of the limiting substrate is added at a rate less than that in options 1 and 2.

This type of fed-batch can be further classified into repeated fed-batch processes or cyclic fed-batch cultures, and single fed-batch processes. In the former, once fermentation reaches a certain stage after which the culture is not effective anymore, a quantity of the culture is removed from the vessel and replaced by a fresh nutrient medium. The decrease in volume results in an increase in the specific growth rate, which is followed by a gradual decrease in volume as the quasi-steady state is established. The latter type refers to a type of fed-batch in which supplementary growth medium is added during fermentation but no culture is removed until the end of the batch. This system presents a disadvantage over the fixed-volume fed-batch and the repeated fed-batch processes. Much of the fermentor volume is not utilized until the end of the batch is reached and, consequently, the duration of the batch is limited by the fermentor volume.

10.8.6.3 Advantages and Disadvantages of Fed-Batch Reactors

Fed-batch fermentation is an intermediary technique between batch and continuous fermentation. A proper feed rate with the right component constitution is required during the process. Fed-batch offers many advantages over batch and continuous cultures. It can be easily concluded that under controllable conditions and with the requisite knowledge of the microorganism involved in the fermentation process, the feed of the required components for growth and/or other substrates required for the production of the product can never be depleted and the nutritional environment can be maintained approximately constant during the course of the batch. The production of by-products that are generally related to the presence of high concentrations of the substrate can also be avoided by limiting its quantity to amounts that are required solely for the production of the biochemical. When high concentrations of the substrate are present, cells get overloaded, that is, their oxidative capacity is exceeded, and due to the Crabtree effect products other than the one of interest are produced, reducing the efficacy of the carbon flux. Moreover, these by-products have even proved to "contaminate" the product of interest, as ethanol does in the production of baker's yeast, and to impair cell growth, thereby reducing fermentation time and related productivity.

Sometimes, controlling the substrate is also important due to catabolic repression. Since this method usually permits the extension of operating time, high cell concentrations, and thereby improved productivity [mass of product/(volume × time)] can be achieved. This aspect is greatly favored in the production of growth-associated products. Additionally, this method allows the replacement of water loss by evaporation and decrease in the viscosity of the broth, such as in the production of dextran and xanthan gum, by the addition of a water-based feed. As previously mentioned, fed-batch might be the only option for fermentations dealing with toxic or low-solubility substrates. When dealing with recombinant strains, fed-batch mode can guarantee the presence of an antibiotic throughout the course of the fermentation process with the intent of keeping the presence of an antibiotic-marked plasmid.

Since growth can be regulated by the feed, and knowing that in many cases a high growth rate can decrease the expression of encoded products in recombinant products, the possibility of having different feeds and feed modes makes fed-batch an extremely flexible tool for control in such cases. Because the feed can also be a multisubstrate, the fermentation environment can still be provided with required protease inhibitors that might degrade the product of interest, metabolites, and precursors that increase the productivity of the fermentation process.

Finally, in fed-batch fermentation no special piece of equipment is required in addition to the one required by batch fermentation, even when the operating procedures for sterilization and prevention of contamination are considered. A cyclic fed-batch culture has an additional advantage: The productive phase of a process may be extended under controlled conditions. Such controlled periodic shifts in growth rate provide an opportunity to optimize product synthesis, particularly if the product of interest is a secondary metabolite whose maximum production takes place during the deceleration in growth.

10.8.6.4 Equipment

10.8.6.4.1 Vessels

Laboratory-scale fermentations are carried out in shaker flasks and flatbed bottles. Large-scale fermentations are carried out in glass or stainless steel tank fermentors. A fermentation vessel should be cheap, not allow contamination of its contents, be nontoxic to the microorganism used

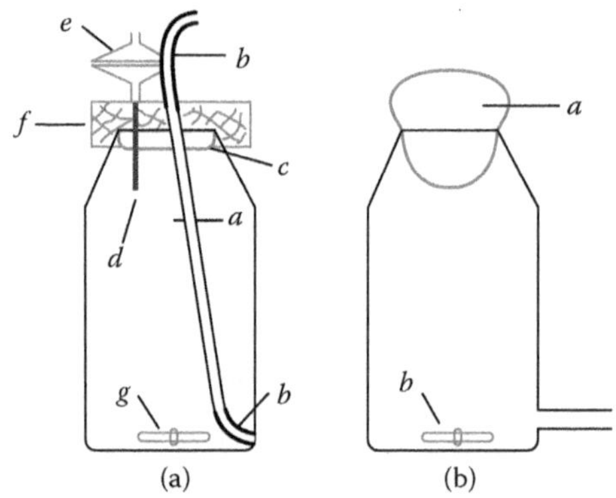

FIGURE 10.20 Fermentation vessels. (a) A holding vessel: A screw-neck borosilicate glass vessel with medium/inoculum addition assembly is shown. In the figure, *a* denotes a stainless steel rod, *b* denotes silicon tubing, *c* denotes a silicon disk, *d* a hypodermic needle, *e* an air vent, *f* a screw cap, and *g* a magnetic bar. (b) An aspirator-type vessel for introducing an inoculum of filamentous fungi into the fermentor is shown. In this figure, *a* denotes a cottonwool plug and *b* denotes a magnetic stirrer bar.

for the process, be easy to sterilize, be easy to operate, be robust and reliable, allow visual monitoring of the fermentation process, allow sampling, and be leak proof (Figure 10.20).

10.8.6.4.1.1 Shaker Flasks

Shaker flasks are conical vessels made of glass and are available in different sizes. The typical volume of these flasks is 250 mL. There are different types of shaker flasks, such as baffled, unbaffled or Erlenmeyer, and flying saucer flasks. Shaker flasks are used for screening of microorganisms as well as their cultivation for inoculation. Shaker beds or shaker tables are used to allow oxygen transfer by their continuous rotary motion. Baffled flasks are used to increase oxygen transfer. Shaker flasks need to be plugged to prevent contamination with other microorganisms. Cottonwool, polyurethane foam, glass, and synthetic plugs are commonly used to make shaker flasks. Fernwald shaker flasks and flatbed Thompson bottles are expensive and are not commonly used (Figure 10.11).

10.8.6.4.2 Pumps

A syringe pump is useful in the case of fed-batch for controlling liquid flow rate. A syringe is filled with liquid and secured onto the pump. Syringe plungers link to movable pistons to firmly secure to the main body of the pump. Regulated piston movement controls the flow rate. This pump is ideal for use with low flow rates.

Peristaltic pumps are also used to control flow rate (Figure 10.15). They control flow rate by squeezing and releasing pulse flow. The tubing through which the liquid flows is housed on a roller. The circular motion (speed) pinches and releases the tubing. The flow is smoother with more rollers. This pump can be controlled by a process controller with inputs of respiratory quotient and optical density. Peristaltic pumps are mostly used for adding acids and bases.

Another type of controlling device is a diaphragm pump (Figure 10.16). The liquid is allowed to flow through a flexible tube that utilizes a piston, and a pump controls the flow. The ball valves present in the flow prevent leakages and control the direction of the flow.

10.8.6.5 Control Techniques for Fed-Batch Fermentations

10.8.6.5.1 Proportional-Integral-Derivative Control Systems

The process parameters measured using probes may be controlled using control loops. A control loop consists of four components: (1) a measuring element, (2) a controller, (3) a final control element, and (4) the process to be controlled. A simple type of control is feedback control, in which the measuring element senses a process property and generates a corresponding output signal. The

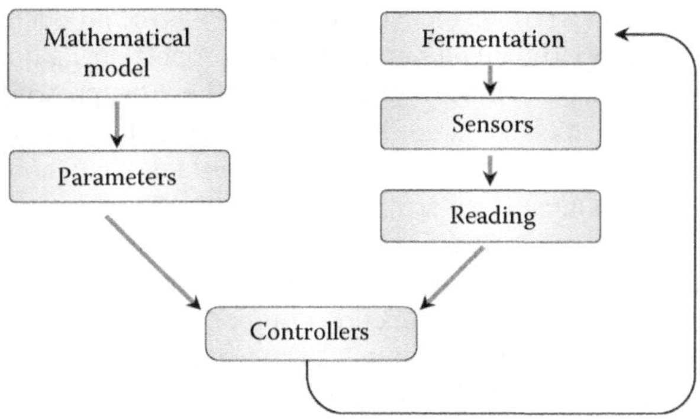

FIGURE 10.21 Control techniques for fed-batch fermentations.

controller compares the measurement signal with a predetermined desired value (setpoint) and produces an output signal to counteract any differences between the two. The final control element receives the control signal and adjusts the process by changing a valve opening or pump speed and causing the controlled process property to return to the set point. In manual control systems, a plant operative is instructed to monitor data throughout the time of operation and take appropriate action if there is any change from the set point (Figure 10.21).

10.8.6.5.2 The Proportional-Integral-Derivative Controller

A proportional-integral-derivative controller (PID controller) is a generic control loop feedback mechanism widely used in industrial control systems. A PID controller attempts to correct the error between a measured process variable and a desired set point by calculating and then outputting a corrective action that can adjust the process accordingly.

The PID controller calculation (algorithm) involves three separate parameters: (1) proportional, (2) integral, and (3) derivative values. The proportional value determines the reaction to the current error, the integral determines the reaction based on the sum of recent errors, and the derivative determines the reaction to the rate at which the error is changing. The weighted sum of these three actions is used to adjust the process through a control element such as the position of a control valve or the power supply of a heating element. By "tuning" the three constants in the PID controller algorithm, the PID can provide the control action designed for specific process requirements.

The response of the controller can be described in terms of the responsiveness of the controller to an error, the degree to which the controller overshoots the setpoint, and the degree of system oscillation. Note that the use of the PID algorithm for control does not guarantee optimal control over the system. Some applications may require the use of only one or two modes to provide appropriate system control. This is achieved by setting the gain of undesired control outputs to zero. A PID controller will be called a PI, PD, P, or I controller in the absence of the respective control actions. PI controllers are particularly common, since derivative action is very sensitive to measurement noise, and the absence of an integral value prevents the system from reaching its target value due to the control action. The PID control scheme is named after its three correcting terms, whose sum constitutes the manipulated variable (*MV*). Hence,

$$MV(t) = P_{\text{out}} + I_{\text{out}} + D_{\text{out}}$$

where P_{out}, I_{out}, and D_{out} are contributions to the output from the PID controller from each of the three terms, as defined next.

The *proportional term* makes a change to the output that is proportional to the current error value. The proportional response can be adjusted by multiplying the error by a constant K_p, which is called proportional gain. The proportional term is given as follows:

$$P_{\text{out}} = K_p e(t)$$

where *SP* is the substrate product ratio in the fermenter and *PV* is the product volume ratio in the fermenter; where P_{out} denotes proportional output; K_p denotes proportional gain, a tuning parameter; e is given by error = $SP - PV$; and t is time or instantaneous time (the present).

A high proportional gain results in a large change in the output for a given change in the error. If the proportional gain is too high, the system can become unstable. In contrast, a small gain results in a small output response to a large input error and a less responsive (or sensitive) controller. If the proportional gain is too low, the control action may be too small when responding to system disturbances. In the absence of disturbances, pure proportional control will not settle at its target value but will retain a steady-state error that is a function of proportional gain and process gain. Despite the steady-state offset, both tuning theory and industrial practice indicate that it is the proportional term that should contribute to the bulk of output change.

The contribution from the *integral term* is proportional to both the magnitude of the error and the duration of the error. Summing the instantaneous error over time (integrating the error) gives the accumulated offset that should have been corrected previously. The accumulated error is then multiplied by integral gain and added to the controller output. The magnitude of the contribution of the integral term to the overall control action is determined by the integral gain, K_i. The integral term is given by

$$I_{out} = K_i \int_0^{\tau} (\tau)\mathrm{d}\tau$$

where I_{out} denotes integral output; K_i is integral gain, a tuning parameter; e is given by error = $SP - PV$; and τ denotes time in the past contributing to the integral response. The integral term (when added to the proportional term) accelerates the movement of the process toward the setpoint and eliminates the residual steady-state error that occurs with a proportional only controller. However, since the integral term is responding to accumulated errors from the past, it can cause the present value to overshoot the set point value (cross over the setpoint and then create a deviation in the other direction).

The *derivative term* is the rate of change of the process error calculated by determining the slope of the error over time (i.e., its first derivative with respect to time) and multiplying this rate of change by the "derivative gain," K_d. The magnitude of the contribution of the derivative term to the overall control action is termed the derivative gain, K_d.

The derivative term is given by

$$D_{out} = K_d \frac{\mathrm{d}e}{\mathrm{d}t}$$

where D_{out} denotes derivative output; K_d denotes derivative gain, a tuning parameter; e is given by error = $SP - PV$; and t is time or instantaneous time (the present).

The derivative term slows down the rate of change of the controller output, an effect that is most noticeable close to the controller setpoint. Hence, derivative control is used to reduce the magnitude of the overshoot produced by the integral component and improve the combined controller-process stability. However, differentiation of a signal amplifies noise in the signal and, thus, this term in the controller is highly sensitive to noise in the error term and can cause a process to become unstable if the noise and the derivative gain are sufficiently large. Outputs from the three terms, proportional, integral, and derivative, are summed to calculate the output of the PID controller. Defining $u(t)$ as the controller output, the final form of the PID algorithm is as follows:

$$MV(t) = K_p e(t) + K_i \int_0^{\tau} e(\tau)\mathrm{d}\tau + K_d \frac{\mathrm{d}e}{\mathrm{d}t}$$

The tuning parameters are as follows:

- Proportional gain (K_p): A large value of K_p typically means a faster response, since the larger the error the larger the feedback to compensate. An excessively large proportional gain will lead to process instability.
- Integral gain (K_i): A large value of K_i implies that steady-state errors are eliminated quicker. The trade-off is larger overshoot: Any negative error integrated during transient response must be integrated away by positive error before reaching steady state.
- Derivative gain (K_d): A large K_d value decreases overshoot, but slows down transient response and may lead to instability.

10.8.7 Continuous Culture Fermentation

In continuous culture fermentation, the exponential growth phase of an organism may be prolonged by the addition of a fresh medium to the vessel. The vessel should be designed in such a way that the added volume displaces an equal volume of culture from the vessel. If a medium is fed continuously to such a vessel at a suitable rate, a steady state is achieved eventually. During steady state, the formation of a new biomass in the vessel is equivalent to the loss of cells from the vessel. The flow of medium into the vessel is related to the total volume of the medium in the vessel; it is expressed as the dilution rate, *D*, which can be expressed in the form of a mathematical equation:

$$D = F/V$$

where *F* is the flow rate (in cubic decimeters per hour) and *V* is the total volume. The net change in cell concentration over a time period may be expressed as follows:

$$dx/dt = \text{growth} - \text{output} \quad \text{or} \quad \mu x - Dx$$

Under steady-state conditions, the cell concentration remains constant; thus, $dx/dt = 0$ and the equation becomes

$$\mu x = Dx$$

and $\mu = D$.

10.8.7.1 Chemostat

Under steady-state conditions, the specific growth rate is controlled by the dilution rate, which is a controllable variable. An important objective of continuous culture operation is to control cell growth at a level at which productivity is optimum. There are several ways this can be achieved. One way is to maintain a constant fermentor volume and to use a flow rate that gives appropriate productivity. In this mode of operation, the fermentor system is known as a chemostat. The chemostat is by far the simplest and most common mode of operation of a continuous culture. Cell growth in chemostat cultures is maintained steady by a constant inflow of a fresh medium consisting of nutrients (nitrogen, phosphorous, glucose) at a concentration that makes it growth limiting. Other constituents of such a medium are present at concentrations higher than the required values. Increases or decreases in the concentration of the growth-limiting factor are correspondingly expressed by increases or decreases in the growth rate of cells.

The volume of the chemostat can be controlled by using either a pump or, where contamination can be a significant problem, a pump-based control system. This setup is commonly used in laboratory investigations and animal cell culture systems. An overflow system has the advantage that only one pump is required. However, as the effluent flow rate is determined by gravity alone, there is a greater possibility of contaminants moving up the effluent tube into the reactor. Overflow systems are widely used in wastewater treatment and have been used in the large-scale continuous culture of bacteria. In other techniques, a fermentor variable, for example, turbidity or pH, will be monitored using an appropriate detector and the liquid flow rate will be automatically adjusted so as to maintain the variable at a constant level. Examples of these types of continuous fermentors are pH-stat, turbidostat, and nutristat. However, apart from the pH-stat, these reactors are rarely used as the necessary measurement-control systems are generally unreliable over long periods of time.

10.8.7.2 Turbidostat

A turbidostat is a continuous culturing method in which the turbidity of the culture is held constant by manipulating the rate at which the medium is fed. If turbidity tends to increase, the feed rate is increased to dilute the turbidity back to its set point. When turbidity tends to fall, the feed rate is lowered so that growth can restore the turbidity to its set point. The most widespread large-scale application of continuous culture reactors is in wastewater treatment. Activated sludge plants, trickle bed filters, anaerobic digester, and ponds all operate in a continuous manner. Cell immobilization is also often used to improve the efficiency of the process. Continuous cultures are well established in the wastewater treatment industry for several reasons:

- Unlike pure-culture microbial and animal cell systems, contamination is not a consideration as the wastewater feed always contains microorganisms.
- Continuous reactors have long been used in waste treatment and their use is not considered a risk.
- Finally, using batch cultures is simply not economically feasible. Wastewater flows are often measured in mega liters per hour and batch reactors simply cannot cope with the load.

10.9 STERILIZATION IN FERMENTATION

It is essential for all fermentation processes to ensure a yield of the desired product as this allows the growth of the desired microorganisms. All the components of fermentation need to be sterilized, including the nutrient medium and other ingredients. The fermentation vessel and other accessory equipment must also be sterilized to ensure pristine conditions and virtually no contamination. Different techniques for sterilization are available based on the properties of the component being sterilized. The most commonly used techniques include sterilization by heat, such as heat sterilization, moist heat sterilization, dry heat sterilization, incineration, and boiling; sterilization by chemicals; sterilization by filtration; and sterilization by radiation. The specific technique of sterilization applied in a fermentation process depends on the following:

- The type of microorganism present: Some microorganisms are more resistant than others and it is very difficult to kill them compared to others.
- The number of microorganisms present: It is much easier to kill a small number of organisms compared to a large number of organisms.
- The amount and type of organic material that protects the microorganisms: Blood or tissue remaining on poorly cleaned instruments acts as a shield to microorganisms during the sterilization process.
- The number of cracks and crevices on an instrument that might harbor microorganisms: Microorganisms collect in and are protected by scratches, cracks, and crevices such as the serrated jaws of tissue forceps.

Sterilization can be achieved in a number of ways, such as moist heat sterilization, dry heat sterilization, incineration, boiling, sterilization by chemicals, filtration, and radiation, as shown in Figure 10.22. The application of heat is considered the most reliable method for sterilization of

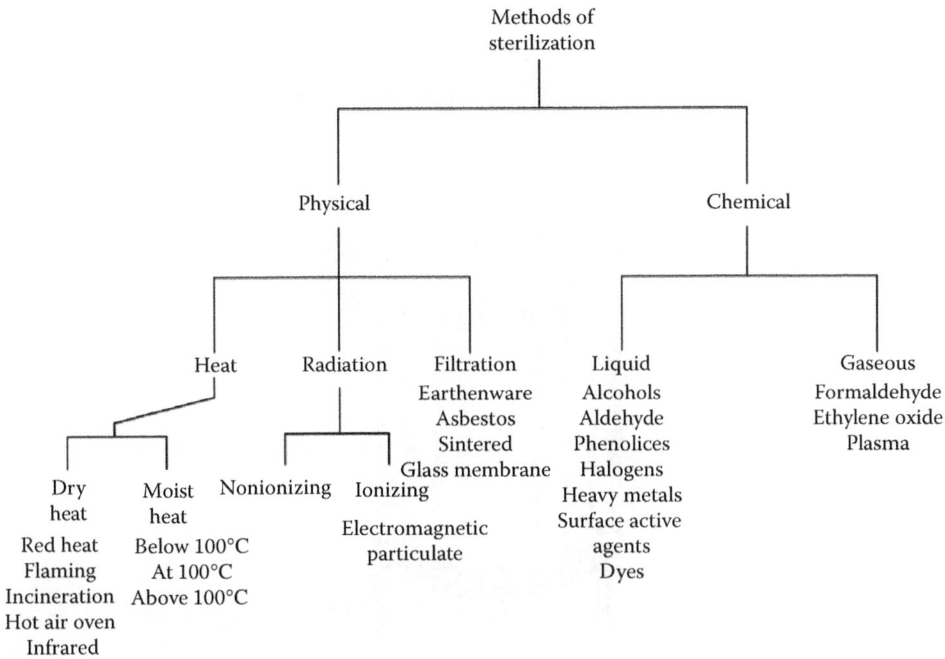

FIGURE 10.22 Various methods of sterilization.

articles that can withstand high temperatures. Heat acts by oxidative effects as well as the denaturation and coagulation of proteins. Articles that cannot withstand high temperatures can still be sterilized at lower temperatures by prolonging the duration of exposure. Factors affecting sterilization by heat are as follows:

- Nature of heat: Moist heat is more effective than dry heat.
- Temperature and time: Temperature and time are inversely proportional to each other. As temperature increases, the time required decreases.
- Number of microorganisms: The higher the number of microorganisms the higher the temperature or longer the duration of exposure required.
- Nature of the microorganisms: The effectiveness of sterilization depends on the species and strain of the microorganism; sensitivity to heat may vary among microorganisms. Spores are highly resistant to heat.
- Type of material: Articles that are heavily contaminated require high temperatures or prolonged exposure. Certain heat-sensitive articles must be sterilized at lower temperatures.
- Presence of organic material: Organic materials such as protein, sugars, oils, and fats increase the time of exposure required.

10.9.1 Moist Heat Sterilization

This is the most commonly used technique for sterilization of a wide variety of fermentation components. Steam is used at high pressure and temperature at 121°C for 15 minutes, causing denaturation of enzymes and degradation of nucleic acids, resulting in the death of the microorganisms present. This process is carried out by using an autoclave in which the components to be sterilized are placed in water and heated (Figure 10.23). Steam is produced and high pressure is developed inside the autoclave. Most of the fermentation components such as the culture medium, small fermentors, and glassware are sterilized by this technique.

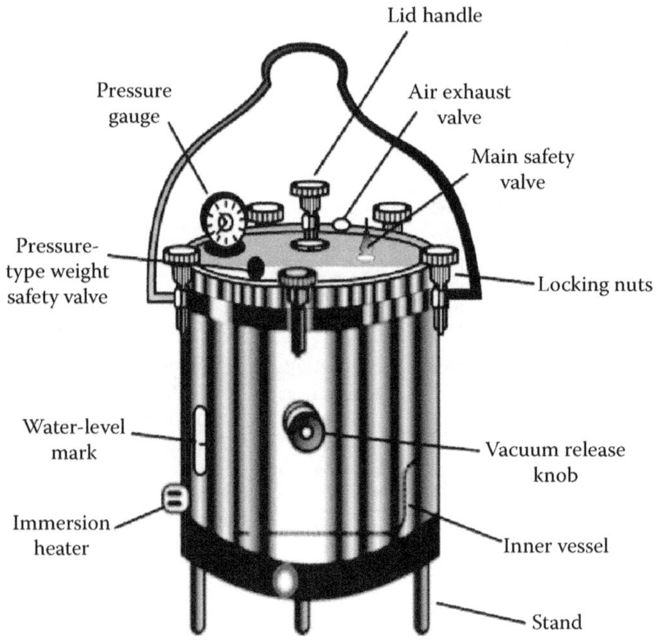

FIGURE 10.23 An autoclave.

Steam sterilization requires four conditions: (1) adequate contact, (2) sufficiently high temperature, (3) correct time, and (4) sufficient moisture. Although all conditions are necessary for sterilization to take place, sterilization failures in clinics and hospitals are most often caused by lack of steam contact or failure to attain adequate temperature. Moist heat acts by coagulation and denaturation of proteins.

10.9.1.1 Moist Heat Sterilization at Temperatures below 100°C

Moist heat sterilization at temperatures below 100°C can occur in the following ways:

- Pasteurization: This process was originally used by Louis Pasteur. Currently, this procedure is used in the food and dairy industry. There are two methods of pasteurization: (1) the holder method (heated at 63°C for 30 minutes) and (2) the flash method (heated at 72°C for 15 seconds followed by quickly cooling to 13°C). Other pasteurization methods include the use of ultra-high temperature (UHT), that is, heating at 140°C for 15 seconds and at 149°C for 0.5 seconds. This method is suitable to destroy most milk-borne pathogens such as *Salmonella, Mycobacteria, Streptococci, Staphylococci*, and *Brucella*; however, *Coxiella* may survive pasteurization. Efficacy is tested by the phosphatase test and the methylene blue test.
- Vaccine bath: The contaminating bacteria in a vaccine preparation can be inactivated by heating the preparation in a water bath at 60°C for 1 hour. Only vegetative bacteria are killed and spores survive.
- Serum bath: The contaminating bacteria in a serum preparation can be inactivated by heating the preparation in a water bath at 56°C for 1 hour on several successive days. Proteins in the serum will coagulate at higher temperatures. Only vegetative bacteria are killed and spores survive.
- Inspissation: This is a technique to solidify as well as disinfect egg- and serum-containing media. The medium containing serum or egg are placed on the slopes of an inspissator and heated at 80°C–85°C for 30 minutes on three successive days. On the first day, the

vegetative bacteria die; spores that germinate by the next day are killed on the following day. The efficacy of the process depends on germination of spores in between inspissation sessions. If the spores fail to germinate, then this technique cannot be considered a sterilization technique.

10.9.1.2 Moist Heat Sterilization at 100°C

Moist heat sterilization at 100°C can occur in the following ways:

- Boiling water: Boiling water (100°C) kills most vegetative bacteria and viruses immediately. Certain bacterial toxins such as staphylococcal enterotoxin are resistant to heat. Some bacterial spores are resistant to boiling and survive; hence, boiling is not a substitute for sterilization. The killing activity can be enhanced by the addition of 2% sodium bicarbonate. When absolute sterility is not required, certain metal articles and glassware can be disinfected by placing them in boiling water for 10–20 minutes. The lid of the boiler must not be opened during the period of exposure.
- Steam at 100°C: Instead of keeping articles in boiling water, they are subjected to free steam at 100°C. Traditionally, Arnold's and Koch's steamers were used. An autoclave (with the discharge tap kept open) can also serve the same purpose. A steamer is a metal cabinet with perforated trays to hold the articles and a conical lid. The bottom of the steamer is filled with water and heated. The steam generated sterilizes the articles when they are exposed for a period of 90 minutes. Media such as TCBS, DCA, and selenite broth are sterilized by steaming. Sugar and gelatin in the medium may get decomposed on autoclaving; hence, they are exposed to free steaming for 20 minutes for three successive days. This process is known as tyndallization (after John Tyndall), fractional sterilization, or intermittent sterilization. The vegetative bacteria in the exposed articles are killed in the first exposure and the spores that germinate by the following day are killed on subsequent days. The success of the process depends on the germination of spores.

10.9.1.3 Moist Heat Sterilization at Temperatures above 100°C

Sterilization can be effectively achieved at a temperature above 100°C using an autoclave. Water boils at 100°C at atmospheric pressure, but if the pressure is raised, the temperature at which water boils also increases. In an autoclave, water is boiled in a closed chamber. As the pressure rises, the boiling point of water increases. At a pressure of 15 lbs inside the autoclave, the temperature is 121°C. Exposure of articles to this temperature for 15 minutes sterilizes them. To destroy the infective agents associated with spongiform encephalopathies (prions), higher temperatures or longer times are used; for this purpose, an application temperature of 135°C or 121°C for at least 1 hour is recommended (Figure 10.23).

Advantages of steam sterilization are as follows: Steam has more penetrative power than dry air, it moistens the spores (moisture is essential for coagulation of proteins), condensation of steam on cooler surfaces releases latent heat, and condensation of steam draws in fresh steam. Exposure to steam is a very effective method of sterilization; it is quicker than using a hot air oven.

Disadvantages of steam sterilization are as follows: Drenching and wetting of articles may occur when they are exposed to steam. The presence of trapped air may reduce the efficacy of sterilization, and articles take a long time to cool down after exposure to steam.

Methods for sterilization control include the following: Physical methods include automatic process control, use of a thermocouple, and use of a temperature chart recorder. Chemical methods include use of Browne's tube No. 1 (black spot) and succinic acid (whose melting point is 121°C) and use of Bowie Dick tape. Bowie Dick tape is applied to articles being autoclaved. If the process is satisfactory, dark brown stripes appear across the tape. Biological methods include using a paper strip containing 106 spores of *Geobacillus stearothermophilus*.

10.9.2 Dry Heat Sterilization

Dry heat acts by protein denaturation, oxidative damage, and toxic effects of elevated levels of electrolytes, whereas moist heat acts by coagulation and denaturation of proteins. Moist heat is superior to dry heat in action. The temperature required to kill microbes by dry heat is more than that required by moist heat. Thermal death time is the minimum time required to kill a suspension of organisms at a predetermined temperature in a specified environment. Dry heat sterilization is performed in the following ways:

- Red heat: Articles such as bacteriological loops, straight wires, tips of forceps, and searing spatulas are sterilized by holding them in a Bunsen flame until they become red hot. This is a simple method for effective sterilization of such articles, but it is limited to those articles that can be heated to redness in flame.
- Flaming: Flaming is the method of passing an article over a Bunsen flame without heating it to redness. Articles such as scalpels, mouths of test tubes, flasks, glass slides, and coverslips are passed through the flame a few times. Even though most vegetative cells are killed by this method, there is no guarantee that spores would die on such short exposures. This method too is limited to articles that can be exposed to a flame. Cracking of glassware may occur as a result of exposure to flame.
- Hot air oven: This method was introduced by Louis Pasteur. Articles to be sterilized are exposed to a high temperature (160°C) for the duration of 1 hour in an electrically heated oven. Since air is a poor conductor of heat, the even distribution of heat throughout the chamber is achieved by using a fan. The heat is transferred to the article by radiation, conduction, and convection. The oven should be fitted with a thermostat control, a temperature indicator, and meshed shelves, and it must have adequate insulation.

Articles such as metallic instruments (e.g., forceps, scalpels, and scissors), glassware (e.g., petri dishes, pipettes, flasks, and all-glass syringes), swabs, oils, grease, petroleum jelly, and some pharmaceutical products are sterilized.

The sterilization process is as follows: Articles to be sterilized must be perfectly dry before placing them inside a hot air oven to avoid breakage. Articles must be placed at sufficient distance from each other so as to allow free circulation of air between them. The mouths of flasks, test tubes, and both ends of pipettes must be plugged with cotton wool. Articles such as petri dishes and pipettes may be arranged inside metal canisters and then placed in the oven. Individual glass articles must be wrapped in kraft paper or aluminum foils.

The sterilization cycle takes into consideration the time taken for articles to reach the sterilizing temperature, the defined period for which the sterilizing temperature is maintained (holding time), and the time taken for the articles to cool down. Different temperature-time relations for holding time are 60 minutes at 160°C, 40 minutes at 170°C, and 20 minutes at 180°C. Increasing the temperature by 10°C shortens the sterilizing time by 50%. The hot air oven must not be opened until the temperature inside falls below 60°C in order to prevent breakage of glassware.

10.9.3 Incineration

Incineration is a method of destroying contaminated materials by burning them in an incinerator. Articles such as soiled dressings, animal carcasses, pathological material, and bedding should be subjected to incineration. This technique results in the loss of the article; hence, it is suitable only for those articles that have to be disposed of. Polystyrene materials emit dense smoke when burned, and should not be incinerated.

Incineration is a waste treatment process that involves the combustion of organic substances contained in waste materials. Incineration and other high-temperature waste treatment systems are described as "thermal treatment." The incineration of waste materials converts the waste into ash, flue gas, and heat. The ash is mostly formed by the inorganic constituents of the waste and may take the form of solid lumps or particulates carried by the flue gas. The flue gases must be cleaned of gaseous and particulate pollutants before they are dispersed into the atmosphere. In some cases, heat produced by incineration can be used to generate electric power.

Incineration with energy recovery is one of several waste-to-energy (WtE) technologies, including gasification, plasma arc gasification, pyrolysis, and anaerobic digestion. Incineration may also be implemented without energy and material recovery. In several countries, there are still concerns from experts and local communities about the environmental impact of incinerators. In some countries, incinerators built just a few decades ago often did not include a materials separation mechanism to remove hazardous, bulky, or recyclable materials before combustion. These facilities tended to risk the health of their workers and the local environment by facilitating inadequate levels of gas cleaning and combustion process control. Most of these facilities did not generate electricity.

Incinerators reduce the solid mass of original waste by 80%–85% and the volume (already compressed somewhat in garbage trucks) by 95%–96%, depending on composition and degree of recovery of materials such as metals from the ash for recycling. This means that although incineration does not completely replace land filling, it significantly reduces the necessary volume for disposal. Garbage trucks often reduce the volume of waste using a built-in compressor before delivering it to the incinerator. Alternatively, the volume of uncompressed garbage is reduced at landfills by approximately 70% using a stationary steel compressor, albeit with significant energy cost. In many countries, simple waste compaction is a common practice for compaction at landfills.

10.9.4 Chemicals

Disinfectants are chemicals that destroy pathogenic bacteria from inanimate surfaces. Some chemicals have a very narrow spectra of activity, whereas some have very wide spectra. Chemicals that can sterilize articles are called chemisterilants. Chemicals that can be safely applied over skin and mucus membranes are called antiseptics. An ideal antiseptic or disinfectant should have the following properties:

- Wide spectrum of activity
- Ability to destroy microbes within practical periods of time
- Active in the presence of organic matter
- Ability to make effective contact and be wettable
- Active in any pH
- Stability
- Long shelf life
- Speed
- High penetrating power
- Nontoxic, nonallergenic, nonirritative, and noncorrosive
- No bad odor
- No nonvolatile residue or stain
- Able to retain efficacy on reasonable dilution
- Inexpensive and easily available

An ideal disinfectant with all the aforementioned properties is not yet available. The level of disinfection achieved depends on contact time, temperature, type and concentration of the active ingredient, presence of organic matter, and type and quantum of microbial load. The chemical disinfectants at working concentrations rapidly lose their strength on standing.

10.9.4.1 Alcohols

Alcohols have the following properties:

- Mode of action: Alcohols dehydrate cells, disrupt membranes, and cause coagulation of proteins.
- Examples: Ethyl alcohol, isopropyl alcohol, and methyl alcohol.
- Application: A 70% aqueous solution is more effective at killing microbes than absolute alcohols. Note that 70% ethyl alcohol (spirit) is used as an antiseptic on skin. Isopropyl alcohol is preferred to ethanol. Isopropyl alcohol can also be used to disinfect surfaces, and is used to disinfect clinical thermometers. Methyl alcohol kills fungal spores; hence, it is useful in disinfecting inoculation hoods.
- Disadvantages: Alcohols are skin irritants, volatile (evaporates rapidly), and inflammable.

10.9.4.2 Aldehydes

Aldehydes have the following properties:

- Mode of action: Aldehydes act through the alkylation of the amino group, carboxyl group, or hydroxyl group and probably damage nucleic acids. They kill all microorganisms, including spores.
- Examples: Formaldehyde and glutaraldehyde.
- Application: Note that 40% formaldehyde (formalin) is used for surface disinfection and fumigation of rooms, chambers, operation theaters, biological safety cabinets, hospital wards, sickrooms, and so on. Fumigation is achieved by boiling formalin, heating paraformaldehyde, or treating formalin with potassium permanganate. Formaldehyde also sterilizes bedding, furniture, and books. Note that 10% formalin with 0.5% tetraborate sterilizes clean metal instruments. Further, 2% gluteraldehyde is used to sterilize thermometers, cystoscopes, bronchoscopes, centrifuges, anesthetic equipments, and so on. An exposure of at least 3 hours at alkaline pH is required for action by gluteraldehyde. Note that 2% formaldehyde at 40°C for 20 minutes is used to disinfect wool and 0.25% at 60°C for 6 hours is used to disinfect animal hair and bristles.
- Disadvantages: Vapors are irritating (they must be neutralized by ammonia). Aldehydes have poor penetration and leave nonvolatile residue. Their activity is reduced in the presence of protein. Glutaraldehyde requires alkaline pH and only those articles that are wettable can be sterilized using gluteraldehyde.

10.9.4.3 Phenol

Phenols have the following properties:

- Mode of action: Phenol acts by disruption of membranes, precipitation of proteins, and inactivation of enzymes.
- Examples: 5% phenol, 1%–5% cresol, 5% lysol (a saponified cresol), hexachlorophene, chlorhexidine, and chloroxylenol (Dettol).
- Applications: Joseph Lister used phenol to prevent infection of surgical wounds. Phenols are coal-tar derivatives. They act as disinfectants at high concentrations and as antiseptics at low concentrations. They are bactericidal, fungicidal, and mycobactericidal, but they are inactive against spores and most viruses. They are not readily inactivated by organic matter. Corrosive phenolics are used for disinfection of ward floors, in discarding jars in laboratories, and for disinfection of bedpans. Chlorhexidine can be used in an isopropanol solution for skin disinfection or as an aqueous solution for wound irrigation. It is often used as an antiseptic hand wash. A 20% chlorhexidinegluconate solution is used for preoperative

hand and skin preparation and for general skin disinfection. Chlorhexidinegluconate is also mixed with quaternary ammonium compounds such as cetrimide to obtain stronger and broader antimicrobial effects (e.g., Savlon). Chloroxylenols are less irritating on skin, can be used for topical purposes, and are more effective against gram-positive bacteria than gram-negative bacteria. Hexachlorophene is chlorinated diphenyl and is much less irritating on skin than chloroxylenols. It has a marked effect on gram-positive bacteria but a poor effect on gram-negative bacteria, mycobacteria, fungi, and viruses. Triclosan is an organic phenyl ether with good activity against gram-positive bacteria, and is effective to some extent against many gram-negative bacteria including *Pseudomonas*. It also has fair activity against fungi and viruses.

- Disadvantages: Phenol is toxic and corrosive and a skin irritant. Chlorhexidine is inactivated by anionic soaps. Chloroxylenol is inactivated by hard water.

10.9.4.4 Halogens

Halogens have the following properties:

- Mode of action: Halogens are oxidizing agents and cause damage by oxidation of the essential sulfhydryl groups of enzymes. Chlorine reacts with water to form hypochlorous acid, which is microbicidal.
- Examples: Chlorine compounds (chlorine, bleach, hypochlorite) and iodine compounds (tincture of iodine, iodophores).
- Applications: Tincture of iodine (2% iodine in 70% alcohol) is an antiseptic. Iodine can be combined with neutral carrier polymers such as polyvinylpyrrolidone to prepare iodophores such as povidone-iodine. Iodophores permit slow release and reduce irritation of the antiseptic. For hand washing, iodophores are diluted in 50% alcohol. Note that 10% povidone-iodine is used in its undiluted form in pre- and postoperative skin disinfection. Chlorine gas is used to bleach water. Household bleach can be used to disinfect floors. In higher concentrations, chlorine is used to disinfect swimming pools. Note that 0.5% sodium hypochlorite is used in serology and virology. It is used at a dilution of 1:10 in decontamination of spillage of infectious material. Mercuric chloride is used as a disinfectant.
- Disadvantages: Halogens are rapidly inactivated in the presence of organic matter. Iodine is corrosive and staining. Bleach solution is corrosive and will corrode stainless steel surfaces.

10.9.4.5 Heavy Metals

Heavy metals have the following properties:

- Mode of action: Heavy metals act by precipitation of proteins and oxidation of the sulfhydryl groups. They are bacteriostatic.
- Examples: Mercuric chloride, silver nitrate, copper sulfate, and organic mercury salts (e.g., mercurochrome and merthiolate).
- Applications: Note that 1% silver nitrate solution can be applied on the eyes as treatment for ophthalmianeonatorum (Crede's method). This procedure is no longer followed. Silver sulfadiazine is used topically to help prevent colonization and infection of burn tissues. Mercurials are active against viruses at dilutions of 1:500 to 1:1000. Merthiolate at a concentration of 1:10,000 is used in the preservation of serum. Copper salts are used as a fungicide.
- Disadvantages: Mercuric chloride is highly toxic; heavy metals are readily inactivated by organic matter.

10.9.4.6 Surface Active Agents

Surface active agents have the following properties:

- Mode of action: Surface active agents concentrate at interfaces between the lipid-containing membrane of a bacterial cell and the surrounding aqueous medium. These compounds have long-chain hydrocarbons that are soluble in fat and charged ions that are soluble in water. Since they contain both hydrocarbons and charged ions, they concentrate on the surface of membranes. They disrupt cell membranes, resulting in leakage of cell constituents.
- Examples: Surface active agents are soaps or detergents. Detergents can be "anionic" or "cationic." Detergents containing negatively charged long-chain hydrocarbons are called anionic detergents. These include soaps and bile salts. If the fat-soluble part is made to have a positive charge by combining it with a quaternary nitrogen atom, it is called a cationic detergent. Cationic detergents are known as quaternary ammonium compounds (or quat). Cetrimide and benzalkonium chloride act as cationic detergents.
- Applications: They are active against vegetative cells, mycobacteria, and enveloped viruses. They are widely used as disinfectants at dilutions of 1%–2% for domestic and hospital uses.
- Disadvantages: Their activity is reduced by hard water, anionic detergents, and organic matter. *Pseudomonas* can metabolize cetrimide, using it as a carbon, nitrogen, and energy source.

10.9.4.7 Dyes

Dyes have the following properties:

- Mode of action: Acridine dyes are bactericidal because of their interaction with bacterial nucleic acids.
- Examples: Examples include aniline dyes such as crystal violet, malachite green, and brilliant green, and acridine dyes such as acriflavine and aminacrine. Acriflavine is a mixture of proflavine and euflavine. Only euflavine has effective antimicrobial properties. The related dye ethidium bromide is also germicidal. It intercalates between base pairs in DNA. Dyes are more effective against gram-positive bacteria than gram-negative bacteria and are more bacteriostatic in action.
- Applications: They may be used topically as antiseptics to treat mild burns. They are used as paint on the skin to treat bacterial skin infections. Dyes are used as selective agents in certain selective media.

10.9.4.8 Hydrogen Peroxide

Hydrogen peroxide has the following properties:

- Mode of action: Hydrogen peroxide acts on microorganisms by releasing nascent oxygen. It produces a hydroxyl-free radical that damages proteins and DNA.
- Applications: It is used at 6% concentration to decontaminate instruments and equipment such as ventilators. A 3% hydrogen peroxide solution is used for skin disinfection and deodorizing wounds and ulcers. Strong solutions are sporicidal.
- Disadvantages: Hydrogen peroxide decomposes in the presence of light and is broken down by catalase; proteinaceous organic matter drastically reduces its activity.

10.9.4.9 Ethylene Oxide

Ethylene oxide (EO) has the following properties:

- Mode of action: EO is an alkylating agent. It acts by alkylating sulfhydryl, amino, carboxyl, and hydroxyl groups.
- Properties: It is a cyclic molecule, which is a colorless liquid at room temperature. It has a sweet ethereal odor, readily polymerizes, and is flammable.

- Application: It is a highly effective chemisterilant capable of killing spores rapidly. Because it is highly flammable, it is usually combined with CO_2 (10% CO_2 + 90% EO) or used as dichlorodifluoromethane. It requires the presence of humidity to be active. EO has good penetration and is well absorbed by porous materials. It is used to sterilize heat-labile articles such as bedding, textiles, rubber, plastics, syringes, disposable petri dishes, and complex apparatus such as the heart-lung machine and respiratory and dental equipment. Efficiency testing is done using *Bacillus subtilis* var. *niger.*
- Disadvantages: EO is highly toxic, irritating to eyes and skin, highly flammable, mutagenic, and carcinogenic.

10.9.4.10 β-Propiolactone

β-Propiolactone (BPL) has the following properties:

- Mode of action: BPL is an alkylating agent and acts through the alkylation of carboxyl and hydroxyl groups.
- Properties: It is a colorless liquid with a pungen, slightly sweetish smell. It is a condensation product of ketone with formaldehyde.
- Application: It is an effective sporicidal agent and has broad-spectrum activity. Note that 0.2% is used to sterilize biological products. It is more efficient in fumigation than formaldehyde. It is used to sterilize vaccines, tissue grafts, surgical instruments, and enzymes.
- Disadvantages: BPL has poor penetrating power and is a carcinogen.

10.9.5 Sterilization Using Filtration

Filtration does not kill microbes; it merely separates them. Membrane filters with pore sizes of 0.2–0.45 μm are commonly used to remove particles from solutions that cannot be autoclaved. Filtration is used to remove microbes from heat-labile liquids such as serum, antibiotic solutions, sugar solutions, and urea solution. Various applications of filtration include removing bacteria from ingredients of culture media, preparing suspensions of viruses and phages free of bacteria, measuring sizes of viruses, separating toxins from culture filtrates, counting bacteria, clarifying fluids, and purifying hydatid fluid. Filtration is aided by the use of either positive or negative pressure provided by vacuum pumps. Older filters made of earthenware or asbestos are called depth filters. Different types of filters are as follows:

- Earthenware filters: These filters are made up of diatomaceous earth or porcelain. They are usually baked into the shape of a candle. Different types of earthenware filters are as follows:
  - Pasteur–Chamberland filter: These candle filters are from France and are made of porcelain (sand and kaolin). A similar filter from Great Britain is a Doulton filter. Chamberland filters are made with various porosities, which are graded as L1, L1a, L2, L3, L5, L7, L9, and L11. Doulton filter porosities are graded as P2, P5, and P11.
  - Berkefeld filter: These filters are made of kieselguhr, a fossilized diatomaceous earth found in Germany. They are available in three grades depending on their porosity (pore size): V (veil), N (normal), and W (wenig). The quality of a V-grade filter is checked using a culture suspension of *Serrtiamarcescens* (0.75 μm).
  - Mandler filter: This filter comes from the United States and is made of kieselguhr, asbestos, and plaster of Paris.
- Asbestos filters: These filters are made from the chrysotile type of asbestos; they chemically comprise magnesium silicate. They are pressed to form disks, which are to be used only once. A disk is held inside a metal mount, which is sterilized by autoclaving. Asbestos

filters are available in the following grades: HP/PYR (for removal of pyrogens), HP/EKS (for absolute sterility), and HP/EK (for clarifying).

- Sintered glass filters: These are made from finely ground glass that is fused sufficiently to make small particles of glass adhere to one another. They are usually available in the form of a disk fused into a glass funnel. Grade 5 filters have average pore diameters of 1–1.5 μm. They are washed in running water in the reverse direction, cleaned with warm concentrated H_2SO_4, and sterilized by autoclaving.
- Membrane filters: These filters are made from a variety of polymeric materials such as cellulose nitrate, cellulose diacetate, polycarbonate, and polyester. The older type of membrane, called "gradocol" (graded colloid ion) membrane, comprised cellulose nitrate. Gradocol membranes have average pore diameters of 3–10 μm. The newer ones comprise cellulose diacetate. These membranes have a pore diameter ranging from 0.015 to 12 μm. These filters are sterilized by autoclaving. Membrane filters are made in two ways: Capillary pore membranes have pores produced by radiation, whereas labyrinthine pore membranes are produced by forced evaporation of solvents from cellulose esters.
- Air filters: Air can be filtered using high-efficiency particle air (HEPA) filters, which are usually used in biological safety cabinets. HEPA filters are at least 99.97% efficient for removing particles greater than 0.3 μm in diameter. Examples of areas where HEPA filters are used include rooms housing severely neutropenic patients and operating rooms designated for orthopedic implant procedures. HEPA filter efficiency is monitored using the dioctylphthalate (DOP) particle test using particles that are 0.3 μm in diameter.

The disadvantages of depth filters are possibility of migration of filter material into the filtrate, absorption or retention of certain volumes of liquid by the filters, pore sizes that are not definite, and the possibility of viruses and mycoplasma passing through the filters. The advantages of membrane filters are known porosity, absence of retention of fluids, ability to reuse after autoclaving, and compatibility with many chemicals. However, membrane filters have little loading capacity and are fragile.

10.9.6 Sterilization Using Radiation

Two types of radiation are used: (1) ionizing and (2) nonionizing. Nonionizing rays are low-energy rays with poor penetrative power, whereas ionizing rays are high-energy rays with good penetrative power. Since radiation does not generate heat, it is termed "cold sterilization." In some parts of Europe, fruits and vegetables are irradiated to increase their shelf life up to 500%.

Rays that are longer than the wavelength of visible light are *nonionizing*. The microbicidal wavelength of UV rays lies in the range 200–280 nm, with 260 nm being the most effective. UV rays are generated using a high-pressure mercury vapor lamp. It is at 260 nm of wavelength that absorption by microorganisms is at its maximum, which results in the germicidal effect of rays. UV rays induce formation of thymine-thymine dimers, which ultimately inhibit DNA replication. UV rays readily induce mutations in cells irradiated with a nonlethal dose. Microorganisms such as bacteria, viruses, and yeast that are exposed to effective UV radiation are inactivated within seconds. Since UV rays do not kill spores, they are considered to be of use in surface disinfection. UV rays are used to disinfect hospital wards, operation theaters, virus laboratories, corridors, and so on. Disadvantages of using UV rays include the low penetrative power of the rays and the limited life of UV bulbs. Some bacteria have DNA repair enzymes that can overcome the damage caused by UV rays. The presence of organic matter and dust prevents UV rays from reaching their destination; UV rays are harmful to skin and eyes. Further, they do not penetrate glass, paper, or plastic.

Ionizing rays are of two types: (1) particulate and (2) electromagnetic rays. They are described as follows:

- Electron beams are particulate in nature, whereas gamma rays are electromagnetic in nature. High-speed electrons are produced by a linear accelerator from a heated cathode. Electron beams are used to sterilize articles such as syringes, gloves, dressing packs, foods, and pharmaceuticals. When using electron beams, sterilization is accomplished in a few seconds. Unlike electromagnetic rays, the instruments can be switched off. Disadvantages of using electron beams include poor penetrative power and the requirement of sophisticated equipment.
- Electromagnetic rays such as gamma rays emanate from the nuclear disintegration of certain radioactive isotopes (Co^{60}, Cs^{137}). They have more penetrative power than electron beams, but they require a longer exposure time. These high-energy radiations damage the nucleic acids of microorganisms. A dosage of 2.5 Mrad kills all bacteria, fungi, viruses, and spores. It is used commercially to sterilize disposable petri dishes, plastic syringes, antibiotics, vitamins, hormones, glassware, and fabrics. Disadvantages of using electromagnetic rays include the following: Unlike electron beams, they cannot be switched off. Glassware tends to become brownish, and there is loss of tensile strength in fabric. Gamma irradiation impairs the flavor of certain foods. *Bacillus pumilus* E601 is used to evaluate the sterilization process.

10.10 PRODUCT RECOVERY

The extraction and purification of fermentation products may be difficult and costly. Ideally, one is trying to obtain a high-quality product as quickly as possible at an efficient recovery rate using minimum plant investment and operation at minimal costs. Unfortunately, the recovery costs of microbial products may vary from as low as 15% to as high as 70% of the total manufacturing costs. Obviously, the chosen process, and therefore its relative cost, will depend on the specific product. The reported percentages of cost are 15% for industrial ethanol, 20%–30% for bulk penicillin G, and up to 70% for enzymes (Figure 10.24). The high cost of downstream processing will affect the overall objective in some fermentation processes. The choice of recovery process is based on the following six criteria:

1. Intracellular or extracellular location of the product
2. Concentration of the product in the fermentation broth

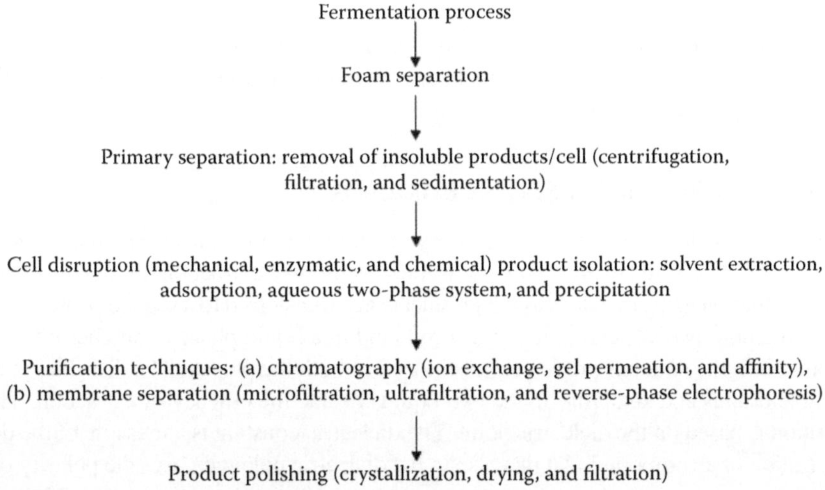

FIGURE 10.24 Flowchart of downstream processing.

3. Physical and chemical properties of the desired product (as an aid to selecting separation procedures)
4. Intended use of the product
5. Minimal acceptable standard of purity
6. Marketable price for the product

The main objective of the first stage in the recovery of an extracellular product is the removal of large solid particles and microbial cells, which is usually done by centrifugation or filtration (Figure 10.24). In the next stage, the broth is fractionated or extracted into major fractions using ultrafiltration, reverse osmosis, adsorption/ion exchange/gel filtration, affinity chromatography, liquid–liquid extraction, two-phase aqueous extraction, or precipitation.

10.10.1 Precipitation

Precipitation may be conducted at various stages of the product recovery process. It is a particularly useful process in that it allows enrichment and concentration in one step, thereby reducing the volume of material for further processing. It is possible to obtain some products (or to remove certain impurities) directly from the broth by precipitation or to use the technique after a crude cell lysate is obtained. Typical agents used in precipitation render the compound of interest insoluble and include the following:

- Acids and bases to change the pH of a solution until the isoelectric point (pI) of the compound is reached and pH equals pI, when there is no overall charge on the molecule and its solubility is decreased.
- Salts such as ammonium and sodium sulfates are used for the recovery and fractionation of proteins. The salts remove water from the surface of a protein, revealing hydrophobic patches that come together and cause the protein to precipitate. The most hydrophobic proteins will precipitate first, allowing fractionation to take place.
- Organic solvents, such as dextrans, can be precipitated out of a broth by the addition of methanol. Chilled ethanol and acetone can be used in the precipitation of proteins mainly because they cause changes in the dielectric properties of the solution.
- Nonionic polymers such as polyethylene glycol (PEG) can be used in the precipitation of proteins and are similar in behavior to organic solvents.
- Polyelectrolytes can be used in the precipitation of a range of compounds, in addition to their use in cell aggregation.
- Protein-binding dyes (triazine dyes) bind to and precipitate certain classes of proteins.
- Affinity precipitation is an area of much current interest in that precipitants are able to bind to, and precipitate, compounds selectively.

10.10.2 Product Recovery by Solvent Extraction

Solvent extraction is a classical method of recovery in addition to the concentration of various products. Solvent extraction has several advantages, such as the selectivity of extraction directly from broth or reaction medium, reduction in product loss as the product is just transferred to a second phase, and easy scale-up. Solvent extraction involves extraction of a compound in a liquid phase to another liquid. The solute originally present in the aqueous phase gets partitioned in both the phases. The distribution between the two immiscible liquids and solubility in the two liquids decide the efficacy of extraction. The choice of solvent selection is based on the dielectric point. The dielectric constant is a measure of the degree of the molar polarization of a compound. An increase in the dielectric pole increases the polarity of the solute:

$$D = C/C_0$$

where D is the dielectric constant, C is electrostatic capacity of a condenser containing the substance between the plates, and C_0 is electrostatic capacity of the same condenser when it is completely evacuated. The final choice of solvent is influenced by the "partition coefficient" (K):

$$K = \text{concentration of solute in the extract/concentration of solute in the raffinate}$$

A high value of k requires single-stage extraction and a low value of k requires multistage extraction. In single-stage batch extraction, the aqueous feed is mixed with the organic solvent, and after equilibration the extract phase containing the desired solute is separated out for further processing. In some cases, single-stage extraction may not be enough and a multistage process is required wherein a fresh volume of solvent is contacted with the raffinate. Continuous extraction can be carried out by cocurrent or countercurrent methods. In cocurrent extraction, there are n mixer vessels in line and the raffinate goes from vessel 1 to vessel n. Fresh solvent is added in each stage and the extracting solvent passes through the cascade of vessels in the same direction. At every stage, the extract is recovered.

In countercurrent extraction, the extracted raffinate passes from vessel 1 to vessel n while the product-enriched solvent flows from vessel n to vessel 1. This is the most efficient method of extraction. Solvent recovery following the extraction process is essential; it is usually done by distillation. Distillation is performed in three stages: (1) evaporation of the solvent into its vapor phase, (2) vapor–liquid separation, and (3) condensation to collect the solvent. The stages are described as follows:

1. Evaporation is the removal of solvent as a vapor from the solution
2. Vapor–liquid separation is performed in a column to separate the lower-boiling (more volatile) component from other less volatile components
3. Condensation of the vapor recovers the more volatile solvent fraction

Evaporation is the removal of a solvent from a solution by the application of heat to the solution. A wide range of evaporators are available. Some are operated on a batch basis, whereas others are operated continuously. Most industrial evaporators use tubular heating surfaces. Circulation of the liquid past the heating surfaces may be induced by boiling or by mechanical agitation. In batch distillation, vapor from the boiler passes up the column and is condensed. Part of the condensate will be returned as the reflux for countercurrent contact; with rising vapor in the column, the distillation is continued until a satisfactory recovery of lower-boiling (more volatile) components is accomplished. The ratio of condensate returned to the column as reflux to condensate withdrawn as product is, along with the number of plates or stages in the column, the major parameter for controlling product purity.

Continuous distillation is started in a similar way as batch distillation, but no condensate is withdrawn initially. There is total reflux of the condensate until ideal operating conditions are established throughout the column. At this stage, the liquid feed is fed into the column at an intermediate level. The more volatile components move upward as vapor and are condensed, followed by partial reflux of the condensate. Meanwhile, the less volatile fractions move down the column to the evaporator. At this part of the bottoms, a fraction is continuously withdrawn and the part is reboiled and returned to the column. Countercurrent contacting of the vapor and liquid streams is achieved by causing the vapor to be dispersed in the liquid phase and the liquid to be dispersed in a continuous vapor phase.

10.10.2.1 Distribution Coefficient

In the typical example of solvent (liquid–liquid) extraction described in Section 10.10.2, the product is a fairly large organic molecule, which one would predict would not be very soluble in water. On the other hand, if the product is a lower-molecular-weight or "small" molecule, one would predict that it might be at least partially soluble in water. Therefore, it might not completely "move" into the organic layer; it might also partially dissolve in the aqueous layer. For water-soluble organic materials, such as acetic acid or sugar, most of the solute will reside in the water phase. A quantitative measure of how an organic compound will distribute itself between aqueous and organic phases is called the "distribution coefficient" or partition coefficient. It is the ratio, K, of the solubility of solute dissolved in the organic layer to the solubility of material dissolved in the aqueous layer (note that K is independent of the actual amounts of the two solvents mixed):

$$K = \text{Distribution coefficient}$$

$$K = \frac{\text{Solubility of organic (g/100 mL)}}{\text{Solubility of water (g/100 mL)}}$$

The constant K is essentially the ratio of the concentrations of the solute in the two different solvents once the system reaches equilibrium. At equilibrium, the molecules naturally distribute themselves in the solvent where they are more soluble. Inorganic and water-soluble materials will stay in the water layer and more organic molecules will remain in the organic layer. By using the correct solvent system, a molecule can be specifically selected and extracted from another solvent.

Since the distribution coefficient is a ratio, unless K is very large not all molecules of a solute will reside in the organic layer in a single extraction. Usually two, three, or four extractions of the aqueous layer with an organic solvent are carried out in sequence in order to remove as much of the desired product as possible from the aqueous layer. The effectiveness of multiple small-volume extractions versus one large-volume extraction can be demonstrated using a simple calculation. Imagine that one extraction can recover 90% of the compound. A second extraction with the same solvent may be able to pull out 90% of the remaining material. Effectively, 99% of the compound is recovered with two extractions. One large extraction can obtain only the initial 90% of the compound. Many smaller extractions are more efficient than one large extraction. This phenomenon can be proved mathematically, using the following equation:

$$\text{Fraction extracted into solvent B} = \left(\frac{1}{1 + \frac{V_B}{V_A n K}} \right)$$

This equation provides the fraction of material extracted by solvent B, where n is the number of extractions performed, K is the distribution coefficient, V_A is the volume of solvent A, and V_B is the volume of solvent B. Distribution coefficients play a large role in the efficacy of a drug. In order for a drug to be absorbed into a brain cell, it must pass through what is called the blood–brain barrier into the brain cell. The drug must have enough water solubility to dissolve in the blood and be carried to the brain. However, to pass through the cell wall, which consists largely of water-insoluble fatty lipids with solubility properties similar to that of an organic solvent, the drug must have reasonable organic solvent solubility. Cell membranes use the same fundamental solubility principles as the extraction process. The cell membrane shown in Figure 10.25 consists of an ionic head and a very nonpolar or hydrophobic center.

The ionic head of the lipid orients itself in aqueous environments, creating a very nonpolar interior. Ions such as K^+ and Ca^{++} cannot traverse the interior of the cell readily because the interior is very nonpolar and does not support these ions. Extraction uses the same partitioning effect to isolate organic compounds. Just as in this biological example, in the extraction process organic compounds choose where to dwell according to the distribution coefficient. Synthetic drug design must take

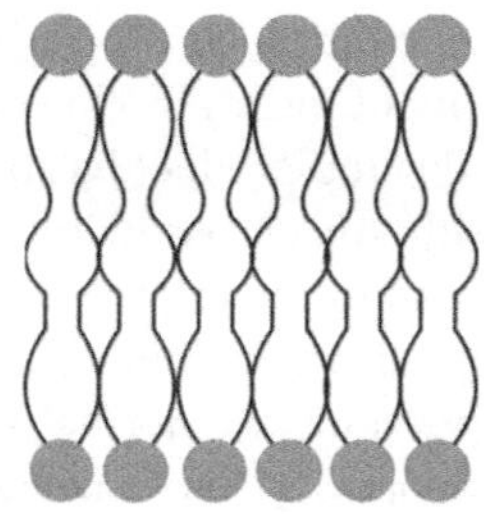

FIGURE 10.25 Simple schematic of a cell membrane.

into account the importance of having a distribution coefficient that allows drug transport both in aqueous blood and through organic membranes.

10.10.2.2 Solvent Selection

One important aspect when choosing a solvent system for extraction is to pick two immiscible solvents. Some common liquid–liquid extraction solvent pairs are water–dichloromethane, water–ether, and water–hexane. Notice that all the combinations include water. Most extractions involve water because it is highly polar and immiscible with most organic solvents. In addition, the compound you are attempting to extract must be soluble in the organic solvent but insoluble in the water layer. An organic compound such as benzene is simple to extract from water, because its solubility in water is very low. However, solvents such as ethanol and methanol will not separate using liquid–liquid extraction techniques, because they are soluble in both organic solvents and water.

There are also practical concerns when choosing extraction solvents. As mentioned previously, the two solvents must be immiscible. The cost, toxicity, and flammability of solvents should be considered. The volatility of the organic solvent is important. Solvents with low boiling points, such as ether, are often used to make isolating and drying the isolate material easier. If ether is used (boiling point = 35°C), then evaporation to collect the solid is fast.

One common mistake when performing an extraction is to mix up the layers and discard the wrong one. The densities of the solvents will predict which solvent is the top or bottom layer. In general, the densities of nonhalogenated organic solvents are less than 1.0 g/mL and the densities of halogenated solvents are greater than 1.0 g/mL. One common solvent pair is dichloromethane and water. The density of dichloromethane is 1.325 g/mL and that of water is 1.000 g/mL. Dichloromethane is more dense that water; therefore, dichloromethane will be the bottom layer and water will be the top layer. Table 10.3 lists the densities of some extraction organic solvents. Although density is the physical property that determines which layer is the top and which is the bottom, a very concentrated solute dissolved in either layer can reverse the order. The best method to avoid making a mistake is a drop test. Add a few drops of water to the layer in question and watch the drop very carefully. If the layer is water, then the drop will mix with the solution. If the solvent

TABLE 10.3
Common Extraction Solvents Listed by Density

Solvent	Density (g/mL)
Hexane	0.695
Ether	0.708
Toluene	0.867
Water	1.000
Dichloromethane	1.325
Chloroform	1.492

is the mistaken organic layer, then the water drop will create a second layer. In general, this method can help to determine the identity of the layer. However, it is still best to keep all the layers until the extraction is complete and the desired product has been isolated.

One significant problem with solvent extraction is that no solvent is completely insoluble in another solvent. In practice, one additional step is usually carried out before evaporating the organic solvent, that is, drying over anhydrous sodium sulfate or some other drying agent. Drying a liquid might seem a peculiar concept, since we normally think of all liquids as being wet; drying an organic liquid in the organic laboratory has a special meaning to chemists. It means to remove all traces of water from the liquid. Even water and hexane are slightly soluble in each other (Table 10.3).

After separating the two solvents, residual water remains in the hexane or ether organic layer. This will remain and stick to the solid product when we remove the more volatile solvent. Therefore, chemists remove the water from the organic layer by adding an insoluble inorganic solid to the solution that absorbs the water, thereby "drying" it. Granular anhydrous sodium sulfate is the drying agent most often used, although other drying agents are also available. All inorganic solids work by reacting with the water to form hydrates, which is their preferred form if water is available:

$$Na_2SO_4 + H_2O \rightarrow Na_2SO_4.5H_2O$$

$$MgSO_4 + H_2O \rightarrow MgSO_4.7H_2O$$

$$CaCl_2 + H_2O \rightarrow CaCl_2.6H_2O$$

$$2CaSO_4 + H_2O \rightarrow (CaSO_4)_2.H_2O$$

The aforementioned compounds will associate or hydrate themselves with water. Table 10.4 lists some common drying agents along with their speed, capacity, and hydration. These drying agents do not dissolve in the solvent they are drying. They may change somewhat, for example, sodium sulfate will clump together as it reacts with water, but they will remain solids in normal extraction solvents. This makes them easy to remove by decantation (pouring off of the liquid) or by gravity filtration. Usually, the organic solvent will go from cloudy to clear in the process of being dried. You should be careful to remove all these solid drying agents before solvent evaporation takes place or you might confuse them with your product. When you take a melting point and the product does not melt by 300°C, you have probably isolated your drying agent. It is recommended that the drying agent chosen is in a granular form. After the drying agent has removed the residual water, it is easier to remove large granular particles. Drying a solvent, however, is not an exact science. An excess of drying agent should be used to ensure that all the water is removed. If water remains after the materials are collected, it could interfere with the analysis. Keep adding the drying agent until there are no more clumps of drying agent stuck to the sides or bottom of the flask. Drying agents should float freely in the beaker, similar to snow.

TABLE 10.4
Hydrated Complexes for Some Common Drying Agents

Drying Agent	Formula	Speed	Capacity[a]	Hydration[b]
Sodium sulfate	Na_2SO_4	Medium	High	7–10
Magnesium sulfate	$MgSO_4$	Fast	High	7
Calcium chloride	$CaCl_2$	Fast	Low	2
Calcium sulfate (Drierite)	$CaCl_2$	Fast	Low	½–2

[a] Capacity refers to the amount of water removed per given weight of the drying agent.
[b] Hydration is the number of water molecules removed per molecule of the drying agent.

10.10.2.3 Acid–Base Extraction

There are three special cases of liquid–liquid extraction that are extremely useful for isolating and purifying amines, carboxylic acids, and phenols. All three of these functional groups can be interconverted from nonionic organic solvent–soluble forms to water-soluble ionic forms by changing the pH:

Amines:	RNH_2	$+ H^+$	$\Leftrightarrow$	RNH_3^+	
Phenols:	PhOH	$+ OH^-$	$\Leftrightarrow$	$PhO^- +$	H_2O
Carboxylic acids:	RCOOH	$+ OH^-$	$\Leftrightarrow$	$RCOO^- +$	H_2O

Solid–liquid or liquid–liquid extractions rely on the solubility of the solute to be extracted. In acid–base extraction, the molecule to be extracted is transformed so that we impose a new solubility on the molecule. One specific example is benzoic acid, which is an organic acid. Benzoic acid is soluble in most organic solvents, including dichloromethane and ether. However, this acid can be easily deprotonated with a base to give a charged ionic species that is readily soluble in water:

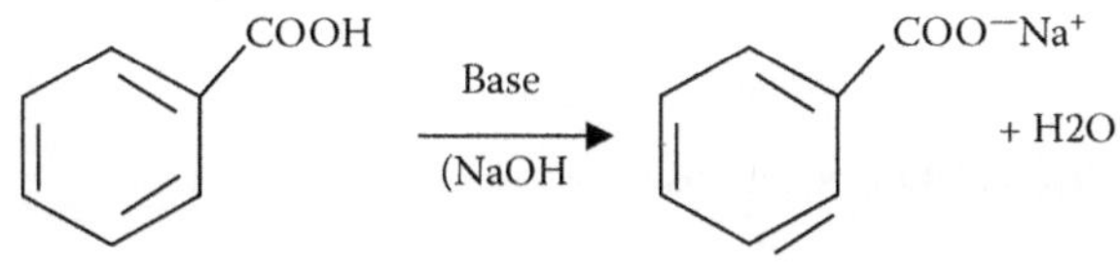

By converting benzoic acid to its sodium salt, the solubility is changed drastically. Now the sodium salt is soluble in water and will migrate to the water layer. Because the solvents chosen are immiscible in each other, the layers can be easily separated. Although the separation is complete, we no longer have benzoic acid. To obtain the original compound, the salt must be protonated with a strong inorganic acid. Once benzoic acid is recovered by adding the inorganic acid, it will precipitate in the water to provide a pure compound. This method works very well with mixtures of strong organic acids, weak organic acids, bases, and neutral compounds. We can use the acid–base functionality to our advantage. By changing the pH of the aqueous phase in a liquid–liquid extraction, the distribution coefficient is drastically changed, pulling the molecules into either an organic layer or an aqueous layer at will. Carboxylic acids, phenols, and amines can be easily separated from neutral components. However, all other common functional groups remain unaffected by changes in aqueous pH, so they will always distribute between layers in the same way because their distribution coefficients are unaffected by pH. Figure 10.26 details the reagents needed to separate benzoic acid, aniline, and phenol.

Whether you use acid–base, solid–liquid, or liquid–liquid, extraction is a useful organic tool to separate a mixture of compounds. From the early drugs that were extracted from trees and plants to modern day pharmacology, extraction is used to separate and purify organic molecules.

10.10.3 Ion Exchange

Ion exchange chromatography is based on synthetic resins of an aromatic nature and were suitable for the separation and purification of inorganic ions and small molecules.

Ion exchange is a reversible chemical reaction in which an ion (an atom or molecule that has lost or gained an electron and has thus acquired an electrical charge) from a solution is exchanged for a similarly charged ion attached to an immobile solid particle. These solid ion-exchange particles are either naturally occurring inorganic zeolites or synthetically produced organic resins. The synthetic organic resins are the predominant type used today because their characteristics can be tailored to specific applications. An organic ion-exchange resin comprises high-molecular-weight polyelectrolytes that can exchange their mobile ions for ions of similar charge from the surrounding medium. Each resin has a distinct number of mobile ion sites that set the maximum quantity of exchanges per

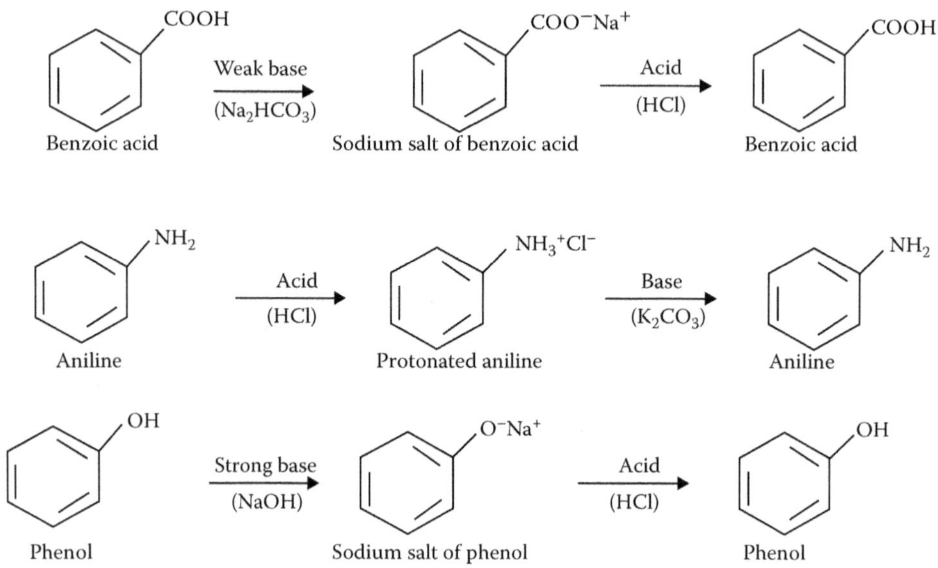

FIGURE 10.26 Acid–base extraction equations.

unit of resin. In most plating processes, water is used to cleanse the surface of the parts after each process bath. To maintain quality standards, the level of dissolved solids in the rinse water must be regulated. Fresh water added to the rinse tank accomplishes this purpose, and the overflow water is treated to remove pollutants and then discharged. As the metal salts, acids, and bases used in metal finishing are primarily inorganic compounds, they are ionized in water and can be removed by contact with ion-exchange resins. In the water-deionization process, the resins exchange hydrogen ions (H^+) for positively charged ions (such as nickel, copper, and sodium ions) and hydroxyl ions (OH^-) for negatively charged sulfates, chromates, and chlorides. Because the quantities of H^+ and OH^- ions are balanced, the result of the ion-exchange treatment is relatively pure, neutral water.

To ensure efficient, reproducible purification that gives the required degree of purity, it is beneficial to develop a multistep process using the purification strategy of capture, intermediate purification, and polishing (CIPP). IEX plays a significant and highly flexible role in most multistep purification schemes. If a specific affinity medium is not available or if little is known about the target molecule, IEX is recommended as the first step to be considered in any purification scheme. The technique can be used for capture, intermediate purification, or polishing, according to the demands of the specific application. Since IEX offers different selectivities (using anion or cation exchangers) and since the pH of the purification can be modified to alter the charge characteristics of the sample components, it is possible to use the technique more than once in the same purification scheme. In addition, IEX can be used with stepwise elution for a rapid capture step or with gradient elution to achieve the highest resolution in a polishing step (Figure 10.27).

Ion-exchange reactions are stoichiometric and reversible, and as such they are similar to other solution phase reactions. For example,

$$NiSO_4 + Ca(OH)_2 = Ni(OH)_2 + CaSO_4$$

In this reaction, the nickel ions of nickel sulfate ($NiSO_4$) are exchanged for the calcium ions of the calcium hydroxide [$Ca(OH)_2$] molecule. Similarly, a resin with hydrogen ions available for exchange will exchange those ions for nickel ions from solution. The reaction can be written as follows:

$$2(R - SO_3H) + NiSO_4 = (R - SO_3)2Ni + H_2SO_4$$

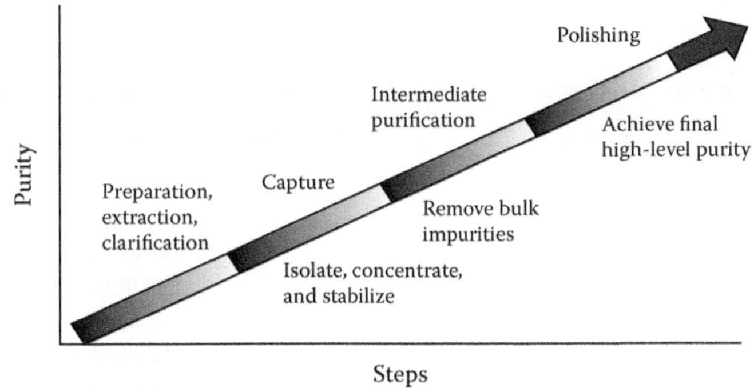

FIGURE 10.27 Various steps involved in ion-exchange purification.

R indicates the organic portion of the resin and SO_3 is the immobile portion of the ion-active group. Two resin sites are needed for nickel ions with a +2 valence (Ni^{+2}). Trivalent ferric ions require three resin sites. As shown, the ion-exchange reaction is reversible. The degree to which the reaction proceeds to the right will depend on the resin's preference or selectivity for nickel ions compared with its preference for hydrogen ions. The selectivity of a resin for a given ion is measured by the selectivity coefficient *K*, which in its simplest form for the reaction

$$R - A^+ + B^+ = R - B^+ + A^+$$

is expressed as

$$K = (\text{Concentration of } B^+ \text{ in resin/concentration of } A^+ \text{ in resin}) \times$$
$$(\text{concentration of } A^+ \text{ in solution/concentration of } B^+ \text{ in solution})$$

The selectivity coefficient expresses the relative distribution of ions when a resin in the A^+ form is placed in a solution containing B^+ ions. Table 10.5 gives the selectivities of strong acid and strong base ion-exchange resins for various ionic compounds. The selectivity coefficient is not constant but varies with changes in solution conditions. It does provide a means to determine what to expect when various ions are involved in a reaction. As indicated in Table 10.5, strong acid resins have a preference for nickel over hydrogen. Despite this preference, the resin can be converted back to the hydrogen form by contact with a concentrated solution of sulfuric acid (H_2SO_4):

$$(R\text{--}SO_4)2Ni + H_2SO_4 \rightarrow 2(R - SO_3H) + NiSO_4$$

This step is known as regeneration. In general terms, the higher the preference a resin exhibits for a particular ion, the greater the exchange efficiency in terms of resin capacity for removal of that ion from solution. Greater preference for a particular ion, however, will result in increased consumption of chemicals for regeneration. Resins currently available exhibit a range of selectivities and thus have broad application. The relative preference for divalent calcium ions (Ca^{++}) over divalent copper ions (Cu^{++}) is approximately 1.5–1. For a heavy metal–selective resin, this preference is reversed and favors copper by a ratio of 2300 to 1.

10.10.3.1 Resin Types

Ion-exchange resins are classified as cation exchangers, having positively charged mobile ions available for exchange, and anion exchangers, whose exchangeable ions are negatively charged. Both anion and cation resins are produced from the same basic organic polymers. They differ in the ionizable group attached to their hydrocarbon networks. It is this functional group that determines

TABLE 10.5
Selectivity of Ion-Exchange Resins in Order of Decreasing Preference

Strong Acid Cation Exchanger	Strong Base Anion Exchanger
Barium	Iodide
Lead	Nitrate
Calcium	Bisulfate
Nickel	Chloride
Cadmium	Cyanide
Copper	Bicarbonate
Zinc	Hydroxide
Magnesium	Fluoride
Potassium	Sulfate
Ammonia–sodium	

the chemical behavior of the resin. Resins can be broadly classified into strong or weak acid cation exchangers and strong or weak base anion exchangers (Table 10.5).

Strong acid cation resins are so named because their chemical behavior is similar to that of a strong acid. The resins are highly ionized in both acid ($R–SO_3H$) and salt ($R–SO_3Na$) forms. They can convert a metal salt to its corresponding acid by the following reaction:

$$2(R - SO_3H) + NiCl_2 \rightarrow (R - SO_4)_2 Ni + 2HCl$$

The hydrogen and sodium forms of strong acid resins are highly dissociated and the exchangeable Na^+ and H^+ are readily available for exchange over the entire pH range. Consequently, the exchange capacity of strong acid resins is independent of the pH of the solution. These resins are used in the hydrogen form for complete deionization; they are used in the sodium form for water softening (removal of calcium and magnesium). After exhaustion, the resin is converted back to the hydrogen form (regenerated) by contact with a strong acid solution, or the resin can be converted to the sodium form with a sodium chloride solution. Hydrochloric acid (HCl) regeneration would result in a concentrated nickel chloride (NiCl) solution.

In a *weak acid cation resin*, the ionizable group is a carboxylic acid group (COOH) as opposed to the sulfonic acid group (SO_3H) used in strong acid resins. These resins behave similarly to weak organic acids that are weakly dissociated. Weak acid resins exhibit a much higher affinity for hydrogen ions than do strong acid resins. This characteristic allows regeneration to the hydrogen form with significantly less acid consumption than is required for strong acid resins. Almost complete regeneration can be accomplished with stoichiometric amounts of acid. The degree of dissociation of a weak acid resin is strongly influenced by the solution's pH. Consequently, resin capacity depends in part on the pH of the solution. A typical weak acid resin has limited capacity below a pH of 6.0, making it unsuitable for deionizing acidic metal finishing wastewater.

Similar to strong acid resins, *strong base anion resins* are highly ionized and can be used over the entire pH range. These resins are used in the hydroxide (OH) form for water deionization. They react with anions in solution and can convert an acid solution to pure water:

$$R - NH_3OH + HCl \rightarrow R - NH_3Cl + HOH$$

Regeneration with concentrated sodium hydroxide (NaOH) converts the exhausted resin to the hydroxide form.

Weak base anion resins are similar to weak acid resins in that the degree of ionization is strongly influenced by the pH of the solution. Consequently, weak base resins exhibit minimum exchange capacity above a pH of 7.0. These resins merely sorb strong acids; they cannot split salts.

In an ion-exchange wastewater deionization unit, the wastewater passes through a bed of strong acid resin first. Replacement of the metal cations (Ni^{++}, Cu^{++}) with hydrogen ions lowers the pH of the solution. The anions (SO_4^{-2}, Cl^-) can then be removed with a weak base resin because the entering wastewater will normally be acidic and weak base resins adsorb acids. Weak base resins are preferred over strong base resins because they require less regenerant chemical. A reaction between the resin in the free base form and Cl proceeds as follows:

$$R - NH_2 + HCl \rightarrow R - NH_3Cl$$

The weak base resin does not have a hydroxide ion form as does the strong base resin. Consequently, regeneration needs to only neutralize the absorbed acid; it need not provide hydroxide ions. Weakly basic reagents such as ammonia (NH_3) or sodium carbonate that are not so expensive can be used.

Heavy metal–selective chelating resins behave similarly to weak acid cation resins but exhibit a high degree of selectivity for heavy metal cations. Chelating resins are analogous to the chelating compounds found in metal finishing wastewater, that is, they tend to form stable complexes with the heavy metals. In fact, the functional group used in these resins is an EDTA–Na compound. The resin structure in the sodium form is expressed as R–EDTA–Na. The high degree of selectivity for heavy metals permits separation of these ionic compounds from solutions containing high background levels of calcium, magnesium, and sodium ions. A chelating resin exhibits greater selectivity for heavy metals in its sodium form than in its hydrogen form. Regeneration properties are similar to those of a weak acid resin; the chelating resin can be converted to the hydrogen form with doses slightly greater than stoichiometric doses of acid because of the fortunate tendency of the heavy metal complex to become less stable under low pH conditions.

Potential applications of chelating resins include polishing to lower the heavy metal concentration in the effluent from a hydroxide treatment process and directly removing toxic heavy metal cations from wastewaters containing a high concentration of nontoxic, multivalent cations. Table 10.6 shows the preference of a commercially available chelating resin for heavy metal cations over calcium ions. (Chelating resins exhibit a similar magnitude of selectivity for heavy metals over sodium or magnesium ions.) The selectivity coefficient defines the relative preference exhibited by the resin for different ions. The preference for copper (Table 10.6) is 2300 times that for calcium. Therefore, when a solution that contains equal molar concentrations of copper and calcium ions is treated, at equilibrium the molar concentration of copper ions in the resin will be 2300 times the concentration of calcium ions. When a solution that contains calcium ions with a molar concentration 2300 times more than that of the copper ion concentration is treated, at equilibrium the resin would hold an equal concentration of the copper and calcium.

TABLE 10.6
Chelating Cation Resin Selectivity for Metal Ions

Metal	KM/Caa
Hg^{+2}	2800
Cu^{+2}	2300
Pb^{+2}	1200
Ni^{+2}	57
Zn^{+2}	17
Cd^{+2}	15
Co^{+2}	6.7
Fe^{+2}	4.7
Mn^{+2}	1.2
Ca^{+2}	1

Experiment: Study Fermentation Process for Lactic Acid Production by *Streptococcus thermophilus* and Cell Biomass

Principle

Fermentation is a biological and enzymatic process of conversion of carbohydrates into lactic acid, alcohols, carbon dioxide, and so on, by some bacteria and fungi. These organisms use fermentation as a method of obtaining energy in the form of ATP. It is a cellular process that transfers the energy in glucose bonds to bonds in ATP.

The energy in ATP can then be used to perform cellular work. Fermentation is an anaerobic (without utilizing oxygen) process; cellular respiration is an aerobic (utilizing oxygen) process. All living organisms, including bacteria, protists, plants, and animals, produce ATP by fermentation or cellular respiration and then use ATP in their metabolism. The process of fermentation followed by *Streptococcus thermophilus* is homolactic fermentation; in this process, one mole of glucose gets converted into two moles of lactic acid:

$$C_6H_{12}O_6 \rightarrow 2\ CH_3CHOHCOOH$$

Before the fermentation starts, a molecule of glucose must split into two moles of pyruvate; this process is known as glycolysis. Fermentation and cellular respiration involve oxidation-reduction reactions (redox reactions). Redox reactions are always defined in terms of electron transfers, with oxidation being the loss of electrons and reduction the gain of electrons. In cellular respiration, two hydrogen atoms are removed from glucose (oxidation) and transferred to a coenzyme called NAD^+, reducing this compound to NADH (Figure 10.28).

Requirements

A small laboratory-grade fermentor, balance, autoclave, cooling centrifuge, pH meter, hot plate.

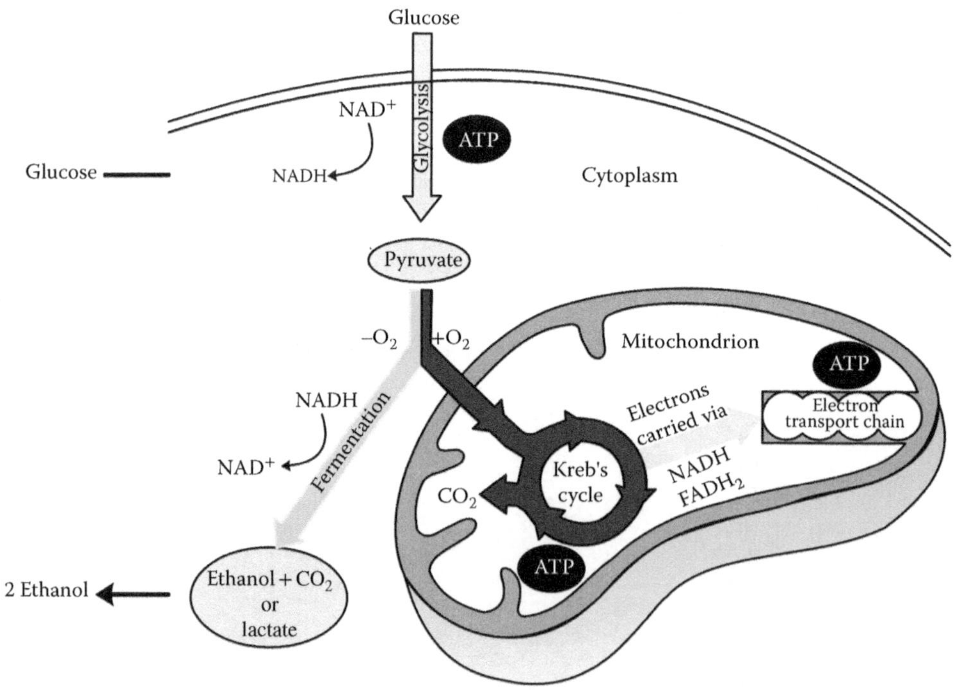

FIGURE 10.28 Stages of cellular respiration and fermentation.

Media Composition

Ingredients	Gram per Liter
Universal peptone	10
Meat extract	5
Yeast extract	5
Glucose	20
Dipotassium hydrogen phosphate	2
Diammonium hydrogen citrate	2
Sodium acetate	5
Magnesium sulfate	0.1
Manganous sulfate	0.05

Final pH = 6.5 ± 0.2 at 37°C.

Procedure

1. Prepare 1 LMRS broth and pour it into a fermentor vessel.
2. Autoclave media at 121 psi for 20 minutes in the vessel and allow it to cool overnight.
3. Inoculate an actively growing culture of *S. thermophilus* to the fermentor vessel and initiate the fermentation process.
4. Allow the process to run for 24 hours.
5. Harvest the culture from the fermentor by centrifugation at 4000 rpm.

Fermentation Process

The inoculation of *S. thermophilus* should always be done carefully under total aseptic conditions through the inoculation port on the top plate of the fermentor with the help of an inoculation syringe. After inoculation, the pH should be adjusted to 7; allow the fermentation process to run for 24 hours for complete cell multiplication, leading to high cell density or mass production. The conditions required for *S. thermophilus* to grow in a uniform manner are as follows:

- Temperature should be not less than 36°C and not more than 41°C.
- The pH of the media should be between 6.5 and 7.2.
- A very low amount of oxygen can be supplied as the organism is a facultative anaerobe.

The pH of the medium starts dropping as the microbe enters the log phase of the bacterial growth curve because of the production of lactic acid, which is an end product of the homofermentative process of the microbe to gain energy for survival (Figures 10.29 and 10.30).

Experiment: Sugar Fermentation by Yeast (*Saccharomyces cerevisiae*) for Production of Ethanol and Cell Biomass

Principle

Yeast can metabolize sugar in two ways: (1) aerobically, with the aid of oxygen, and (2) anaerobically, without the aid of oxygen. When yeasts respire aerobically, oxygen gas is consumed at the same rate that CO_2 is produced; there will be no change in gas pressure in the test tube. However, when yeasts ferment the sugars anaerobically, CO_2 production will cause a change in the pressure of a closed test tube since no oxygen is being consumed. This pressure change monitors the fermentation rate and metabolic activity of the organism. The fermentation of glucose can be described by the following equation (Figure 10.31):

$$\underset{\text{Glucose}}{C_6H_{12}O_6} \rightarrow 2\,\underset{\text{Ethanol}}{CH_3CH_2OH} + 2\,\underset{\text{Carbon dioxide}}{CO_2} + \text{ENERGY}$$

Note that alcohol is a by-product of this fermentation.

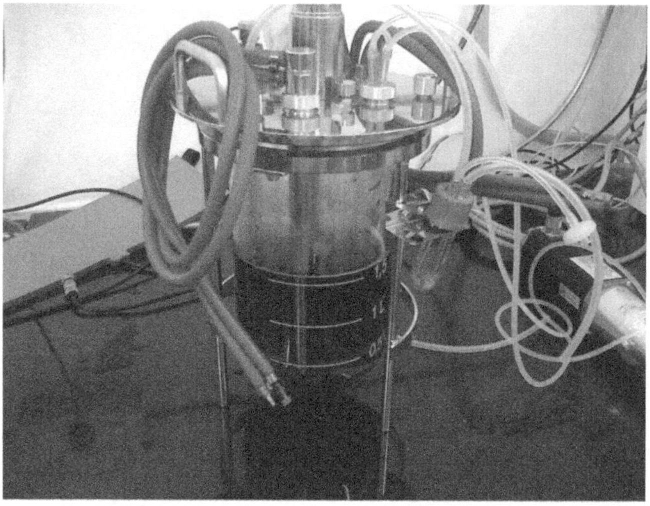

FIGURE 10.29 **(See color insert.)** A laboratory-grade fermentor of capacity 2 L containing MRS broth.

FIGURE 10.30 **(See color insert.)** A fermentor after 24 hours of incubation. The visible color difference in the media shows a high cell density of *S. thermophilus.*

Requirements

A 50-mL centrifuge tube, glucose reagent, yeast inoculum, distilled water.

Procedure

1. Take two centrifuge tubes and add glucose to each centrifuge tube. Add 1, 2, 3, and 4 g of glucose to achieve 5%, 10%, 15%, and 20% w/v glucose concentrations, respectively.
2. Add distilled water to increase the volume to 20 mL.
3. Add 150 μL of yeast inoculum to the glucose solution. Yeast inoculum is prepared by incubating active dry yeast with distilled water for 20 minutes in a water bath at 32°C at 120 rpm.
4. Weigh the tubes. This is the zero-hour reading.
5. Loosen the caps of the centrifuge tubes. Keep them in a water bath at 32°C with shaking at 100 rpm.
6. Weigh the tubes again and record the time.

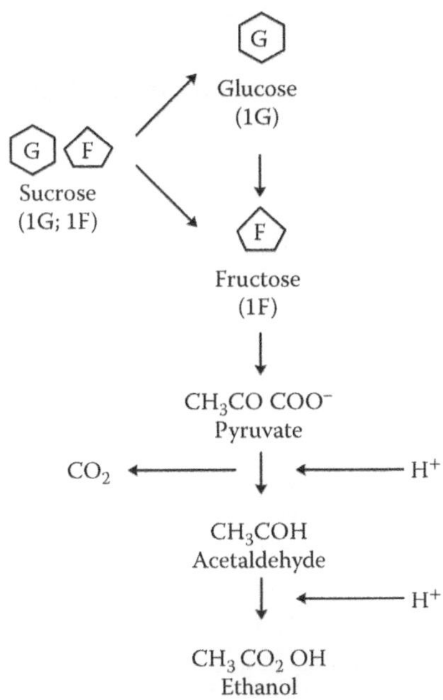

FIGURE 10.31 Sugar fermentation process for the production of ethanol and cell biomass.

7. On the next day, weigh the tubes after 24 hours (according to the fermentation start time). Necessary equations:

$$W_{co_2} = W_t - W_0$$

$$m_{EtoH} = m_{CO_2} = W_{CO_2} / M_{CO_2}$$

$$A_{EtoH} = m_{EtoH} \times M_{EtoH}$$

In these equations, W_{CO_2} is mass of CO_2 produced at a particular time period, m_{CO_2} denotes moles of CO_2 produced, m_{EtOH} denotes moles of ethanol produced, M_{CO_2} is the molecular weight of CO_2, M_{EtOH} is the molecular weight of ethanol, and A_{EtOH} is amount of ethanol produced in grams.

Data Analysis Procedure

1. Determine the amount of ethanol produced (in grams) at respective time periods for four different glucose concentrations in the fermentation media.
2. Plot the amount of ethanol produced versus the fermentation time for the four glucose concentrations (Table 10.7).
3. Discuss the correlation of total ethanol produced with glucose concentrations.

Yeast Count Exercise

Perform yeast count using hemocytometer technique.

Equipment and Materials

Dry yeast, distilled water, methylene blue solution

TABLE 10.7
Effect of Different Concentration of Glucose on Ethanol Production by Yeast

Glucose Concentration (% w/v)	Time (hours)	Weight of Centrifuge Tube (W_{time})	Calculated Loss in Weight (W_t–W_0)	Amount of Ethanol Produced (g)
5	0	W_0=		
	8	W_8=		
	24	W_{24}=		
	40	W_{40}=		
10	0	W_0=		
	8	W_8=		
	24	W_{24}=		
	40	W_{40}=		
15	0	W_0=		
	8	W_8=		
	24	W_{24}=		
	40	W_{40}=		
20	0	W_0=		
	8	W_8=		
	24	W_{24}=		
	40	W_{40}=		

Procedure

1. Yeast inoculum is prepared by adding 5 g of dry yeast and 25 mL of distilled water.
2. This solution is incubated in a water bath at 32°C at 120 rpm for 20 minutes.
3. After incubation, the yeast solution is diluted 1000 times.
4. To 1000 μL of diluted yeast solution, methylene blue dye is added. Dead yeast cells become blue in color, whereas live cells retain the original mustard yellow color.
5. Pour 100 μL of the aforementioned solution into the hemocytometer well.
6. Count the yeast cells in the four corner squares of the hemocytometer.
7. Yeast concentration will be determined based on the dimensions of the hemocytometer square and cells counted.

Calculations

$$\text{Total live cell concentration in cells per milliliter} = \frac{\text{number of live cells} \times \text{dilution factor}}{\text{volume of corner square}}$$

$$\text{Total dead cell concentration in cells per milliliter} = \frac{\text{number of dead cells} \times \text{dilution factor}}{\text{volume of corner square}}$$

$$\text{Volume of the corner square per milliliter} = 16 \times \text{volume of the smallest square}$$

Experimental Procedures

A. Determining Cell Concentration

1. Zeroing the spectrophotometer: Set the wavelength to 630 nm. Using the knob on the left, set the reading to 0% transmission when the chamber is empty; using the knob on the right, set the reading to 100% transmission when the chamber contains a test tube with about 4 mL of pure medium.

2. Determining cell concentration: First, flush out the sample tube for your group's fermentor by taking an 8–10 mL sample, which you will then discard. Take another 8–10 mL sample from your group's fermentor and gently mix it. Take about 4 mL from your sample tube and transfer it to a glass test tube. Clean the outside of the test tube with ethanol, insert it into the spectrophotometer, and record the absorbance reading. If the absorbance reading is greater than 0.25, which is a typical limit of linear correlation between absorbance and cell mass concentration, dilute the sample with a known amount of pure medium and measure the absorbance again to check if the absorbance reading is on the linear portion of the calibration curve. Record time for the sample taken along with the absorbance reading in the linear range, as well as the dilution factor.

B. Glucose Measurement

Take your undiluted yeast sample tube and insert into it the thin plastic sample tubing from the glucose analyzer. If the analyzer is in "Standby" mode, hit the "Run" key; after the machine is calibrated, hit the "Sample" key. If the analyzer is not in the standby mode, just hit the "Sample" key. The glucose concentration in the medium from your sample will be printed out in about 1 minute. Record this value.

C. Ethanol Assay

The ethanol concentration is determined using a simple and quick chemical assay. This assay is based on the following reaction:

$$\text{Ethanol} + \text{NAD} \xrightarrow{\text{alcohol dehydrogenase}} \text{acetaldehyde} + \text{NADH}$$

The amount of NADH produced in this reaction is proportional to the amount of ethanol added as a substrate. This quantity of NADH is determined spectrophotometrically at 340 nm. The ethanol assay reagent for this assay contains NAD and alcohol dehydrogenase.

1. Take 0.5 mL from your undiluted sample, pipette it into an Eppendorf tube, and spin at 14,000 rpm for 5 minutes.
2. Add 2 mL of ethanol assay reagent to each of three cuvettes.
3. Add 10 μL of pure medium to the first cuvette; this is your blank. Cover with a parafilm and mix gently.
4. Add 10 μL of the ethanol standard (0.08% w/v) to the second cuvette; this is your standard. Cover with a parafilm and mix gently.
5. Add 10 μL of sample (from the supernatant from step 4) to the third cuvette; this is your sample. Cover with a parafilm and mix gently.
6. Incubate the cuvettes at room temperature for 10 minutes.
7. Clean the outside of the cuvettes, and then read the absorbance of each cuvette at 340 nm; use the blank to zero the spectrophotometer. If the sample absorbance is greater than 1.7, dilute the supernatant from your sample to 1:4 and perform the assay again using this diluted supernatant. Do not redo the blank and the standard.
8. Calculate the ethanol concentration for your sample using the following formula:

$$\text{Ethanol (in grams per liter)} = \frac{A(\text{sample})}{A(\text{standard})} \times 0.8 \text{ (for the undiluted samples)}$$

Record this value.

SUGGESTED READING

Bailey, J. E., and D. Ollis. 1986. *Biochemical Engineering Fundamentals*, 2nd ed. New York: McGraw-Hill.

Casida, L. E. 1968. *Industrial Microbiology*, 460. Hoboken, NJ: John Wiley & Sons.

Creuger, W., and A. Crueger. 2004. *Biotechnology: A Textbook of Industrial Microbiology*, 2nd ed., 260. Sutherland, MA: Sinaur Assoc., Inc.

Doelle, H. W. 1994. *Microbial Process Development*. Singapore: World Scientific Publ.

Glazer, A. N., and H. Nikaido. 2007. *Microbial Biotechnology-Fundamentals of Applied Microbiology*, 2nd ed., 541. Cambridge: Cambridge University Press.
Prescott, S. C. 2007. *Industrial Microbiology*. India: Agrobios.
Stanbury, P. F., A. Whitaker, and S. J. Hall. 1999. *Principles of Fermentation Technology*, 376. Oxford, UK: Butterworth-Heinemann.

IMPORTANT LINKS

1. Fermentor: http://www.gelifesciences.com/aptrix/upp01077.nsf/Content/na_blotting_site~nadetection~na_autoradiography_cassettes?opendocument&cmpid=ppcaw100
2. ALF: http://www.electrolab.biz/Product_List/FerMac_Air_Lift_Vessel.aspx
3. Fixed-bed fermentor: http://www.environmental-expert.com/products/fixed-bed-fermentor-101742/view-comments
4. Incinerator: http://www.nikaengg.com/incinerators.htm

11 Conductivity Meter

11.1 INTRODUCTION

Conductivity is the ability of a solution, metal, or gas—in brief, all materials—to pass an electric current. In solutions, the current is carried by cations and anions, whereas in metals it is carried by electrons. How well a solution conducts electricity depends on a number of factors, namely concentration, mobility of ions, valence of ions, and temperature. All substances possess some degree of conductivity. In aqueous solutions, the level of ionic strength varies from the low conductivity of ultra pure water to the high conductivity of concentrated chemical samples.

11.2 PRINCIPLES

Conductivity is typically measured in aqueous solutions of electrolytes. Electrolytes are substances containing ions, that is, solutions of ionic salts or of compounds that ionize in a solution. The ions formed in a solution are responsible for carrying the electric current. Electrolytes include acids, bases, and salts, and they can be either strong or weak. Most conductive solutions measured are aqueous solutions, as water has the capability of stabilizing the ions formed by a process called solvation (see Figure 11.1).

11.2.1 Strong Electrolytes

Strong electrolytes are substances that are fully ionized in a solution. As a result, the concentration of ions in a solution is proportional to the concentration of the electrolyte added. They include ionic solids and strong acids, for example, HCl. Solutions of strong electrolytes conduct electricity because positive and negative ions can migrate largely independently under the influence of an electric field.

11.2.2 Weak Electrolytes

Weak electrolytes are substances that are not fully ionized in solution. For example, acetic acid partially dissociates into acetate ions and hydrogen ions, so that an acetic acid solution contains both molecules and ions. A solution of a weak electrolyte can conduct electricity, but usually not as well as a strong electrolyte because there are fewer ions to carry the charge from one electrode to the other.

11.3 COMMON DEFINITIONS

11.3.1 Resistance

This is the reciprocal of the conductivity value and is measured in ohm-cm. It is generally limited to the measurement of ultrapure water, the conductivity of which is very low. The resistance of the solution (R) can be calculated using Ohm's law ($V = R \times I$):

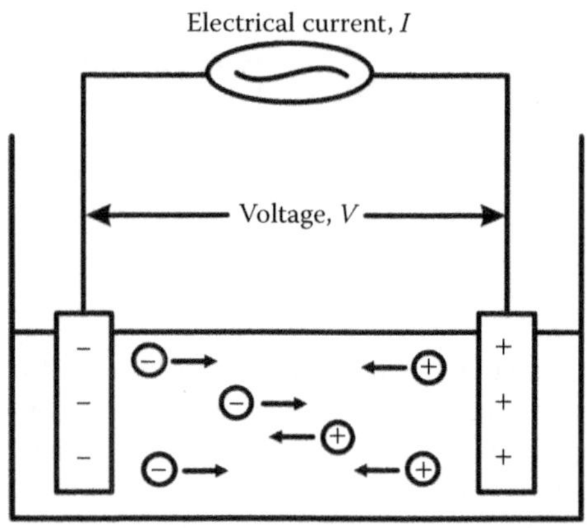

FIGURE 11.1 Migration of ions in a solution.

$$R = V/I$$

where

V = voltage (volts)
I = current (amperes)
R = resistance of the solution (ohms)

11.3.2 Conductance

Electricity is the flow of electrons. This indicates that ions in a solution will conduct electricity. Conductivity is the ability of a solution to pass an electric current. The conductivity reading of a sample will change with temperature. Conductance (G) is defined as the reciprocal of the electrical resistance (R) of a solution between two electrodes:

$$G = 1/R(S)$$

$$\kappa = G \bullet K$$

where

κ = conductivity (S/cm)
G = conductance (S), where $G = 1/R$
K = cell constant (cm^{-1})

The principle by which instruments measure conductivity is simple: two plates are placed in a sample, a potential is applied across the plates (normally a sine wave voltage), and the current is measured. Conductivity (G), the inverse of resistivity (R), is determined from the voltage and current values according to Ohm's law:

$$G = I/R = I(\text{amps})/E(\text{volts})$$

The conductivity meter measures the conductance and displays the reading converted into conductivity. Since the charge on ions in a solution facilitates the conductance of an electric current, the conductivity of a solution is proportional to its ionic concentration. In some situations, however,

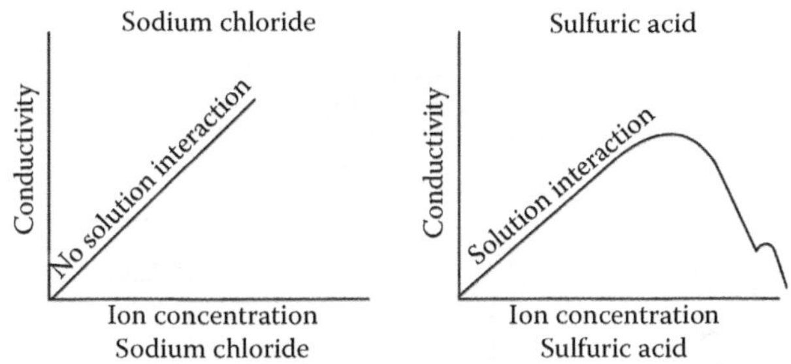

FIGURE 11.2 Conductivity of sulfuric acid and sodium chloride/ion concentration.

TABLE 11.1
Conductivity of Some Common Solutions

Solution	Conductivity
Absolute pure water	0.055 μS/cm
Power plant boiler water	1.0 μS/cm
Good city water	50 μS/cm
Ocean water	53 mS/cm
Reverse osmosis water	50–100 μS/cm
Domestic tap water	500–800 μS/cm

conductivity may not correlate directly to concentration. The graphs in Figure 11.2 illustrate the relationship between conductivity and ion concentration for two common solutions. Notice that the graph is linear for sodium chloride solution, but not for highly concentrated sulfuric acid. Ionic interactions can alter the linear relationship between conductivity and concentration in some highly concentrated solutions.

The water-softening industry refers to "grains" of hardness and uses TDS (total dissolved solids) as a measurement scale. While TDS is really a gravimetric measurement, since in a solution the solids are predominately present in ionic form, they can be approximated with conductivity. The TDS scale uses 2 μS/cm = 1 ppm (part per million as $CaCO_3$). It is also expressed as 1 mg/L TDS. While the method of measurement is the same, some conductivity meters can make the conversion and express the results of a measurement in many different units. This is helpful for users who are accustomed to one particular unit of measurement (Table 11.1).

11.4 CONDUCTIVITY METER

A typical conductivity meter applies an alternating current (I) at an optimal frequency* to two active electrodes and measures the potential (V). Both the current and the potential are used to calculate the conductance (I/V). The conductivity meter then uses the conductance and cell constant to display the conductivity (Figure 11.3):

$$\text{Conductivity}^{\dagger} = \text{cell constant} \times \text{conductance}$$

Note that the current source is adjusted so that the measured potential (V) is equal to the reference potential ($\mathbf{E}_r$) (approximately ±200 mV).

* The measuring frequencies of a typical conductivity meter are 94 Hz in the 4.000–40.00 μS range and 46.9 kHz in the 400.0 mS–2.000 S range.

† At sample temperature.

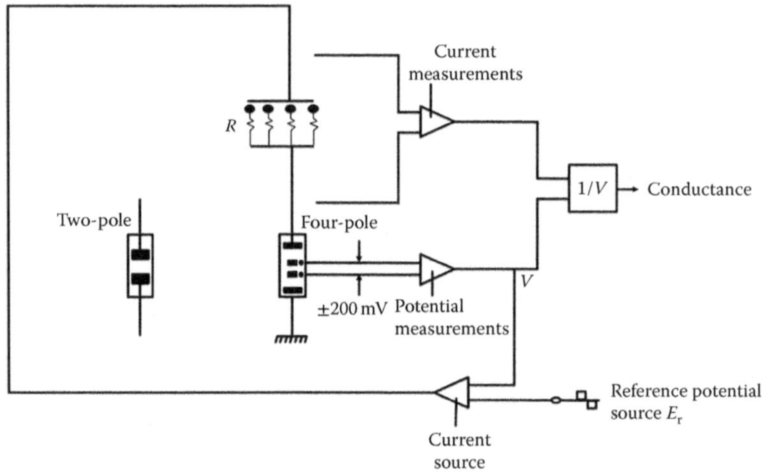

FIGURE 11.3 Diagram of a simplified conductivity meter.

TABLE 11.2
Optimum Conductivity Range for Cells of Three Different Constants

Cell Constant	Optimum Conductivity Range (μS/cm)
0.1	0.5–400
1.0	10–2000
10.0	1000–200,000

TABLE 11.3
Common Alpha Values of Some Common Solutions

Substance at 25°C	Concentration (wt%)	α
HCl	10	1.56
KCl	10	1.88
H_2SO_4	50	1.93
NaCl	10	2.14
HF	1.5	7.20
HNO_3	31	31.0

11.4.1 Cell Constant

This is the ratio of the distance (d) between the electrodes to the area (a) of the electrodes (see Table 11.2):

$$K = d/a$$

where

K = cell constant (cm^{-1})
a = effective area of the electrodes (cm^2)
d = distance between the electrodes (cm)

Common alphas (α) are listed in Table 11.3. To determine α of other solutions, simply measure conductivity at a range of temperatures and graph the change in conductivity versus the change in temperature. Divide the slope of the graph by $G_{t_{cal}}$ to get α.

11.5 CONDUCTIVITY CELLS

11.5.1 Two-Pole Cell

In a traditional two-pole cell (Figure 11.4), an alternating current is applied between the two poles and the resulting voltage is measured. The aim is to measure the solution resistance (R_{sol}) only. However, the resistance (R_{el}) caused by the polarization of the electrodes and the field effect interferes with the measurement, and both R_{sol} and R_{el} are measured.

11.5.2 Three-Pole Cell

The three-pole cell (Figure 11.5) is not as popular now as it has been replaced by the four-pole cell. The advantage of this design was that the third pole, which was linked to pole 1, allowed the field lines to be guided and confined in an optimal manner, limiting dispersion in the measurement and minimizing influences on the measurement such as beaker volume and the position of the cell in the beaker (field effect). It guarantees a better reproducibility when determining the cell constant and therefore more reproducible results.

11.5.3 Four-Pole Cell

In a four-pole cell (Figure 11.5), a current is applied to the outer rings (1 and 4) in such a way that a constant potential difference is maintained between the inner rings (2 and 3). As this voltage measurement takes place with a negligible current, these two electrodes are not polarized ($R_2 = R_3 = 0$). The conductivity will be directly proportional to the applied current. The geometry of four-pole cells with outer tubes minimizes the beaker field effect, due to the measurement volume being well defined within the tube. The position of the conductivity cell in the measuring vessel or the sample volume therefore has no influence on the measurement.

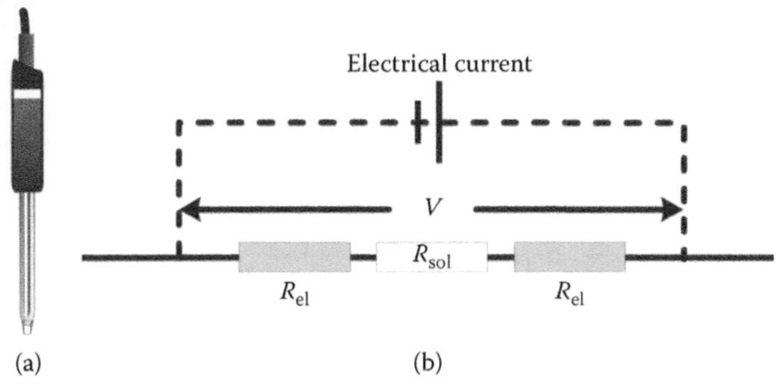

FIGURE 11.4 (a) Two-pole conductivity cell and (b) simplified diagram of a two-pole conductivity cell.

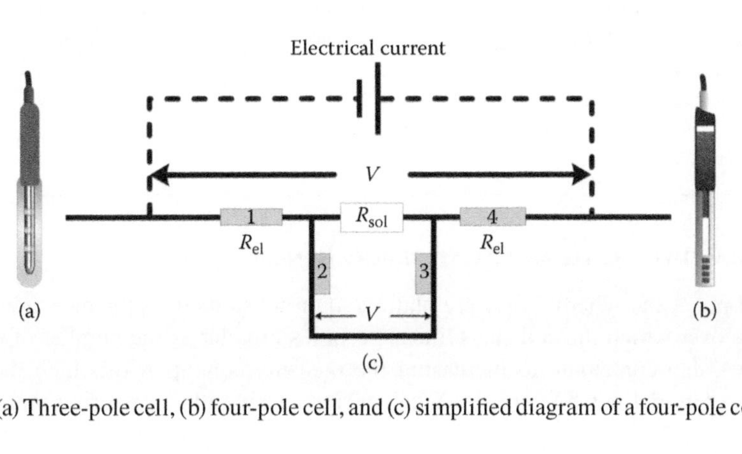

FIGURE 11.5 (a) Three-pole cell, (b) four-pole cell, and (c) simplified diagram of a four-pole conductivity cell.

11.5.4 Platinized Cells

Covering the cell poles (plates or rings) with a layer of platinum black is another way to minimize polarization effects and avoid error in measurements. The surface of the pole is increased and the current density is decreased; therefore, the polarization effect is less. Consequently, the cell constant is linear over 2–3 decades toward the higher conductivity range. The platinum black must not be damaged or scratched, as this will modify the surface of the poles and therefore the cell constant. However, one minor disadvantage of platinized cells is that the cell constant tends to drift faster than the constant of nonplatinized cells. It is advisable to use platinized cells only in nonviscous samples, without suspensions and to perform frequent calibrations.

11.5.5 Flow-Through Cell

Flow-through-type conductivity cells (Figure 11.6) are designed for flow measurements and measurements in small sample volumes. These measurements can be performed in a closed liquid system protected from the air. If a measurement is to be performed in pure water, it is necessary to use a flow cell. Contact with air must be avoided. The reason for this is that the carbon dioxide in the air forms hydrogen carbonate ions in water and leads to a change in the conductivity. A circulation cell can be used in two ways: (1) circulation, where the solution flows nonstop during the measurement, and (2) pipette, where a quantity of solution is drawn into the cell. This technique is ideal for small sample volumes.

11.5.6 Advantages and Disadvantages of Two-Pole and Four-Pole Cells

Two-Pole Cell

Advantages	Disadvantages
Easier to maintain	Field effects—cell must be positioned in the centre of the measuring vessel
Use with sample changer (no carryover)	Only cells with no bridge between the plates
Economical	Polarization in high-conductivity samples
Recommended for viscous media or samples with suspension	Calibrate using a standard with a value close to measuring value accurate over 2 decades

Four-Pole Cell

Advantages	Disadvantages
Linear over a very large conductivity range	Unsuitable for micro samples; depth of immersion 3–4 cm
Calibration and measurement in different ranges	Unsuitable for use with a sample changer
Flow-through- or immersion-type cells	
Ideal for high-conductivity measurements	
Can be used for low-conductivity measurements if the cell capacitance compensated	

11.5.7 Conductivity Cells and Measurement Ranges

The number of poles and whether they are platinized or not influences the measurement. The measurement range over which the cell stays linear becomes broader as the number of poles increases. Platinized poles also contribute to increasing the measurement span in which the cell is linear. Figure 11.7 shows guidelines for choosing a cell corresponding to a measuring range.

FIGURE 11.6 Flow-through cell.

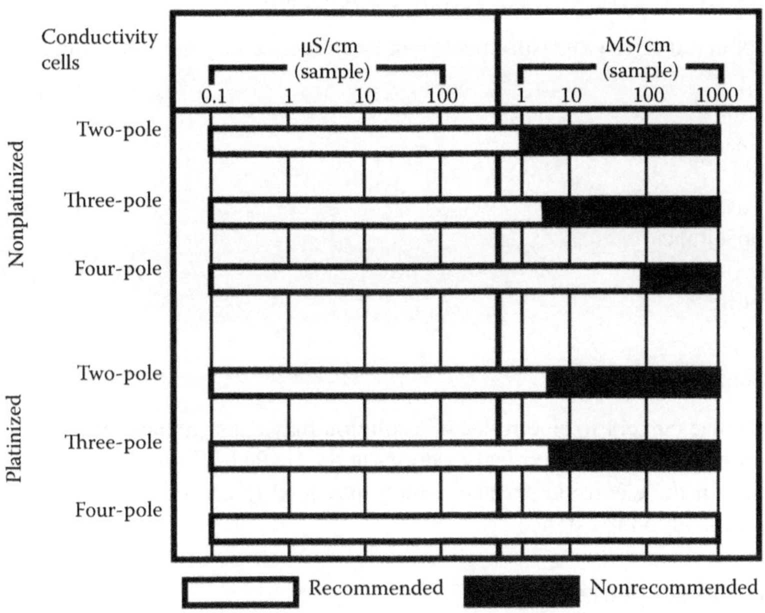

FIGURE 11.7 Guidelines for choosing a cell corresponding to a measuring range.

11.5.8 Calibration

Determination of the cell constant is required to convert conductance readings into conductivity results.

11.5.9 Standard Solution

A standard solution of known conductivity is used to calibrate the conductivity-measuring chain.

11.5.10 Reference Temperature

Conductivity readings are often referenced to a specific temperature, typically 20°C or 25°C, for comparative purposes.

11.5.11 Automatic Temperature Correction

Algorithms for automatic conversion of sample conductivity to a reference temperature are as follows:

Conductivity Values at 25°C	
Pure water	0.055 µS/cm
Deionized water	1 µS/cm
Rainwater	50 µS/cm
Drinking water	500 µS/cm
Industrial wastewater	5 mS/cm
Seawater	50 mS/cm
1 mol/L NaCl	85 mS/cm
1 mol/L HCl	332 mS/cm

11.6 FACTORS INFLUENCING THE MEASUREMENT

The accuracy of conductivity measurements can be influenced by the following factors:

- Polarization
- Contamination
- Geometry
- Cable resistance
- Cable capacitance
- Frequency change
- Temperature

11.6.1 Polarization

Applying an electric current to electrodes in a solution may cause an accumulation of ionic species near the electrode surfaces and chemical reactions at the surfaces (Figure 11.8). Thus, a polarization resistance arises on the electrode surface, which may lead to erroneous results as it is a parasitic component to the solution resistance.

11.6.1.1 Preventing Polarization

Polarization effects can be reduced or prevented by the following:

- Applying an alternating current: The measuring current will flow through the double layer capacitance (C_{dl}) of the electrodes instead of building up a voltage drop across the electrode surface due to the solution resistance (R_{sol}). R_{el} is then much smaller than R_{sol}.
- Adjusting the measuring frequency: The frequency has to be adapted to the conductivity of the sample. Low frequencies are applied at low conductivities, where polarization is negligible compared to the solution resistance (R_{sol}). High frequencies are applied at high conductivities, where the solution resistance (R_{sol}) is low in order to minimize the polarization resistance (R_{el}) (Figure 11.9).
- Optimizing the electrode areas: Increasing the active surface area of the electrodes with a layer of platinum black reduces the current density and consequently the polarization effect.
- Using a four-pole conductivity cell: The polarization resistance has no influence on the measurement.

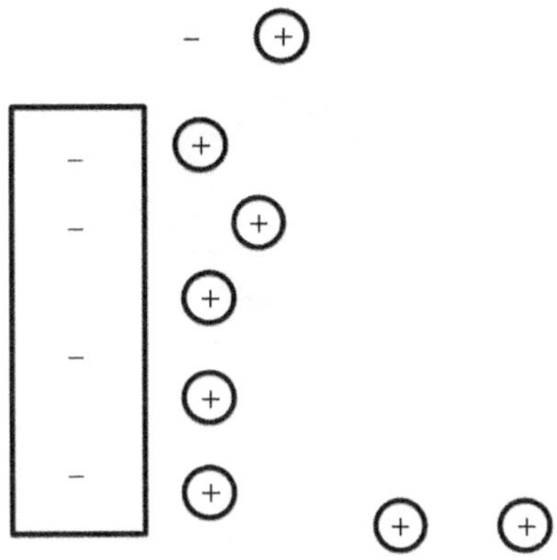

FIGURE 11.8 Accumulation of ionic species at the electrode surface.

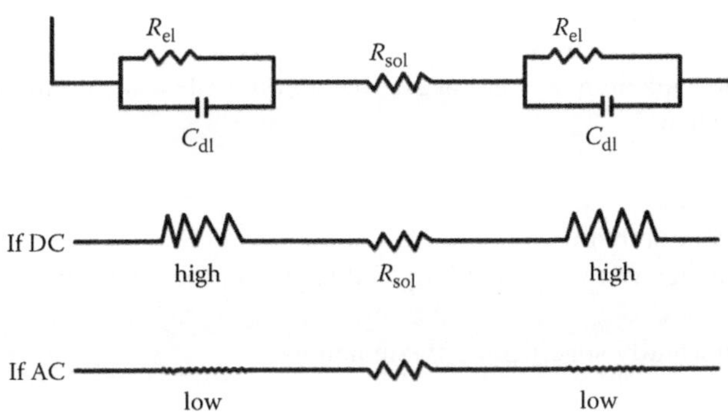

FIGURE 11.9 Polarization resistance.

11.6.2 Contamination of Electrode Surfaces

Deposits on the electrode surface of a two-pole cell have an effect similar to polarization errors, that is, they cause the conductivity reading to be lower than usual. These effects can also be prevented with a four-pole conductivity cell.

11.6.3 Geometry-Related Errors: Field Effects

Errors are also caused by field effects—the part of the measuring field that falls outside the geometric space of the two-pole cell (Figure 11.10). These field lines can affect the measurement if something interferes with the field lines, such as the beaker walls. Three- and four-pole conductivity cells are designed to minimize this effect. If the entire measuring field is contained within the body of the cell, then field errors cannot be caused by the beaker wall.

11.6.4 Cable Resistance

A cable has a given length, and therefore a given resistance. The cable resistance induces error on the result and must be taken into account. Compensate the cable resistance when

- Solution resistance is low (below 50 Ω), that is, for high-conductivity measurements
- Performing measurements using two- or three-pole cells

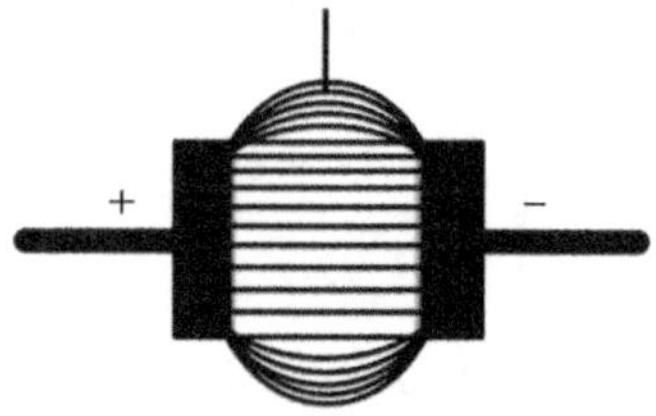

FIGURE 11.10 Geometry-related errors field effects.

The cable resistance of a cell is normally specified by the manufacturer. Note that for four-pole cells, the cable resistance has no influence. If a value is demanded during programming of the conductivity meter, enter zero.

11.6.5 Cable Capacitance

A shielded cable of a given length has a given capacity. When the measured conductance is low (below 4 μS), the cable capacitance is not negligible and must be taken into account. Compensate the cable capacitance when

- Using a four-pole cell
- Measuring low conductivities
- The cable capacitance of the conductivity cell is greater than 350 pF

The cable capacity is normally specified by the manufacturer.

11.6.6 Frequency Change

Low frequencies are applied at low conductivities, where the polarization resistance is negligible when compared to the resistance of the solution. They also contribute to reducing the cable capacitance effect, which is greater when the conductivity is low (high solution resistance).

High frequencies are applied at high conductivities, where the solution resistance is low. In most conductivity meters, the frequency is automatically increased with increasing conductance of the sample, to avoid polarization errors at high conductivity.

11.6.7 Temperature Effect

Conductivity measurements are temperature dependent: if temperature increases, conductivity increases. For example, the conductivity measured in a 0.01 D KCl solution at 20°C is 1.273 mS/cm, whereas at 25°C, it is 1.409 mS/cm. The concept of reference temperature was introduced to allow the comparison of conductivity results obtained at different temperatures. The reference temperature is usually 20°C or 25°C. The conductivity meter measures the actual conductivity and temperature and then converts it to the reference temperature using a temperature correction function and displays the conductivity at the reference temperature. It is mandatory to always associate the temperature with a conductivity result. If no temperature correction is applied, the conductivity is the value taken at the measurement temperature. For temperature correction, different options can be selected:

- Linear function
- Nonlinear function for natural waters according to ISO/DIN7888
- No correction

To perform correct conductivity measurements, use a temperature sensor or a conductivity cell with a built-in temperature sensor. For high-accuracy measurement, thermostat samples so that the same temperature is used for the calibration and measurement.

11.6.7.1 Linear Temperature Correction

In moderately and highly conductive solutions, temperature correction can be based on a linear equation involving a temperature coefficient (θ). The coefficient is usually expressed as a conductivity variation in %/°C. Linear temperature correction is used, for example, for saline solutions, acids, and leaching solutions:

$$\kappa_{T_{ref}} = \frac{100}{100+\theta(T-T_{ref})} \cdot \kappa_T$$

where

$\kappa_{T_{ref}}$ = conductivity at T_{ref}
κ_T = conductivity at T
T_{ref} = reference temperature
T = sample temperature
θ = temperature coefficient

Note that the correction is accurate only within a limited temperature range around T_1 and T_2. Figure 11.11 shows $T_1 = 26°C$, $T_2 = 14°C$, and $T_{ref} = 25°C$. The greater the difference between T and T_{ref}, the higher the risk of error.

11.6.7.2 Determination of the Temperature Coefficient (θ)

By measuring the conductivity of a sample at temperature T_1 close to T_{ref} and another temperature T_2, you can calculate the temperature coefficient by using the following equation:

$$\theta = \frac{(\kappa_{T_2} - \kappa_{T_1}) \cdot 100}{(T_2 - T_1) \cdot \kappa_{T_1}}$$

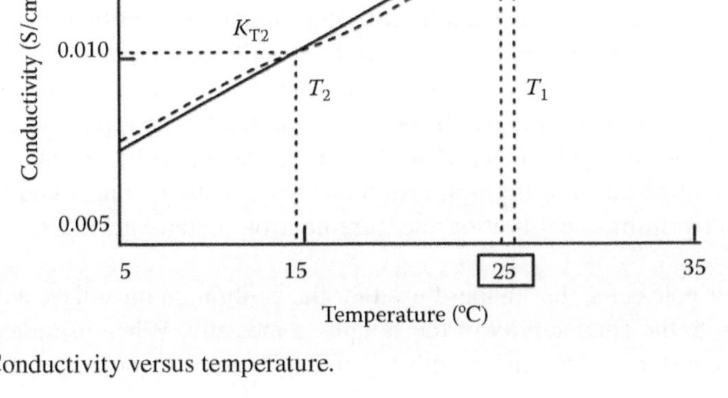

FIGURE 11.11 Conductivity versus temperature.

T_2 should be selected as a typical sample temperature and should be approximately 10°C different from T_1. The temperature coefficients of the following electrolytes generally fall into the ranges shown below

- Acids: 1.0–1.6%/°C
- Bases: 1.8–2.2%/°C
- Salts: 2.2–3.0%/°C
- Drinking water: 2.0%/°C
- Ultrapure water: 5.2%/°C

11.6.7.3 Nonlinear Temperature Correction

Linear temperature correction is not suitable for many aqueous liquids under test, and the temperature dependency can only be described by nonlinear functions such as the nonlinear function for natural waters, that is, for ground water, surface water, drinking water, and wastewater. The principle of this correction is that the conductivity measured at the sample temperature is corrected to 25°C to give K_{25}:

$$K_{25} = f_{25}(T) \cdot K_{\mathrm{T}}$$

$f_{25}(T)$ is the temperature correction factor used for the conversion of the conductivity values of natural water from T to 25°C. The conductivity meter calculates $f_{25}(T)$ from a four-degree polynomial equation. This equation fits conductivity variations against temperature for natural water stated in Natural Water Temperature Correction (ISO/DIN 7888). The nonlinear correction is defined by ISO/DIN 7888 standard and is applicable for measurements between 0°C and 35.9°C.

11.7 MEASURING TECHNIQUES

11.7.1 Determination of the Cell Constant

Calibration is important because the result is the correct value of the cell constant in your working conditions. The cell constant is a factor that is used to convert the measured conductance to conductivity:

$$\text{Conductivity}\left(\mathrm{S \cdot cm^{-1}}\right) = \text{cell constant}\left(\mathrm{cm^{-1}}\right) \times \text{conductance}\left(\mathrm{S}\right)$$

The cell constant is determined by the geometry of the cell but in practical terms can only be measured using a standard of known conductivity, for example, KCl 0.01 D solution. The cell constant changes with time. Some modifications can occur due to contamination or due to physical–chemical modification, in the case of platinized cells. It is therefore recommended to calibrate the cell at least once a week. However, if you use a platinized cell, it is advisable to perform a daily conductivity measurement in a standard. If the result obtained is in accordance with the theoretical value, continue your measurements. If not, your cell needs to be cleaned (if nonplatinized) or replatinized (if platinized). For high-precision measurements, it is necessary to determine the cell constant by performing a calibration measurement on a standard thermostat at the desired temperature.

Note that for two-pole cells, the standard used for the calibration must have a conductivity value as close as possible to the conductivity of the sample to measure. When using a two-pole cell, the choice of the cell constant value varies with the linear measurement range of the cell selected.

Typically, a cell with $K = 0.1\ cm^{-1}$ is chosen for pure water measurements, while for environmental water and industrial solutions a cell with $K = 0.4–1\ cm^{-1}$ is used. Cells with up to $K = 10\ cm^{-1}$ are best for very-high-conductivity samples. In the case of a four-pole cell, the cell constant value is generally included in the range $0.5–1.5\ cm^{-1}$.

11.7.2 Conductivity Measurements

Conductivity measurements can only be performed after calibration, as the cell constant value is used to calculate the conductivity.

11.7.2.1 Low-Conductivity Measurements (Pure Water)

One of the main applications of low-conductivity measurements is checking the quality of pure water. Pharmaceutical laboratories are obliged to respect regulations laid down by national pharmacopoeias; for example, the fifth supplement of the United States Pharmacopoeia (USP) lays down rules for checking the quality of pure water or fully deionized water used for the production of injection products.

11.7.2.2 Principle of Pure Water Measurements

11.7.2.2.1 USP Standards

The conductivity partly depends on the pH, the temperature, and the amount of atmospheric carbon dioxide, which has been dissolved in the water to form ions (intrinsic conductivity). The conductivity also depends on the chloride, sodium, and ammonium ions considered as water impurities (extraneous conductivity). The conductivity (intrinsic and extraneous) of the water is measured and compared to values listed in a table to evaluate if the studied water is suitable for use in pharmaceutical applications. If the sample fails stage 1, additional tests have to be performed (stages 2 and 3) in order to determine if the excessive conductivity value is due to intrinsic factors or extraneous ions. The main requirement is that the cell constant be known with an uncertainty better than ±2%.

11.7.2.2.2 European Pharmacopoeia Standards

The cell consists of two parallel platinum plates at a defined distance. It is confined within a glass jacket with two pipe connectors enabling measurement in flow mode. This cell is calibrated using a conductivity standard of 26.6 μS/cm at 20°C, which is traceable to NIST (National Institute of Standards and Technology). All measurements are made with a precision conductivity meter using the AC current at a low frequency.

11.7.2.3 High-Conductivity Measurements

High-conductivity measurements are performed when samples are water solutions, highly concentrated in dissociated species, that is, salts, acids, and bases. The conductivity method is used to control the concentration of pure or mixed solutions. For example, the conductivity method is preferred to pH measurements in production plants of concentrated acids or bases.

11.7.3 Contacting Conductivity

Contacting conductivity uses a cell with two metal or graphite electrodes in contact with the electrolyte solution. An AC current is applied to the electrodes by the conductivity meter, and the resulting AC voltage is measured. This technique can measure down to pure water conductivity. Its main drawback is that the cell is susceptible to coating and corrosion, which drastically decreases the reading. In strongly conductive solutions, there can also be polarization effects, which result in nonlinearity of measurements.

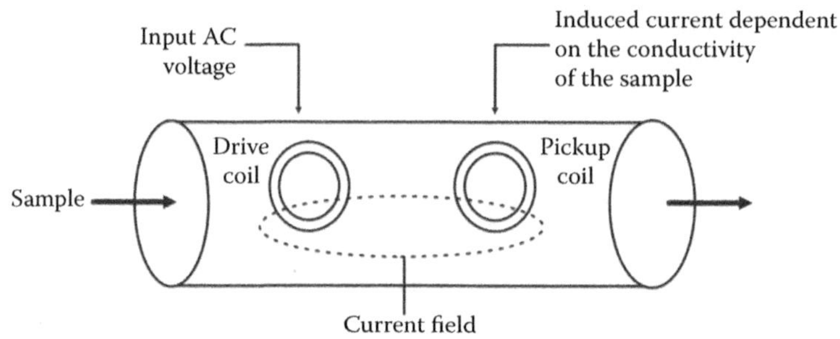

FIGURE 11.12 Toroidal conductivity AC current passing through a toroidal drive coil.

11.7.4 Toroidal "Inductive" Conductivity

A toroidal conductivity measurement is made by passing an AC current through a toroidal drive coil, which induces a current in the electrolyte solution (see Figure 11.12). This induced solution current, in turn, induces a current in a second toroidal coil, called the pickup toroid. The amount of current induced in the pickup toroid is proportional to the solution conductivity.

The main advantage of toroidal conductivity is that the toroidal coils are not in contact with the solution. They are either encased in a polymeric material or are external to a flow-through cell. One of the main disadvantages of toroidal conductivity measurements is that it lacks the sensitivity of contacting measurement. Toroidal sensors are also typically larger than contacting sensors, and the solution current induced by the toroid occupies a volume around the sensor. Hence, toroidal sensors need to be mounted in a larger pipe.

11.8 RELIABLE MEASUREMENTS

Reliable measurement results can be obtained when you follow the simple rules discussed in this section.

11.8.1 Calibrate Frequently

The cell constant value is an important factor of conductivity measurements. You must be sure of the cell constant value before starting measurements. It is impossible to recommend a given calibration interval for conductivity cells because the interval will depend on the application, the samples, and the operating conditions in use. Check the cell constant regularly, especially when using a platinized cell, due to increased risk of contamination or physical–chemical modification of the platinum layer.

11.8.2 Temperature and Stirring Conditions

Thermostat if you want accurate measurements. The temperature and stirring conditions during calibration must be as close as possible to the conditions used during the measurement. Note that for conductivity measurements, the conductivity reading can be expressed at the measuring temperature or at a reference temperature using a temperature correction factor.

11.8.3 Position of Conductivity Cell

Make sure that all the poles of the conductivity cell are completely covered by the sample. Always position a two-pole cell in the center of the measuring vessel.

11.8.4 Low-Conductivity Measurements

- Use a flow-through cell, to avoid atmospheric contamination from carbon dioxide.
- Use cells with a low cell constant, 1 cm^{-1} or lower.
- Use nonplatinized cells for easier cleaning and faster response.
- Make sure that your instrument is able to apply an appropriate measuring frequency.

11.8.5 High-Conductivity Measurements

- Use platinized cells to avoid polarization errors, preferably four-pole cells.
- Use cells with a high cell constant (1 cm^{-1} or higher if possible).
- Do not dilute samples in order to bring them into a measuring range. Conductivity is not proportional to concentration at high levels.
- Make sure that your instrument is able to apply an appropriate measuring frequency.

11.8.6 Traceability

Using one of Radiometer Analytical's Demal range of certified conductivity standards to calibrate your measuring chain guarantees that your conductivity measurements are traceable to NIST.

11.9 MAINTENANCE AND STORAGE

11.9.1 Make Sure That Your Cell Is Clean

Between measurements, rinse the cell with deionized water. If the cell has been exposed to a solvent immiscible with water, clean it with a solvent miscible with water, for example, ethanol or acetone, and rinse carefully with water. In the event of deposits on the cell, dip the cell in normal cleaning solution for approximately 1 hour. Rinse thoroughly with distilled water after cleaning.

11.9.2 Store the Conductivity Cell Carefully

Before storing the cell, rinse it carefully in deionized water.

- Short-term storage: in deionized water
- Long-term storage: in deionized water or store dry
- After long-term storage, condition the cell for 8 hours in deionized water, before use

11.9.3 Handle Platinized Cells Carefully

Do not touch the black platinum layer of the platinized cells. To rinse the platinized cells, dip the cell several times in a beaker of demineralized water. Be careful when rinsing platinized cells using deionized water from a water bottle. The force of the water could remove some of the platinum layer and consequently modify the cell constant.

11.10 APPLICATIONS

11.10.1 Conductivity Measurements

Conductivity measurements simply detect the presence of ions in a solution and are therefore non-specific measurements. Conductivity applications encompass, for instance, the monitoring of water purity, drinking water, and process water quality. Conductivity measurement is also a rapid and inexpensive way of determining the ionic strength of a solution. The conductivity κ is calculated using the conductance G and the cell constant K:

$$\kappa = G \cdot K(\mathrm{S/cm})$$

11.10.2 Resistivity Measurements

Resistivity measurements are used as a reliable indicator of ionic water quality, especially for ultrapure water and more generally when a resistivity value is preferred to a conductivity value, for example, when checking for water contamination in organic solvents. The resistivity of a solution is calculated on the basis of the conductance G compensated for the cable resistance, cable capacitance, and cell constant of the conductivity cell used. The resistivity ρ is calculated as follows:

$$\rho = \frac{1}{\kappa} \Omega \cdot \mathrm{cm}^{-1}$$

11.10.3 TDS Measurements

TDS measurements are the measure of the total concentration of ionic species of a sample. The magnitude is relative to the standard solution used to calibrate the meter. TDS measurements in the pulp and paper industry measure accurately and easily the total organic and inorganic dissolved solids in water.

11.10.3.1 What Is TDS and How Is It Measured?

The TDS corresponds to the total weight of cations, anions, and the undissociated dissolved species in one liter of water. The standard method (Arnold et al. 1992) to determine TDS is to evaporate a measured sample of water to dryness at 180°C, under strict laboratory conditions, and carefully weigh the amount of dry solids remaining. The precision of the standard method depends on the nature of the dissolved species. The TDS method in a typical conductivity meter offers a quicker and easier way of determining TDS by measuring the conductiviy than using a conversion factor to give TDS readings.

11.10.3.2 Determination of the TDS Factor

Conductivity readings are converted to TDS readings by multiplication with a known mathematical factor. The factor depends on the reference material used to prepare the standard. Perform a calibration using a standard of known TDS(STD). The TDS factor is calculated as follows:

$$\text{TDS factor} = \frac{\text{TDS(STD)}}{\kappa_{18}(\text{STD})}$$

TDS(STD) is expressed in milligrams per liter. κ_{18}(STD) is the conductivity of the standard corrected to 18°C (in microsiemens per centimeter). The conductivity of the standard measured is corrected to 18°C using the corresponding temperature correction table.

The TDS factor calculated by the conductivity meter also provides information about the qualitative ionic composition of the water sample. If the TDS factor is outside the 0.55–0.7 range, the TDS calibration should be considered suspect and must be repeated. If a TDS factor below 0.55 is confirmed, the sample probably contains a significant concentration of a constituent that cannot be measured (e.g., ammonia or nitrite). A TDS factor above 0.8 may indicate the presence of a large amount of poorly dissociated calcium and sulfate ions.

The sample conductivity is measured at the sample temperature (0°C–99°C) and corrected to 18°C. The sample TDS, TDS (SMP), is calculated from the sample conductivity corrected to 18°C, κ_{18} (SMP):

$$\text{TDS (SMP) (in milligram per liter)} = \text{TDS factor} \times k_{18} \text{ (SMP)}.$$

TDS values between 4 and 20,000 mg/L can be displayed. To obtain the most accurate measurements, perform the standard and sample measurements at the same temperature. Such TDS measurements are accurate as long as the composition of the samples varies only slightly.

11.10.4 Concentration Measurements

Since the charge of the ions in solution facilitates the conductance of electric current, the conductivity of a solution is highly (but not totally) proportional to its ion concentration. As conductivity is a nonspecific technique, concentration calculation using conductivity measurements is valid for samples containing only the species of interest. The first step to measuring concentration is to know the conductivity of the solution as a function of the concentration of the species of interest. This data can come from published conductivity versus concentration curves for electrolytes, or from laboratory measurements. Over large conductivity ranges, conductivity will increase with concentration, but may reach a maximum and then decrease with increasing concentration. When using conductivity measurements to determine the concentration, it is important to work at a constant temperature for calibration and measurements because the shape of the conductivity versus concentration curve will change with the temperature.

11.10.4.1 Necessary Conditions

With the help of conductivity versus concentration data, the following conditions are necessary to perform a concentration measurement:

- There must be a measurable change in conductivity over the desired concentration range.
- Conductivity must either increase or decrease with concentration. If a maximum or minimum is present in the desired concentration range, concentration cannot be measured (Figure 11.13).

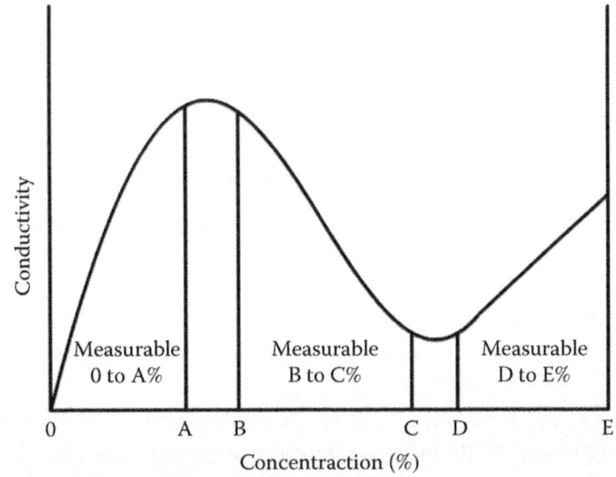

FIGURE 11.13 Conductivity versus concentration.

To calculate a sample concentration, the conductivity meter must be calibrated against 1–3 standards of known concentration. At the end of the calibration, a formula expressing the concentration as a function of the conductivity measured is determined. The sample concentration is calculated from the sample conductivity measured and the coefficients obtained during the calibration.

11.10.4.2 Determination of Concentration Coefficients

A calibration using 1–3 standards of known concentration is performed. For each standard

- The conductivity of the standard is measured at the standard's temperature.
- The conductivity of the standard is corrected to a reference temperature T_{ref} (linear correction or natural water correction as for a conductivity measurement).

At the end of the calibration, the conductivity meter calculates and displays the A_0, A_1, and A_2 coefficients of the following equation:

$$C = A_0 + A_1 \bullet \kappa_{T_{ref}} + A_2 \bullet (\kappa_{T_{ref}})^2$$

where

C = concentration
$\kappa_{T_{ref}}$ = conductivity corrected to the T_{ref} reference temperature

In one-point calibration, A_1 is calculated, and A_0 and A_2 are equal to 0. In two-point calibration, A_0 and A_1 are calculated, and $A_2 = 0$.

All the measurements of the standards must be performed at the same temperature. To obtain the most accurate measurements, thermostat the standards and the samples at the reference temperature T_{ref}.

11.10.4.3 Determination of the Sample Concentration

The conductivity of a sample is measured at the sample's temperature, then corrected to a reference temperature T_{ref} to give $\kappa_{T_{ref}}$ (linear correction or "natural water" correction as for a conductivity measurement). The sample concentration C_{SMP} is then calculated using the following equation:

$$C_{SMP} = A_0 + A_1 \times \kappa_{T_{ref}} + A_2 \times (\kappa_{T_{ref}})^2$$

A_0, A_1, and A_2 are the coefficients determined during the calibration.

11.10.4.4 Limitations of the Concentration Method

For some samples with high concentrations, the conductance = f (concentration) curve may show a maximum. This can lead to major measurement errors as shown in Figure 11.14 for the example of HCl. For HCl, a maximum around 860 mS is obtained for a concentration of about 19% (see the curve in Figure 11.14). Two calibrations have been performed using three standards each. The two calibration curves obtained are represented as dashed lines on the conductance = f (concentration) curve.

When performing measurements on a 32% HCl sample using calibration no. 1, the conductivity meter measures a conductance of 720 mS and finds a concentration of 10% instead of 32%. The correct result (32%) is obtained if sample measurements are performed using calibration no. 2.

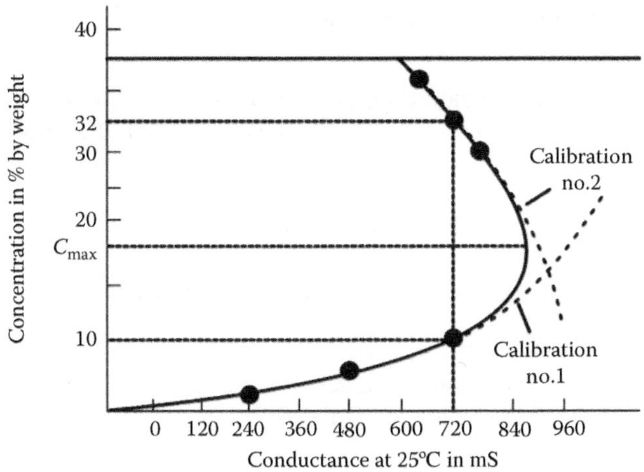

FIGURE 11.14 Concentration versus conductance for a HCl solution.

11.10.4.5 Recommendations

To obtain accurate measurements in concentrated samples, some precautions must be taken:

- For the sample analyzed, you must know if the conductance = *f* (concentration) curve shows a maximum for the conductance. If this is the case, the concentration value C_{max} must be known.
- A calibration on three standards must be performed with the following recommendation: the three standard concentrations must be greater or less than the C_{max} value.
- Make sure that the change in conductivity is only due to the species of interest. As conductivity is a nonselective measurement, conductivity change due to any other species would lead to misinterpretation of the concentration results.

11.10.5 Salinity Measurements

Salinity is a measurement without units that corresponds to the weight of dissolved salts in seawater. The salinity is calculated from an empirical relationship between the conductivity and salinity of a seawater sample. Oceanographic tables and standards endorsed by UNESCO/SCOR/ICES/IAPSO are used for the calculation. Salinity measurements are performed with no direct temperature correction. The calculation is valid for salinity values in the range 2–42 at a sample temperature of −2°C to +35°C:

$$G_t = G_{t_{cal}}\{1+\alpha(T-T_{cal})\}$$

where

G_t = conductivity at any temperature T in °C
$G_{t_{cal}}$ = conductivity at calibration temperature T_{cal} in °C
α = temperature coefficient of the solution at T_{cal} in °C

To determine α of other solutions, simply measure the conductivity at a range of temperatures and graph the change in conductivity versus the change in temperature. Divide the slope of the graph by $G_{t_{cal}}$ to get α.

11.10.5.1 Determination of the Sample Salinity

11.10.5.1.1 Calibration

Calibration is carried out using a standard seawater solution κ_{15} (STD) (salinity = 35, conductivity = 42.896 mS/cm at 15°C). The conductivity of the sample is measured at the sample temperature *T*. The conductivity of standard seawater κ_T (STD) is calculated from the following equation:

$$k_T(\text{STD}) = f(T)\,\kappa_{15}(\text{STD})$$

The conversion factor $f(T)$ is calculated from a four-degree polynomial formula.

11.10.5.1.2 Salinity of the Sample

At the sample temperature *T*, the sample conductivity measured is κ_T (SMP). Salinity is calculated from the equation below combined with a five-degree polynomial formula:

$$R = \kappa_T(\text{SMP})/\kappa_T\,(\text{STD})$$

11.10.5.1.3 Ensure Correct Cell Positioning and Reproducible Stirring Conditions

The positioning of a two-pole conductivity cell in the beaker is essential for accurate results due to the influence of the field lines.

11.10.5.1.4 Select the Appropriate Cell for Application

Different types of conductivity cells exist; the following criteria will help to determine the selection. Regarding the measuring range (high or low conductivity), a four-pole cell covers a wide measurement range, whereas a platinized cell is recommended for high-conductivity measurements.

11.10.5.1.5 Use Certified Conductivity Standards

No conductivity measurement can be more precise than the standards used for the calibration. To ensure optimum accuracy time after time, calibrate your system with solutions that are fully traceable to national standards. Two types of conductivity standards exist, certified or economical. If a laboratory is involved in an accreditation, certification, or quality procedure, use the certified standards.

11.10.5.2 Demal Solution

11.10.5.2.1 Definition

In 1940, Jones and Bradshaw determined the demal definition as the weight of KCl/weight of the solution.

11.10.5.2.2 KCl Demal Solutions

The mass of KCl/1000 g of solution as the molar weight is redefined from time to time; there is a need to redefine the conductivity for a given molarity of KCl solution. Using a given mass in a defined mass of water means that the determined conductivity of these solutions will not change over time (Table 11.4).

TABLE 11.4
Mass of KCl/1000 g of Solution in Demal at 25°C

Solution Type at 25°C	Mass of KCl/1000 g of Solution (g)	Conductivity
1 D	71.1352	111.3 mS/cm
0.1 D	7.4191	12.85 mS/cm
0.01 D	0.745263	1408 μS/cm

The conductivity of the demineralized water used must not exceed 2 μS/cm. Corrections for air buoyancy must be applied to the weighing. For the preparation of standards, refer to OIML (the International Organisation of Legal Metrology) Recommendation no. 56 (OIML 1991).

11.10.5.2.3 U.S. Pharmacopoeia and European Pharmacopoeia Standards

USP <645> with stage 1, 2, and 3 compliance is required for purified water and water for injection. Only a few resistivity/conductivity meters conform to these requirements. Some of these requirements are:

- Resolution of 0.1 μS/cm or greater
- Accuracy at 1.3 μS/cm of 0.1 μS/cm
- Must be able to read with or without automatic temperature compensation
- Verifiable cell constant ±2%

11.10.5.2.4 Correct Reading to a Reference Temperature

Conductivity measurements are strongly temperature dependent. Add a temperature sensor to your system for automatic temperature measurements and temperature correction of the conductivity to a reference temperature (Tables 11.5 through 11.8).

TABLE 11.5
Conductivity (in mS/cm) of Demal Concentrations of 1, 0.1, and 0.01 D KCl Solutions

Temperature (°C)	KCl[a] 1 D	KCl[b] 0.1 D	KCl[b] 0.01 D
0	65.14	7.13	0.773
1	66.85	7.34	0.796
2	68.58	7.56	0.82
3	70.32	7.77	0.843
4	72.07	7.98	0.867
5	73.84	8.2	0.891
6	75.62	8.42	0.915
7	77.41	8.64	0.94
8	79.21	8.86	0.965
9	81.03	9.08	0.989
10	82.85	9.31	1.014
11	84.68	9.54	1.039
12	86.54	9.76	1.065
13	88.39	9.99	1.09
14	90.26	10.22	1.116
15	92.13	10.46	1.142
16	94.02	10.69	1.168
17	95.91	10.93	1.194
18	97.81	11.16	1.22
19	99.72	11.4	1.247
20	101.63	11.64	1.273
21	103.56	11.88	1.3
22	105.49	12.12	1.327
23	107.42	12.36	1.354
24	109.36	12.61	1.381
25	111.31	12.85	1.409
26	113.27	13.1	1.436
27	115.22	13.35	1.464
28		13.59	1.491
29		13.84	1.519

(*Continued*)

TABLE 11.5 (*Continued*)
Conductivity (in mS/cm) of Demal Concentrations of 1, 0.1, and 0.01 D KCl Solutions

Temperature (°C)	KCl[a] 1 D	KCl[b] 0.1 D	KCl[b] 0.01 D
30		14.09	1.547
31		14.34	1.575
32		14.59	1.603
33		14.85	1.632
34		15.1	1.66
35		15.35	1.688
36		15.61	1.717
37		15.86	1.745
38		16.12	1.774
39		16.37	1.803
40		16.63	1.832
41		16.89	1.861
42		17.15	1.89
43		17.4	1.919
44		17.66	1.948
45		17.92	1.977
46		18.18	2.007
47		18.44	2.036
48		18.7	2.065
49		18.96	2.095
50		19.22	2.124

[a] For the Demal KCl solution: extrapolation of the NIST results from OIML conductivity values at 0°C, 18°C and 25°C.

[b] "Organisation Internationale de Métrologie Légale (OIML)", Recommendation no. 56 and "The National Institute of Standards and Technology (NIST)", Journal of Solution Chemistry, Vol. 20, no. 4 1991.

TABLE 11.6
Conductivity (in mS/cm) of Various Molar Concentrations of KCl Solutions

Temperature (°C)	KCl 1 M	KCl 10^{-1} M	KCl 2×10^{-2} M	KCl 10^{-2} M
0	65.41	7.15	1.521	0.776
1	67.13	7.36	1.566	0.8
2	68.86	7.57	1.612	0.824
3	70.61	7.79	1.659	0.848
4	72.37	8	1.705	0.872
5	74.14	8.22	1.752	0.896
6	75.93	8.44	1.8	0.921
7	77.73	8.66	1.848	0.945
8	79.54	8.88	1.896	0.97
9	81.36	9.11	1.945	0.995
10	83.19	9.33	1.994	1.02
11	85.04	9.56	2.043	1.045
12	86.89	9.79	2.093	1.07
13	88.76	10.02	2.142	1.095

TABLE 11.6 (*Continued*)
Conductivity (in mS/cm) of Various Molar Concentrations of KCl Solutions

Temperature (°C)	KCl 1 M	KCl 10^{-1} M	KCl 2×10^{-2} M	KCl 10^{-2} M
14	90.63	10.25	2.193	1.121
15	92.52	10.48	2.243	1.147
16	94.41	10.72	2.294	1.173
17	96.31	10.95	2.345	1.199
18	98.22	11.19	2.397	1.225
19	100.14	11.43	2.449	1.251
20	102.07	11.67	2.501	1.278
21	104	11.97	2.553	1.305
22	105.94	12.15	2.606	1.332
23	107.89	12.39	2.659	1.359
24	109.84	12.64	2.712	1.386
25	111.8	12.88	2.765	1.413
26	113.77	13.13	2.819	1.441
27	115.74	13.37	2.873	1.468
28		13.62	2.927	1.496
29		13.87	2.981	1.524
30		14.12	3.036	1.552
31		14.37	3.091	1.581
32		14.62	3.146	1.609
33		14.88	3.201	1.638
34		15.13	3.256	1.667
35		15.39	3.312	
36		15.64	3.368	

TABLE 11.7
Average Temperature Coefficients of Standard Electrolyte Solutions Expressed as %/°C of the Conductivity Value at 25°C

Temperature Range °C	KCl 1 M	KCl 0.1 M	KCl 0.01 M	Saturated NaCl
15–25	1.725	1.863	1.882	1.981
15–25–35	1.73 (15 °C–27°C)	1.906	1.937 (15°C–34°C)	2.041
25–35	1.762 (25°C–27°C)	1.978	1.997 (25°C–34°C)	2.101

TABLE 11.8
Conductivity (in μS/cm) Values of a 0.05% NaCl Solution

Temperature (°C)	Conductivity	Temperature (°C)	Conductivity
0	540.4	27	1056.53
1	557.73	28	1077.54
2	575.2	29	1098.67
3	592.79	30	1119.92
4	610.53	31	1141.3
5	628.4	32	1162.8
6	646.4	33	1184.41

(*Continued*)

TABLE 11.8 (*Continued*)
Conductivity (in μS/cm) Values of a 0.05% NaCl Solution

Temperature (°C)	Conductivity	Temperature (°C)	Conductivity
7	664.55	34	1206.15
8	682.83	35	1228
9	701.26	36	1249.96
10	719.82	37	1272.03
11	738.53	38	1294.21
12	757.37	39	1316.49
13	776.36	40	1338.89
14	795.48	41	1361.38
15	814.74	42	1383.97
16	834.14	43	1406.66
17	853.68	44	1429.44
18	873.36	45	1452.32
19	893.18	46	1475.29
20	913.13	47	1498.34
21	933.22	48	1521.48
22	953.44	49	1544.71
23	973.8	50	1568.01
24	994.28		
25	1014.9		
26	1035.65		

Experiment: Measurement of Soil Electrical Conductivity

Objectives

Define and understand the meanings of the term soil electrical conductivity (EC)

Principles

Salinity affects the physical and chemical characteristics of soil aggregates. These properties may be influenced by parent materials, rainfall, organic materials, plants, and various manmade soil amendments.

High salt concentrations, particularly those containing K^+ and Na^+, that have a weak charge and large hydrated radii increase the swelling and dispersion of clay particles in soil. This effect decreases with increased charge and decreased hydrated radii of base cations. Swelling and dispersion (puddling) both cause a loss of pore space and permeability. This occurs due to clay particles expanding in physical size and a loss of aggregation. The tendency to cause swelling and dispersion can be shown as follows:

$$Ca^{2+} < Mg^{2+} < K^+ < Na^+$$

Salinity may be measured in multiple ways. The EC of soil to determine relative salinity is an easy and quick method. EC measures the ability of soil to conduct an electric charge across a distance. This measurement is recorded as decisiemens per meter. Keep in mind that porosity, water content, temperature, and cation exchange capacity also affect the EC of a soil. Specific base cation content, such as Na^+, may also be quantified to determine salinity using an atomic adsorption or mass spectrometer.

Requirements

Beaker, conductivity meter, deionized water, balance, pipette, spatula, soil sample

Method

1. Place five scoops of a soil sample into a beaker.
2. Add five scoops of deionized water to the beaker and stir. Mix the soil and water thoroughly. About 30 strokes is adequate.
3. Let the solution sit for 5 minutes.
4. Rinse the probes with deionized water.
5. Place the probes into the soil–water paste.
6. Press the measurement button on the conductivity instrument.
7. The measurement is stable when the measure icon stops flashing.
8. Record the results displayed on the screen. The EC has been temperature-corrected by the instrument.
9. To proceed with measuring another sample, repeat from step 1.

Results

Record the EC for all soil samples and sampling depths. Repeat readings for mulch and roadside soils.

Discussion

1. Compare the change of EC with the depth of your soil. What is the trend and why?
2. How do your EC values compare with mulch and roadside soil collections and with soils of other groups?

SUGGESTED READING

"Conductivity Theory and Practice." Radiometer Analytical SOS, 72 rue d' Alsace, 69627, Villeurbanne Cedex, France, p. 50.

Greenberg, A E., L S. Clesceri, and A D. Eaton, eds. *Standard Methods for the Examination of Water and Wastewaters*, 18th ed., part 2540C, pp. 2–55. Washington, DC: American Public Health Association, 1992.

Hewitt, G. F. *Chemical Engineering Division*. Harwell, U.K: U.K.A.E.A Research Group Atomic Energy Research Establishment, 1960.

OIML (Organisation Internationale de Métrologie Légale). "Recommendation No. 56 and The National Institute of Standards and Technology (NIST)." *Journal of Solution Chemistry* 20, no.4 (1991).

Wagner, R. "Temperaturkorrekturfaktoren für die elektrische Leitfähigkeit von Wässern." Zeitschrift für Wasser und Abwasserforschung 2 (1980).

IMPORTANT LINKS

1. **Conductivity meter:** http://www.radiometer-analytical.com
2. http://www.eutech.nl

Glossary

α-helix A common motif in the secondary structure of proteins.

β-pleated sheet (β sheet) The second form of regular secondary structure in proteins.

Abbe condenser A lens that is specially designed to mount under the stage of a microscope and typically moves in a vertical direction.

Absorbance Refers to the mathematical quantity of light.

Absorption Refers to the physical process of absorbing light.

Acid fast stain A staining property of the *Mycobacterium* species such that cells stained with hot carbol fuchsin will not decolorize with acid–alcohol.

Acidity Ability of a substance to react with a base.

Adsorption The process of interaction between a solute and the surface of an adsorbent.

Adsorption chromatography A chromatography technique in which we make one solvent immobile (by adsorption on a solid support matrix) and another mobile.

Aerobe Microbes growing in the presence of oxygen.

Affinity chromatography A method of separating biochemical mixtures based on a highly specific interaction such as the interaction between antigen and antibody, enzyme and substrate, or receptor and ligand.

Agarose gel electrophoresis Technique used for the separation of DNA fragments ranging from 50 base pairs to several megabases (millions of bases) using a specialized apparatus.

Aggregate A clustered mass of individual cells that is solid, fluffy, or pelletized, which can clog the pores of filters or other fermentation apparatus.

Alkalinity Alkalinity is the acid neutralizing capacity of solutes in a water sample, reported in milliequivalents per liter and consists of the sum of titratable carbonate and noncarbonate chemical species in a filtered water sample.

Alpha particle A positively charged particle that consists of two protons and two neutrons bound together.

Alternate current A type of current in which the direction of the current flowing in the circuit is constantly reversed back and forth.

Amino acids A class of 20 hydrocarbon molecules that combine to form proteins in living things.

Amplifier A device for increasing the power of a signal.

Anaerobe Microbes growing in the absence of air or oxygen.

Analog A voltage or current signal that is a continuous function of the measured parameter.

Analog meter A scale and pointer meter capable of indicating a continuous range of values.

Analytical ultracentrifuge An ultracentrifuge that uses one of the three optical systems (schlieren, Rayleigh, or absorption) for the accurate determination of sedimentation velocity or equilibrium.

Angular aperture The apparent angle of a lens aperture as seen from the focal point.

Angular velocity A vector quantity that specifies the angular speed of an object and the axis about which the object is rotating.

Anion An ion in which the number of electrons is greater than the number of protons; the particle is a negative ion, also called an anion.

Anode An electrode through which electric current flows into a polarized electrical device.

Antibodies A protein present in serum or other body fluids that combines specifically with antigens.

Antifoam agent A chemical added to the fermentation broth to reduce surface tension and counteract the foaming (bubbles) caused by mixing, sparging, or stirring.

Antigen A substance, usually macromolecular, that induces specific antibody formation.

Ascending chromatography A chromatography technique in which the solvent is moving against the gravitational force.

Aseptic Sterile atmosphere free from bacteria, viruses, and contaminants such as foreign DNA.

Atom A basic unit of matter that consists of a dense central nucleus surrounded by a cloud of negatively charged electrons.

Atomic absorption spectroscopy (AAS) A spectroanalytical procedure for the qualitative and quantitative determination of chemical elements using the absorption of optical radiation (light) by free atoms in a gaseous state.

Attenuated Weakened (attenuated) bacteria and viruses are often used as vaccines; they can no longer produce diseases, but they can still stimulate a strong immune response similar to natural microbes.

Autoradiography A process by which radioactive materials, although not often exclusively incorporated into cell structures, are located by exposure to a photographic emulsion forming a pattern on a film corresponding to the location of the radioactive compounds within the cell.

Backflushing A column-switching technique in which a four-way valve placed between the injector and the column allows the mobile phase to flow in either direction.

Bacterial capsule A compact layer of polysaccharide exterior to the cell wall in some bacteria.

Bacteriophage Viruses that infect bacteria; sometimes used as a vector.

Bandwidth Also known as spectral bandwidth or bandpass; relates to the physical size of the slit from which light passes out from a monochromator.

Base pair Two bases on different strands of a nucleic acid that join together; in DNA, cytosine (C) always pairs with guanine (G), and adenine (A) always links with thymine (T). In RNA molecules, adenine pairs with uracil (U).

Baseline The line drawn by a data system when the only signal from the detector is from the mobile phase.

Batch culture Large-scale cell culture in which the cell inoculum is cultured to maximum density in a tank or air-lift fermentor, harvested, and processed as a batch.

Beer–Lambert law (or Beer's law) The linear relationship between the absorbance and concentration of an absorbing species.

Beta particle High-speed electron or positron, especially one emitted in radioactive decay.

Binocular microscope A microscope with a head that has two eyepiece lenses.

Bioactivity Ability of a molecule to function correctly after it has been delivered to the active site of the body (in vivo).

Bioavailability Measure of the true rate and the total amount of drug that reaches the target tissue after administration.

Biologic A therapeutic agent derived from living things.

Biopharmaceutical A therapeutic product created through the genetic manipulation of living things, including (but not limited to) proteins and monoclonal antibodies, peptides, and other molecules that are not chemically synthesized, along with gene therapies, cell therapies, and engineered tissues.

Bioprocessing Organisms or biologically derived macromolecules that are used to carry out enzymatic reactions or manufacture products

Bioreactor A vessel capable of supporting a cell culture in which a biological transformation takes place (also called a fermentor or reactor).

Bright field illumination Lighting that shines through a transparent specimen that appears dark against a bright background; also called transmitted light illumination. It is the opposite of dark field illumination.

Broth The contents of a microbial bioreactor—cells, nutrients, waste, and so on.

Buffer Measure of the ability of a solution to resist pH change when a strong acid or base is added.

Capacitance The ability of a capacitor to store energy in an electric field.

Capillary gel electrophoresis Bundle of extremely narrow gel-filled tubes used to significantly speed up the electrophoresis process of separating biological materials.

Carrier Term used most often in affinity chromatography, which refers to the support that is used to carry the active ligand, usually by a covalent bond. It can also refer to the support in other chromatographic modes.

Carrier ampholytes Small, soluble molecules that are capable of acting as an acid or a base in standard 2D gel electrophoresis.

Cascade effects A series of events that result from one initial cause.

Catabolites Waste products of catabolism by which organisms convert substances into excreted compounds.

Cathode An electrode through which an electric current flows out of a polarized electrical device.

Cation An ion in which the number of electrons is less than the number of protons; the particle is a positive ion, also called a cation.

Cell constant The ratio of distance between conductance titration electrodes to the area of the electrodes. It is measured from the determined resistance of a solution of known specific conductance.

Cell culture Cells taken from a living organism and grown under controlled conditions (in culture). It is a method used to maintain cell lines or strains.

Cell density Measurement of the absorbance (also known as optical density or OD) of a suspension of cells at 600 nanometers (OD 600 nanometers), which can give an indication of when cells are ready for harvesting.

Cell line When cells from the first culture (taken from the organism) are used to make subsequent cultures, a cell line is established.

Cell organelle A specialized subunit within a cell that has a specific function and is usually separately enclosed within its own lipid bilayer.

Centrifugal field A space in which centrifugal forces may be detected.

Centrifugation A process that involves the use of centrifugal force for the sedimentation of mixtures using a centrifuge.

Centrifuge An apparatus, generally driven by an electric motor (some older models were spun by hand), that puts an object in rotation around a fixed axis by applying a force perpendicular to the axis.

Chemostat A growth chamber that keeps a bacterial culture at a specific volume and rate of growth by limiting the nutrient medium and removing the spent culture.

Chromatic aberration (CA) A type of distortion in which a lens fails to focus all colors to the same convergence point. It occurs because lenses have different refractive indexes for different wavelengths of light (dispersion of the lens). The refractive index decreases with increasing wavelength.

Chromatography A separation technique based on the differential distribution of constituents of a mixture between two phases, one of which moves relative to the other.

Circular dichroism (CD) The differential absorption of left and right circularly polarized light.

Clone To duplicate something exactly, whether a gene or a whole organism; or, an organism that is a genetically identical copy of another organism.

Cloning vectors Methods of transferring desired genes to organisms that will be used to express them. Cloning vectors are used to make recombinant organisms.

Column A tube that contains the stationary phase. The stationary phase differentially interacts with the sample's constituent compounds as they are carried along in the mobile phase.

Column chromatography Any form of chromatography that uses a column to hold the stationary phase.

Compound microscope A microscope with more than one objective lens.

Concentration using a factor The concentration of a sample is determined by multiplying the absorbance value by a specific factor.

Condenser A lens that concentrates light on a specimen and increases the resolution.

Conductance An expression of the ease with which an electric current flows through a substance.

Conductivity The ability of a material to conduct an electric current.

Conductivity meter A device to measure conductivity.

Contrast The difference in visual properties that makes an object (or its representation in an image) distinguishable from other objects and the background.

Countercurrent chromatography (CCC) or **partition chromatography** A category of chromatography techniques.

Counting chamber A device used for counting the cell number.

Coverslip A thin square or circular piece of transparent glass or plastic placed over the specimen on a microscope slide. It flattens out liquid samples and helps single-plane focusing.

cpm Counts per minute; used for measuring radioactivity.

Cryopreservation Maintenance of frozen cells, usually in liquid nitrogen.

Cyanobacteria A phylum of bacteria that obtain their energy through photosynthesis.

Cyclotrons Circular particle accelerators in which charged subatomic particles generated at a central source are accelerated spirally outward in a plane perpendicular to a fixed magnetic field by an alternating electric field.

Cytoplasm The protoplasm of a cell outside the nucleus.

Dalton The unit of molecular weight equal to the weight of a hydrogen atom.

Dark field microscopy Technique used to enhance the contrast in unstained specimens.

Dark field plate Circular iris that sits on the base of the microscope above the light source and reflects the light horizontally to the specimen, thereby achieving lateral illumination.

Decay The disintegration of the nucleus of an unstable nuclide by the spontaneous emission of charged particles, photons, or both.

Degassing Process of removing dissolved gas from the mobile phase before or during use.

Density A physical property of matter; each element or compound has a unique density associated with it.

Density gradient centrifugation Sedimentation based on a spatial variation in density over an area.

Descending chromatography A chromatography technique in which the solvent moves toward the gravitational force.

Detector An electronic device that quantitatively discerns the presence of separated components as they elute in a separation process.

DGGE (denaturing gradient gel electrophoresis) A technique for screening single nucleotide polymorphisms (SNPs).

Differential centrifugation Sedimentation based on differences in the size and density of particles.

Diffraction Various phenomena that occur when a wave encounters an obstacle.

Digital signal An output signal that represents the size of an input in the form of a series of discrete quantities.

DNA (deoxyribonucleic acid) The nucleic acid based on deoxyribose (a sugar) and the nucleotides G, A, T, and C. It occurs in a corkscrew-ladder shape; it is the primary component of chromosomes, which thus carry the inheritable characteristics of life.

DNA fingerprinting Sequences of nucleic acids in specific areas (loci) on a DNA molecule are polymorphic, which means that the genes in these locations may differ from person to person. The DNA fragments can be cut from such sequences using restriction enzymes. Fragments from various samples can be analyzed to determine whether they are from the same person. The technique of analyzing restriction fragment length polymorphism (RFLP) is called DNA typing or DNA fingerprinting.

DNA vaccine A nucleic acid vaccine; genes coding for specific antigenic proteins are injected to produce those antigens and trigger an immune response.

Double beam A configuration in which a light source is split in two. One beam illuminates the reference cell holder and the other illuminates the sample. This configuration enables reference corrections to be applied continually throughout a measurement and is commonly used when the sample and the reference change over time, for example, in kinetics measurements.

Downstream processing Bioprocessing steps following fermentation or cell culture or both; a sequence of separation and purification activities needed to obtain the required product at the necessary level of purity.

dpm Disintegrations per minute of an isotope in terms of efficiency.

Efficacy The ability of a substance (such as a protein therapeutic) to produce a desired clinical effect; its strength and effectiveness.

Efficiency Measurement of the sharpness of an eluting band or peak; high efficiencies indicate low diffusion and dispersion, which are major contributors to peak broadening.

Effluent The mobile phase after it has passed through the analytical column.

Electrode An electrical conductor used to make contact with a nonmetallic part of a circuit (e.g., a semiconductor, an electrolyte, or a vacuum).

Electrolyte Substance that dissociates into ions when dissolved in a suitable medium or melted and thus form conductors of electricity.

Electromagnetic field (EMF) Physical field produced by moving electrically charged objects.

Electromagnetic interference (EMI) Disturbance that affects an electrical circuit due to either electromagnetic induction or electromagnetic radiation (EMR) emitted from an external source.

Electromagnetic radiation (EMR) Form of energy that exhibits wavelike behavior as it travels through space. EMR has both electric and magnetic field components, which oscillate in phase perpendicular to each other and perpendicular to the direction of energy propagation.

Electromagnetic technique Active method that uses an electromagnetic signal to detect variations in subsurface conductivity.

Electromotive force (emf) The maximum potential difference between two electrodes of a galvanic or voltaic cell.

Electron A negatively charged subatomic particle. It can be either free (not attached to any atom) or bound to the nucleus of an atom.

Electron beam A beam of electrons that illuminates the specimen and produces a magnified image.

Electron capture detector (ECD) A very sensitive detector used in GLC/GC for analysis of halogenated compounds. As electronegative species pass through the detector, they capture low-energy electrons, causing a decrease in cell current.

Electron microscope The type of microscope that uses electrons rather than light to create an image of the target.

Electron spin resonance (ESR) A sophisticated spectroscopic technique that detects free radicals or inorganic complexes in chemical and biological systems.

Electrophoresis Separation technique based on the motion of colloidal particles in an electric field.

ELISA (enzyme-linked immunosorbent assay) A test to measure the concentration of antigens or antibodies.

Eluant The mobile phase in a chromatographic separation.

Elution order The order in which components pass out of the column in the separation process.

Emission flame spectroscopy A flame photometry technique in which the solution containing the sample to be analyzed is optically excited in an oxyhydrogen or oxyacetylene flame.

Emission spectrum The spectrum of frequencies of EMRs emitted by an element's atoms or a compound's molecules when they return to a lower energy state.

Endogenous Growing or developing from a cell or organism, or arising from causes within an organism.

Endonuclease A restriction enzyme that breaks up nucleic acid molecules at specific sites along their length; such enzymes are naturally produced by microorganisms as a defense against foreign nucleic acids.

Endoplasmic reticulum An extensive array of internal membranes in eukaryotes.

Endotoxin A poison in the form of a fat–sugar complex (lipopolysaccharide) that forms a part of the cell wall of some types of bacteria; it is released only when the cell is ruptured and can cause septic shock and tissue damage.

Enzymes Functional proteins that catalyze biochemical reactions by causing or speeding up reactions without getting changed in the process.

Epithelium (epithelial) The layers of cells between an organism and its tissues or organs and their environment (skin cells, inner linings of lungs or digestive organs, outer linings of kidneys, etc.).

Eukaryotic Organism whose cells contain complex structures enclosed within membranes.

European Pharmacopoeia (EP) Guidelines for product quality and measurement consistency.

Exclusion chromatography (size-exclusion chromatography) A chromatographic method in which molecules in a solution are separated by their size and, in some cases, molecular weight.

Exogenous Developing from outside; originating externally.

Express To translate a cell's genetic information, which is stored in its DNA (gene), to a specific protein.

Expression system Organism chosen to manufacture (by expression) a protein of interest through recombinant DNA technology.

Expression vector A way of delivering foreign genes to a host, creating a recombinant organism that will express the desired protein.

Eyepiece Referred to as an ocular, the eyepiece is the lens nearest to your eye.

Eyepiece reticles Sometimes also called "eyepiece micrometers"; they are clear circular glass inserts with a scale inscribed on them.

Fermentation The process by which microorganisms break down complex organic substances, generally in the absence of oxygen, to produce alcohol and carbon dioxide.

Fermentor A bioreactor used to grow bacteria or yeasts in liquid culture.

Filar eyepiece A micrometer is a combination of a focusing eyepiece and a reticle, but it features a moveable line rule.

Fine focus A knob used to fine-tune the focus of a specimen in conjunction with coarse focus.

Flame ionization detector (FID) A common type of detector used in gas chromatography (GC) and supercritical fluid chromatography.

Flame photometer An instrument that uses a flame to evaporate a solvent, sublimate and atomize metal, and then excite a valence electron to an upper energy state.

Flame photometric detector (FPD) Sulfur- and phosphorus-containing hydrocarbons burn in a flame, producing chemiluminescent species that are monitored at selective wavelengths.

Floc A fluffy aggregate that resembles a woolly cloud.

Flow rate (*F*) The volumetric rate of flow of a mobile phase through an LC column.

Fluorescence The ability of a substance to emit light of a certain wavelength when it is activated by the light of another wavelength.

Fluorescence detectors Fluorescence detectors project a specified wavelength of light into the sample, causing the component of interest to fluoresce; thus the emitted light is detected.

Fluorescence illumination A technique that provides cooler and brighter light than tungsten. This is beneficial when viewing slides for long periods of time or observing live specimens, such as protozoa.

Fluorescence microscope An optical microscope used to study the properties of organic or inorganic substances using the phenomena of fluorescence and phosphorescence instead of, or in addition to, reflection and absorption.

Fluorescence spectroscopy A type of electromagnetic spectroscopy that analyzes fluorescence from a sample.

Fluorescent confocal microscopes Usually laser based; designed to provide high-resolution images without interference from out-of-focus portions of the sample. They can also view deeper into a specimen when compared with conventional epifluorescence microscopes.

Focal length A measure of how strongly the system converges (focuses) or diverges (defocuses) light.

Focal planes The plane that is perpendicular to the axis of a lens or mirror and passes through the focal point.

Focus The ability to achieve a clear image, which is typically achieved by moving either the eyepiece tubes or the stage.

Gamma particle Denoted as γ; EMR of high frequency (very short wavelength) with energies in the range 10^4–10^7 eV.

Gas ionization The condition of being dissociated into ions (as by the action of heat, radiation, chemical reaction, or electrical discharge); "the ionization of a gas."

Gas–liquid chromatography (GLC) The mobile phase (or moving phase) is a carrier gas, usually an inert gas such as helium or an unreactive gas such as nitrogen. The stationary phase is a microscopic layer of liquid or polymer on an inert solid support inside a piece of glass or metal tubing called a column (a homage to the fractionating column used in distillation). It is also sometimes known as GC, vapor-phase chromatography (VPC), or gas–liquid partition chromatography (GLPC). These alternative names, as well as their respective abbreviations, are frequently found in scientific literature. Strictly speaking, GLPC is the most correct terminology and is thus preferred by many authors.

Gas–solid chromatography (GSC) A subclassification of GC in which the mobile phase is a gas and the stationary phase is a solid.

Geiger–Mueller counter An instrument that detects and measures the intensity of radiation, such as particles from a radioactive material, consisting of a Geiger tube and associated electronic equipment.

Gene The unit of inheritance consisting of a sequence of DNA, occupying a specific position within the genome. Three types of genes have been identified: (1) structural genes encoding particular proteins, (2) regulatory genes controlling the expression of other genes, and (3) genes for transfer RNA (tRNA) or ribosomal RNA (rRNA) instead of proteins.

Genetic engineering Altering the genetic structure of an organism (adding foreign genes, removing native genes, or both) through technological means rather than traditional breeding.

Genotype The genetic composition of an organism (including expressed and nonexpressed genes), which may not be readily apparent.

Germ cell The "sex cells" in higher animals and plants that carry only half of the organism's genetic material and can combine to develop into new living things.

Glass electrode A type of ion-selective electrode made of a doped glass membrane that is sensitive to a specific ion. It is an important part of the instrumentation for chemical analysis and physicochemical studies.

Glass probe Sensors or elements used to measure temperature by correlating the resistance. Mostly they are consist of a length of fine coiled wire wrapped around a ceramic or glass core.

Glycosylation Adding one or more carbohydrate molecules onto a protein (a glycoprotein) after it has been built by the ribosome; a posttranslational modification.

GMPs Good manufacturing practices required by FDA regulations.

Golgi body A cell organelle consisting of stacked membranes where posttranslational modifications of proteins are performed; also called a Golgi apparatus.

Gram stain A method of differentiating bacterial species into two large groups (gram-positive and gram-negative groups).

Growth hormone A protein produced in the pituitary gland to control cell growth.

Guard column A small column placed between the injector and the analytical column. It protects the analytical column against contamination by sample particulates and strongly retained species.

Hemocytometer A device originally designed for counting blood cells. Today, it is also used to count other types of cells as well as other microscopic particles.

Hemoglobin The iron-containing oxygen transport metalloprotein in the red blood cells of all vertebrates with the exception of the fish family Channichthyidae, as well as the tissues of some invertebrates.

Half-life The period of time it takes for the amount of a substance undergoing decay to decrease by half.

Halogen Provides the brightest illumination. The best microscopes have halogen lighting; stereo microscopes with top lighting also use halogen lighting.

Hanging drop method One of the most common methods for demonstrating bacterial motility.

High-performance liquid chromatography (HPLC) This is sometimes referred to as high-pressure liquid chromatography. A chromatographic technique that can separate a mixture of compounds and is used in biochemistry and analytical chemistry to identify, quantify, and purify the individual components of a mixture.

High-performance thin-layer chromatography (HPTLC) An enhanced form of thin-layer chromatography (TLC). A number of enhancements can be made to the basic method of TLC to automate the different steps, increase the resolution achieved, and allow more accurate quantitative measurements.

Hormone A protein released by an endocrine gland to travel in the blood and act on tissues at another location in the body.

Huygenian eyepiece Eyepiece consisting of two plano-convex lenses positioned with both convex lenses facing toward the object.

Hybridoma An immortalized cell line (usually derived by fusing β-lymphocyte cells with myeloma tumor cells) that secretes desirable antibodies.

Illumination system The light source on light microscopes; it is typically mounted under the stage.

Image An artifact, for example, a two-dimensional picture, that has a similar appearance to some subject, usually a physical object or a person.

Immersion oil A special oil used with the 100× objective in order to concentrate light rays and increase the resolution of an image.

Immortalize To alter cells (either chemically or genetically) so that they can reproduce indefinitely.

Immunoelectrophoresis A method for the separation and characterization of proteins based on electrophoresis and reaction with antibodies.

Immunofluorescence A technique used for light microscopy with a fluorescence microscope; used primarily on biological samples.

In vitro Performed in the laboratory rather than in a living organism (in vivo).

In vivo In the body; in a living organism.

Incident ray A ray of light that strikes a surface. The angle between this ray and the perpendicular or normal to the surface is the angle of incidence.

Infrared (IR) spectroscopy A type of spectroscopy that deals with the IR region of the electromagnetic spectrum, that is, light with a longer wavelength and lower frequency than visible light.

Injector A mechanism for accurately injecting a predetermined amount of sample into the mobile phase stream.

Inlet The initial part of a column where the solvent and the sample enter.

Inoculate To introduce cells into a culture medium.

Inoculum Material (usually cells) introduced into a culture medium.

Interference A phenomenon in which two waves superpose to form a resultant wave of greater or lower amplitude.

Interference microscope Uses two separate light beams with much greater lateral separation than those used in phase contrast microscopy or in differential interference microscopy (DIC).

Interferon A cytokine that inhibits virus reproduction.

Ion An atom or a group of atoms in which the number of electrons is different from the number of protons.

Ion chamber A gas-filled enclosure containing positive and negative electrodes that measures the amount of radiation passing through the enclosure according to the degree of ionization caused by the radiation.

Ion-exchange chromatography (or ion chromatography) A process that allows the separation of ions and polar molecules based on their charge.

Ionization (or ionizing) radiation Radiation composed of particles that individually have sufficient energy (or the ability to liberate sufficient energy) to remove an electron from an atom or molecule.

Iris diaphragm A diaphragm found on high-power microscopes under the stage; the diaphragm is typically a five-hole disk; each hole has a different diameter.

Isoelectric focusing Protein separation technique in which a mixture of protein molecules is resolved into its components by subjecting the mixture to an electric field in a supporting gel having a previously established pH gradient.

Isocratic Constant-composition mobile phase used in liquid chromatography.

Isopycnic centrifugation The word "isopycnic" means "equal density." It is also known as density gradient centrifugation or equilibrium sedimentation. It is a technique used to separate molecules on the basis of buoyant density.

Isotachophoresis (ITP) A nonlinear electrophoretic technique used in the separation of a variety of ionic compounds ranging from small molecules, like metal ions, to large molecules, like proteins.

Isotope A chemical element with the same atomic number as another element (i.e., the same number of nuclear protons) but a different atomic mass (i.e., a different number of nuclear neutrons).

Kilodalton (kDA) Unified atomic mass unit (also known as amu; symbol: u) or dalton (symbol: Da) that is used for indicating mass on an atomic or molecular scale.

Kinetic measurements The absorbance of a sample at a given wavelength is measured over a specified time. The change in absorbance and slope of this plot provides details about the rate of reaction.

Lens Standard achromatic lenses help to prevent color distortion.

Ligase An enzyme that causes fragments of DNA or RNA to link together; used with restriction enzymes to create recombinant DNA.

Light-emitting diode (LED) Produces light at a single wavelength without the need for a monochromator. Lamp life is almost infinite and LED sources are stable with little variation in bandwidth, making them an attractive, low-cost solution for simple applications.

Light microscopes Any microscope that uses a source of light to create an image of the specimen; essentially includes all compound and stereo microscopes.

Linear accelerator (LINAC) An apparatus for accelerating charged subatomic particles; used in radiotherapy, physics research, and the production of radionuclides.

Linearity A measurement process that ensures accurate quantitation.

London dispersion forces (LDFs) Forces named after the German–American physicist Fritz London; they are weak intermolecular forces that arise from the interactive forces between instantaneous multipoles in molecules without permanent multipole moments.

Lymphocytes White blood cells that produce antibodies.

Lysosomes Cell organelles containing enzymes; responsible for degrading proteins and other materials ingested by the cell.

Macrokinetics Movement of whole cells and their media within a bioreactor.

Magnification The essence of a microscope is its ability to magnify a specimen.

Mechanical stage A flat mechanism that sits on top of the stage and allows the viewer to move a specimen through small distances, a task that is otherwise difficult at high magnifications.

Media A (usually sterile) preparation made for the growth, storage, maintenance, or transport of microorganisms or other cells.

Membrane Any thin sheet or layer.

Metabolites Chemical by-products of metabolism, which is the chemical process of life.

Microanalysis The chemical identification and quantitative analysis of very small amounts of chemical substances (generally less than 10 mg or 1 mL) or very small surfaces of materials (generally less than 1 cm^2).

Microbiology The study of microscopic life, such as bacteria, viruses, and yeast.

Microbore A column with a smaller-than-usual internal diameter (<2 mm) used in HPLC.

Microcarrier A microscopic particle (often, a 200-μm polymer bead) that supports cell attachment and growth in a suspension culture.

Microencapsulated Surrounded by a thin, protective layer of biodegradable substance referred to as a microsphere.

Microheterogeneity Slight differences in the amino acid sequence of a protein.

Microinjection Manually using tiny needles to inject microscopic material (such as DNA) directly into cells or cell nuclei; computer screens provide a magnified view.

Microkinetics Movement of chemicals into, out of, and within a cell.

Micrometer Also called a micron; it is the metric linear measurement used in microscopy. There are 1000 microns in a millimeter.

Microorganism A microbe; a living thing too small to be seen by the naked eye.

Microscope An instrument used to see objects that are too small to be seen by the naked human eye. The science of investigating small objects using such an instrument is called microscopy.

Microtubules Cellular organelles common in microorganisms; they are thin tubes that make structures involved in cellular movement.

Mirror An instrument that allows you to direct ambient light up through the hole in the stage of a microscope and illuminate a specimen.

Mitochondria Cell organelles; seat of respiration and electron transport. Often referred to as the powerhouse of a cell.

Mobile phase A solvent that moves the solute through the column.

Molecule The result of two or more atoms combining by chemical bonding.

Monoclonal antibody (MAb) A highly specific, purified antibody that recognizes only a single antigen.

Monocular microscope A compound microscope with a single eyepiece.

Multicellular Organisms comprising more than one cell; they often contain billions of cells, arranged in various organs, tissues, and systems.

Multiwavelength measurement Sample absorbance is recorded at multiple wavelengths.

Mutagen An agent (chemicals, radiation) that causes mutations in DNA.

Mutation A permanent change in DNA sequence or chromosomal structure.

Mycoplasma A genus of bacteria that lack a cell wall. Without a cell wall, they are unaffected by many common antibiotics.

Myeloma Lymphocytic cancer; a malignancy normally found in bone marrow.

Myoglobin An iron- and oxygen-binding protein found in the muscle tissue of vertebrates in general and in almost all mammals.

Normal-phase chromatography A mode of chromatography carried out with a polar stationary phase and a nonpolar mobile phase.

Nuclear magnetic resonance (NMR) A physical phenomenon in which magnetic nuclei in a magnetic field absorb and reemit EMR.

Nuclear reactor A device used to generate power in which nuclear fission takes place as a controlled chain reaction, producing heat energy that is generally used to drive turbines and provide electric power. Nuclear reactors are used as a source of power in large power grids and in submarines.

Nucleic acids DNA or RNA: long, chain-like molecules comprising nucleotides.

Nucleotides Molecules comprising a nitrogen-rich base, phosphoric acid, and a sugar. The bases can be adenine (A), cytosine (C), guanine (G), thymine (T), or uracil (U).

Nucleus A specialized, usually spherical mass of protoplasm encased in a double membrane, and found in most living eukaryotic cells, directing their growth, metabolism, and reproduction, and functioning in the transmission of genetic characters.

Numerical aperture (NA) A measure of the diameter of the aperture compared to the focal length of a lens and, ultimately, of the resolving power of a microscope.

Objective lens The lens closest to the specimen that first receives the rays from the specimen (the object) and forms the image in the focal plane of the eyepiece.

Ocular meter An ocular micrometer serves as a scale or rule. This is simply a disk of glass on which equally spaced divisions are etched. The rule may be divided into 50 subdivisions or, more rarely, into 100 subdivisions.

Oncogene A gene that, when expressed as a protein, can lead a cell to become cancerous, usually by removing the normal constraints on its growth.

Organelle A structurally discrete component that performs a certain function inside a eukaryotic cell.

Organism A single, autonomous living thing. Bacteria and yeasts are organisms, whereas mammalian and insect cells used in a culture are not.

Oxygen electrode Also referred to as the Clark electrode, it measures oxygen on a catalytic platinum surface using the net reaction.

Paper chromatography A technique that involves placing a small dot or line of a sample solution on a strip of chromatography paper. The paper is placed in a jar containing a shallow layer of solvent and the jar is sealed. As the solvent rises through the paper, it meets the sample mixture, which starts to travel up the paper with the solvent. This paper is made of cellulose, a polar substance, and the compounds within the mixture travel farther if they are nonpolar. Substances that are more polar bond with the cellulose paper more quickly and, therefore, they do not travel as far.

Particle size (dp) The average particle size of the packing in an LC column.

Partition chromatography A method of chromatography using the partition of the solutes between two liquid phases (the original solvent and the film of solvent on the adsorption column).

Partition coefficient (K_d) A measure of how a compound distributes itself between two immiscible solvent phases.

Peak When the detector registers the presence of a compound, the normal baseline signal it sends to the data system changes. This results in a deflection from the baseline called a peak.

Peak identification Performed by comparing the sample chromatogram to a chromatogram of a standard solution separated under the same conditions.

Peak shape The profile of a chromatographic peak.

Peptides Proteins consisting of fewer than 40 amino acids.

pH meter An electronic instrument used for measuring the pH (acidity or alkalinity) of a liquid (although special probes are sometimes used to measure the pH of semisolid substances).

Phase contrast A contrast-enhancing microscopic technique that shifts the light phase wavelength, thereby causing the light deviated by a specimen to appear dark on a light background.

Phenotype The physical appearance of an organism as distinguished from its genetic makeup.

Photocell A resistor whose resistance decreases with increasing incident light intensity; it can also be referred to as a photoconductor.

Photodiode array (PDA) PDA detectors are UV/Vis detectors that record the absorbance of light at many different wavelengths simultaneously.

Photomultiplier An electronic sensing device used to detect the electromagnetic energy of a wide range of frequencies and intensity levels.

Photon An elementary particle, the quantum of an electromagnetic interaction, and the basic unit of light and all other forms of EMR.

Pilot plant A medium-scale bioprocessing facility used as an intermediate in scaling up processes from the laboratory to commercial production.

Plasmid Hereditary material that is not part of a chromosome; plasmids are circular and self-replicating and are found in the cytoplasm of cells (naturally in bacteria and some yeast). They can be used as vectors for introducing up to 10,000 base pairs of foreign DNA into recipient cells.

Polyacrylamide gel electrophoresis (PAGE or SDS PAGE) Electrophoresis in which a polyacrylamide gel is used as the diffusion medium.

Polymerase An enzyme that catalyzes the production of nucleic acid molecules.

Polymerase chain reaction (PCR) A method of duplicating genes exponentially.

Posttranslational modifications Protein processing done by Golgi bodies after proteins are constructed by ribosomes.

Potentiometer An instrument for measuring the potential (voltage) in a circuit.

Preparative chromatography The process of using liquid chromatography to isolate a sufficient amount of material for other experimental or functional purposes.

Prions Resembling viruses, these pathogens comprise only protein, with no detectable nucleic acid.

Prokaryotes Simple organisms with no cell organelles.

Proteins Macromolecules whose structures are coded in an organism's DNA. Each is a chain of more than 40 amino acids folded back upon itself in a particular way.

Proteolytic Capable of lysing (denaturing or breaking down) proteins.

Pulsed-field gel electrophoresis (PGFE) Electrophoresis in which the direction of the electric field is changed periodically.

Radioactive decay Spontaneous decomposition of the nuclei of atoms of radioactive substances; measured as the proportion of atoms in a radionuclide that decompose per unit of time, usually stated as the half-life of that particular isotope.

Radioactive fallout Dissemination of radioactive substances through the atmosphere and deposition on the environment generally; causes radiation injury.

Radioactive isotope One of two or more atoms with the same number of protons but different number of neutrons with a nuclear composition.

Radioactive tracer A radioactive isotope that replaces a stable chemical element in a compound and can thus be followed or tracked through one or more reactions or systems; generally an isotope that is introduced into and followed through the body.

Radioimmunoassay (RIA) A technique in radiology used to determine the concentration of an antigen, antibody, or other protein in the serum. The technique involves the injection of a known amount of a radioactively labeled substance that reacts with the protein in question.

Radioisotopes A version of a chemical element that has an unstable nucleus and emits radiation during its decay to a stable form.

Radiotherapy The treatment of diseases with ionizing radiation.

Ramsden eyepiece Comprises two plano-convex lenses with the same focal length and glass placed less than one focal length apart; a design created by the astronomical and scientific instrument maker Jesse Ramsden in 1782.

Recombinant Containing genetic material from another organism. Genetically altered microorganisms are usually referred to as recombinant, whereas plants and animals so modified are called transgenic.

Reflection The change in direction of a wave front at an interface between two different media so that the wave front returns into the medium from which it originated.

Refraction The change in direction of a wave due to a change in its speed. It is essentially a surface phenomenon.

Refractive index Expressed as the ratio of the speed of light in a vacuum relative to that in the considered medium.

Reference electrode An electrode that has a stable and well-known electrode potential.

Relative centrifugal force (RCF) The measurement of force applied to a sample within a centrifuge is traditionally known as RCF.

Resistance The property of a component that restricts the flow of an electric current.

Resolution The ability of a lens to distinguish the fine details of the specimens being viewed.

Resolving power The ability of an imaging device to separate (i.e., to see as distinct) points of an object that are located at a small angular distance.

Restriction enzyme Bacterial enzyme that cuts DNA molecules at the location of particular sequences of base pairs.

Retina A translucent filmy piece of tissue lining the back of the eyeball.

Reversed-phase chromatography (RPC) The most common HPLC mode; uses hydrophobic packings such as octadecyl or octysilane phases bonded to silica or neutral polymeric beads. The mobile phase used is usually water and a water-miscible organic solvent such as methanol or acetonitrile.

R_f (value retention factor) Defined as the ratio of the distance traveled by the substance to the distance traveled by the solvent.

Ribosome Cell organelles that translate RNA to build proteins.

RNA Ribonucleic acid; similar to DNA but is based on ribose sugar and has the base uracil (U) in place of thymine (T). Various forms of RNA are found, such as messenger RNA (mRNA), tRNA, and rRNA. Most RNA molecules are single-stranded molecules, although they can form double-stranded units.

Roller bottle A container with large growth surfaces in which cells can be grown in a confluent monolayer. The bottles are rotated or agitated to keep the cells in suspension, but they require extensive handling, labor, and media.

Rotor Rotating part of a mechanical device.

Sample capacity The amount of sample that can be injected onto an LC column without overloading it.

Sampling rate The frequency with which a detector checks the flow cell. Sampling rate is often called the time constant.

Saturated calomel electrode (SCE) A reference electrode based on the reaction between elemental mercury and mercury chloride.

Scale up To take a biopharmaceutical manufacturing process from the laboratory scale to a scale at which it is commercially feasible.

Scanning electron microscope (SEM) A type of electron microscope that scans the surface of a sample using a beam of electrons. These systems are characterized by producing high-resolution images with a large depth of field and are capable of reaching extremely high magnifications.

Scintillation counter A measuring instrument for counting individual ionizing events.

Seed stock The initial inoculum, or the cells placed in growth medium from which other cells will grow.

Sensitivity The sensitivity setting is the line between normal background noise and a true peak.

Sequence The precise order of bases in a nucleic acid or amino acids in a protein.

Serum The watery portion of an animal or plant fluid (such as blood) remaining after coagulation. When cheese is made, whey is the milk serum that is left behind.

Shorter wavelength Electromagnetic radiation with shorter wavelengths may be called "millimeter waves", terahertz radiation or even T-rays.

Single beam All light passes through the sample holder. Measurements are made by placing a reference in the sample holder, which is measured to standardize the instrument. This reference value is subtracted from subsequent sample measurements to remove effects from the solvent and the cell.

Single wavelength measurement Also known as a fixed wavelength measurement. This is the simplest application of a spectrophotometer and allows the measurement of absorbance or transmittance (percentage of transmission [%T]) at a specified wavelength.

SIP Steam in place or sterilize in place.

Slide A flat, rectangular glass plate on which a specimen may be placed.

Somatic cell In higher organisms, a cell that (unlike germ cells) carries the full genetic makeup of an organism.

Sparge To spray. A sparger is the component of a fermentor that sprays air into the broth.

Spherical aberration An optical effect observed in an optical device (lens, mirror, etc.) that occurs due to the increased refraction of light rays when they strike a lens or the reflection of light rays when they strike a mirror near its edge in comparison with light rays that strike nearer to the center.

Split beam Light from a source is split into two paths with approximately 70% of the energy from the monochromator passing through the sample and the rest going to a separate feedback detector. This corrects variations in energy and improves measurement stability and reproducibility.

Stable isotope One that does not transmute into another element with emission of corpuscular or EMRs.

Stage The platform on which slides and specimens are placed for viewing.

Stage micrometer Simply a microscope slide with a scale etched on the surface.

Staining An auxiliary technique used in microscopy to enhance contrast in a microscopic image.

Stationary phase A solid, or a liquid supported on a solid.

Strain A population of cells all descended from a single cell.

Substage The part of a microscope below the stage, including the illumination system.

Substrate Reactive material; the substance on which an enzyme acts.

Substratum The solid surface on which cells move or grow.

Supernatant Material floating on the surface of a liquid mixture (often the liquid component that has the lowest density).

Support The solid particles in a column. The support can be naked, coated, or have a chemically bonded phase in HPLC.

Surfactant Any substance that changes the nature of a surface, such as lowering the surface tension of water.

Suspension Particles floating in (not necessarily on) a liquid medium, or the mix of particles and the liquid itself.

Symbiotic Living together for mutual benefit.

Synthesis Creating products through chemical and enzymatic reactions.

Tailing A peak with an asymmetrical factor greater than 1. An asymmetrical peak is the result of a component that is excessively retarded in eluting.

Teflon membranes They are chemically stable and inert, and suitable for applications involving aggressive organic solvents, strong acids, and alkalis.

Thermal conductivity detector (TCD) A filament's temperature increases as analytes present in the carrier gas pass over it, causing its resistance to increase.

Thin-layer chromatography (TLC) Involves a stationary phase of a thin layer of an adsorbent, such as silica gel, alumina, or cellulose, on a flat, inert substrate. Compared to paper chromatography, this technique has the advantages of faster runs, better separations, and the choice between different adsorbents. For even better resolution and to allow for quantification, high-performance TLC can be used.

Tissue culture Growing plant or animal tissues outside the body, as in a nutrient medium in a laboratory; similar to cell culture, but cells are maintained in their structured tissue form.

Titer A measured sample. (Drawing a measured, representative sample from a larger amount is to titrate.)

Total dissolved solids (TDS) A measure of the combined content of all inorganic and organic substances contained in a liquid in molecular, ionized, microgranular (colloidal sol), or suspended form.

Tracer carbon dating A process in which a naturally radioactive carbon isotope with an atomic mass of 14 and a half-life of 5730 years is used in determining the age of ancient organic, geologic, or archeological specimens.

Transgenics The alteration of plant or animal DNA so that it contains a gene from another organism.

Translation The process by which information transferred from DNA by RNA specifies the sequence of amino acids in a polypeptide (protein) chain.

Transmission electron microscope (TEM) A microscope in which a beam of electrons is transmitted through an ultrathin specimen, which interacts with the specimen as it passes through.

Transmittance The fraction of incident light (or other EMR) at a specified wavelength that passes through a sample.

Trypsin, tryptic digestion Trypsin, from the pancreas, is an enzyme that helps in the digestion of proteins. A tryptic digest is accomplished simply by taking a pure, whole protein and subjecting it to certain enzymatic activity that breaks down the protein into specific peptides and or peptide chains.

Tryptic fragment analysis Quantitating the resultant fragments caused by tryptic digestion.

Tungsten The least expensive source of illumination. It is hotter but less bright than other kinds of illumination.

Turbidostat A variation of a chemostat; a turbidostat is designed to keep organisms at a constant concentration. A turbidity sensor measures the concentration of organisms in a culture and adds additional medium when a preset value is exceeded.

Turbulent flow field The state that results from mixing the contents of a fermentor or bioreactor to provide oxygen to the cells. This must be balanced against the shear that causes cell damage and death.

Two-dimensional chromatography A type of chromatographic technique in which the injected sample is separated by passing it through two different separation stages.

Two-dimensional difference gel electrophoresis A method of comparing two or more samples on the same two-dimensional gel electrophoresis gel.

Two-dimensional gel electrophoresis Electrophoresis in which a second perpendicular electrophoretic transport is performed on the separate components resulting from the first electrophoresis. This technique is usually performed on polyacrylamide gels.

Ultracentrifuge A centrifuge optimized for spinning a rotor at very high speeds; capable of generating accelerations as high as 2,000,000 *g*.

Ultraviolet (UV) Portion of the electromagnetic spectrum below blue light (380 nm).

UV/Vis detector The tunable- or variable-wavelength UV/Vis detector is the most popular form of detector. For methods involving organic compounds in aqueous mobile phases, the UV/Vis detector takes advantage of compounds' varying absorptivities of UV and visible light.

Unicellular Composed of only a single cell.

Vaccine Preparation of an antigen from a killed or modified organism that elicits an immune response (the production of antibodies) to protect a person or animal from a disease-causing agent.

Vacuolation Term used in cell and tissue culture. Excess fluid, debris (aggregates), or gas (from sparging) can form inside a cell vacuole. A vacuole is a cavity within a cell that can be relatively clear and filled with fluid, gas (as in a number of blue-green algae), or food (as in protozoa).

Van der Waals forces Refers to forces including attractions between atoms, molecules, and surfaces, as well as other intermolecular forces.

Vector The plasmid, virus, or some other vehicle used to carry a DNA sequence into the cell of another species.

Velocity Speed in a given direction.

Vessel jacket A temperature control method consisting of a double wall outside the main vessel wall. Liquid or steam flows through the jacket to heat (or cool) the fluid in the vessel.

Viability Life and health; the ability to grow and reproduce. A measure of the proportion of live cells in a population.

Virus The simplest form of life. It comprises RNA or DNA wrapped in a shell of protein; a small infectious agent that can replicate only inside the living cells of organisms.

Viscosity A measure of the resistance of a fluid, which is being deformed by either shear or tensile stress.

Visible light Electromagnetic radiation that is visible to the human eye and is responsible for the sense of sight.

Voltage Electric potential energy per unit charge; measured in joules per coulomb (unit is volts).

Water for injection Very pure water that is suitable for medical uses.

Wavelength scanning Also known as spectral scanning or wave scan measurements.

Xenon Xenon flash lamps provide high-energy light sources with short warm-up times and long lamp lives. They measure in both the UV and visible regions of the electromagnetic spectrum from 190–1100 nanometers.

X-radiation (x-ray) A form of EMR with a wavelength in the range of 0.01–10 nanometers, corresponding to frequencies in the range of 30 PHz to 30 EHz (3×10^{16} to 3×10^{19} Hz) and energies in the range of 120 eV to 120 keV. They are shorter in wavelength than UV rays and longer than gamma rays.

X-ray diffraction (XRD) A versatile, nondestructive technique that reveals detailed information about the chemical composition and crystallographic structure of natural and manufactured materials.

Yeast A single-celled fungus that feeds on sugar and is used to make bread rise and for alcoholic fermentation.

Index

D

G

H

I

J

K

L

M

N

Q

R

S

For Product Safety Concerns and Information please contact our EU representative GPSR@taylorandfrancis.com Taylor & Francis Verlag GmbH, Kaufingerstraße 24, 80331 München, Germany

Batch number: 10731069

Printed by Printforce, the Netherlands